AF678175

Intelligent Sensors for Structural Health Monitoring and Mechanical Fault Diagnosis

Intelligent Sensors for Structural Health Monitoring and Mechanical Fault Diagnosis

Guest Editors

Zijian Qiao
Ke Feng
Zhihui Lai

MDPI

Basel • Beijing • Wuhan • Barcelona • Belgrade • Novi Sad • Cluj • Manchester

Guest Editors
Zijian Qiao
Faculty of Mechanical Engineering and Mechanics
Ningbo University
Ningbo
China

Ke Feng
School of Mechanical Engineering
Xi'an Jiaotong University
Xi'an
China

Zhihui Lai
College of Mechatronics and Control Engineering
Shenzhen University
Shenzhen
China

Editorial Office
MDPI AG
Grosspeteranlage 5
4052 Basel, Switzerland

This is a reprint of the Special Issue, published open access by the journal *Sensors* (ISSN 1424-8220), freely accessible at: https://www.mdpi.com/journal/sensors/special_issues/192H3X3UC3.

For citation purposes, cite each article independently as indicated on the article page online and as indicated below:

Lastname, A.A.; Lastname, B.B. Article Title. *Journal Name* **Year**, *Volume Number*, Page Range.

ISBN 978-3-7258-4965-9 (Hbk)
ISBN 978-3-7258-4966-6 (PDF)
https://doi.org/10.3390/books978-3-7258-4966-6

© 2025 by the authors. Articles in this book are Open Access and distributed under the Creative Commons Attribution (CC BY) license. The book as a whole is distributed by MDPI under the terms and conditions of the Creative Commons Attribution-NonCommercial-NoDerivs (CC BY-NC-ND) license (https://creativecommons.org/licenses/by-nc-nd/4.0/).

Contents

sensors

MDPI

Article

An Adaptive Signal Denoising Method Based on Reweighted SVD for the Fault Diagnosis of Rolling Bearings

Baoxiang Wang [1] and Chuancang Ding [2,*]

1 School of Mechanical Engineering, Suzhou University of Science and Technology, Suzhou 215009, China; wangbaoxiang@usts.edu.cn
2 School of Rail Transportation, Intelligent Urban Rail Engineering Research Center of Jiangsu Province, Soochow University, Suzhou 215131, China
* Correspondence: ccding@suda.edu.cn

Highlights:

What are the main findings?

- This paper proposed an adaptive signal denoising method based on frequency domain multipoint kurtosis (FDMK) and singular value decomposition (SVD).
- FDMK is used to identify sensitive singular components that contain fault-related information, and an estimation process is proposed to calculate the fault characteristic frequency.

What is the implication of the main finding?

- The proposed method FDMK-SVD can effectively extract fault features from raw vibration signals, even when faced with significant background noise and other interferences, enabling accurate fault diagnosis of rolling bearings.

Abstract: Due to the harsh and complex operating conditions, rolling element bearings (REBs) are prone to failures, which can result in significant economic losses and catastrophic breakdowns. To efficiently extract weak fault features from raw signals, singular value decomposition (SVD)-based signal denoising methods have been widely adopted in the field of rolling bearing fault diagnosis. In traditional SVD-based methods, singular components (SCs) with significant singular values are selected to reconstruct the denoized signal. However, this approach often overlooks low-energy SCs that contain important fault information, leading to inaccurate diagnosis. To address this issue, we propose a new selection scheme based on frequency domain multipoint kurtosis (FDMK), along with a reweighting strategy based on FDMK to further emphasize weak fault features. In addition, the estimation process of fault characteristic frequency is introduced, allowing FDMK to be calculated without prior information. The proposed FDMK-SVD can adaptively extract periodic fault features and accurately identify the health condition of REBs. The effectiveness of FDMK-SVD is validated using both simulated and experimental data obtained from a locomotive bearing test rig. The results show that FDMK-SVD can effectively extract fault features from raw vibration signals, even in the presence of severe background noise and other types of interferences.

Keywords: singular value decomposition; frequency domain multipoint kurtosis; signal denoising; rolling element bearings; fault diagnosis

1. Introduction

Rolling element bearings (REBs) are essential components of rotating machinery and play a significant role in the modern industry. Due to their harsh operating conditions, REBs are prone to unexpected failures, which can lead to machine breakdowns and substantial economic losses [1,2]. Consequently, the condition monitoring of REBs has become crucial, attracting significant attention from both academia and industry [3,4].

Since the vibration signal contains abundant fault information that can effectively reflect the health conditions of REBs, vibration analysis-based diagnostic methods have been widely applied for fault diagnosis [5]. When a fault occurs in a bearing, a series of repetitive impulses with a unique period are produced in the vibration signal and excite the resonances of the bearing and mechanical system [6,7]. Therefore, fault diagnosis of REBs can be achieved by detecting the periodical fault impulses in the time domain waveform or fault characteristic frequency (FCF) in the envelope spectrum domain. However, fault-induced repetitive transients are often submerged by strong background noise and other interferences, making fault diagnosis in real-world applications challenging. Therefore, extracting the repetitive impulses from the raw vibration signal is a critical and difficult task in the fault detection of REBs.

In recent years, many researchers have worked to address this issue, proposing various diagnostic methods, including spectral kurtosis [8,9], wavelet transform (WT) [10], empirical mode decomposition (EMD) [11,12], minimum entropy deconvolution (MED) [13–15], and singular value decomposition (SVD). Among these methods, SVD has garnered significant attention due to its ability to perform adaptive decomposition and its high computational efficiency.

Due to the abovementioned merits, SVD-based signal denoising methods have been widely employed in condition-based maintenance (CBM). Zhao et al. [16] illustrated the similarities between SVD and WT when the Hankel matrix is selected to convert one-dimension discrete signal into two-dimension matrix. On the basis of the above findings, a singular value decomposition packet (SVDP) algorithm was developed for decomposing vibration signals and extracting weak fault features [17]. Moreover, SVD has been employed to enhance the signal-to-noise ratio (SNR) in both the time and frequency domains [18], which efficiently promotes the improvement of the reliability of fault detection. Cong et al. presented a hybrid method consisting of SVD, singular value ratio, and short-time matrix series for the vibration signal analysis [19], which achieves an accurate diagnosis of rolling bearings. Duan et al. [20] proposed an impulsive feature extraction with improved SVD and sparsity-closing morphological analysis, in which A length factor is designed to optimize the construction of the matrix. Jiang et al. [21] combined SVD with the whale optimization algorithm for the quantitative detection of multiple damages in turbine blades. In addition to the SVD-based signal denoising method, SVD can also be combined with other fault diagnosis methods, where SVD is utilized to extract the fault feature for further signal processing. Jiang et al. [22] proposed a novel method combining SVD with a continuous hidden Markov model, achieving high classification accuracy in rolling bearing diagnosis. In addition, Kang et al. [23] presented an SVD-based method for fault classification, in which the features extracted using SVD were adopted as the input of the multi-layer support vector machines. Muruganatham et al. [24] took the singular value or the energy of SCs as the input features for the neural network and successfully achieved the automated fault diagnosis of REBs.

The key to the SVD-based signal denoising method is to determine an appropriate criterion for selecting the SC for subsequent signal reconstruction. A difference spectrum [25] was proposed to capture the sudden change point, which reflects the boundary between the signal and noise. These change points are then used to select effective singular values. A similar approach based on the correlation coefficient was introduced in [26]. However, periodicity and impulsiveness are not taken into account in the above two methods, which may cause effective singular values to be ignored and even incorrect diagnostic results. To address this issue, a novel information index, the periodic modulation intensity (PMI) [27], was used to evaluate whether the SCs contained diagnostic information. This index helps to accurately select SCs that carry fault-related information, thereby improving the reliability of fault diagnosis. However, one challenge of PMI is that the fault period of interest must be predetermined, which can hinder its application in real-world situations in which prior knowledge of the fault period is not always available.

To overcome the drawbacks of the abovementioned SVD-based signal denoising method, this paper proposes a new selection scheme based on frequency domain multi-point kurtosis (FDMK) to select the effective SCs. FDMK evaluates SCs based on their periodicity and impulsiveness, ensuring the accurate selection of SCs that contain fault-related information, unlike traditional SVD-based methods that retain only large-energy SCs. Additionally, a reweighting strategy based on FDMK is introduced to enhance weak fault features during the reconstruction of the denoized signal. In addition, to solve the problem of the fault characteristic frequency or fault period not being given in advance, the process of estimating the fault period was incorporated, which allowed the proposed FDMK-SVD to operate effectively without prior knowledge of the fault period or characteristic frequency. Therefore, the proposed FDMK-SVD can adaptively achieve fault diagnosis and health assessment of REBs without prior information.

The remainder of this paper is organized as follows. In Section 2, the principle of the SVD-based signal denoising method is briefly introduced. In Section 3, the proposed FDMK-SVD for fault diagnosis of REBs is presented. Specifically, Sections 3.1 and 3.2 introduce the definition of FDMK and the estimation process of fault characteristic frequency (FCF), respectively, and the detailed flowchart of the FDMK-SVD is illustrated in Section 3.3. Section 4 presents the preliminary validation of the FDMK-SVD using a simulated signal. Subsequently, in Section 5, the vibration signal acquired from the experiment is used to further confirm the effectiveness of FDMK-SVD. Finally, the conclusions are summarized in Section 6.

2. Principle of SVD-Based Signal Denoising Approach

For a real matrix $A \in R^{m \times n}$, the SVD of the matrix A can be defined as follows [16,25].

$$A = \boldsymbol{U\Sigma V}^T \tag{1}$$

where the matrices $\boldsymbol{U} = [u_1, u_2, \ldots, u_m] \in R^{m \times m}$ and matrix $\boldsymbol{V} = [v_1, v_2, \ldots, v_n] \in R^{n \times n}$ are orthogonal to each other. The column vectors of matrices $\boldsymbol{U}$ and $\boldsymbol{V}$ are the eigenvectors of $\boldsymbol{AA}^T$ and $\boldsymbol{A}^T\boldsymbol{A}$, respectively. $\boldsymbol{\Sigma} = [\mathrm{diag}(\sigma_1, \sigma_2, \ldots, \sigma_k), \boldsymbol{0}] \in R^{\mathrm{m} \times \mathrm{n}}$ is a diagonal matrix, where $\boldsymbol{0}$ is a zero matrix, $k = \min(m, n)$, σ is the singular value of matrix $\boldsymbol{A}$, and $\sigma_1 \geq \sigma_2, \ldots, \geq \sigma_{\mathrm{k}}$.

As illustrated in [25,27,28], SVD is effectively used for vibration signal denoising, which can be achieved through the following steps:

Step 1: Construction of Hankel matrix

In practice, the vibration signal is usually a one-dimensional time series rather than a matrix; thus, converting the one-dimensional signal into a matrix A is indispensable. Some matrix forms have been presented, such as the Toeplitz matrix, the cycle matrix, and the Hankel matrix. Because the Hankel matrix has zero phase-shifting characteristics and wavelet-like properties, the Hankel matrix is employed in [16].

The Hankel matrix of the discrete signal $x = [x(1), x(2), \ldots, x(N)]$ can be written as [27]

$$A = \begin{bmatrix} x(1) & x(2) & \cdots & x(n) \\ x(2) & x(3) & \cdots & x(n+1) \\ \vdots & \vdots & \ddots & \vdots \\ x(m) & x(m+1) & \cdots & x(N) \end{bmatrix} \tag{2}$$

where $1 < n < N$ and $m + n - 1 = N$. Parameter m is usually selected less than n, and it is suggested to set as 3 times the number of sub-components of the vibration signal [29]. This choice of m also determines the number of SCs after performing SVD.

Step 2: Signal decomposition and reconstruction

Once matrix A is constructed, it can be expressed as the product of matrices U, Σ, and V^T according to Equation (1). This means matrix A can be decomposed into a sum of unit-rank elementary matrices that are mutually orthogonal, which can be expressed as

$$\begin{aligned} A &= [u_1, u_2, \ldots, u_m] \begin{bmatrix} \sigma_1 & 0 & \cdots & 0 & \mathbf{0} \\ 0 & \sigma_2 & \cdots & 0 & \mathbf{0} \\ \vdots & \vdots & \ddots & 0 & \mathbf{0} \\ 0 & 0 & \cdots & \sigma_m & \mathbf{0} \end{bmatrix} \begin{bmatrix} v_1^T \\ v_2^T \\ \vdots \\ v_n^T \end{bmatrix} \\ &= \sigma_1 u_1 v_1^T + \sigma_2 u_2 v_2^T + \ldots + \sigma_m u_m v_m^T \\ &= A_1 + A_2 + \ldots + A_m \end{aligned} \tag{3}$$

where $u_i \in R^{m\times 1}$ and $v_i \in R^{m\times 1}$ are the ith column vectors of matrices U and V, respectively.

As illustrated in Equation (3), matrix A can be decomposed into the sum of sub-matrix A_i, and each singular component x_i corresponding to sub-matrix A_i can be obtained using two methods, as demonstrated in Figure 1. In the direct method, the singular component x_i can be obtained by concatenating the first row and sub-column vector in the last column of A_i, which can be written as

$$x_i = \left[R_{i,1}, C_{i,n}^T\right] \tag{4}$$

where $R_{i,1} \in R^{n\times 1}$ and $C_{i,n} \in R^{(m-1)\times 1}$. Another way to obtain the singular component x_i is by diagonal averaging, where each element of x_i can be obtained by averaging along the anti-diagonals, as shown in Figure 1b. Interestingly, regardless of any of the methods mentioned above being used, the raw signal can be perfectly reconstructed by adding all decomposed SCs, as shown in Equation (5).

$$x = \sum_{i=1}^{m} x_i \tag{5}$$

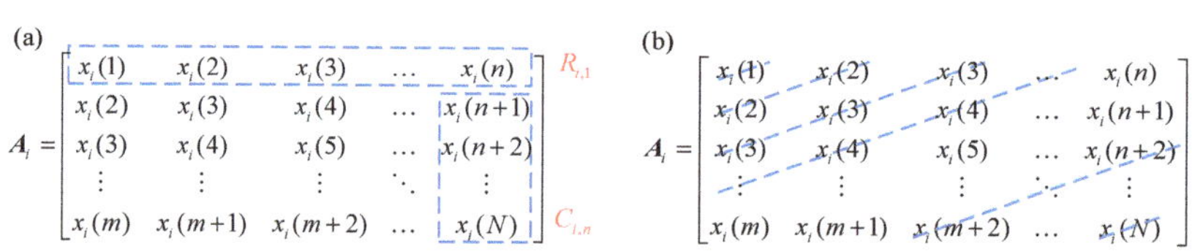

Figure 1. Obtaining x_i from the sub-matrix A_i. (**a**) direct method; and (**b**) anti-diagonals averaging method.

Step 3: Signal denoising

As illustrated in Equations (3) and (5), each singular value is associated with an SC. This means that the selection of effective SCs can be transformed into the selection of the corresponding singular values. Therefore, the SVD-based signal denoising method can be achieved by selecting some effective SCs and simply adding their corresponding component signals together. In general, the SCs with large singular values are considered as the effective components and would be retained [25], which can be expressed as

$$\widetilde{A} = [u_1, u_2, \ldots, u_m] \begin{bmatrix} \sigma_1 & 0 & \cdots & 0 & \mathbf{0} \\ 0 & \sigma_2 & \cdots & 0 & \mathbf{0} \\ \vdots & \vdots & \ddots & \vdots & \vdots \\ 0 & 0 & \cdots & \sigma_k & \mathbf{0} \\ \mathbf{0} & \mathbf{0} & \cdots & \mathbf{0} & \mathbf{0} \end{bmatrix} \begin{bmatrix} v_1^T \\ v_2^T \\ \vdots \\ v_n^T \end{bmatrix} = \sum_{i=1}^{k} \sigma_i u_i v_i^T = \sum_{i=1}^{k} A_i \Rightarrow \widetilde{x} = \sum_{i=1}^{k} x_i \quad (6)$$

where the first k largest singular values are added together; $\widetilde{A}$ and $\widetilde{x}$ represent the denoized matrix and signal, respectively.

In the abovementioned SVD-based signal processing approach, it is important to determine the proper threshold k to guarantee denoising performance. The selection of k has been investigated in several studies, such as the difference spectrum [25], correlation coefficients [26], and the median value of singular values [28]. Among these methods, the difference spectrum is frequently adopted because of its easy implementation and good performance in applications. More details about the difference spectrum can be found in [28].

3. Reweighted Singular Value Decomposition Based on Frequency Domain Multipoint Kurtosis

3.1. Frequency Domain Multipoint Kurtosis

In the fault diagnosis of REBs, many indicators can be used to detect whether the signal contains fault information and quantify the extent of this information. Common indicators include Shannon entropy [30,31], Gini index [32], kurtosis [33], and l_2/l_1 norm [34]. However, most of these indicators mentioned above are introduced from the field of statistics, economics, and information theory, and the periodicity of the defect impulses caused by defective rotating machinery is not considered. In order to take into account the periodicity of the defect impulses, several other indicators, such as correlated kurtosis [35], envelope harmonic-to-noise ratio (EHNR) [36], and multipoint kurtosis (MK) [37], have been proposed in the literature. In this paper, the multipoint kurtosis that considers the periodicity of the defect impulses is utilized to select the sensitive SCs containing the information related to the health conditions of REBs.

The multipoint kurtosis of M-shift of zero-mean data $\boldsymbol{y}$ is expressed as [37]

$$MK(\vec{y}, \vec{t}) = \frac{\left(\sum_{n=1}^{N} t_n^2\right)^2}{\sum_{n=1}^{N} t_n^8} \frac{\sum_{n=1}^{N} (t_n y_n)^4}{\left(\sum_{n=1}^{N} y_n^2\right)^2} \quad (7)$$

where N is the length of the original signal $\boldsymbol{y}$ and $\vec{t}$ is the target vector. For example, $\vec{t}$ can be defined as

$$\vec{t} = \begin{bmatrix} 0 & 0 & 0 & 1 & 0 & 0 & 0 & 1 & 0 & \ldots & 0 & 1 \end{bmatrix}^T \quad (8)$$

The above $\vec{t}$ controls the locations of the interested multiple points and the interval between adjacent peaks or impulses, where the interval is four points and the first impulse is located at the 4th point.

Although the MK can extract repetitive impulses, a major drawback still exists that hinders its application. As demonstrated in Equation (8), the locations of the impulses are assumed to be known in advance. However, it is impossible to determine the locations of fault impulses in actual situations, which makes it unrealistic to calculate the MK directly in the time domain. As illustrated in the previous literature, the position of the frequency lines associated with the FCF and its harmonics is constant in the envelope spectrum for periodic fault transients, regardless of their positions in the time waveform. Therefore, it is practical to calculate the MK in the frequency domain (envelope spectrum), and FDMK can be defined by

$$FDMK(\vec{y},\vec{t}) = \frac{\left(\sum\limits_{n=1}^{N} t_n^2\right)^2 \sum\limits_{n=1}^{N} (t_n S_n)^4}{\sum\limits_{n=1}^{N} t_n^8 \left(\sum\limits_{n=1}^{N} S_n^2\right)^2} \tag{9}$$

where S represents the envelope spectrum of the signal and $\vec{t}$ represents the target vector. According to this definition, it can be observed that the FCF and sampling frequency f_s must be predetermined to construct the target vector $\vec{t}$. However, it is sometimes impossible to obtain the FCF in advance, or the theoretical FCF deviates from the real value, which leads to an inaccurate FDMK and even incorrect diagnostic results. To address this issue, a new strategy that can adaptively estimate the fault characteristic frequency is presented in the next section.

3.2. The Estimation of Fault Characteristic Frequency and Construction of Target Vector $\vec{t}$

As demonstrated in [38,39], the autocorrelation function can be utilized to estimate the period of the fault impulses. The definition of autocorrelation function of $x(t)$ is expressed as follows

$$r_x(\tau) = \int x(t)x(t+\tau)dt \tag{10}$$

As illustrated in Figure 2, τ is the time lag, and the autocorrelation function of $x(t)$ reaches a local maximum when $\tau = \tau_{\max}$. Therefore, $\tau_{\max}$ is considered the fault period.

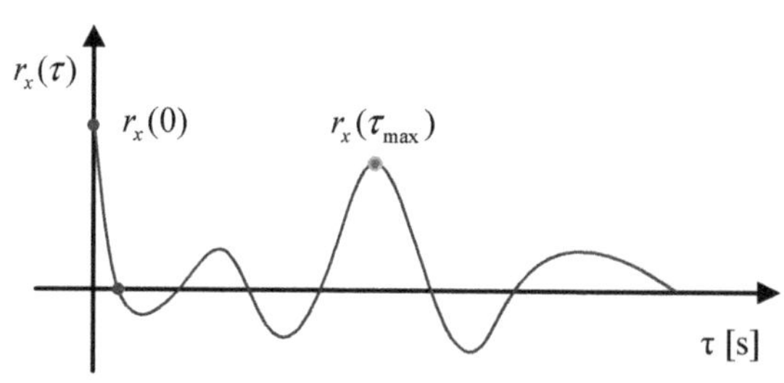

Figure 2. The autocorrelation function of $x(t)$.

However, the above method for estimating the period or FCF through the envelope autocorrelation function of the raw vibration signal may fail when the defect impulses are submerged by harmonics or severe background noise. To overcome these limitations, a new strategy based on SVD to estimate the period or fault characteristic frequency is proposed in this paper.

Step 1: Obtain the singular components using SVD.

In real applications, defect impulses are usually overwhelmed by harmonic components and background noise, leading to inaccurate estimates of the fault period or FCF from the raw signal. Therefore, SVD is adopted to separate the components in the raw signal, and SCs are obtained.

Step 2: Calculate the Gini index (GI) of each SC and then calculate the envelope harmonic-to-noise ratio (EHNR) to select the most sensitive SC.

In this part, GI and EHNR are introduced to select the most sensitive SC containing the diagnostic information. The definitions of GI and EHNR are provided below.

The Gini index (GI) was originally used to represent the wealth distribution of a nation's population. In recent years, the GI has been increasingly used as a measure of signal sparsity [32,40,41]. The GI values range from 0 to 1, and a large GI value indicates that the signal is sparser. For a real signal indexed in non-decreasing order, its GI can be given below

$$GI(x) = 1 - 2\sum_{i=1}^{N} \frac{|\mathbf{x}_i|}{\|\mathbf{x}\|_1}\left(\frac{N - i + 0.5}{N}\right) \tag{11}$$

where $||x||_1$ represents the l_1-norm of x and N is its length.

EHNR is proposed for detecting periodic impulses [39,42], and the formula is defined as follows:

$$EHNR = \frac{r_{E[x(t)]}(\tau_{\max})}{r_{E[x(t)]}(0) - r_{E[x(t)]}(\tau_{\max})} \tag{12}$$

where $E[x(t)]$ represents the envelope of signal$x$, and the autocorrelation function of $E[x(t)]$ reaches a local maximum when $\tau = \tau_{\max}$.

After separating the components contained in the raw signal, the GI value of each singular component is calculated, and the EHNR of the first several SCs with large GI values are further calculated. In this way, the periodicity and sparsity of the defect impulse are considered at the same time. The SC with the maximum ENHR is considered the most sensitive one, which is related to the fault feature.

Step 3: Estimate the fault period or fault characteristic frequency

The envelope autocorrelation function of the selected SC is calculated, and the time lag $\tau_{\max}$ that causes the envelope autocorrelation function to reach a local maximum is considered the fault period. Correspondingly, the fault characteristic frequency is $1/\tau_{\max}$.

3.3. The Proposed FDMK-SVD for Fault Detection of REBs

The framework of the proposed FDMK-SVD for bearing fault detection is shown in Figure 3, and the concrete steps are presented as follows:

Step 1: Signal decomposition

A one-dimensional signal must first be converted into a Hankel matrix, as illustrated in Section 2 for SVD. When a relatively large m is selected, the raw signal, consisting of fault impulse components, harmonic components, and other interferences, can be separated, and the computational cost increases as m increases. According to the analysis of the simulated signal and experimental signals, m = 30 is appropriate for separating the components contained in the vibration signal under different situations.

Step 2: Estimate the fault characteristic frequency

After obtaining SCs, GI and EHNR are set as the criteria to select the most sensitive SC, followed by the estimation of the fault period or FCF using the envelope autocorrelation function of the selected SC.

Step 3: Construct the target vector $\vec{t}$ After obtaining the characteristic frequency, the corresponding target vector $\vec{t}$ can be generated. Because the estimated fault characteristic has a slight deviation from the real value, the locations of '1' in the target vector $\vec{t}$ are determined by seeking the locations of the local maximum around the estimated value, which indicates that '1' is not equally distributed.

Step 4: Calculate the FDMK of each singular component

In this step, the FDMK of each singular component is calculated to evaluate whether the diagnostic information is included.

Step 5: Denoized signal reconstruction

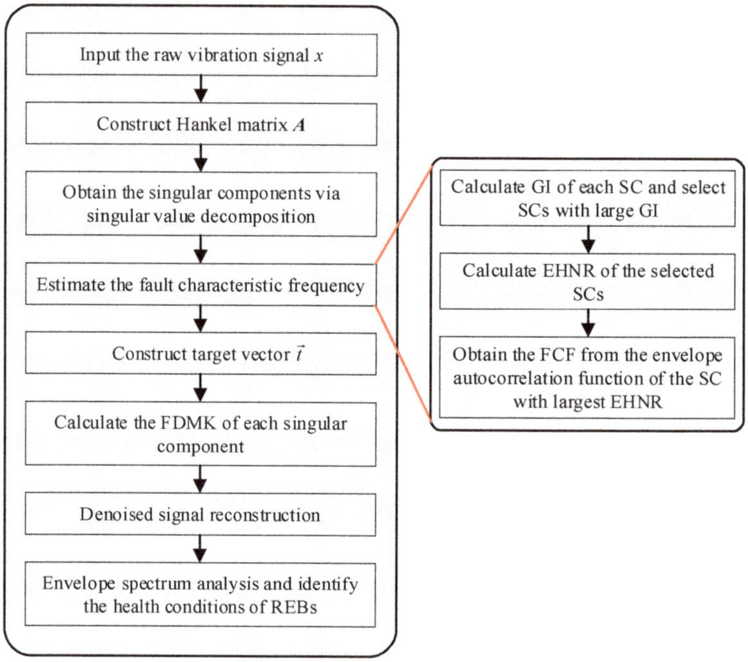

Figure 3. The flowchart of the proposed FDMK-SVD.

The denoized signal is obtained by assigning a new weighting coefficient to each SC and then adding the weighted SCs, as given in Equations (13) and (14).

$$\tilde{X} = w(1)X_1 + w(2)X_2 + w(3)X_3 + \cdots + w(m)X_m \quad (13)$$

$$w(i) = \begin{cases} 0 & mk_i \leq threshold \\ mk_i/\sum_i mk_i & mk_i > threshold \end{cases} \quad (14)$$

In Equation (14), the weighting function $w(i)$ determines whether each SC is highlighted or suppressed when reconstructing the denoized signal. The greater the weight value of each SC, the greater its contribution to reconstructing the denoized signal. For each SC, if its corresponding $mk_i \leq threshold$, then the SC is considered to contain no fault feature information, and it is culled by assigning it a weight of zero when reconstructing the denoized signal. Conversely, if the SCs with $mk_i > threshold$, then it is considered to contain fault information and should be preserved. Furthermore, since different SC contain different levels of information, SC is reweighted by the normalized FDMK, as shown in Equation (14). This ensures that SCs with larger FDMK can be highlighted more effectively in the reconstruction of the denoized signal. In this paper, the threshold is empirically set to 0.02.

Step 6: Envelope spectrum analysis and condition identification

Finally, the envelope spectrum of the reconstructed denoized signal is analyzed to detect the fault characteristic frequencies. By identifying the presence of these frequencies in the envelope spectrum, the health condition of the REBs can be accurately assessed.

4. Simulation Analysis

In this section, a simulated signal of a faulty rolling bearing is constructed to verify the effectiveness of the proposed FDMK-SVD. As mentioned above, the bearing vibration signal collected from an actual industrial environment contains multiple signal components while being submerged in heavy background noise. To create a more realistic signal model, we incorporate four common signal components into the simulated signal of a faulty rolling bearing, as described below [43].

$$y(t) = \underbrace{\sum_n A_n \cos(2\pi n f_0 t + a_n)}_{Rotorvibration} + \underbrace{\sum_n B_n \cos(2\pi n T f_0 t + \beta_n)}_{Gearmeshing} \\ + \underbrace{\sum_i D_i S_d(t - iT_d - \tau_i)}_{Defectimpulses} + \underbrace{\sum_i R_i S_r(t - T_i)}_{Randomshocks} + n(t) \tag{15}$$

The first part is the vibration from the rotor or shaft, where f_0 is the rotation frequency, A_n and a_n represent the amplitude and phase of the nth harmonic, respectively. The vibration signal originating from gear meshing is represented by the second part, where the number of gear teeth is denoted as T, and the amplitude and phase of the nth meshing frequency are represented by B_n and β_n. The third part represents the defect impulses caused by the failure of the rolling bearing, where the amplitude of the ith impulse is represented by D_i, and T_d represents the interval time between the two adjacent impulses. Moreover, τ_i is used to simulate the random slip of the bearing, which is usually set to 1~2% of the fault period.

In general, the defect impulses can be represented as an exponentially decaying sinusoid, as follows:

$$S_d(t) = e^{(-\alpha_r t)} sin(2f_r t) \tag{16}$$

where f_r is the resonant frequency of the impulse response and α_r is the damping ratio of the impulses.

The fourth part is the external random shock, where R_i and T_i denote the amplitude and occurrence time of the ith random shock, respectively. Furthermore, $n(t)$ represents white Gaussian noise.

Table 1 lists the parameters of the simulated signal. In the simulation, the time length of the simulated signal and sampling frequency are set as 1 s and 10,000 Hz, respectively. Figure 4a–d display the time waveforms of the rotor vibration, gear meshing, defective impulses and random shocks, respectively. After adding white Gaussian noise with an SNR of −3 dB, the synthetic signal is shown in Figure 4e, and no obvious defect impulses can be recognized. The envelope spectrum of the simulated signal is presented in Figure 4f, where the frequency lines of the BPFO are ambiguous, and no signature indicates the existence of a bearing fault. Here, the BPFO represents the FCF of the outer race.

Table 1. The parameters of the simulated signal.

Rotor and Shaft			Gear Meshing					Defect Impulses				Random Shocks	
f_0	A_1	α_1	T	B_1	β_1	B_2	β_2	D_1	T_d	f_r	α_r	f_r	α_r
14	0.1	$\pi/2$	30	0.2	$\pi/2$	0.08	$\pi/2$	0.3	1/50	2400	200	4000	700

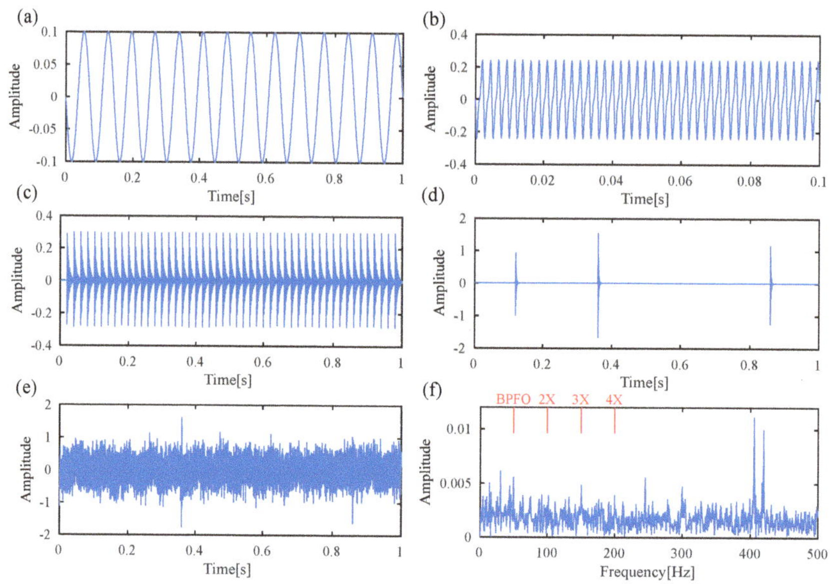

Figure 4. Simulated signal. (**a**) rotor vibration; (**b**) gear meshing; (**c**) defect impulses; (**d**) random shocks; (**e**) noise-added mixture signal; (**f**) envelope spectrum.

Next, the traditional DS-SVD [25] and the proposed FDMK-SVD are applied to analyze the simulated vibration signal. In traditional DS-SVD, the threshold k is determined by the difference spectrum (DS), and the fault information is considered to be contained in the first k SCs. Figures 5 and 6 show the results obtained using DS-SVD. According to the difference spectrum shown in Figure 5, the maximum peak is observed at $k = 2$, which means that only the first two singular components are selected to reconstruct the denoized signal. Figure 6a presents the denoized signal, and no obvious defect impulses are observed. The envelope spectrum of the reconstructed signal is displayed in Figure 6b, and it also fails to provide useful information for fault diagnosis. To further investigate the reason for this phenomenon, the decomposition results obtained using SVD are shown in Figure 7. As shown in Figure 7, the first two SCs selected for reconstructing the denoized signal based on DS-SVD are composed of the harmonic components originating from gear meshing, and the defect impulses are ignored in the DS-SVD. The above results indicate that the selection scheme of traditional DS-SVD is inappropriate when the diagnostic information is very weak and disturbed by other large-energy components.

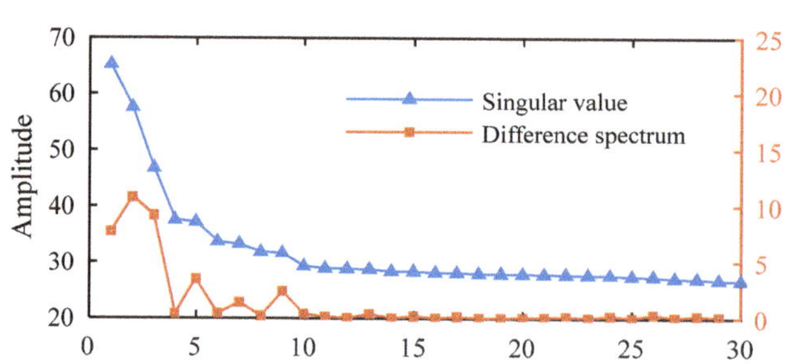

Figure 5. Singular values and difference spectra.

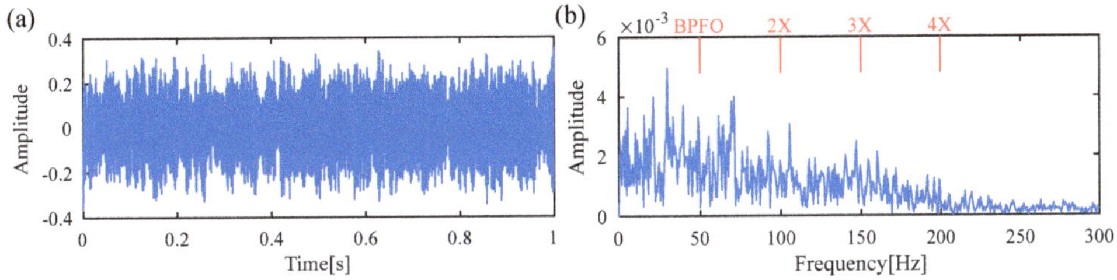

Figure 6. The results obtained by DS-SVD for simulated signal. (**a**) denoized signal; (**b**) envelope spectrum.

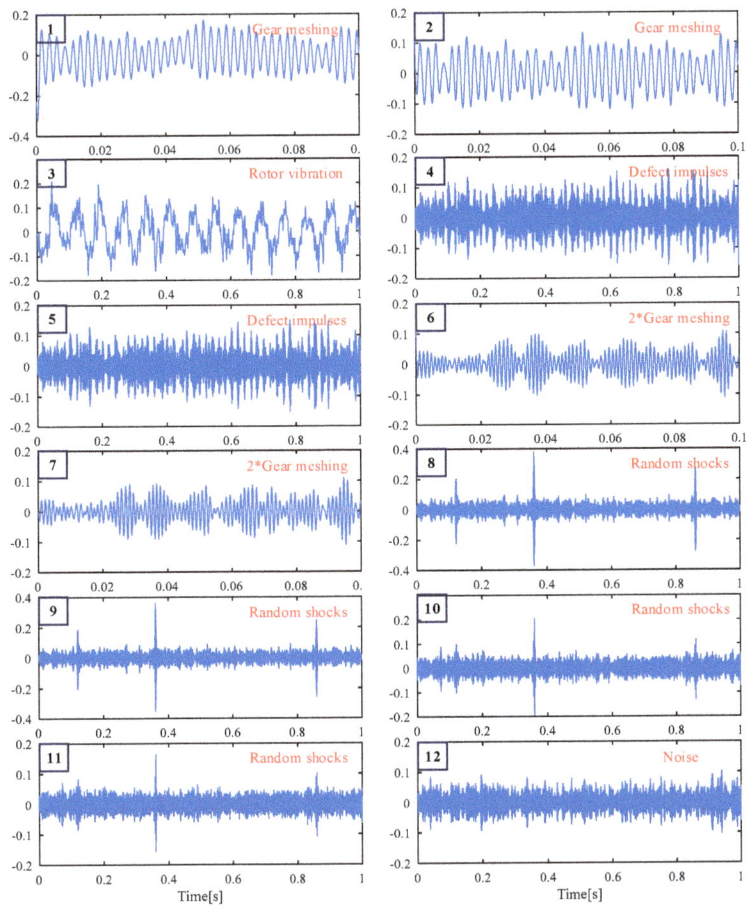

Figure 7. The first 12 singular components obtained using SVD.

For comparison, the same simulated signal is processed using the proposed FDMK-SVD. The envelope autocorrelation function of the selected most sensitive SC and raw signal is displayed in Figure 8. As shown in Figure 8a, the local maximum is located at τ_{max} = 0.0202 s, and its corresponding frequency is 49.505 Hz, which has a slight deviation from the real value. On the contrary, the fault period is masked by the harmonic period (τ_{max} = 0.0024 s) corresponding to gear meshing when the raw signal is used to estimate the period or FCF. The reason for this phenomenon is that the energy of the defect impulses is less than that of the harmonic components. After obtaining the estimated fault characteristic frequency, the target vector $\vec{t}$ is constructed, and the FDMK value of SC is calculated and is displayed in Figure 9. As shown in Figure 9, although the first three SCs have large energies, their FDMKs are very small since they are not relevant to the bearing fault. On the contrary, the FDMK of the 4th and 5th components are significantly larger than those of the others. By checking the decomposition results obtained by SVD shown in Figure 7, the 4th and 5th SCs correspond to the defect impulses, which is significant information for identifying the health conditions of bearings. The time waveforms of the denoized signal and the corresponding envelope spectrum are displayed in Figure 10. According to Figure 10, the defect impulses are successfully extracted in the time domain, and BPFO and its harmonics are clearly recognized in the envelope spectrum. There is no doubt that the fault is successfully detected by the proposed FDMK-SVD.

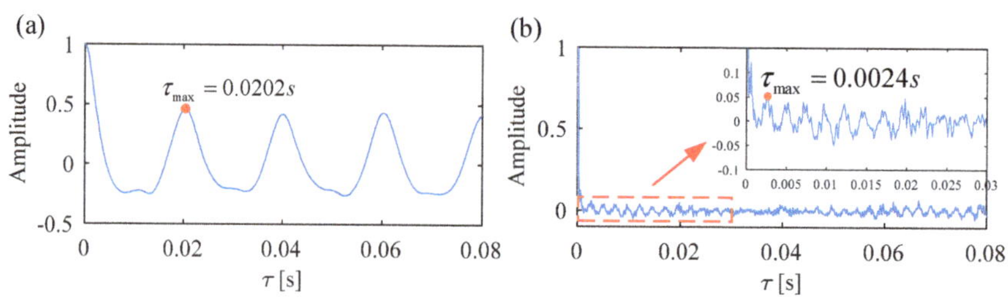

Figure 8. Envelope autocorrelation function of (**a**) selected singular component and (**b**) raw signal.

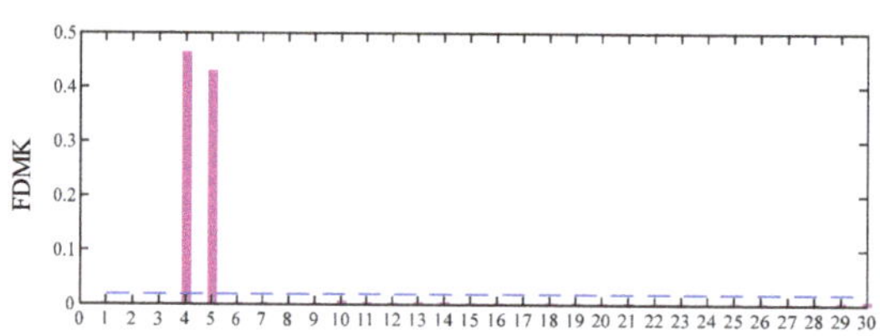

Figure 9. FDMK of each singular component for simulated signal.

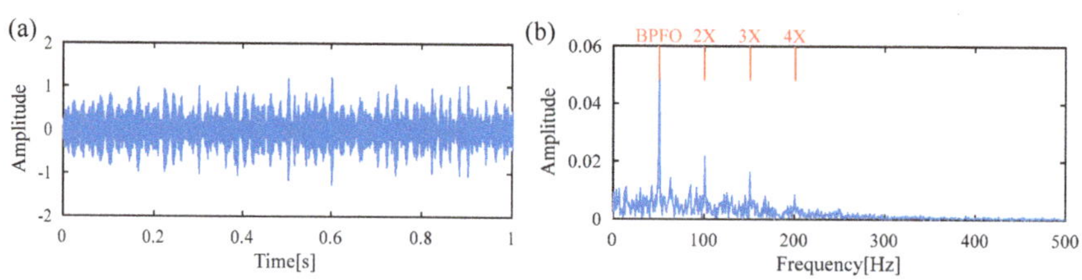

Figure 10. The results obtained by FDMK-SVD for simulated signal. (**a**) denoized signal; (**b**) envelope spectrum.

To evaluate the computational efficiency of the proposed method, the execution time is analyzed on a laptop equipped with a 12th Gen Intel® Core™ i7-12700H processor (2.7 GHz) and 16 GB DDR4 RAM, running MATLAB R2022a. The execution time for analyzing the simulated signal was 1.563 s. This relatively short processing time highlights the proposed method's suitability for real-time applications, including online monitoring. Therefore, the proposed method not only provides accurate fault diagnosis but also ensures computational efficiency, making it ideal for condition monitoring of bearings.

5. Experimental Verification

In this section, an experimental test is conducted on a locomotive bearing test rig to verify the validity of the proposed FDMK-SVD. The test rig consists of a driving wheel, a hydraulic motor, and a locomotive wheel, as shown in Figure 11. Notably, in this setup, the outer race of the bearing is driven by the hydraulic motor, while the inner race remains fixed. As a result, when a fault occurs in the outer race, amplitude modulation occurs, and sidebands are expected to appear in the envelope spectrum. In contrast, no amplitude modulation is observed for faults in the inner race. The vibration signal is collected by a three-axis PCB accelerometer with a sensitivity of 100 mV/g, which is placed at the end of the shaft. The sampling frequency is set to 76,800 Hz, and the data length is 1 s. Moreover, the speed information of the driving wheel is also acquired by a tachometer. In the experiment, the bearing under test is a tapered roller bearing, model 197,726. The geometric parameters of the locomotive bearings are listed in Table 2. To simulate real-world scenarios, the test rig was subjected to various noise sources, including mechanical vibrations from the surrounding machinery and environmental noise.

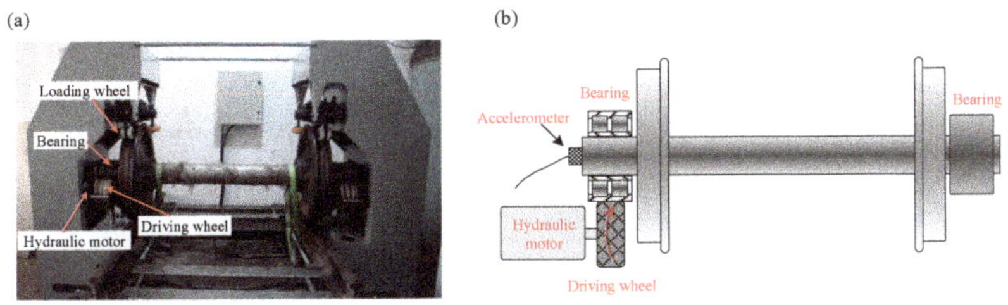

Figure 11. (**a**) Experiment setup of the test rig; (**b**) schematic view of the locomotive bearing test rig.

Table 2. Geometric parameters of the locomotive bearing.

Number of Rollers	Pitch Diameter (mm)	Roller Diameter (mm)	Contact Angle (Degree)
20	180	23.775	9

5.1. Experiment 1: The Bearing with Inner-Race-Fault

In this case, in order to further validate the validity of the proposed FDMK-SVD, the vibration signal from a bearing with an inner-race fault is processed. Table 3 lists the FCFs of the tested inner-race-fault bearing, where f_r specifies the rotating frequency of the outer race, and BPFO, BPFI, and BSF represent the FCF of the outer race, inner race, and rolling element, respectively.

Table 3. Bearing characteristic frequencies (Hz) in Experiment 1.

f_r	BPFO	BPFI	BSF
4.6183	40.1584	52.2083	17.1851

Figure 11 displays the raw vibration signal and its envelope spectrum. Owing to the heavy noise, the periodic defect impulses cannot be obviously observed from the time waveform, as shown in Figure 12a. Meanwhile, no significant BPFI or its harmonics are observed in the envelope spectrum, as shown in Figure 12b. To extract the weak fault feature from the raw vibration signal, DS-SVD is applied, and the corresponding denoized signal is illustrated in Figure 13. However, useful information related to the inner-race fault cannot be observed from the waveform of the denoized signal or its envelope spectrum. Moreover, it can be seen that the rotating frequency f_r is dominant in the envelope spectrum, which illustrates that the DS-SVD method, which relies on an energy-oriented selection scheme, fails to effectively select the sensitive SCs containing diagnostic information when harmonic components dominate the signal.

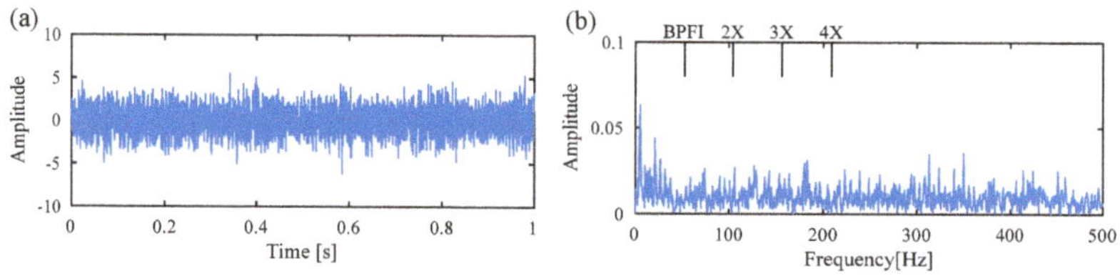

Figure 12. (**a**) Raw signal of the inner race fault bearing; (**b**) envelope spectrum.

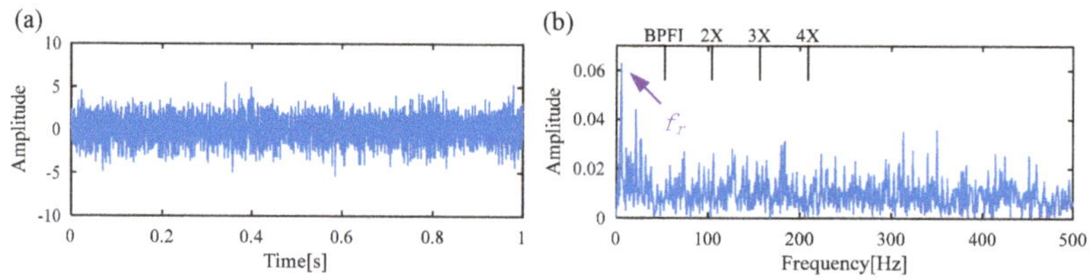

Figure 13. The results obtained by DS-SVD for the signal of the inner race fault bearing. (**a**) denoized signal; (**b**) envelope spectrum.

Spectral kurtosis (SK), a widely used and state-of-the-art method, is also adopted to extract the fault feature from the raw signal. Figure 14a presents the Kurtogram of the raw vibration signal, from which the resonance band with maximum spectral kurtosis (center frequency: 21,600 Hz; Bandwidth: 600 Hz) is easily determined. A band-pass filter with a center frequency of 21,600 Hz and a bandwidth of 600 Hz is designed to filter the vibration signal, and the filtered signal is displayed in Figure 14b. However, only several random impulses can be observed, and the periodic fault impulses are not detected. Similarly, the envelope spectrum in Figure 14c shows no obvious fault information. This failure results from the assumption that fault impulses occur in the frequency band with the largest kurtosis value. However, kurtosis is only a statistical indicator of signal sparsity, and the periodicity of the defect impulses is not taken into consideration. Therefore, it is susceptible to interference from non-Gaussian signals and random noise.

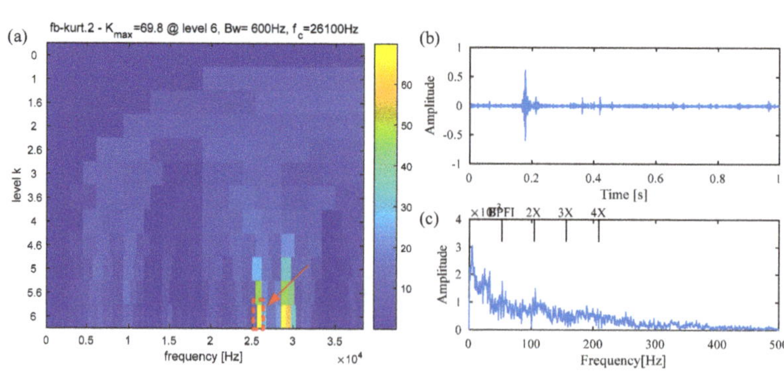

Figure 14. The results obtained by the SK method for the signal of the inner race fault bearing. (**a**) Kurtogram; (**b**) filtered signal; (**c**) envelope spectrum.

For comparison, the proposed FDMK-SVD is conducted on the same vibration signal. The FDMK value of each SC is calculated and is displayed in Figure 15. The denoized signal and its envelope spectrum are displayed in Figure 16. As expected, defect impulses are effectively detected, which illustrates the effectiveness of FDMK-SVD in extracting fault features and suppressing noise interference. Moreover, the BPFI and its harmonics are clearly observed, which provides evidence of the presence of an inner-race fault. To verify the results of the above analysis, the test bearing is disassembled, and the picture is illustrated in Figure 17, from which the inner-race fault of the bearing is clearly identified.

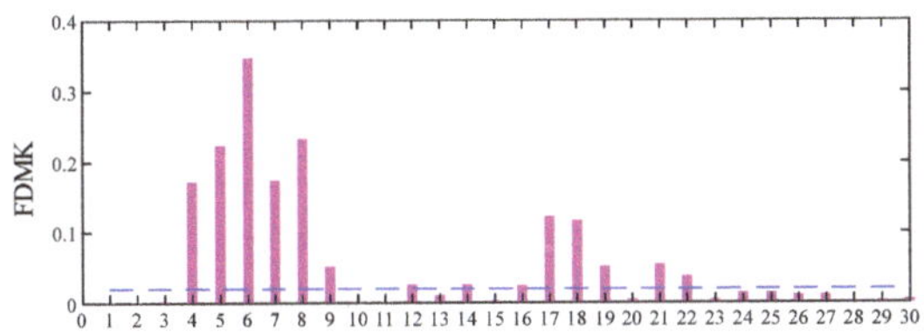

Figure 15. FDMK of each singular component for the signal of the inner race fault bearing.

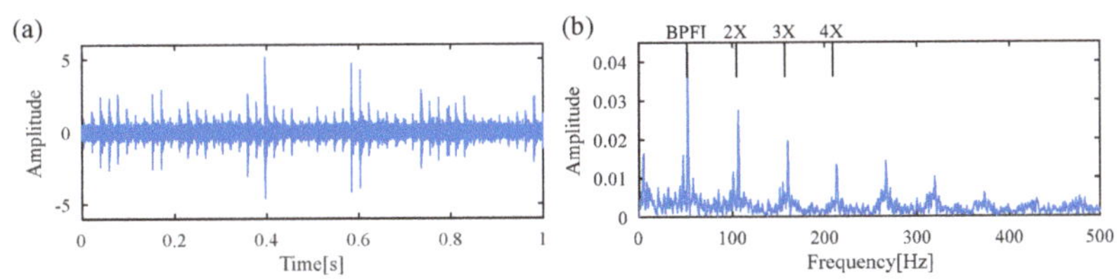

Figure 16. The results obtained by FDMK-SVD for the signal of the inner race fault bearing. (**a**) denoized signal; (**b**) envelope spectrum.

Figure 17. The fault on the inner race.

5.2. Experiment 2: The Bearing with Outer-Race Fault

In this case, the vibration signal from a bearing with an outer-race fault is used to further verify the effectiveness of FDMK-SVD in detecting bearing faults. Table 4 lists the fault characteristic frequencies for the outer-race fault.

Table 4. Bearing characteristic frequencies (Hz) in Experiment 2.

f_r	BPFO	BPFI	BSF
5.2747	45.8657	59.6281	19.6274

Figure 18 presents the time waveform of the raw vibration signal and its envelope spectrum, from which no periodic fault impulses or fault characteristic frequency can be observed. Then the DS-SVD is performed on the vibration signal, and the results are displayed in Figure 19. Similar to Experiment 1, it can be seen that the rotating frequency and its multipliers are dominant in the envelope spectrum, while the BPFO and its harmonics are not prominent, which implies that the suitable SCs with abundant information are not identified. Figure 20a shows the Kurtogram of the signal, where the resonance band is determined to be at the center frequency of 282,000 Hz with a bandwidth of 1200 Hz. The filtered signal obtained through the corresponding band-pass filter is presented in Figure 20b, where only random shocks are extracted. Moreover, there is no obvious fault characteristic frequency or its multiplier in the envelope spectrum. Finally, the proposed FDMK-SVD is applied to the same signal, and the results are shown in Figures 21 and 22. According to Figure 21, it can be seen that the FDMK of several SCs exceeds the threshold line, which implies that these SCs may contain useful diagnostic information. To validate the above analysis, a reweighting scheme is adopted, and the reconstructed signal is shown in Figure 22. Periodic fault impulses are effectively extracted, which verifies the effectiveness of FDMK-SVD in extracting weak fault features and suppressing noise interference. Moreover, the BPFO and its harmonics are obvious in the envelope spectrum, which indicates that the outer race of the rolling bearing is faulty. Then the locomotive bearing is disassembled, from which the outer-race fault of the bearing is clearly recognized, and the picture is illustrated in Figure 23.

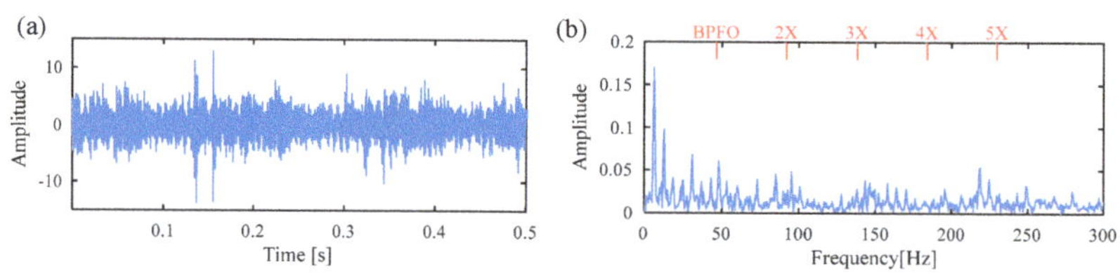

Figure 18. (**a**) Raw signal of the outer race fault bearing; (**b**) envelope spectrum.

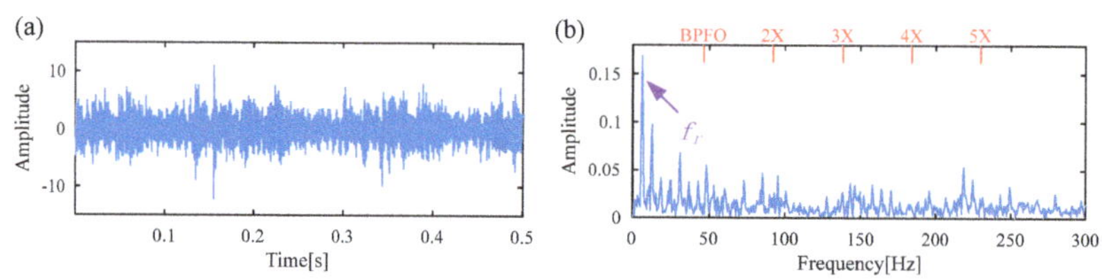

Figure 19. The results obtained by DS-SVD for the signal of the outer race fault bearing. (**a**) denoized signal; (**b**) envelope spectrum.

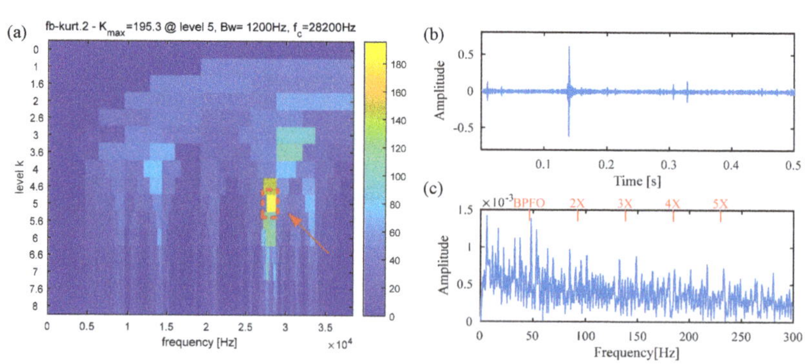

Figure 20. The results obtained by the SK method for the signal of the outer race fault bearing. (**a**) Kurtogram; (**b**) filtered signal; (**c**) envelope spectrum.

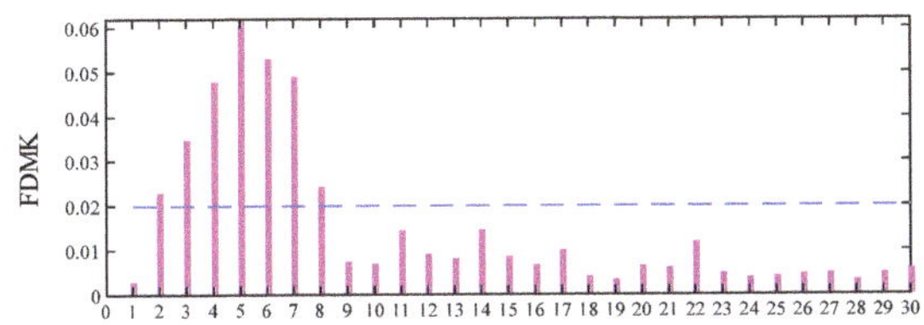

Figure 21. FDMK of each singular component for the signal of the outer race fault bearing.

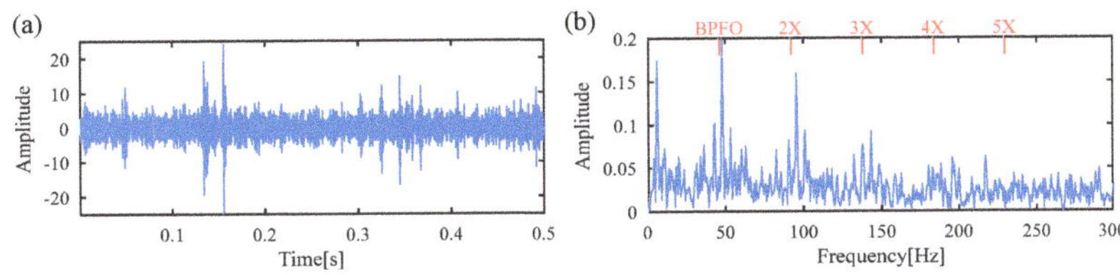

Figure 22. The results obtained by FDMK-SVD for the signal of the outer race fault bearing. (**a**) denoized signal; (**b**) envelope spectrum.

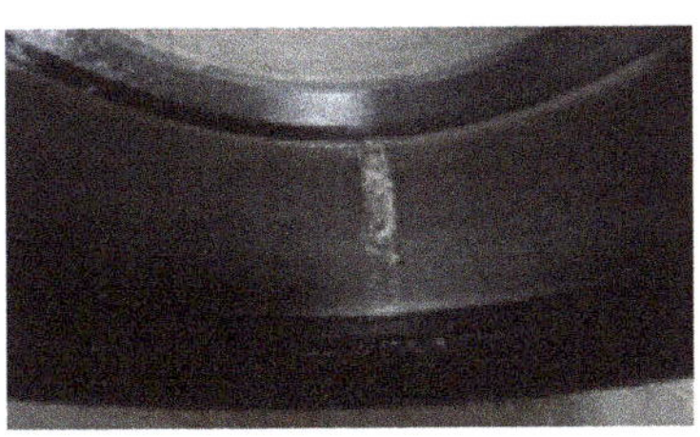

Figure 23. The fault on the outer race of the bearing.

6. Conclusions

This paper proposes a novel method, FDMK-SVD, for the diagnosis of rolling element bearings. In this work, the vibration signal is firstly decomposed via SVD, and then the sensitive SCs containing diagnostic information are selected and reweighted based on the FDMK index to reconstruct the denoized signal. Meanwhile, the estimation process of fault characteristic frequency is incorporated into the proposed method to address the issue that prior information about FCF is not predetermined in practice. Using the proposed method, weak fault features indicating the occurrence of bearings are successfully extracted. Moreover, the effectiveness and advantages of FDMK-SVD in detecting bearing faults are validated by comparing it with DS-SVD and Kurtogram. The results demonstrate that FDMK-SVD is a promising method for the condition monitoring of bearings, even in the presence of heavy noise and strong random interference.

The proposed FDMK-SVD has been validated for a constant-speed operation; however, real-world machinery often operates at varying speeds. Future work will focus on adapting SVD-based signal analysis to variable-speed conditions.

Author Contributions: Conceptualization, B.W.; methodology, B.W.; software, B.W.; writing—original draft preparation, B.W.; writing—review and editing, C.D.; and supervision, C.D. All authors have read and agreed to the published version of the manuscript.

Funding: This research was supported by the National Natural Science Foundation of China under Grant numbers. 52405124, 52205119, 52275157, the open project of the Intelligent Urban Rail Engineering Research Center of Jiangsu Province under Grant SDGC2414, and the Natural Science Foundation of Jiangsu Province under Grant BK20220497.

Institutional Review Board Statement: Not applicable.

Informed Consent Statement: Not applicable.

Data Availability Statement: Data are unavailable due to privacy restrictions.

Conflicts of Interest: The authors declare no conflicts of interest.

References

1. Que, H.; Liu, X.; Jin, S.; Huo, Y.; Wu, C.; Ding, C.; Zhu, Z. Partial transfer learning method based on inter-class feature transfer for rolling bearing fault diagnosis. *Sensors* **2024**, *24*, 5165. [CrossRef] [PubMed]
2. Xu, Y.; Li, S.; Feng, K.; Sun, B.; Yang, X.; Kou, L.; Zhao, Z.; Huang, G.Q. Multiperspective temporal pooling convolutional neural networks for fault diagnosis of mechanical transmission systems. *IEEE Trans. Instrum. Meas.* **2025**, *74*, 1–12. [CrossRef]
3. Shi, M.; Ding, C.; Chang, S.; Shen, C.; Huang, W.; Zhu, Z. Cross-domain class incremental broad network for continuous diagnosis of rotating machinery faults under variable operating conditions. *IEEE Trans. Ind. Inform.* **2024**, *20*, 6356–6368. [CrossRef]
4. Bhattacharjee, B.; Chakraborti, P.; Choudhuri, K. Selection of suitable lubricant for sliding contact bearing and the effect of different lubricants on bearing performance: A review and recommendations. *Tribol.-Finn. J. Tribol.* **2020**, *37*, 13–25. [CrossRef]
5. Ding, C.; Huang, W.; Shen, C.; Jiang, X.; Wang, J.; Zhu, Z. Synchroextracting frequency synchronous chirplet transform for fault diagnosis of rotating machinery under varying speed conditions. *Struct. Health Monit.* **2024**, *23*, 1403–1422. [CrossRef]
6. Ambrozkiewicz, B.; Litak, G.; Georgiadis, A.; Syta, A.; Meier, N.; Gassner, A. Effect of radial clearance on ball bearing's dynamics using a 2-DOF model. *Int. J. Simul. Model.* **2021**, *20*, 513–524. [CrossRef]
7. Dadouche, A.; Kerrouche, R. Bearing skidding detection for high speed and aero engine applications[C]//Turbo Expo: Power for Land, Sea, and Air. *Am. Soc. Mech. Eng.* **2021**, *85024*, V09AT24A014.
8. Zhang, B.; Miao, Y.; Lin, J.; Li, H. Weighted envelope spectrum based on the spectral coherence for bearing diagnosis. *ISA Trans.* **2022**, *123*, 398–412. [CrossRef]
9. Mauricio, A.; Gryllias, K. Cyclostationary-based multiband envelope spectra extraction for bearing diagnostics: The combined improved envelope spectrum. *Mech. Syst. Signal Process.* **2021**, *149*, 107150. [CrossRef]
10. Zhang, Q.; Ding, J.; Zhao, W. An adaptive boundary determination method for empirical wavelet transform and its application in wheelset-bearing fault detection in high-speed trains. *Measurement* **2021**, *171*, 108746. [CrossRef]
11. Sun, Y.; Li, S.; Wang, Y.; Wang, X. Fault diagnosis of rolling bearing based on empirical mode decomposition and improved manhattan distance in symmetrized dot pattern image. *Mech. Syst. Signal Process.* **2021**, *159*, 107817. [CrossRef]
12. Zheng, J.; Su, M.; Ying, W.; Tong, J.; Pan, Z. Improved uniform phase empirical mode decomposition and its application in machinery fault diagnosis. *Measurement* **2021**, *179*, 109425. [CrossRef]
13. Miao, Y.; Li, C.; Zhang, B.; Lin, J. Application of a coarse-to-fine minimum entropy deconvolution method for rotating machines fault detection. *Mech. Syst. Signal Process.* **2023**, *198*, 110431. [CrossRef]
14. Xiong, M.; Lv, Y.; Dang, Z.; Yuan, R.; Song, H. Early fault diagnosis of rolling bearings based on parameter-adaptive multipoint optimal minimum entropy deconvolution adjusted and dynamic mode decomposition. *Meas. Sci. Technol.* **2022**, *33*, 125101. [CrossRef]
15. López, C.; Wang, D.; Naranjo, Á.; Moore, K.J. Box-cox-sparse-measures-based blind filtering: Understanding the difference between the maximum kurtosis deconvolution and the minimum entropy deconvolution. *Mech. Syst. Signal Process.* **2022**, *165*, 108376. [CrossRef]
16. Zhao, X.; Ye, B. Similarity of signal processing effect between Hankel matrix-based SVD and wavelet transform and its mechanism analysis. *Mech. Syst. Signal Process.* **2009**, *23*, 1062–1075. [CrossRef]
17. Zhao, X.; Ye, B. Singular value decomposition packet and its application to extraction of weak fault feature, Mechanical Systems and Signal Processing. *Mech. Syst. Signal Process.* **2016**, *70*, 73–86. [CrossRef]
18. Golafshan, R.; Sanliturk, K.Y. SVD and Hankel matrix based de-noising approach for ball bearing fault detection and its assessment using artificial faults. *Mech. Syst. Signal Process.* **2016**, *70*, 36–50. [CrossRef]
19. Cong, F.; Chen, J.; Dong, G.; Zhao, F. Short-time matrix series based singular value decomposition for rolling bearing fault diagnosis. *Mech. Syst. Signal Process.* **2013**, *34*, 218–230. [CrossRef]
20. Duan, R.; Liao, Y. Impulsive feature extraction with improved singular spectrum decomposition and sparsity-closing morphological analysis. *Mech. Syst. Signal Process.* **2022**, *180*, 109436. [CrossRef]
21. Jiang, H.; Jiang, Y.; Xiang, J. Method using singular value decomposition and whale optimization algorithm to quantitatively detect multiple damages in turbine blades. *Struct. Health Monit.* **2024**, *23*, 1025–1036. [CrossRef]
22. Jiang, H.; Chen, J.; Dong, G.; Liu, T.; Chen, G. Study on Hankel matrix-based SVD and its application in rolling element bearing fault diagnosis. *Mech. Syst. Signal Process.* **2015**, *52*, 338–359. [CrossRef]

23. Kang, J.; Kim, M. Singular value decomposition based feature extraction approaches for classifying faults of induction motors. *Mech. Syst. Signal Process.* **2013**, *41*, 348–356. [CrossRef]
24. Muruganatham, B.; Sanjith, M.A.; Krishnakumar, B.; Murty, S.A.V.S. Roller element bearing fault diagnosis using singular spectrum analysis. *Mech. Syst. Signal Process.* **2013**, *35*, 150–166. [CrossRef]
25. Zhao, X.; Ye, B. Selection of effective singular values using difference spectrum and its application to fault diagnosis of headstock. *Mech. Syst. Signal Process.* **2011**, *25*, 1617–1631. [CrossRef]
26. Qiao, Z.; Pan, Z. SVD principle analysis and fault diagnosis for bearings based on the correlation coefficient. *Mech. Syst. Signal Process.* **2015**, *26*, 085014. [CrossRef]
27. Zhao, M.; Jia, X. A novel strategy for signal denoising using reweighted SVD and its applications to weak fault feature enhancement of rotating machinery. *Mech. Syst. Signal Process.* **2017**, *94*, 129–147. [CrossRef]
28. Zhao, X.; Ye, B.; Chen, T. Difference spectrum theory of singular value and its application to the fault diagnosis of headstock of lathe. *J. Mech. Eng.* **2010**, *46*, 100–108. [CrossRef]
29. Li, Q.; Weaver, B.; Chu, F.; Wang, W. A fast damping ratio identification method for rotating machinery based on the singular value of a directional power spectra density function matrix. *J. Sound Vib.* **2020**, *488*, 115630. [CrossRef]
30. Su, W.; Wang, F.; Zhu, H.; Zhang, Z.; Guo, Z. Rolling element bearing faults diagnosis based on optimal Morlet wavelet filter and autocorrelation enhancement. *Mech. Syst. Signal Process.* **2010**, *24*, 1458–1472. [CrossRef]
31. Kim, T.; Lee, G.; Youn, B.D. PHM experimental design for effective state separation using Jensen–Shannon divergence. *Reliab. Eng. Syst. Saf.* **2019**, *190*, 106503. [CrossRef]
32. Meng, Z.; Cui, Z.; Liu, J.; Li, J.; Fan, F. Maximum cyclic Gini index deconvolution for rolling bearing fault diagnosis. *IEEE Trans. Instrum. Meas.* **2023**, *72*, 1–12. [CrossRef]
33. He, W.; Jiang, Z.N.; Feng, K. Bearing fault detection based on optimal wavelet filter and sparse code shrinkage. *Measurement* **2009**, *42*, 1092–1102. [CrossRef]
34. Wang, D. Some further thoughts about spectral kurtosis, spectral L2/L1 norm, spectral smoothness index and spectral Gini index for characterizing repetitive transients. *Mech. Syst. Signal Process.* **2018**, *108*, 360–368. [CrossRef]
35. Zhang, C.; Liu, Y.; Wan, F.; Chen, B.; Liu, J.; Hu, B. Adaptive filtering enhanced windowed correlated kurtosis for multiple faults diagnosis of locomotive bearings. *ISA Trans.* **2020**, *101*, 421–429. [CrossRef]
36. Xu, X.; Zhao, M.; Lin, J.; Lei, Y. Envelope harmonic-to-noise ratio for periodic impulses detection and its application to bearing diagnosis. *Measurement* **2016**, *91*, 385–397. [CrossRef]
37. McDonald, G.L.; Zhao, Q. Multipoint optimal minimum entropy deconvolution and convolution fix: Application to vibration fault detection. *Mech. Syst. Signal Process.* **2017**, *82*, 461–477. [CrossRef]
38. Miao, Y.; Zhao, M.; Lin, J.; Lei, Y. Application of an improved maximum correlated kurtosis deconvolution method for fault diagnosis of rolling element bearings. *Mech. Syst. Signal Process.* **2017**, *92*, 173–195. [CrossRef]
39. Xu, X.; Zhao, M.; Lin, J.; Lei, Y. Periodicity-based kurtogram for random impulse resistance. *Meas. Sci. Technol.* **2015**, *26*, 085011. [CrossRef]
40. Loza, C.A.; Principe, J.C. Transient model of EEG using Gini index-based matching pursuit. In Proceedings of the 2016 IEEE International Conference on Acoustics, Speech and Signal Processing (ICASSP), Shanghai, China, 20–25 March 2016; pp. 724–728.
41. Zonoobi, D.; Kassim, A.A.; Venkatesh, Y.V. Gini index as sparsity measure for signal reconstruction from compressive samples. *IEEE J. Sel. Top. Signal Process.* **2011**, *5*, 927–932. [CrossRef]
42. Miao, Y.; Zhao, M.; Lin, J. Periodicity-impulsiveness spectrum based on singular value negentropy and its application for identification of optimal frequency band. *IEEE Trans. Ind. Electron.* **2018**, *66*, 3127–3138. [CrossRef]
43. Ding, C.; Zhao, M.; Lin, J. Sparse feature extraction based on periodical convolutional sparse representation for fault detection of rotating machinery. *Meas. Sci. Technol.* **2020**, *32*, 015008. [CrossRef]

Disclaimer/Publisher's Note: The statements, opinions and data contained in all publications are solely those of the individual author(s) and contributor(s) and not of MDPI and/or the editor(s). MDPI and/or the editor(s) disclaim responsibility for any injury to people or property resulting from any ideas, methods, instructions or products referred to in the content.

sensors

MDPI

Article

Static Calibration of a New Three-Axis Fiber Bragg Grating-Based Optical Accelerometer

Abraham Perez-Alonzo [1,*], Luis Alvarez-Icaza [1] and Gabriel E. Sandoval-Romero [2]

[1] Instituto de Ingeniería, Universidad Nacional Autónoma de México, Mexico City 04510, Mexico; alvar@pumas.iingen.unam.mx

[2] Instituto de Ciencias Aplicadas y Tecnología, Universidad Nacional Autónoma de México, Mexico City 04510, Mexico; eduardo.sandoval@icat.unam.mx

* Correspondence: aperezal@iingen.unam.mx

Abstract: Optical sensors are a promising technology in structural and health monitoring due to their high sensitivity and immunity to electromagnetic interference. Because of their high sensitivity, they can register the responses of buildings to a wide range of motions, including those induced by ambient noise, or detect small structural changes caused by aging or environmental factors. In previous work, an FBG-based accelerometer was introduced that is suitable for use as an autonomous unit since it does not make use of any interrogator equipment. In this paper, we present the results of the characterization of this device, which yielded the best precision and accuracy. The results show the following: (i) improvements in the orthogonality of the sensor axes, which impact their cross-axis sensitivity; (ii) reductions in the electronic noise, which increase the signal-to-noise ratio. The results of our static characterization show that, in the worst case, we can obtain a correlation coefficient R^2 of 0.9999 when comparing the output voltage with the input acceleration for the X- and Y-axes of the sensor. We developed an analytical, non-iterative, 12-parameter matrix calibration approach based on the least-squares method, which allows compensation for different gains in its axes, offset, and cross-axis. To improve the accuracy of our sensor, we propose a table with correction terms that can be subtracted from the estimated acceleration. The mean error of each estimated acceleration component of the sensor is zero, with a maximum standard deviation of 0.018 m/s^2. The maximum RMSE for all tested positions is 6.7×10^{-3} m/s^2.

Keywords: sensor characterization; static calibration; FBG-based accelerometer; overlapping interrogation method

1. Introduction

In this work, we focus on the full characterization of the accelerometer developed in [1] for seismic applications, which is a fiber Bragg grating (FBG)-based triaxial accelerometer that uses the overlapping interrogation method for the demodulation of the Bragg wavelength of the gratings. The common approach is to use two FBGs for each sensing axis; however, we require just one FBG for each sensor axis and one reference FBG. Thus, this device only needs four FBGs in total.

The relevance of accelerometers in the field of structural health monitoring lies in the fact that they are part of a system that measures different vibration parameters, such as strain, where FBGs can be directly embedded in buildings, as studied in [2]. Although acceleration can be calculated from displacement or velocity measurements and vice versa, the derivation or integration process causes large errors; thus, the direct measurement of the desired magnitude is preferred. Due to their good response, many researchers are working on the development of triaxial FBG-based accelerometers; for example, in [3], the authors developed a triaxial device with a measurement range between 0.1 Hz and 20 Hz and a high sensitivity of 606.2 pm/g; in [4], a triaxial sensor with an operating frequency between 0.1 Hz and 50 Hz was developed, achieving a sensitivity between 395 pm/g and 590 pm/g on its axes, and with a minimum detectable acceleration of 0.01 g; in both [3,4], the authors presented applications to seismic detection, similar to [5–7] and many other authors. These devices, and almost all FBG-based accelerometers, use an FBG demodulator (FBG interrogator) to measure the changes in the FBG. This FBG interrogator turns the sensor into a more complex and expensive device. We found that there is a lack of research into FBG sensors that do not use interrogators.

For the work presented here, we developed a new triaxial sensor assembled with new fiber Bragg gratings. Manufacturing helps to improve the orthogonality in its axes. This new sensor is based on the same operation principle, in which we do not use an interrogator to demodulate the FBG changes and use only one FBG for each axis and one additional reference FBG for all the axes. This approach was already employed, with satisfactory results, in the biaxial accelerometer developed in [8].

The characterization we perform in this work is of vital importance because it allows us to prove that our new device is suitable for the proposed application, and ensures that its measurements are reliable. Calibration methods are classified as static and dynamic characterizations. We selected a static characterization, which is performed by leaving the accelerometer fixed in different positions for a short period of time and recording the sensor output for each position; this procedure is the fundamental first step in the characterization of an accelerometer, since it allows us to determine the non-orthogonalities, the sensitivity in each axis, and the effects of temperature on the measurements.

Static calibration can be performed by iterative or analytical methods. Iterative methods have the advantage of not requiring one know the orientation of the accelerometer, as is the case with Newton's method [9], the Levenberg–Marquardt algorithm, or the Thin-Shell algorithm [10], which try to minimize the calibration error in each iteration until it falls to a value equal to or smaller than a previously established one. In analytical methods, there is no need for iterations; however, the tilt angles of the accelerometer for each position must be known.

The analytical method was chosen in this work, which allows the linearity of the sensor to be determined in each of its axes by considering that its measurement range is less than ±0.5 g (±4.89 m/s^2 in Mexico City), which can be provided by a tilt table [11]. The principle of this static calibration method is that the sum of three acceleration components in the accelerometer should be equal to the gravitational field vector [12].

On the other hand, since Bragg gratings are sensitive to both temperature and strain [13], we need to decouple the strain from temperature. In theory, the overlapping interrogation method of FBGs allows us to intrinsically compensate for temperature because the reference FBG does not undergo any change due to vibrations, only due to temperature changes; then, when each of the spectra of the sensing FBGs overlap with the spectrum of the reference FBG, the changes in the FBGs due to temperature are compensated. In practice, this compensation is not complete; in our prototype it is between 50% and 60%.

This incomplete compensation behavior has also been reported in other works, such as [14], where two FBGs that should have had the same sensitivity to temperature showed that they had different sensitivities in experiments: 23 pm/°C in one and 28 pm/°C in the other. Another example is [15], where the temperature coefficients of the two FBGs were

16.9 pm/°C and 18.2 pm/°C; thus, although the temperature dependence was reduced to 1.3 pm/°C, there was still some temperature dependence.

This imperfect temperature compensation approach has limited effects on the measurement of vibrations due to seismic motion, since the signal variation with respect to temperature is very small compared to the signal variations due to vibration. The longest period of the signal that our sensor is intended to measure is ten seconds; thus, the temperature variations are negligible, since they normally have a period of 24 h when the sensor is in a static position. Therefore, we can decouple the vibration signals from the temperature signals.

2. Materials and Methods

The static characterization was performed on the hand-made tilting platform shown in Figure 1, whose parameters can be seen in Table 1. This platform allows us to change the pitch and roll angle by rotating the corresponding bolt nuts to slide one end of the tilting table. The hinges are tightened in such a way that there is no gap between their parts, allowing only rotational movement.

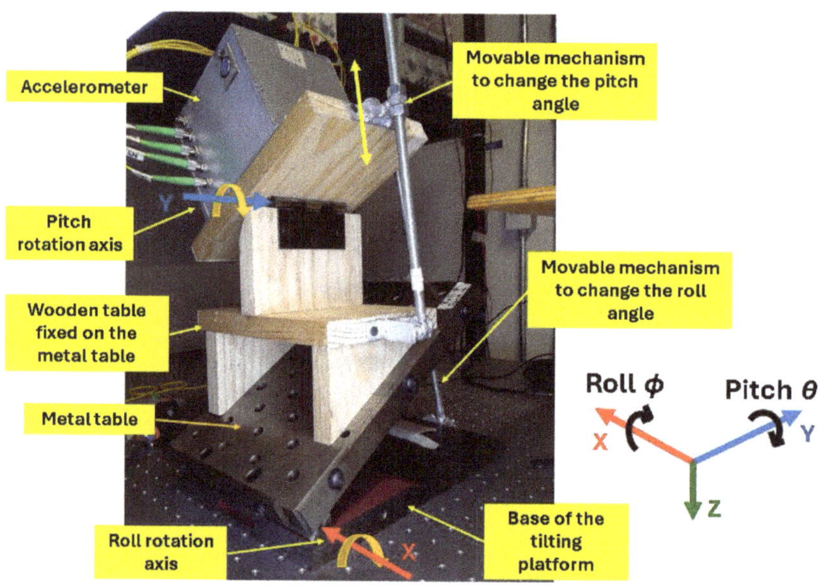

Figure 1. Tilting platform in pitch and roll rotations. The optical sensor is placed on top of the platform. The coordinate system is shown on the right of the figure.

Table 1. Characteristics of the tilting platform.

Parameter	Range	Resolution
Pitch rotation, θ (Rotation about the Y-axis)	±60 deg	0.1 deg
Roll rotation, ϕ (Rotation about the X-axis)	±30 deg	0.1 deg

The magnitude of the gravitational acceleration varies with latitude, altitude, gravity anomaly, and tidal effect. We use the International Gravity Formula derived in [9], which depends on latitude and altitude, as shown in Equation (1).

$$g(\gamma, h) = 9.780327\left[1 + 5.3024 \times 10^{-3}\sin^2(\gamma) - 5.08 \times 10^{-6}\sin^2(\gamma)\right] - 3.086 \times 10^{-6}h \ \left[\mathrm{m/s^2}\right], \tag{1}$$

where γ is the latitude, and h is the height (in meters) with respect to sea level. Substituting γ = +19.32078°, which corresponds to Mexico City's latitude, and h = 2293 m according to [16], results in a local acceleration value of 9.7789 $\mathrm{m/s^2}$; this value will be used in the rest of the article as the magnitude of gravity.

As our accelerometer has a high sensitivity and a low range ($\pm$4.89 $\mathrm{m/s^2}$), it is enough to rotate its horizontal axes between a range of $\pm$30 degrees. In the case of the vertical axis, to change the Z component of the gravity vector by 4.89 $\mathrm{m/s^2}$, it is necessary to tilt the platform within a $\pm$60 degrees range. Therefore, the characteristics of the tilting platform are sufficient to characterize our sensor. The tilt angle of the platform was measured with a digital protractor model Pro 360 950-315 from the Japanese manufacturer Mitutoyo, which has a tilt range of 360 degrees and a resolution of 0.1 degree.

When the accelerometer is at rest in the horizontal position, the acceleration on the vertical axis of the sensor is 9.78 $\mathrm{m/s^2}$ downward, but since we want to measure the acceleration relative to the gravitational field, we take the acceleration as 0 $\mathrm{m/s^2}$ in all three axes at rest position. Another thing to note in the vertical axis is that the changes in acceleration are symmetric regardless of whether the platform is tilted in the positive or negative sense. Thus, while the horizontal axes are tested over a range of $\pm$4.89 $\mathrm{m/s^2}$, the vertical axis is only tested from 0 to $-$4.89 $\mathrm{m/s^2}$. The accelerometer cannot be turned over (rotated 180 degrees about the X- or Y-axis) because it is outside the measurement range of the sensor.

Static Calibration Procedure

The most general linear calibration of the accelerometer has a total of 12 parameters [17], where a recalibration is performed over the previous factory calibration. We use this same procedure to perform the first calibration of our sensor. This is a non-iterative method, which is based on the linear least-squares optimization.

At first, the general linear model that relates the dependent variable y to the N inputs x_0 to x_{N-1}, through N coefficients β_0 to β_{N-1}, is

$$y = \beta_0 x_0 + \beta_1 x_1 + \beta_2 x_2 + \ldots + \beta_{N-1} x_{N-1}. \tag{2}$$

The N model parameters for each sensor axis β_0 to β_{N-1} are determined from the M measurements from i = 0 to i = $M - 1$, where $M \geq N$. Equation (2), written in vector form for all M measurements, is

$$\begin{pmatrix} y[0] \\ y[1] \\ \vdots \\ y[M-1] \end{pmatrix} = \begin{pmatrix} x_0[0] & x_1[0] & \cdots & x_{N-1}[0] \\ x_0[1] & x_1[1] & \cdots & x_{N-1}[1] \\ \vdots & \vdots & \ddots & \vdots \\ x_0[M-1] & x_1[M-1] & \cdots & x_{N-1}[M-1] \end{pmatrix} \begin{pmatrix} \beta_0 \\ \beta_1 \\ \vdots \\ \beta_{N-1} \end{pmatrix}. \tag{3}$$

The vector $\boldsymbol{Y}$ of the dependent variables for each sensor axis is defined as

$$\boldsymbol{Y} = \begin{pmatrix} y[0] \\ y[1] \\ \vdots \\ y[M-1] \end{pmatrix}. \tag{4}$$

The matrix $\boldsymbol{X}$ of the measurements of the independent variables is defined as

$$\boldsymbol{X} = \begin{pmatrix} x_0[0] & x_1[0] & \cdots & x_{N-1}[0] \\ x_0[1] & x_1[1] & \cdots & x_{N-1}[1] \\ \vdots & \vdots & \ddots & \vdots \\ x_0[M-1] & x_1[M-1] & \cdots & x_{N-1}[M-1] \end{pmatrix}. \tag{5}$$

The solution vector β of the N model parameters for each sensor axis is defined as

$$\beta = \begin{pmatrix} \beta_0 \\ \beta_1 \\ \vdots \\ \beta_{N-1} \end{pmatrix}. \tag{6}$$

Using Equations (4)–(6), Equation (3) can be rewritten as

$$\boldsymbol{Y} = \boldsymbol{X}\beta. \tag{7}$$

In the linear least-squares optimization, a residuals vector is defined as

$$\boldsymbol{r} = \boldsymbol{Y} - \boldsymbol{X}\beta. \tag{8}$$

The optimal least-squares fit for β is the one that minimizes the performance function P defined as the squared modulus of the residuals vector [17]. Finally, the commonly known normal equations for least squares optimization result in:

$$\beta = \left(\boldsymbol{X}^T\boldsymbol{X}\right)^{-1}\boldsymbol{X}^T\boldsymbol{Y}. \tag{9}$$

The matrix $\boldsymbol{X}$ of measurements is reused in the calculation of the parameter of the three axes; the vectors $\boldsymbol{Y}$ and β are different for each axis. The calibrated accelerometer output $\boldsymbol{G}$ is defined by

$$\boldsymbol{G} = \begin{pmatrix} G_x \\ G_y \\ G_z \end{pmatrix} = \boldsymbol{W}\boldsymbol{V}_0 + \boldsymbol{V} = \begin{pmatrix} W_{xx} & W_{xy} & W_{xz} \\ W_{yx} & W_{yy} & W_{yz} \\ W_{zx} & W_{zy} & W_{zz} \end{pmatrix} \begin{pmatrix} V_{ox} \\ V_{oy} \\ V_{oz} \end{pmatrix} + \begin{pmatrix} V_x \\ V_y \\ V_z \end{pmatrix}, \tag{10}$$

where $\boldsymbol{W}$ is the gain matrix, which has nine independent elements. It includes all possible cross-axis interactions and any internal rotation of the sensor structure with respect to its case. $\boldsymbol{V}_0$ are the output voltages of the sensor, and $\boldsymbol{V}$ is the offset vector at 0 m/s^2. Using Equation (9), the optimum least squares solution for β_x of the X-axis is

$$\beta_x = \begin{pmatrix} W_{xx} \\ W_{xy} \\ W_{xz} \\ V_x \end{pmatrix} = \left(\boldsymbol{X}^T\boldsymbol{X}\right)^{-1}\boldsymbol{X}^T\boldsymbol{Y}_x = \left(\boldsymbol{X}^T\boldsymbol{X}\right)^{-1}\boldsymbol{X}^T \begin{pmatrix} -\sin\theta[0] \\ -\sin\theta[1] \\ \vdots \\ -\sin\theta[M-1] \end{pmatrix}. \tag{11}$$

Similarly, the optimum least-squares solution for the remaining eight calibration parameters for the Y- and Z-axes is

$$\beta_y = \begin{pmatrix} W_{yx} \\ W_{yy} \\ W_{yz} \\ V_y \end{pmatrix} = \left(X^T X\right)^{-1} X^T Y_y = \left(X^T X\right)^{-1} X^T \begin{pmatrix} \cos\theta[0]\sin\phi[0] \\ \cos\theta[1]\sin\phi[1] \\ \vdots \\ \cos\theta[M-1]\sin\phi[M-1] \end{pmatrix}, \quad (12)$$

$$\beta_z = \begin{pmatrix} W_{zx} \\ W_{zy} \\ W_{zz} \\ V_z \end{pmatrix} = \left(X^T X\right)^{-1} X^T Y_z = \left(X^T X\right)^{-1} X^T \begin{pmatrix} \cos\theta[0]\cos\phi[0] \\ \cos\theta[1]\cos\phi[1] \\ \vdots \\ \cos\theta[M-1]\cos\phi[M-1] \end{pmatrix}. \quad (13)$$

The measurement matrix $\boldsymbol{X}$ is common to all three Equations (11)–(13). The angle θ is the pitch rotation about the Y-axis, the angle ϕ is the roll rotation about the X-axis, and M is the number of different sensor orientations used to find the calibration parameters.

3. Results

In this section, we show the dependence of the Bragg wavelength of the four FBGs used in the sensor on temperature changes, and how to compensate for them. We show that the reference FBG is not affected by vibrations, only by temperature. The response of our sensor to input acceleration, in wavelength and voltage, is shown, along with the 12-parameter model calibration and error analysis.

3.1. *Bragg Wavelength Dependence of FBGs with Temperature*

As mentioned in the introduction, the reference FBG acts as a temperature compensation device, which in theory must be 100%. In practice, all FBGs have different sensitivity coefficients to a greater or lesser degree. To characterize the dependence on temperature changes in all four FBGs used in our sensor, we kept the sensor stable at a fixed position and changed the temperature of the sensor. We placed an incandescent lamp over the sensor at a certain distance from it, which was periodically reduced to increase the sensor temperature. The internal temperature of the sensor, measured using a DHT22 sensor, was changed from 24 °C to 31 °C. The Bragg wavelength was recorded with a device, the Micron Optics—Optical Sensing Interrogator sm130, with a sampling rate of 100 Hz. To reduce noise, 1000 samples were averaged for each data point [18]. Although the external temperature was changed from 24 °C to about 50 °C, the internal temperature did not reach the same value, which means that the sensor case helps to reduce the variation of the output voltages due to temperature. The total time to change the temperature from 24 °C to 31 °C was 3 h and 25 min.

Figure 2a shows the Bragg wavelengths of all FBGs used in our sensor with respect to its internal temperature. To compare the responses of the gratings, they must have the same initial Bragg wavelength; therefore, an offset was applied to the Bragg wavelength of the sensing FBGs. Then, the Bragg wavelength of the sensing FBGs was adjusted by a factor using Equation (14), which changes the temperature sensitivity. The reference FBG has the lowest sensitivity to temperature; thus, the adjustment factors found for the sensing gratings were less than 1—0.51, 0.51, and 0.59 for the X-, Y-, and Z-axes, respectively. The temperature adjustment is then of the form

$$\lambda_{BS_{adj},i} = \lambda_{BS,0} + (\lambda_{BS,i} - \lambda_{BS,0}) f_a + O \quad (14)$$

where $\lambda_{BS_{adj},i}$ is the *i*-th adjusted Bragg wavelength of the *i*-th measurement point, $\lambda_{BS,0}$ is the Bragg wavelength of the first measurement point, $\lambda_{BS,i}$ is the Bragg wavelength of the *i*-th measurement point, f_a is the adjustment factor, and O is the offset. Equation (14) is applied for each of the sensing axes.

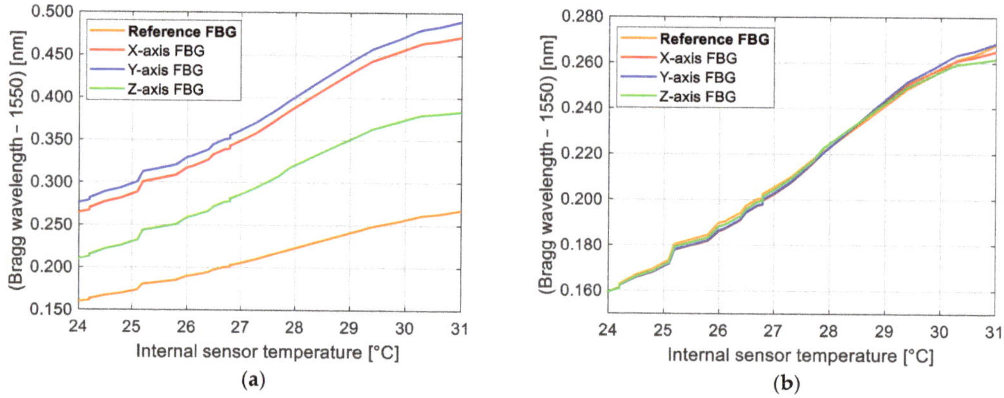

Figure 2. Dependence of the FBGs of our prototype on temperature: (**a**) apart from an offset, the sensitivity in all FBGs is different; (**b**) the Bragg wavelength of sensing FBGs is adjusted by a factor and an offset; now all FBGs show the same trend.

In Figure 3, we can see the responses of the FBGs of our sensor, which was left recording in the laboratory for 17 h. The accelerometer was left on an optical table with active pneumatic isolation; thus, it would not suffer from the ambient vibrations of the building. The changes in the Bragg wavelength of the FBGs are due to temperature, in the same way as in Figure 2. This figure was added since the accelerometer is not intended to work at a controlled ambient temperature, and the temperature surrounding it will change in a similar way as in Figure 3. The trend in this figure confirms once again that the Bragg wavelength of all FBGs shows the same tendency with time, as the ambient temperature changes. The initial Bragg wavelength of the FBGs may vary depending on the sensor calibration.

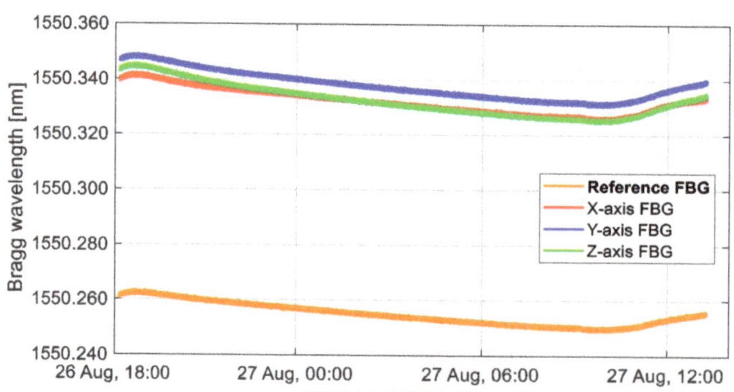

Figure 3. Changes in Bragg wavelength due to temperature over 17 h at laboratory room temperature. Measurements were made with a sampling period of 0.1 s, and a simple moving average with a window size of 5 was applied to the data.

Therefore, we can state that the reference FBG, when used in the overlapping interrogation method, helps to decrease the influence of temperature by reducing the drift of the sensor output. By using the adjustment factors to change the sensitivity of sensing FBGs, we can almost eliminate the signal variation. In addition, the signal variation due to temperature is very small compared to the variations due to vibrations. This is because the temperature changes at a very slow rate compared to the period of vibrations.

3.2. Inmunity of Reference FBG to Vibrations

The reference FBG has two purposes in our sensor: the first is to complete the demodulation scheme for the accelerometer, and the second is to compensate for temperature. Therefore, it is very important that the reference FBG is not affected by vibrations. In our prototype sensor, this fiber grating is fixed in an aluminum clamp with a set screw for pretension [1]. This grating is not connected to the inertial mass as the sensing gratings are; thus, this FBG is not affected by vibrations as demonstrated in Figure 4. In Figure 4a, we apply a directional vibration by hitting the tilting platform to which the accelerometer is attached. It can be seen that the reference FBG is not affected by these hits. In Figure 4b, we apply aleatory shocks, and we can observe that all three axes are affected, but, again, the reference FBG is not affected by these impulses.

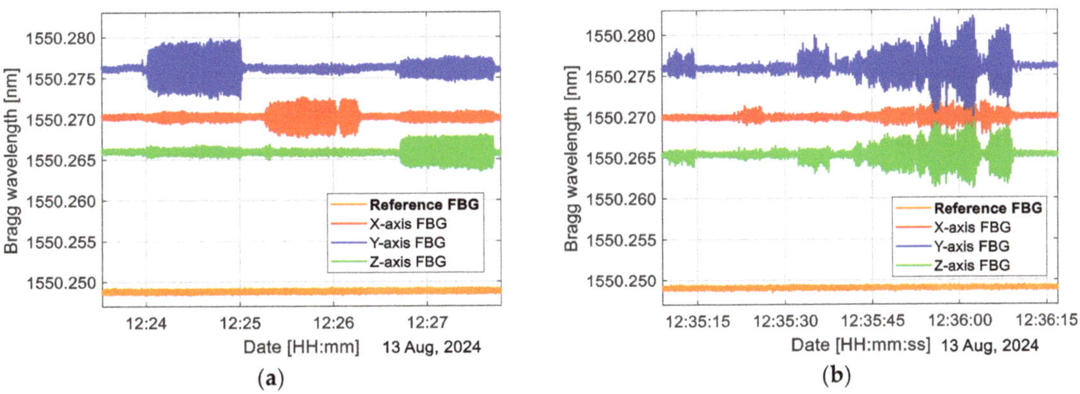

Figure 4. Vibration is applied to the tilting platform to which our sensor is attached; it can be noted that the reference FBG is not affected by such vibration: (**a**) vibration applied in a preferred direction; the X-, Y-, or Z-axis; (**b**) vibration applied in arbitrary directions.

We performed a simple moving average with a window size of 10 on data in Figure 4 that were sampled at 1 kHz.

The reference FBG is attached to the sensor structure by two points with a 25 mm clearance between the fixing points; this clearance is the same as that of the sensing grating; therefore, the immunity of the reference FBG to vibrations is proof that the possible bending of optical fibers under the influence of acceleration is decreased by the pre-strain applied to the fiber at the time of its fixing, as reported in [19]. Although the pre-strain limits the maximum measurable acceleration, since in our case the maximum acceleration to be measured is only $\pm 4.89\ \mathrm{m/s^2}$, this limitation does not affect it. Due to the pre-strain, the fiber is never slack, but its tension is higher or lower depending on the direction of motion; therefore, the bending of the fiber is decreased, and its effect is not seen in the result of the experiment.

3.3. Static Characterization of the Accelerometer

3.3.1. Wavelength Characterization

As described in the materials and methods section, the X-axis was tilted in a ± 60 degrees range to obtain the response of the vertical axis to -0.5 g, while the Y-axis was only tilted in a ± 30 degrees range. The step in the measurements was 5 degrees.

In Figure 5, we can see a little offset between the forward and the backward sweep measurements due to temperature changes; this offset is clearer in Figure 5b. Before analyzing this data, we performed compensation based on the changes of the reference FBG using Equation (15).

$$\lambda_{BS_{adj},i} = \lambda_{BS,i} - (\lambda_{BR,i} - \lambda_{BR,0}) f_{ar}, \tag{15}$$

where $\lambda_{BS_{adj},i}$ is the i-th adjusted Bragg wavelength of the i-th measurement point, $\lambda_{BS,i}$ is the Bragg wavelength of the i-th measurement point, $\lambda_{BR,i}$ is the Bragg wavelength of the i-th measurement point of reference FBG, $\lambda_{BR,0}$ is the Bragg wavelength of the initial measurement point of the reference FBG, and f_{ar} is the adjustment factor. Equation (15) applies to each of the sensing axes. The adjustment factors used in the compensation were 1.15, 0.90, and 1.00 for the X-, Y-, and Z-axes, respectively. The results are shown in Figure 6, where the response of reference FBG is shown without compensation.

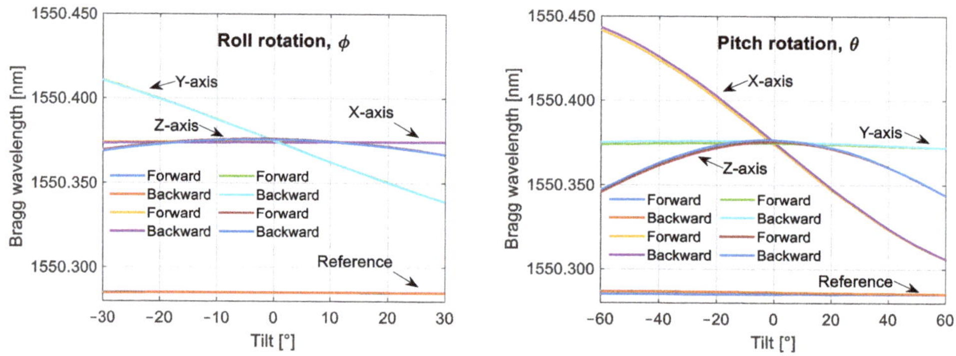

(a) (b)

Figure 5. Response in wavelength of the FBGs used in our prototype with respect to tilt. The tilt is referenced to the horizontal plane; two sweeps were performed: one forward and one backward. (**a**) The sensor is rotated about its X-axis; (**b**) the sensor is rotated about its Y-axis.

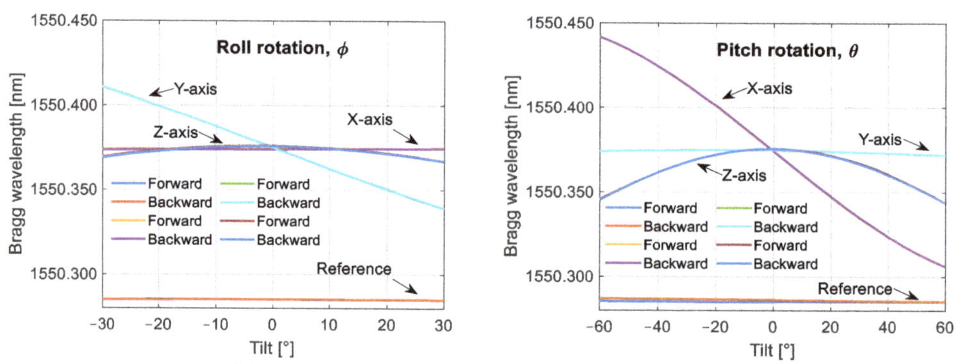

(a) (b)

Figure 6. Response in wavelength of the FBGs used in our prototype with temperature compensation. The tilt is referenced to the horizontal plane; the temperature compensation was performed with respect to the reference FBG: (**a**) the sensor is rotated about its X-axis; (**b**) the sensor is rotated about its Y-axis.

After temperature compensation, the forward and backward sweep data are averaged for each sensor orientation and plotted with the horizontal axis in units of acceleration. The acceleration was calculated with Equation (16) by applying the sine function to the tilt angles in the X- and Y-axes and the cosine function for the vertical axis of the sensor. The results are shown in Figure 7.

$$\begin{aligned} a_x &= -\sin\theta \\ a_y &= \cos\theta \sin\phi \\ a_z &= \cos\theta \cos\phi \end{aligned} \quad (16)$$

where a_x, a_y, and a_z are the components of the gravity vector coupled to the X-, Y-, and Z-axes of the sensor, θ is the pitch rotation about the Y-axis, and ϕ is the roll rotation about the X-axis of the sensor.

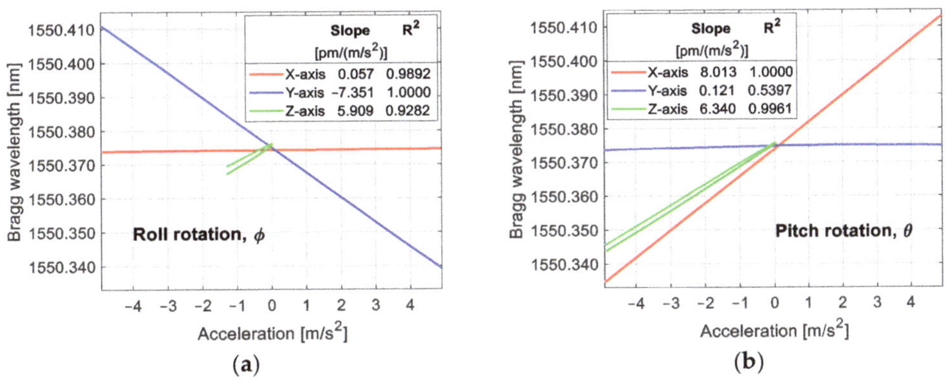

Figure 7. Response of the sensing FBG of our sensor, the horizontal axis of the plot was converted to acceleration by using the sine function of the tilt angle for the X- and Y-axes and the cosine function for the Z-axis. In the Z-axis, the magnitude of the gravity vector is subtracted from the measurements: (**a**) the sensor is rotated about its X-axis; (**b**) the sensor is rotated about its Y-axis.

For Figure 7a, where the sensor rotates about its X-axis, after a linear fit, we obtained a sensitivity of -7.351 pm/(m/s^2) for the Y-axis with an R-squared (correlation factor) of 1.0000, while we have a cross-sensitivity of 0.057 pm/(m/s^2) on the X-axis. This means a cross-sensitivity 129 times lower than the corresponding axis sensitivity. In the second case, where the sensor is rotated about its Y-axis, we obtained a sensitivity of 8.013 pm/(m/s^2) for the X-axis with an R-squared value of 1.0000, and 0.121 pm/(m/s^2) on the Y-axis; thus, we have a cross-sensitivity 66 times lower than the corresponding axis.

For the Z-axis, where the acceleration is symmetrical with respect to both senses of rotation of the horizontal axes of the sensor, the ideal would be to obtain a line that goes from -5 to 0 m/s^2, then from 0 to -5 m/s^2, arriving at the same place; however, in Figure 7a,b, we can see that its sensitivity is somewhat different depending on the sense of rotation of the sensor. If we analyze Figure 6, the difference in sensitivity can be attributed to a slight misalignment of the Z-axis with respect to the vertical. This is because the maximum wavelength of the Z-axis FBG is not obtained when the sensor is in a flat position, but when it is somewhat tilted. The average sensitivity of this axis, when the sensor is rotated about the X-axis, is 5.909 pm/(m/s^2), with R-squared equal to 0.9282, and 6.340 pm/(m/s^2) when the sensor is rotated about the Y-axis, with R-squared equal to 0.9961.

3.3.2. Voltage Characterization

The sensitivities and response curves we obtained earlier were at the wavelength scale with respect to the acceleration experienced in each axis. These values help to reveal the response of our sensor to temperature and acceleration during characterization. However, one of the main advantages of our sensor is that we do not need interrogation equipment; instead, the reference FBG allows us to demodulate the Bragg wavelength of the sensing fiber gratings into optical power changes, which are measured with simple photodetectors. The output of the photodetectors is a current, which is proportional to the input optical power. Hence, the next step is to obtain the response of our sensor by measuring the voltage at the output of the electronic stage. The way of characterization is the same as in the wavelength characterization; the sweeps are in the same range, and the results are shown in Figure 8.

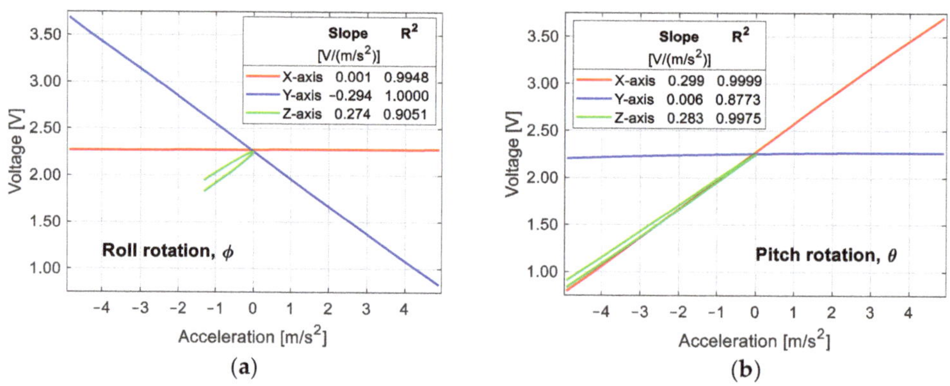

Figure 8. Sensor output in voltage at the final sensor stage with respect to the input acceleration: (**a**) rotation about the X-axis; (**b**) rotation about the Y-axis.

According to Figure 8a, the sensitivity of the Y-axis when the sensor rotates about the X-axis (roll rotation) is -0.294 V/(m/s^2) with a correlation coefficient of 1.0000, and 0.001 V/(m/s^2) for the X-axis, which means a cross-sensitivity 267 times lower (0.34%) than that of the corresponding Y-axis. When the sensor rotates about the Y-axis (pitch rotation), the sensitivity for the X-axis is 0.299 V/(m/s^2) with R-squared equal to 0.9999, and a cross-sensitivity of 0.006 V/(m/s^2) for the Y-axis, i.e., a cross-sensitivity 48 times lower (2.08%) than that of the X-axis. For the Z-axis, the average sensitivity when the sensor rotates about the X-axis is 0.274 V/(m/s^2) and 0.283 V/(m/s^2) when it rotates about the Y-axis. The cross-axis interference for our sensor is comparable to that of some commercial accelerometers, such as the SF3000L calibrated in [12].

The overlapping interrogation method has a nonlinear response because the output power of the method is a convolution of the two FBG spectra, modeled as Gaussians [20]. There is a region where the response of this method is nearly linear; this region can be achieved with a suitable separation between the Bragg wavelength of the reference FBG and that of the sensing FBG. The correlation coefficient that we found in the sensor for the X- and Y-axes is very close to unity, which means that the FBGs are working in the linear region. The proper separation between the reference FBG and the sensing FBGs found experimentally in our sensor is shown in Table 2.

Table 2. Optimal separation between the Bragg wavelength of sensing FBGs and the reference FBG.

FBG_1	FBG_2	Bragg Wavelength Separation (pm)
Reference	X-axis FBG	102 ± 2
Reference	Y-axis FBG	98 ± 2
Reference	Z-axis FBG	95 ± 2

3.3.3. The 12-Parameters Calibration Model

Along with the voltage characterization, we performed the calibration model described in Equation (10). The sensor orientations used as input for parameter estimation are shown in Table 3, where θ is the angle of rotation about the Y-axis of the sensor from the horizontal plane, and ϕ is the angle of rotation about the X-axis of the sensor from the horizontal plane too. For the estimation of the X- and Y-axes parameters, the positions 0, 1, 2, 4, 5, and 6 were used. For the Z-axis, all eight positions were used.

Table 3. Orientations for sensor calibration.

Orientation Number	Pitch, θ [deg]	Roll, ϕ (deg)
0	0	−30
1	0	0
2	0	+30
3	−60	0
4	−30	0
5	0	0
6	+30	0
7	+60	0

The 12-parameter model includes three main gains, cross-axis gains, sensor axes misalignment, and offsets for zero input. The model parameters' values found for our sensor are shown in Equation (17).

$$G = \begin{pmatrix} G_x \\ G_y \\ G_z \end{pmatrix} = \begin{pmatrix} 3.3610 & 0.0171 & -0.1107 \\ 0.0719 & -3.4242 & 0.0844 \\ -0.0608 & -0.1500 & 3.5454 \end{pmatrix} \begin{pmatrix} V_{ox} \\ V_{oy} \\ V_{oz} \end{pmatrix} + \begin{pmatrix} -7.4219 \\ 7.3733 \\ -7.5029 \end{pmatrix}. \quad (17)$$

In the estimated parameters, the vertical acceleration is considered relative to the magnitude of the gravity field; then, at the rest position of the sensor, all three axes have a zero value. G_x, G_y, and G_z are the acceleration components in m/s^2 for the X-, Y-, and Z-axes, respectively. V_{ox}, V_{oy}, and V_{oz} are the output voltages of the accelerometer electronic stage in volts.

3.3.4. Errors in the 12-Parameter Calibration Model

We verified the estimated calibration model parameters for our sensor by introducing the sensor output voltages from each of the sensor orientations obtained during the sweeps, as shown in Figure 5, and comparing the model output to each acceleration component exerted at each position. We calculate the full-scale error (FSE) as a percentage for all measurements with Equation (18).

$$\text{FSE}\ [\%] = \frac{\text{measured value} - \text{reference value}}{\text{full} - \text{scale acceleration}} \times 100\%. \quad (18)$$

The maximum full-scale errors are 0.58%, 0.48%, and 0.39% for the X-, Y-, and Z-axes, respectively. In absolute values, the maximum acceleration errors were 0.057 m/s^2, 0.047 m/s^2 and 0.038 m/s^2 for the X-, Y-, and Z-axes, respectively. These errors found are comparable to the errors found after calibration in [12].

In the estimation of the calibration parameters, the 38 recorded positions were also considered to see how much the maximum error decreased, finding that there was no improvement. Therefore, we can conclude that using eight orientations to estimate the model parameters is a good compromise between the number of positions and the error obtained.

In Figure 9, we show the errors present after calibration for each sensor orientation in each axis. The largest errors are present at the extremes of the sensor range, i.e., near $\pm 4.89\ \mathrm{m/s^2}$. The plot has the horizontal axis in units of angles because this unit is common to the horizontal and vertical axes of the sensor, but acceleration is not. The errors are not totally random, as they show a tendency to increase or decrease systematically.

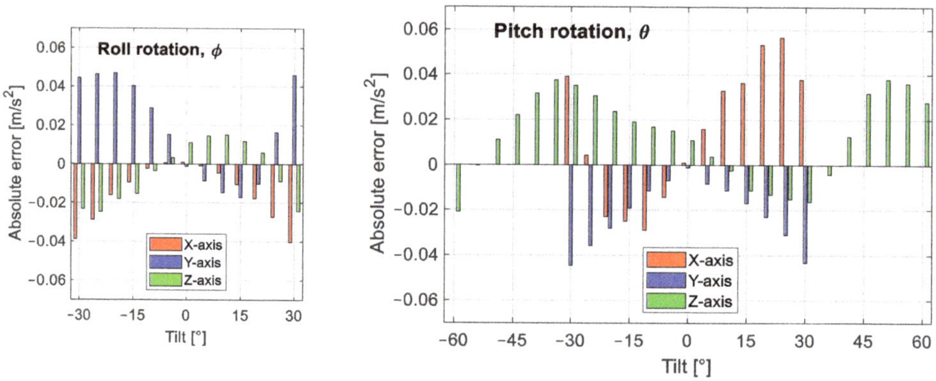

(a) (b)

Figure 9. Absolute acceleration error comparing the theoretical acceleration due to each tilt angle with the estimated acceleration using the sensor output voltages: (**a**) the sensor is rotated about its X-axis; (**b**) the sensor is rotated about its Y-axis.

To further reduce errors, we propose to subtract one term from each estimated component of acceleration. To make sure our proposal is adequate and that the errors decrease, we performed the sweep for the 38 positions five times. We then subtracted the mean error of the five measurements at each position. In Table 4, we can see the terms to be subtracted from the estimated acceleration together with the standard deviation of the error at each orientation. The remaining errors are plotted in Figure 10, where we group the errors for each position and each axis. Now, the errors have a zero mean and show a random behavior.

Table 4. Correction terms to be subtracted from estimated accelerations and the respective standard deviation (SD) of the zero-mean error.

Ref. Angle θ; ϕ [deg]	Correction Term C_{Gx}, C_{Gy}, C_{Gz} (m/s^2)	Error SD σ_{Ex}, σ_{Ey}, σ_{Ez} (m/s^2)	Ref. Angle θ; ϕ (deg)	Correction Term C_{Gx}, C_{Gy}, C_{Gz} (m/s^2)	Error SD σ_{Ex}, σ_{Ey}, σ_{Ez} (m/s^2)
0; −30	0.0388; −0.0443; 0.0233	0.0047; 0.0106; 0.0027	−30; 0	−0.0391; 0.0446; −0.0351	0.0154; 0.0018; 0.0049
0; −25	0.0288; −0.0461; 0.0247	0.0025; 0.0058; 0.0027	−25; 0	−0.0043; 0.0359; −0.0305	0.0116; 0.0019; 0.0053
0; −20	0.0160; −0.0468; 0.0181	0.0054; 0.0055; 0.0039	−20; 0	0.0230; 0.0281; −0.0236	0.0144; 0.0028; 0.0040
0; −15	0.0095; −0.0401; 0.0153	0.0024; 0.0072; 0.0031	−15; 0	0.0250; 0.0192; −0.0191	0.0086; 0.0021; 0.0055
0; −10	0.0024; −0.0288; 0.0036	0.0019; 0.0041; 0.0012	−10; 0	0.0290; 0.0115; −0.0170	0.0105; 0.0014; 0.0034
0; −5	−0.0006; −0.0151; −0.0033	0.0028; 0.0066; 0.0020	−5; 0	0.0143; 0.0068; −0.0152	0.0107; 0.0024; 0.0052
0; 0	−0.0010; 0.0012; −0.0110	0.0000; 0.0000; 0.0000	0; 0	−0.0010; 0.0012; −0.0110	0.0000; 0.0000; 0.0000
0; 5	0.0008; 0.0086; −0.0145	0.0025; 0.0064; 0.0012	5; 0	−0.0159; 0.0082; −0.0037	0.0037; 0.0032; 0.0101
0; 10	0.0046; 0.0147; −0.0150	0.0025; 0.0057; 0.0008	10; 0	−0.0329; 0.0114; 0.0024	0.0072; 0.0031; 0.0100
0; 15	0.0105; 0.0173; −0.0116	0.0032; 0.0037; 0.0009	15; 0	−0.0364; 0.0169; 0.0114	0.0120; 0.0014; 0.0067
0; 20	0.0178; 0.0102; −0.0060	0.0048; 0.0041; 0.0017	20; 0	−0.0533; 0.0230; 0.0132	0.0102; 0.0015; 0.0062
0; 25	0.0274; −0.0164; 0.0088	0.0077; 0.0016; 0.0026	25; 0	−0.0567; 0.0301; 0.0153	0.0095; 0.0041; 0.0069
0; 30	0.0403; −0.0459; 0.0244	0.0073; 0.0065; 0.0029	30; 0	−0.0379; 0.0432; 0.0164	0.0112; 0.0025; 0.0064
−60; 0	-----; -----; 0.0209	-----; -----; 0.0106	35; 0	-----; -----; −0.0172	-----; -----; 0.0071
−55; 0	-----; -----; 0.0002	-----; -----; 0.0066	40; 0	-----; -----; −0.0339	-----; -----; 0.0076
−50; 0	-----; -----; −0.0111	-----; -----; 0.0076	45; 0	-----; -----; −0.0532	-----; -----; 0.0099
−45; 0	-----; -----; −0.0221	-----; -----; 0.0023	50; 0	-----; -----; −0.0592	-----; -----; 0.0072
−40; 0	-----; -----; −0.0315	-----; -----; 0.0025	55; 0	-----; -----; −0.0574	-----; -----; 0.0089
−35; 0	-----; -----; −0.0375	-----; -----; 0.0037	60; 0	-----; -----; −0.0491	-----; -----; 0.0180

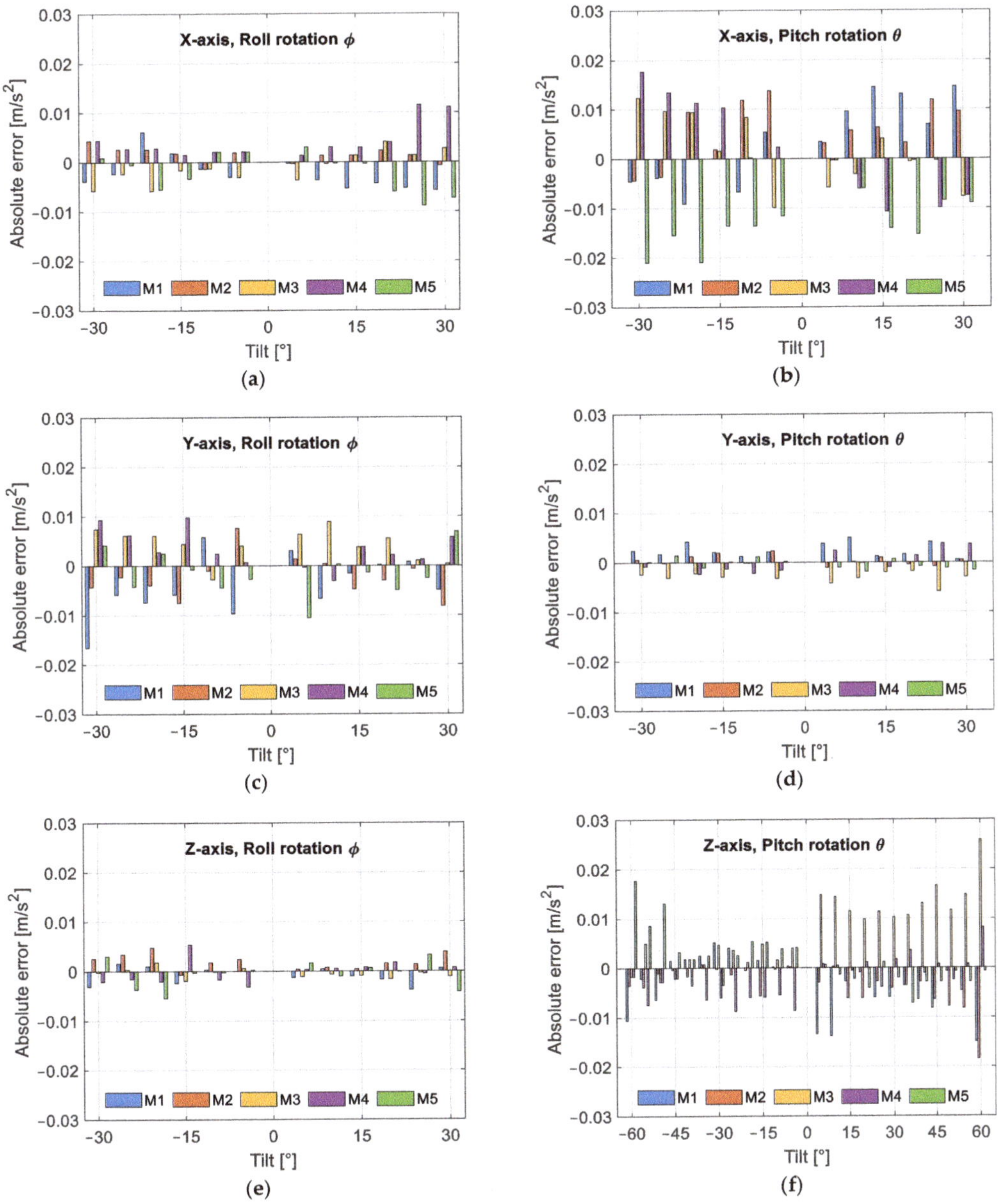

Figure 10. Absolute acceleration error after subtracting the mean error for each sensor orientation: (**a**) X-axis response to roll rotation; (**b**) X-axis response to pitch rotation; (**c**) Y-axis response to roll rotation; (**d**) Y-axis response to pitch rotation; (**e**) Z-axis response to roll rotation; (**f**) Z-axis response to pitch rotation.

From Figure 10a,d,e, we see that the errors are below ± 0.01 m/s^2, while in Figure 10b,c, the maximum errors are around ± 0.02 m/s^2, with the largest maximum error in Figure 10f being between +0.03 and -0.02 m/s^2. With the correction term applied to the estimated acceleration components, the maximum errors for all sweeps are as follows: for the X-axis ± 0.02 m/s^2, for the Y-axis between +0.01 and -0.02 m/s^2, and for the Z-axis between +0.03, and -0.02 m/s^2. In percentage, the maximum error was reduced by 58%, 58%, and 54% for the X-, Y-, and Z-axes, respectively.

Figure 11 shows the error between the gravity magnitude of the calibrated accelerometer output and the magnitude of the gravity field. Figure 11a shows the errors in each sensor orientation for each of the five sweeps and the root mean square error (RMSE), with the maximum value being 0.025 m/s^2 before error correction. From Figure 11b, we can see that the RMSE decreases to 0.007 m/s^2; therefore, the RMSE is reduced by 72%. In Table 5, we can see the pitch and roll angle for each of the positions in Figure 11.

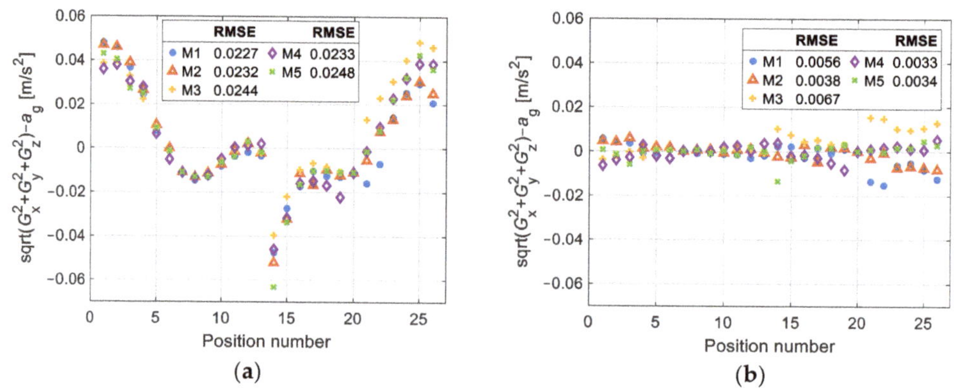

Figure 11. Absolute acceleration error, between the magnitude of the calibrated output of the accelerometer and the magnitude of the gravity field, and their respective RMSEs: (**a**) errors before subtracting the correction term; (**b**) errors after subtracting the correction term.

Table 5. Pitch and roll angle for each of the positions in Figure 11.

Orientation	Pitch, θ (deg)	Roll, ϕ (deg)	Orientation	Pitch, θ (deg)	Roll, ϕ (deg)	Orientation	Pitch, θ (deg)	Roll, ϕ (deg)
1	0	−30	10	0	15	19	−5	0
2	0	−25	11	0	20	20	0	0
3	0	−20	12	0	25	21	5	0
4	0	−15	13	0	30	22	10	0
5	0	−10	14	−30	0	23	15	0
6	0	−5	15	−25	0	24	20	0
7	0	0	16	−20	0	25	25	0
8	0	5	17	−15	0	26	30	0
9	0	10	18	−10	0	-	-	-

Now, we compare the errors obtained during calibration and the noise present in the circuit. The root mean square (RMS) values of the noise, which we can see in Figure 12, are 9.8×10^{-4} V, 5.3×10^{-4} V, and 4.9×10^{-4} V for the X-, Y-, and Z-axes, respectively. These values are converted to acceleration using the sensitivities shown in Figure 8, and result in 0.0033 m/s^2, 0.0018 m/s^2 and 0.0017 m/s^2, for the X-, Y- and Z-axes, respectively. The cutoff frequency is 20 Hz, and the window size is 58 s. The noise we see in Figure 12 is the sum of all noise sources present in the electronic circuit, which includes the thermal noise within the analog-to-digital converter and quantization noise due to the analog-to-digital conversion process [21]. The SNR (signal-to-noise ratio) is calculated using

$$\mathrm{SNR} = 20\log\left(\frac{\text{full} - \text{scale input}}{\text{noise}}\right). \quad (19)$$

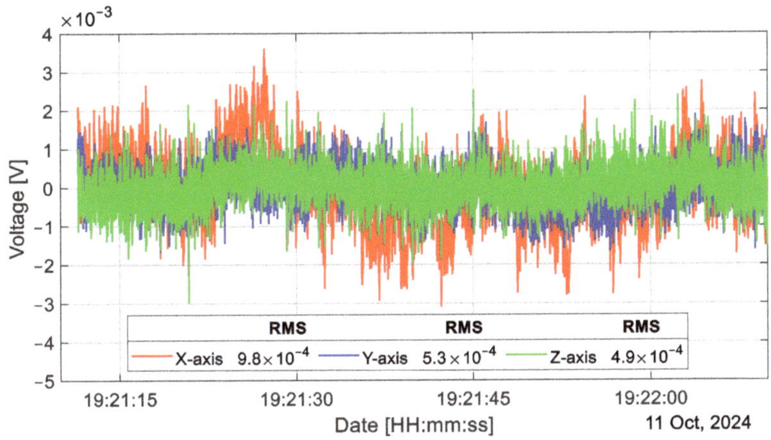

Figure 12. Electrical noise present at the output of the accelerometer electronic stage in a 60 s window.

The full-scale input of our sensor is 9.7789 m/s^2, which, when converted to RMS, is 3.4574 m/s^2. The full-scale input (or range) is converted to RMS for consistency with the RMS noise used in the calculation. Using Equation (19), the SNR values for the X-, Y-, and Z-axes are 60.40 dB, 65.7 dB, and 66.2 dB, respectively. On the other hand, the dynamic range (DR) is defined as

$$DR = 20\log\left(\frac{\text{range}}{2 \times \text{noise}}\right). \tag{20}$$

DR is found to be 54.4 dB, 59.6 dB, and 60.1 dB for the X-, Y-, and Z-axes, respectively. We consider the minimum detectable acceleration as twice the noise value, i.e., the minimum detectable acceleration is 6.6×10^{-3} m/s^2, 3.6×10^{-3} m/s^2, and 3.4×10^{-3} m/s^2.

4. Discussion and Conclusions

The characterization of a new triaxial FBG-based accelerometer has been performed. This triaxial accelerometer uses only a total of four fiber gratings. This has been achieved by reusing the reference FBG, which is not part of any sensor axis, with each of the sensing axes. We need to confirm that the Bragg wavelength of the reference FBG is not affected by vibrations, and that we can reuse it for demodulating the Bragg wavelength of the three sensing axes without loss of linearity. Therefore, the first experiment was to apply acceleration on the sensor axes, where the results show that the reference FBG is immune to vibrations. The reference FBG is oriented along the X-axis of the sensor, but because it is vibration-insensitive, its orientation has no effect on the responses of the sensing FBGs.

We have shown that all FBGs exhibit similar behaviors with respect to temperature, and that although the temperature compensation in the sensing FBGs is not at 100%, the reference FBG helps to compensate between 50% and 60%. The remaining temperature dependence can be neglected because the signal variation due to temperature is much slower than the signal variation due to vibration. Over a short period of time, the signal variation is also very small.

The overlapping interrogation method is nonlinear; then, the linearity of the output power from the overlapping interrogation method with respect to the input acceleration is another issue. A nearly linear operating range can be used with a suitable Bragg wavelength difference between the sensing FBG and the reference grating. The optimal spacing for the FBGs used in our sensor was found to be 102 ± 2 pm, 98 ± 2 pm, and 95 ± 2 pm for the X-, Y-, and Z-axes, respectively. Since the correlation factors for the X- and Y-axes are 1.0000 and 0.9999, respectively, we are operating within the nearly linear operating range, and the linear model of the sensor could be constructed.

We chose an analytical calibration method instead of an iterative one in order to determine the linearity of our sensor and to make the necessary adjustments to the Bragg wavelength of the gratings until we were operating in the center of the linear range of the demodulation scheme. We performed a 12-parameter calibration based on the linear least squares optimization by placing the sensor in seven different positions. The full-scale errors found were all under 1%—0.58%, 0.48%, and 0.39% for the X-, Y-, and Z-axes, respectively. We performed five sweeps in all sensor orientations, and the errors in the estimated acceleration components were repetitive; thus, we proposed to subtract a term, equal to the average of the five sweeps, from the estimated acceleration. After error compensation, the values obtained were 0.24%, 0.20%, and 0.18% for the X-, Y-, and Z-axes, which means reductions of 58%, 58%, and 54%, respectively.

The dynamic range found in the sensor was 54.4 dB, 59.6 dB, and 60.1 dB for the X-, Y-, and Z-axes, respectively. The minimum detectable acceleration was 6.6×10^{-3} m/s^2, 3.6×10^{-3} m/s^2, and 3.4×10^{-3} m/s^2 for the X-, Y-, and Z-axes, respectively. The minimum detectable acceleration was small enough to detect a seismic event of magnitude less than 3.0, and the errors in the measurements were all low enough to estimate the amplitude of the motions accurately. The larger cross-sensitivity value for our sensor was found to be 1.51%.

The characteristics of our sensor are comparable to those of the high-figure-of-merit, low-cross-sensitivity accelerometer developed in [3], where the minimum detectable acceleration was 1.67×10^{-3} m/s^2 with a transverse sensitivity of 2.3%. Although the accelerometer in [3] showed a higher sensitivity, 606.2 pm/g, compared to 78.4 pm/g, 71.9 pm/g, and 57.8 pm/g for the X-, Y-, and Z- axes of our sensor, we did not have the limitation of the interrogator resolution, and this has allowed us to achieve similar minimum detectable acceleration. Our sensor improved by 15-fold the minimum detectable acceleration when compared to that in [4], the value of which was 9.8×10^{-2} m/s^2. The low-frequency accelerometer developed in [22] had a sensitivity of 681.7 pm/g and a transverse interference of less than 4.9%, and, considering the errors in the sensor, the minimum acceleration amplitude that it could measure was around 2.9×10^{-1} m/s^2. In [23], a high sensitivity of 763.2 pm/g was achieved, but again, the minimum detectable acceleration was limited by the resolution of the interrogator used, which in this case was 1 pm, meaning that its minimum detectable acceleration was about 1.3×10^{-2} m/s^2, and its cross-sensitivity was 5%. It is worth mentioning that the accelerometers mentioned above used an interrogator. Therefore, it is demonstrated that our sensor, which does not need an interrogator and uses only one FBG for each sensing axis, shows similar or better characteristics than those that do.

Our accelerometer can be used to measure vibration response in buildings, and help preserve structures and human lives. This device does not require an interrogator and can be operated as an autonomous unit. As future work, since the sensor sensitivity may vary with frequency, a dynamic characterization of the sensor is needed. The reduction in its size will improve internal temperature uniformity and further reduce the possibility of optical fiber bending with vibration. Further, the construction of the sensor frame in one piece will prevent possible resonances after prolonged use. This developed accelerometer prototype is suitable for field applications, and is not only a laboratory experiment.

Supplementary Materials: The following supporting information can be downloaded at: https://www.mdpi.com/article/10.3390/s25030835/s1, Table S1: F2-Temperature Dependence; Table S2: F3-Peaks_20240826180838; Table S3: F4a-Peaks_20240813122330; Table S4: F4b-Peaks_20240813123508; Table S5: F5to7-Caract_est_for_art_X; Table S6: F5to7-Caract_est_for_art_Y; Table S7: F8-Caract_est_for_art_X_Volt_avg; Table S8: F8-Caract_est_for_art_Y_Volt_avg; Table S9: F9-Positions_Voltaje15nov2024; Table S10: F10-Positios_Voltaje15nov2024_Sev_meas; Table S11: F12-Noise_data.

Author Contributions: Conceptualization, A.P.-A., L.A.-I. and G.E.S.-R.; methodology, A.P.-A. and G.E.S.-R.; software, A.P.-A.; validation, L.A.-I. and G.E.S.-R.; formal analysis, A.P.-A.; investigation, A.P.-A.; resources, L.A.-I. and G.E.S.-R.; data curation, A.P.-A.; writing—original draft preparation, A.P.-A.; writing—review and editing, A.P.-A., G.E.S.-R. and L.A.-I.; visualization, A.P.-A.; supervision, L.A.-I. and A.P.-A.; project administration, A.P.-A., G.E.S.-R. and L.A.-I.; funding acquisition, L.A.-I. All authors have read and agreed to the published version of the manuscript.

Funding: This work was supported by UNAM Posdoctoral Program, UNAM-PAPIIT grant IT100623, and UNAM-PAPIIT grant IT104123.

Institutional Review Board Statement: Not applicable.

Informed Consent Statement: Not applicable.

Data Availability Statement: The original contributions presented in this study are included in the article/Supplementary Materials. Further inquiries can be directed to the corresponding author.

Conflicts of Interest: The authors declare no conflicts of interest.

References

1. Perez-Alonzo, A.; Velazquez-Carreon, F.; Sandoval-Romero, G.E. High sensitivity three-axis optical vibration sensor for low-frequency measurement. In Proceedings of the 2023 World Congress on Advances in Structural Engineering and Mechanics (ASEM23), Seoul, Republic of Korea, 16–18 August 2023; GECE: Seoul, Repulic of Korea, 2023; pp. 1–11.
2. Zabihollah, A.; Hajyalikhani, P.; Vuddandam, R. Evaluation of embedded FBG sensors for strain monitoring of residential timber buildings under various wind speeds. *Struct. Monit. Maint.* **2024**, *11*, 263–275.
3. Qiu, Z.; Zhang, Y. Three-Dimensional Low-Frequency Earthquake Monitoring Vibration Sensor Based on FBG. *IEEE Trans. Instrum. Meas.* **2024**, *73*, 1–9. [CrossRef]
4. Zhang, B.; Ning, T.; Pei, L.; Liu, S.; Teng, Y. A Triaxial Dual FBG Accelerometer That Can Be Used for Seismic Wave Detection. *IEEE Sens. J.* **2024**, *24*, 10031–10038. [CrossRef]
5. Reghuprasad, A.E.; Colombero, C.; Godio, A. Serially Connected Cantilever Beam-Based FBG Accelerometers: Design, Optimization and Testing. *Sensors* **2023**, *23*, 3188. [CrossRef] [PubMed]
6. Hong, L.; Mai, M.; Zhang, R.; Luo, Q.; Wang, X. Rotating cantilever-based low-frequency double-fiber grating acceleration sensor. *Rev. Sci. Instrum.* **2023**, *94*, 065004. [CrossRef] [PubMed]
7. Hong, L.; Zhang, J.; Zhang, R.; Wu, P.; Duan, L. Research on low-frequency FBG accelerometer based on double curved beam reed. *Opt. Commun.* **2023**, *543*, 1–9. [CrossRef]
8. Perez-Alonzo, A.; Velazquez-Carreon, F.; Sandoval-Romero, G.E. Biaxial FBG Vibration Sensor With a Single Edge Filter and Matching Demodulation. *IEEE Access* **2022**, *10*, 127798–127805. [CrossRef]
9. Frosio, I.; Pedersini, F.; Borghese, N.A. Autocalibration of MEMS Accelerometers. In *Advanced Mechatronics and MEMS Devices*; Zhang, D., Ed.; Springer: New York, NY, USA, 2013; pp. 53–88. [CrossRef]
10. Sipos, M.; Paces, P.; Rohac, J.; Novacek, P. Analyses of Triaxial Accelerometer Calibration Algorithms. *IEEE Sens. J.* **2012**, *12*, 1157–1165. [CrossRef]
11. Wielandt, E. Seismic Sensors and their Calibration. In *New Manual of Seismological Observatory Practice 2 (NMSOP-2)*; Deutsches GeoForschungsZentrum GFZ: Potsdam, Germany, 2012. [CrossRef]
12. Won, S.P.; Golnaraghi, F.A. Triaxial Accelerometer Calibration Method Using a Mathematical Model. *IEEE Trans. Instrum Meas.* **2010**, *59*, 2144–2153. [CrossRef]
13. Hisham, H. Sensor Mechanisms in Fiber Bragg Gratings. In *Fiber Bragg Grating Sensors: Development and Applications*; CRC Press: Boca Raton, FL, USA, 2020; pp. 49–54.
14. Sun, L.; Liang, T.; Sun, X.; Li, C.; Zhang, C. Temperature self-compensating and high-sensitivity FBG inclination sensor based on the sliding mass principle. *Opt. Fiber Technol.* **2023**, *81*, 103539. [CrossRef]

15. Fan, W.; Zhou, R.; Li, H.; Yu, D.; Qiao, X. Low-frequency temperature compensation fiber Bragg grating accelerometer based on double-arched beam. *Opt. Commun.* **2023**, *528*, 129058. [CrossRef]
16. Mapas Topográficos. Available online: https://es-mx.topographic-map.com/ (accessed on 2 November 2024).
17. Pedley, M. *High Precision Calibration of a Three-Axis Accelerometer*; Freescale Semiconductor: Austin, TX, USA, 2013; pp. 1–41.
18. Forsberg, T.; Grip, N.; Sabourova, N. Non-iterative calibration for accelerometers with three non-orthogonal axes, reliable measurement setups and simple supplementary equipment. *Meas. Sci. Technol.* **2013**, *24*, 1–14. [CrossRef]
19. Basumallick, N.; Chatterjee, I.; Biswas, P.; Dasgupta, K.; Bandyopadhyay, S. Fiber Bragg grating acceleromter with enhanced sensitivity. *Sens. Actuators A Phys.* **2012**, *173*, 108–115. [CrossRef]
20. Lobo Ribeiro, A.B.; Fereira, L.A.; Santos, J.L.; Jackson, D.A. Analysis of the reflective-matched fiber Bragg grating sensing interrogation scheme. *Appl. Opt.* **1997**, *36*, 934–939. [CrossRef] [PubMed]
21. McCarthy, M. *Peak-to-Peak Resolution Versus Effective Resolution*; Analog Devices: Wilmington, MA, USA, 2003; pp. 1–2.
22. Qiu, Z.; Su, K.; Zhang, J.; Teng, Y. A miniaturized low-frequency FBG accelerometer based on symmetrical cantilever beam. *IEEE. Sens. J.* **2023**, *23*, 5831–5840. [CrossRef]
23. Liu, Q.; Liu, W.; Wang, C.; Zhang, R.; Fan, W.; Yu, D.; Qiao, X. High figure of merit and low cross sensitivity fiber Bragg grating accelerometer based on double grid-diaphragms. *IEEE. Sens. J.* **2021**, *21*, 27503–27509. [CrossRef]

Disclaimer/Publisher's Note: The statements, opinions and data contained in all publications are solely those of the individual author(s) and contributor(s) and not of MDPI and/or the editor(s). MDPI and/or the editor(s) disclaim responsibility for any injury to people or property resulting from any ideas, methods, instructions or products referred to in the content.

sensors

MDPI

Article

A Comprehensive Approach for Detecting Brake Pad Defects Using Histogram and Wavelet Features with Nested Dichotomy Family Classifiers

Sakthivel Gnanasekaran [1,2], Lakshmi Pathi Jakkamputi [2], Jegadeeshwaran Rakkiyannan [2,*], Mohanraj Thangamuthu [3,*] and Yogesh Bhalerao [4]

1 School of Mechanical Engineering, Vellore Institute of Technology, Chennai 600127, India; sakthivel.g@vit.ac.in
2 Centre for Automation, Vellore Institute of Technology, Chennai 600127, India; lakshmipati@vit.ac.in
3 Department of Mechanical Engineering, Amrita School of Engineering, Amrita Vishwa Vidyapeetham, Coimbatore 641112, India
4 Department of Mechanical Engineering and Design, School of Engineering, University of East Anglia, Norwich Research Park, Norwich NR4 7TJ, UK; y.bhalerao@uea.ac.uk
* Correspondence: jegadeeshwaran.r@vit.ac.in (J.R.); t_mohanraj@cb.amrita.edu (M.T.)

Abstract: The brake system requires careful attention for continuous monitoring as a vital module. This study specifically focuses on monitoring the hydraulic brake system using vibration signals through experimentation. Vibration signals from the brake pad assembly of commercial vehicles were captured under both good and defective conditions. Relevant histograms and wavelet features were extracted from these signals. The selected features were then categorized using Nested dichotomy family classifiers. The accuracy of all the algorithms during categorization was evaluated. Among the algorithms tested, the class-balanced nested dichotomy algorithm with a wavelet filter achieved a maximum accuracy of 99.45%. This indicates a highly effective method for accurately categorizing the brake system based on vibration signals. By implementing such a monitoring system, the reliability of the hydraulic brake system can be ensured, which is crucial for the safe and efficient operation of commercial vehicles in the market.

Keywords: histogram feature; wavelet feature; nested dichotomy; class-balanced nested dichotomy; data-near-balanced nested dichotomy

1. Introduction

The hydraulic brake system is a critical component of an automobile and is responsible for controlling the vehicle's speed and reducing the stopping distance. Any malfunction in the brake system can have a detrimental impact on the vehicle's stability and the safety of its passengers. According to the National Motor Vehicle Causation Survey, it has been revealed that 22% of accidents occur as a result of brake system failures [1]. In 2007, BMW recalled over 155,000 Sports Utility Vehicles (SUVs) due to brake fluid issues. This recall was initiated to address concerns regarding the performance and functionality of the brake system in these vehicles. The recall aimed to ensure the safety of the vehicles and their occupants by rectifying the brake-fluid-related issues. Similarly, Chrysler also faced a recall in 2007 involving approximately 60,000 vehicles. The recall was due to braking issues, specifically, a lack of success in the braking system. Chrysler undertook this recall to address the identified deficiencies and ensure the proper functioning of the brakes in the affected vehicles [2].

These recalls demonstrate the commitment of automotive manufacturers to prioritize safety and take prompt action to rectify any issues related to braking systems. This statistic underscores the significance of maintaining a reliable and properly functioning brake system to prevent accidents and ensure the safety of both drivers and passengers. Regular maintenance, inspections, and testing are crucial for vehicle owners, manufacturers, and regulatory authorities to mitigate the risk of brake system failures. By implementing effective monitoring systems and taking preventive measures, potential issues can be identified early on, allowing for timely corrective actions and reducing the occurrence of accidents stemming from brake system failures. Examining such systems for defects will decrease the total crashes and enhance safety to a certain extent. Currently, a specific method for monitoring brake system failure is unavailable. Therefore, it is essential to develop a suitable method to monitor failures related to the brake system. The brake system may become flawed owing to pad wear, mechanical fade, inadequate oil pressure in the control cylinder, and oil leakage. If these flaws are not observed, accidents will take place. Hence, proactive maintenance is recommended to prevent accidents and can be attained through continuous monitoring [3,4].

Condition monitoring (CM) is continuously observing a system's operational attributes to initiate necessary steps before the occurrence of a breakdown. Many CM methods utilize devoted sensors and data assessment tools to examine a specific type of disparity in functional attributes [5]. Thomas et al. outlined a new communications module for identifying brake faults through sensors. The module detects the malfunction through the recorded signals from the brake system [6]. A method was patented for monitoring the vehicle condition and generating a maintenance plan according to the vehicle condition [7–9]. A new approach for monitoring the aircraft brake was proposed to determine the brake condition [10].

Nevertheless, to the best knowledge of the authors, there is no available solution for monitoring the brake pad conditions. Hence, a new method is required to diagnose brake pad-related failure. Fault diagnosis is a sub-area of CM in which specific system conditions are continuously studied. Fault identification can be performed in numerous ways, like visual testing, acoustic emission (AE) [11], thermal imagers, ultrasonic, and vibration signals [12]. Visual testing is the most commonly used technique in the industry. However, visual inspection has drawbacks since it requires skilled laborers who can identify the problems through experience. Hence, it is unsuitable for monitoring hidden components such as the brake system [13].

A passive infrared thermal imager was utilized to monitor the weld quality of the specimen, but sophisticated equipment was needed to measure the temperature variance [14]. Since it is an offline process, it is also unsuitable for brake fault diagnosis. AE can be used to monitor the machining process [15]. The bearing fault in a helicopter gearbox was identified by measuring the AE and vibration signals [16]. The spur gear condition was monitored using AE and vibration signals [17]. The study proved that vibration analysis is more appropriate for fault than AE signal analysis. However, the AE signal acquisition is complicated in machine components such as a brake.

Recently, most of the fault diagnostic studies focused on the vibration signal. The vibration signals are examined through frequency domain analysis, wavelet analysis, waveform examination, spectral analysis, statistical learning, histogram learning, etc. These examinations gave the data needed to decide on the maintenance plan. The consequences of such examinations are utilized for fault investigation to determine the fault's original cause. The brake system produces vibrations under various working circumstances. The measured vibration signals correlate precisely with the fault conditions, which are used to identify the defects in the system. Therefore, vibration signals can be considered appropriate for condition monitoring through feature-based analysis.

The feature is a specific, measurable characteristic of an anomaly being examined. Selecting data, separating, and categorizing independent features is essential in classification, pattern recognition, and regression. Numerous features, specifically statistical [18], histogram [19], and wavelet [20], can be obtained from the acquired signals. The application of a wavelet plays a crucial role in CM. The main advantage of the wavelet transform (WT) is to denoise or compress a signal without reducing the original signal and computational time. WT could be successfully applied to the different fault diagnosis studies. Hence, WT has been used for diagnosing brake failure in this study.

After extracting the features, the most significant features have to be selected. The attribute evaluator, the influence of the number of features, and Decision Tree (DT) are the three most important methods for selecting the features. DT utilizes a graphical structure to select the most significant features [21]. The attribute evaluator calculates the value of a characteristic subset by studying each feature's analytical capacity and severance level. The influence of the number of features is one of the most helpful techniques in feature selection. Instead of taking all the features, the performance of the specific features is tested using this examination. The influence of several features study was applied to diagnose the brake faults and identify the best feature combination that produces the maximum classification accuracy [22].

Feature classification is the last phase in the classifier. The selected features should be classified for predicting performance. The commonly used classifiers are Decision Tree [23,24], Fuzzy [25], Artificial Neural Network [20], Bayes Net [26], Naive Bayes [20], Support Vector Machine (SVM) [27], Convolutional Neural Network (CNN) [28] and the Hidden Markov Model [29]. Different classifiers monitored the tool wear using vibration signals with the discrete wavelet feature [20]. So, there is considerable scope for the wavelet features with other machine learning algorithms in the brake monitoring system. The nested dichotomy (ND) classifier was used for fault diagnosis in the hydraulic brake system. They proved that the class-balanced nested dichotomy (CBND) prediction accuracy was more than that of the SVM and DT classifiers. The authors used the statistical feature alone for classification. The other features can also be considered to make a more detailed study.

To the authors' best knowledge, there is no literature on CM of a brake system using histogram and wavelet features. Hence, histogram and wavelet features are considered to monitor the hydraulic brake system. An attribute evaluator is used for feature selection. The feature selection through the attribute evaluator is validated using the influence of the number of features studied. The preferred wavelet and histogram features were categorized using the family of ND classifiers, and the results were compared and discussed.

2. Experimental Procedure

Experimentation was conducted to supervise the brake system using histogram and wavelet features extracted from the vibration signals. A brake system test rig was used to perform the experiments. The experimental test rig and procedure are briefed in the subsequent sections.

2.1. Experimental Test Rig

A commercial vehicle, as depicted in Figure 1, was utilized for the experiment. Its drive shaft consistently operated at 331 rpm on a test bed equipped with rollers. A piezo-electric uni-axial accelerometer (Make: Dytran, Sensitivity: 10 mV/g) was securely positioned at the center of the flat surface of the brake caliper, as illustrated in Figure 1c. As the drive shaft rotated, various dynamic forces and loads impacted the brake pad and caliper assembly. These forces resulted from the interaction between the brake pad and the brake rotor, leading to vibrations within the brake pad and caliper assembly. As the brake pad wears, it undergoes changes in both rigidity and mass distribution, which, in turn, cause fluctuations in the assembly's stiffness and mass. These variations in vibration within the brake pad assembly, associated with the brake pad's stiffness, are crucial parameters for understanding and identifying brake pad defects. The attached piezo-electric accelerometer captures these variations in vibration signals, which are then processed using the NI 9234 data acquisition (DAQ) system, converting them into digital data. The encoded data from the DAQ system is subsequently processed and presented in the LabVIEW 2021 software, providing results in the time domain as presented in Figure 1b. The experiments were performed with different brake pad conditions, as depicted in Figure 1d.

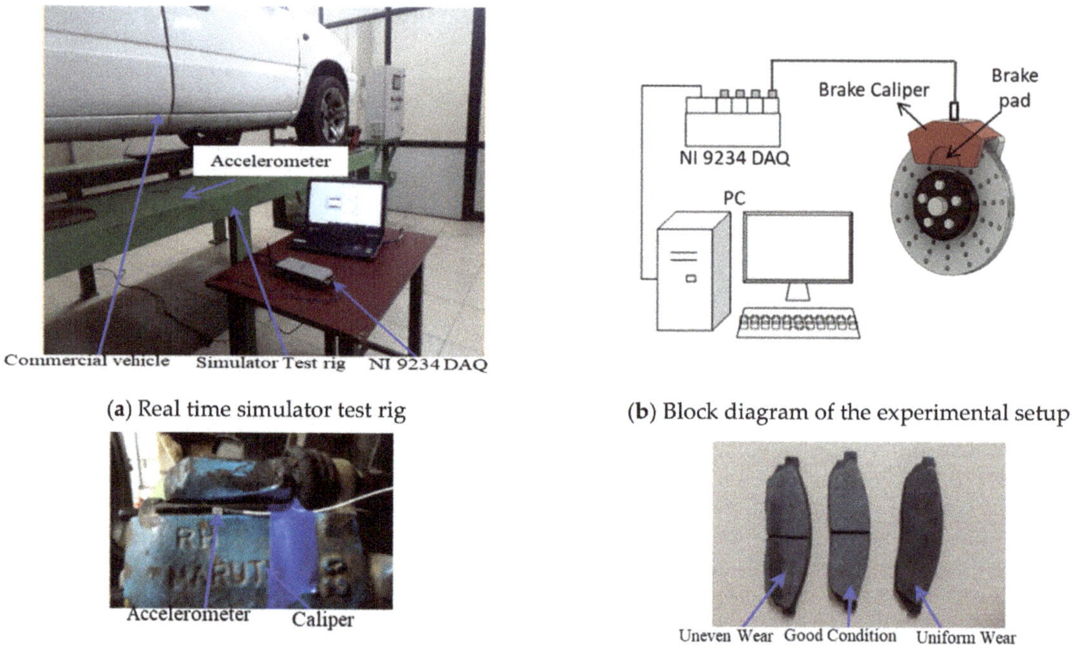

(**a**) Real time simulator test rig

(**b**) Block diagram of the experimental setup

(**c**) Exploded view of experiment setup

(**d**) Brake pad conditions

Figure 1. Experimental test rig.

2.2. Experimental Procedure

The frequently occurring failures, such as Brake Oil Spill (BOS), Drum Brake Mechanical Fade (DBM), Disc Pad Wear (Uneven) Inner (DBP(UE)I), Disc Pad Wear (Uneven) Inner and Outer (DBP(UE)IO), Disc Pad Wear (Even) Inner (DBPI), Disc Pad Wear (Even) Inner and Outer (DBPIO), Drum Brake Pad Wear (DBPW), Air in Brake Fluid (ABF), and Reservoir Leak (RL) were considered for simulating the fault conditions. The vibration signals were acquired from the test setup with good and various simulated faulty brake conditions. The LabVIEW program used for vibration signal acquisition is shown in Figure 2. The data acquisition was performed with the following specifications:

➢ Sample length and no. of samples: 2^{13} and 55 (arbitrarily chosen);
➢ Sampling frequency: 24 kHz (as per the Nyquist sampling theorem);
➢ Wheel speed and brake force: 331 rpm (~30 KMPH) and 68.7 N.

The wheel speed, which was 331 rpm, was measured and controlled using a tachometer. The brakes were applied gradually and evenly to prevent RPM fluctuations, and, importantly, as the brakes were applied, minor manual adjustments were made to the throttle to ensure a consistent RPM was maintained.

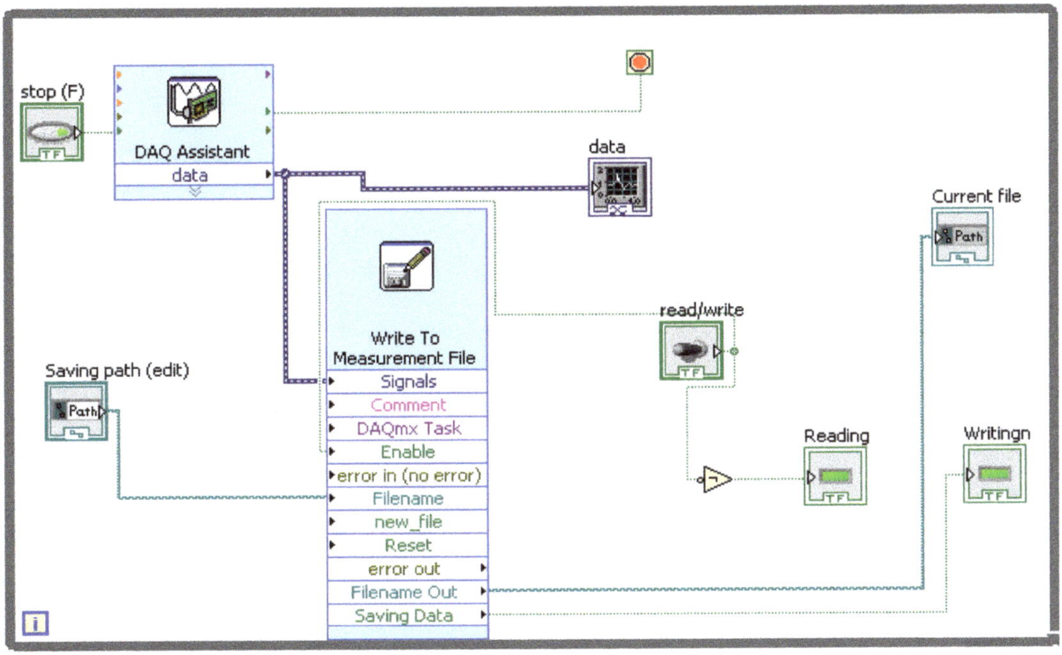

Figure 2. LabVIEW program for data acquisition.

3. Histogram Features

The histogram data is used as a feature in the brake CM. The magnitude range of the vibration is split into various sub-ranges (bins) from the smaller to the larger value of the vibration signal. The quantity of information focuses on whose magnitude of vibration value cataracts inside a specific bin are calculated. The bin and count frame the histogram's x and y axes. The histogram for each defect is plotted utilizing vibration motions as an independent graph. The goal is to discover the bins whose y-axis values are the same for a specific defect but not the same for another defect. The histogram for a specific brake defect might be small; however, it might be huge for other defects. The bin range chosen should suit all defects. The strategy for estimating the base of least magnitude is as follows:

a. Compute the least magnitude of each signal in a specific condition;
b. Repetition of step (i) for all conditions;
c. Identify the least of those least magnitudes.

Similarly, the base of the largest magnitude is computed. The bin width should be fixed so that the bin height is diverse for various brake conditions. It requires not to be valid for all widths of the bin, but rather, at least a few of them should be taken after this measure is utilized as a feature for recognizing different conditions. This study used the bin width and range to plot the histogram, as shown in Figure 3a. The vibration value between the larger and smaller values was divided by a number of bins. The larger and smaller value range is divided into 69 bins (2 to 70). Figure 3a–j shows the histogram obtained from the vibration signals.

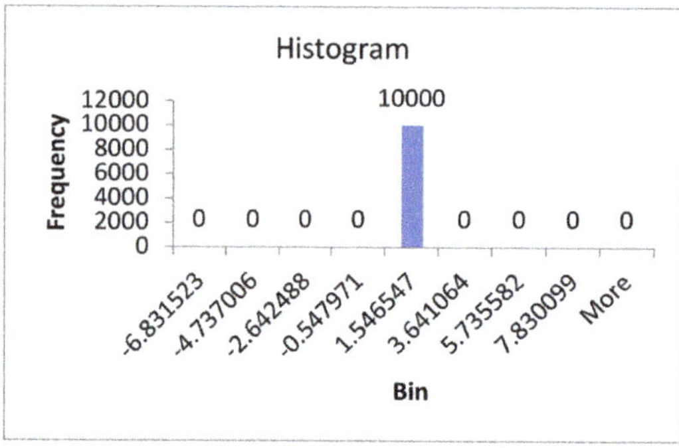

(**a**) GOOD

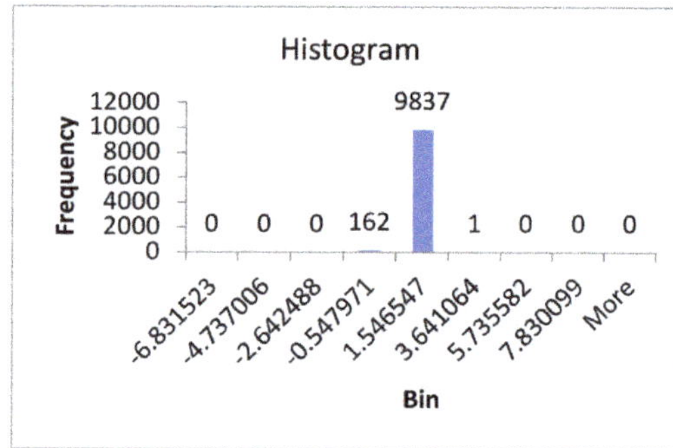

(**b**) ABF

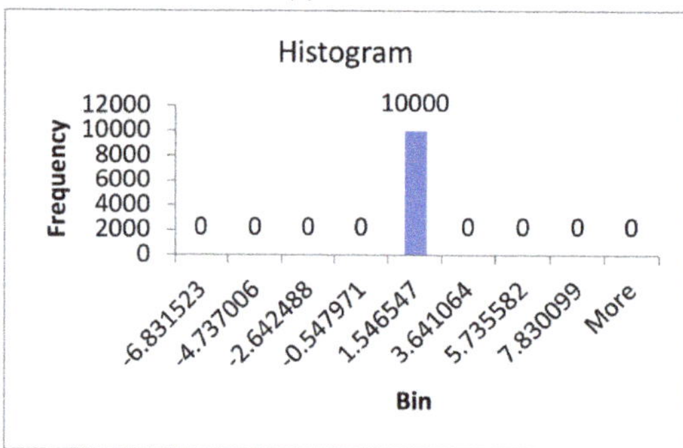

(**c**) BOS

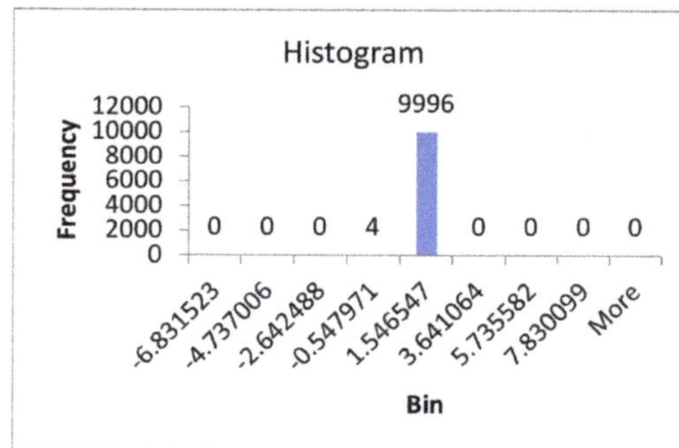

(**d**) DBPI

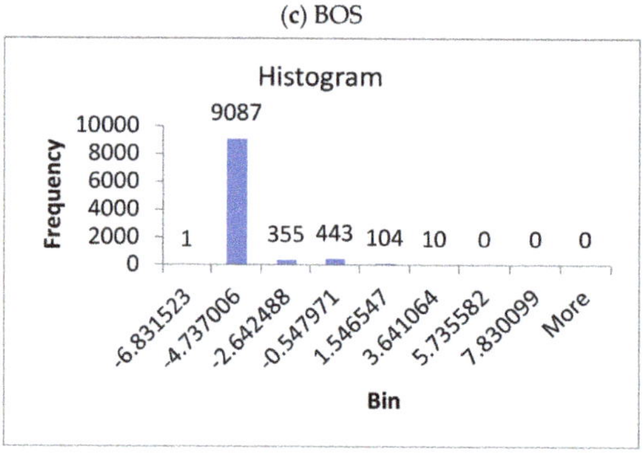

(**e**) DBPIO

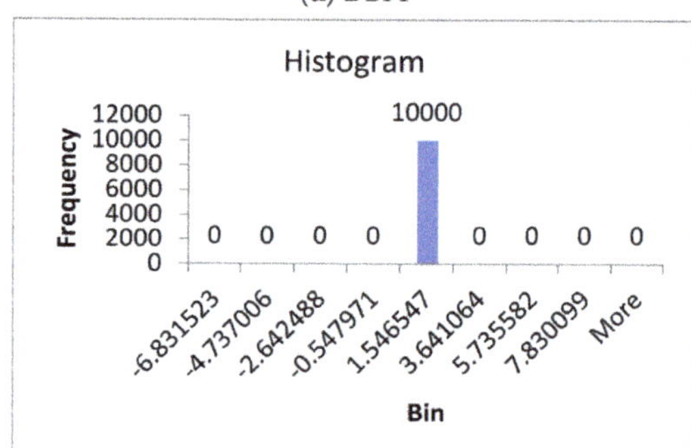

(**f**) DBP(UE) I

Figure 3. *Cont.*

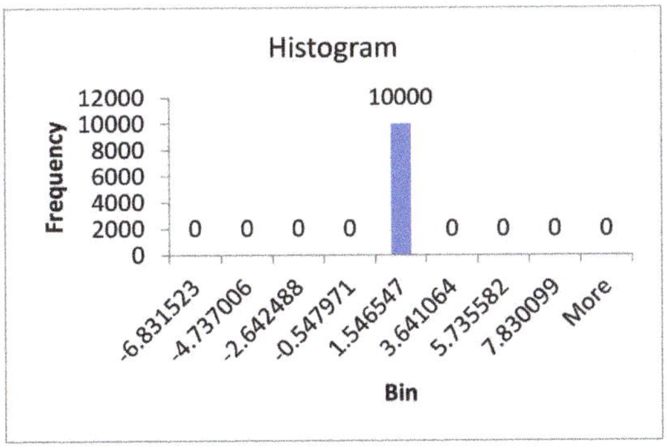

(**g**) DBP(UE) IO

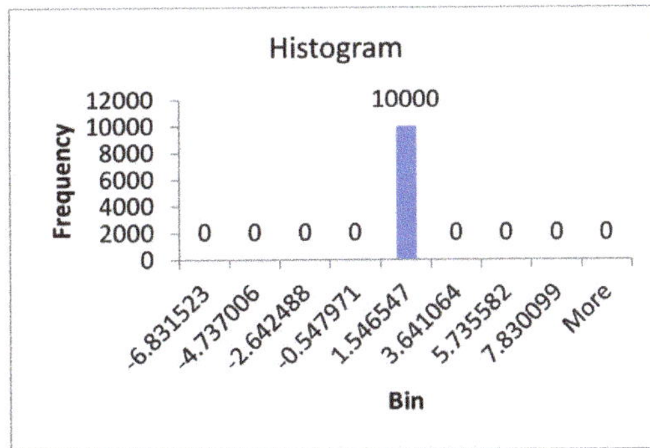

(**h**) DBPW

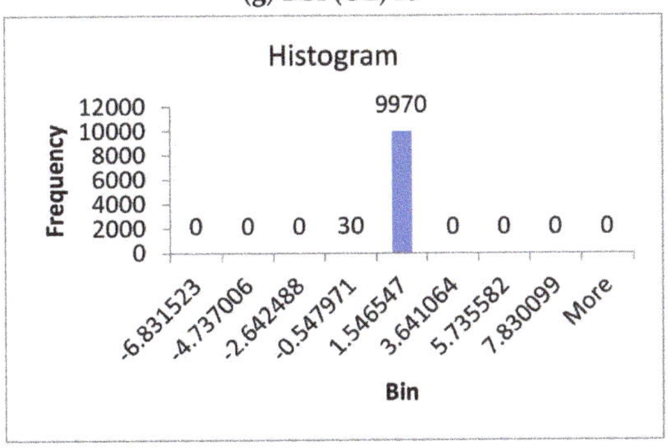

(**i**) DBM

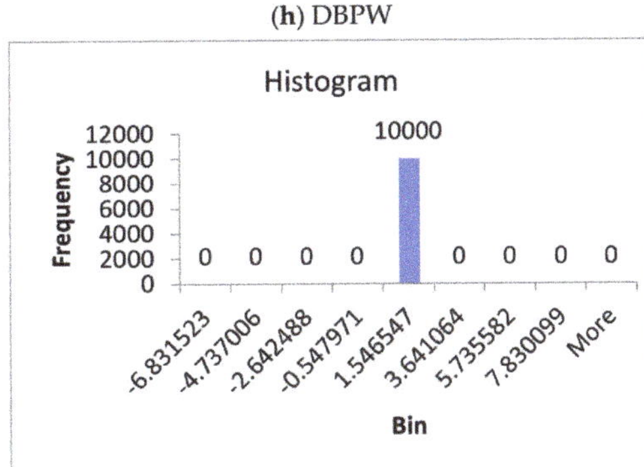

(**j**) RL

Figure 3. Histogram of different faulty conditions for 7 bin range.

4. Wavelet Filter

The wavelet transform (WT) is a prevailing time and frequency domain signal tool. WT can be continuous or discrete, permitting the investigation of fleeting, irregular, or local parts. The continuous WT can expose more insights about a signal but is computationally intensive. For fault analysis, discrete wavelet transform (DWT) aims to characterize a signal proficiently with fewer parameters and computation time. DWT is an appropriate method for this study [30].

4.1. Wavelet Transform

The WT is drawn as

$$\begin{aligned} WT_f(p,q) &= f(t).\psi_{p,q}(t) \\ WT_f(p,q) &= \frac{1}{\sqrt{p}}\int_0^\infty f(t)\psi\left(\frac{t-q}{p}\right)dt \end{aligned} \tag{1}$$

where $\psi_{p,q}(t)$ is the alternative of the "Mother wavelet", which is considered as

$$\psi_{p,q}(t) = \frac{1}{\sqrt{p}}\int_0^\infty \psi\left(\frac{t-q}{p}\right) \tag{2}$$

where t, p, $q \in R$, and $p \neq 0$ are continuously varying quantities, p is the scale factor, q is a shift factor, and $\frac{1}{\sqrt{p}}$ is used for energy conservation. If p is small, extreme-frequency components are analyzed, and vice versa.

4.2. Discrete Wavelet Transform

DW performs scaling and translation procedures in distinct phases, restraining the selection of scales (a) and translation (τ) to specific numbers. However, the examination is satisfactorily precise and articulated in the following equation:

$$\psi_{j,k}(t) = \frac{1}{\sqrt{s_0^j}}\psi\left(\frac{t - k\tau_0 s_0^j}{s_0^j}\right) \quad (3)$$

where j and k are integers.

The time scale is tested at a distinct interval due to discretizing the wavelets, called decomposition, and succeeds a sequence of wavelet coefficients [31]. The wavelet features were obtained for every condition utilizing the Daubechies wavelets "*db1*" to "*db10*". The wavelets studied for this examination are provided below:

- Haar—'haar';
- Daubechies—'db';
- Symlet—'sym';
- Coiflets—'coif';
- Biorthogonal—'bior';
- Reverse biorthogonal—'rbio';
- Discrete approximation of Meyer—'dmey';
- Fejer–korovkin—'fk'.

The WT removed the noise from the raw vibration signal and reconstructed the signals. The successive approximation filters certain more high-frequency data from the raw signal. Using this method, substantial data loss is possible. Therefore, a new method is required for denoising. Optimal denoising demands a more sensitive approach known as thresholding, which includes discarding a small number of progressive elements that exceed a specific threshold. Thresholding is reformed from level to level. For the current study, level 5 approximation was applied. The eight types of wavelet features were extracted for every condition. Figure 4a–j shows the denoised sample signal attained through DWT for the Coiflets wavelet. From the denoised signals, relevant statistical features were collected.

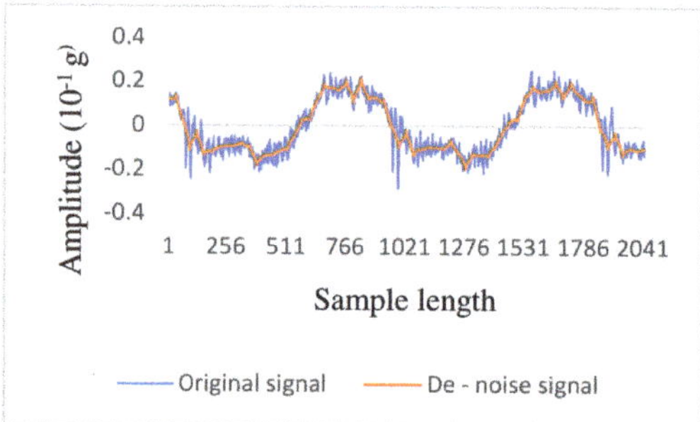

(**a**) Good

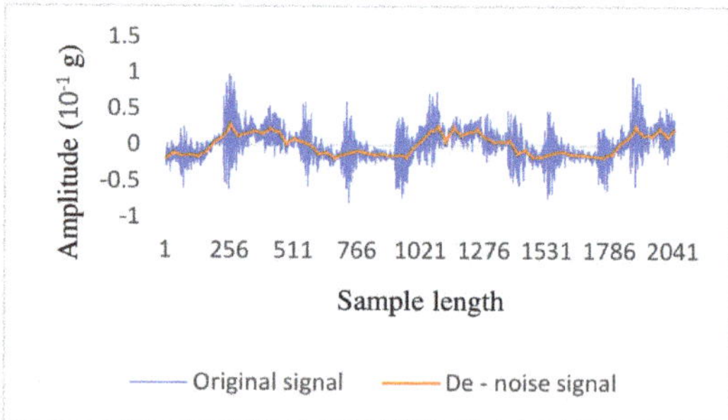

(**b**) ABF

Figure 4. *Cont.*

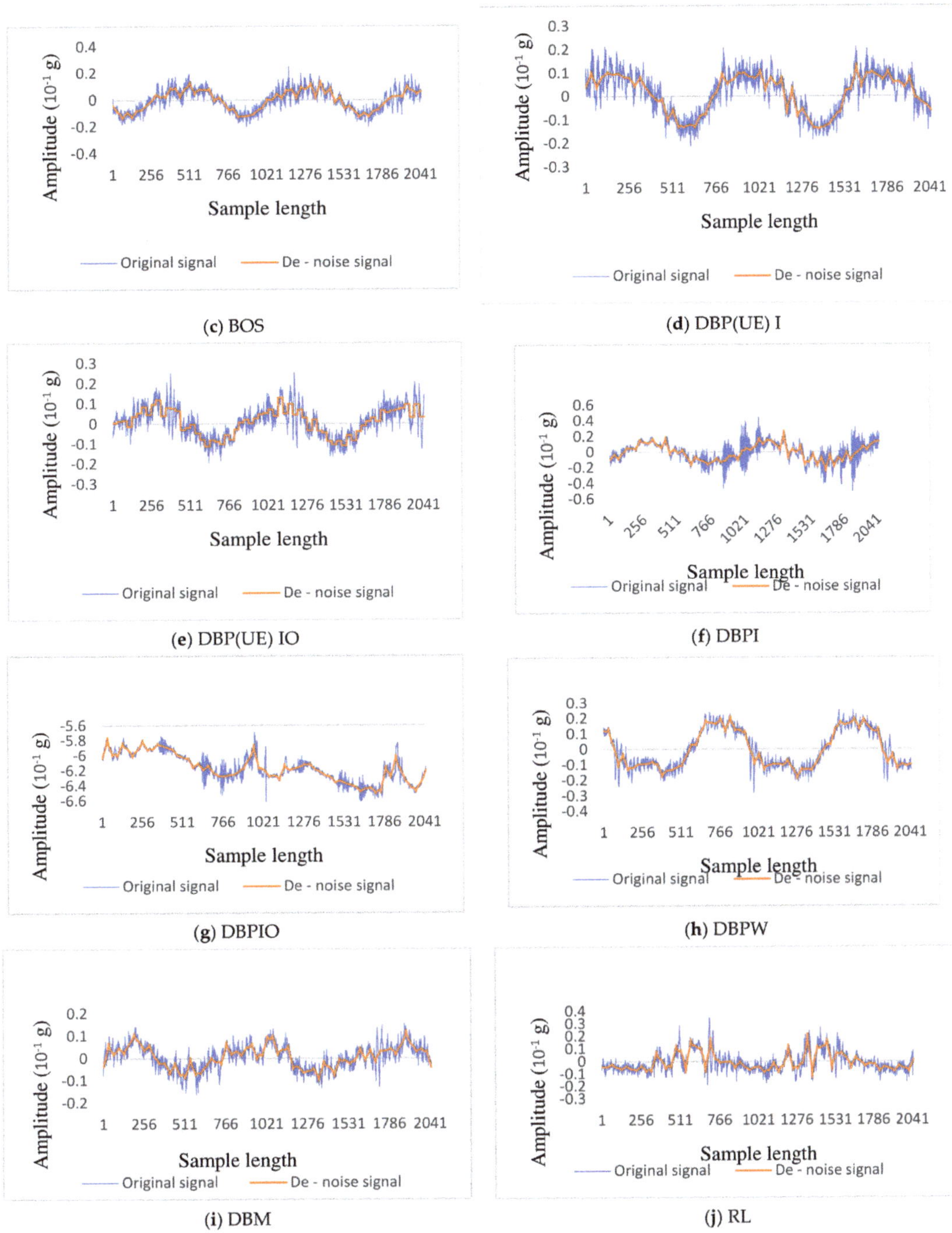

(**c**) BOS (**d**) DBP(UE) I (**e**) DBP(UE) IO (**f**) DBPI (**g**) DBPIO (**h**) DBPW (**i**) DBM (**j**) RL

Figure 4. Original and denoised signals for various faulty conditions.

5. Feature Selection

The attribute evaluator, DT, and the influence of the number of features studied were used to select the most significant features. This process is called feature selection. The attribute evaluator was utilized to find the best feature set from the obtained features. In this study, the two different search methods, Correlation-based Feature Selection Subset Eval and Best First Search were applied to select the sequence of the significant features. The chosen feature was tested through the effect of the number of features studied. The first element suggested by the attribute evaluator was classified first, and the accuracy was recorded. Then, the second feature recommended by the evaluator was combined with the first feature and classified. The classification accuracy with two feature combinations was also recorded. The process was repeated until all the features were combined and classified. The best feature combination that only provides maximum accuracy was selected for further study. The DT was used for choosing the most significant features from the histogram feature.

6. Feature Classification

The selected features should be classified to obtain the classification result. A system of ND is employed to disintegrate a multinomial problem into multiple binary problems. The disintegration is described using a binary tree. A set of class labels is stored in each tree node, and a binary classifier from the training data. The training data is divided into two subsets conforming to the two meta-classes, and one subset of training data is considered a true example.

In contrast, the other subset of testing data is considered a false example. The two next-in-line root nodes receive two subsets of the true class labels with their conforming training data. A tree is constructed by continuously employing this process. Lastly, this process succeeds a leaf node if the node has only one label. The choice of tree structure significantly impacts the accuracy.

The CBND method depends on regulating the number of conditions at every node. As an alternative to examining each possible tree's space, CBND examined the space of each balanced tree. The quantity of possible trees in CBND is lesser than the entire tree of ND. The detailed algorithm for developing a system of CBND is given in [32]. The usual difficulty with the CBND method is that few multinomial issues are not balanced and more populated than others. A CBND does not have additional data balanced. This can undesirably affect the computational time of the algorithm. DNBND arbitrarily assigns the classes to two subsets until the training data size in one subset surpasses 50% of training data at the node. With the DNBND algorithm, assortment suffers when class dispersal is disturbed. The number of classes depends on the class distribution in the data.

7. Results and Discussion

The wavelet and histogram features were extracted, and the most significant features were chosen. The chosen features were classified using ND family classifiers.

7.1. Histogram Feature Classification Using Nested Dichotomy Family Classifiers

The DT algorithm chose the most significant features [23]. All the 69 bins ranging from 2 to 70 were classified using the DT algorithm one after another. The classification accuracy for different bin ranges is shown in Figure 5. The ND classifier delivers a maximum classification accuracy for the 59th bin of 97.64%. The algorithm uses decision rules for classification. These decision rules are graphically represented as a DT, as shown in Figure 6. Hence, the graphical DT shows the contributing bin ranges. Even though 59 bin ranges were used as input in the algorithm, it uses only 7 for classification. Figure 6 selected the contributing bin range (59th) alone which have the maximum classification accuracy (97.64%). The features were categorized using various ND-family algorithms, including ND, CBND, and DNBND. The parameters used for the classification are given below:

- Random no. seed: 1;

- No. of trees: 10;
- Base learner: best first tree;
- Tree depth: unlimited.

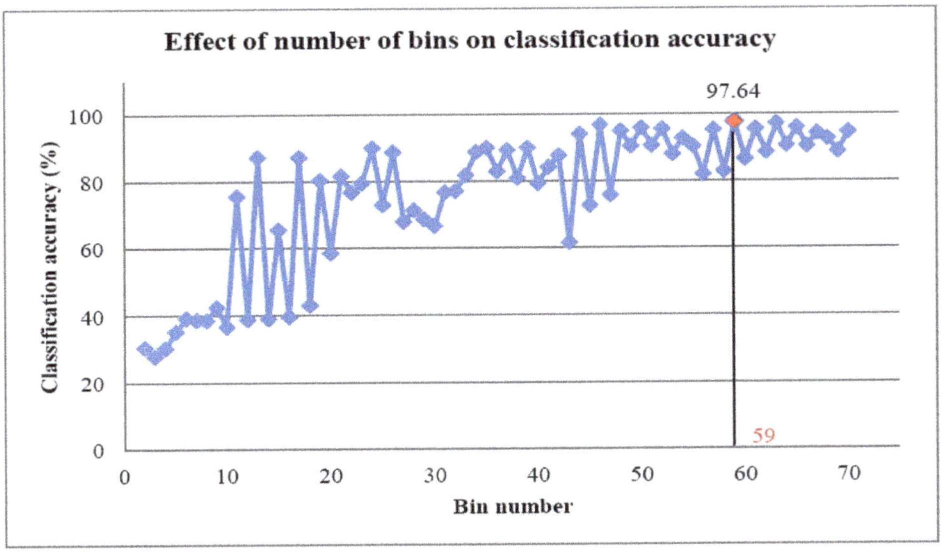

Figure 5. Classification accuracy vs. number of bins.

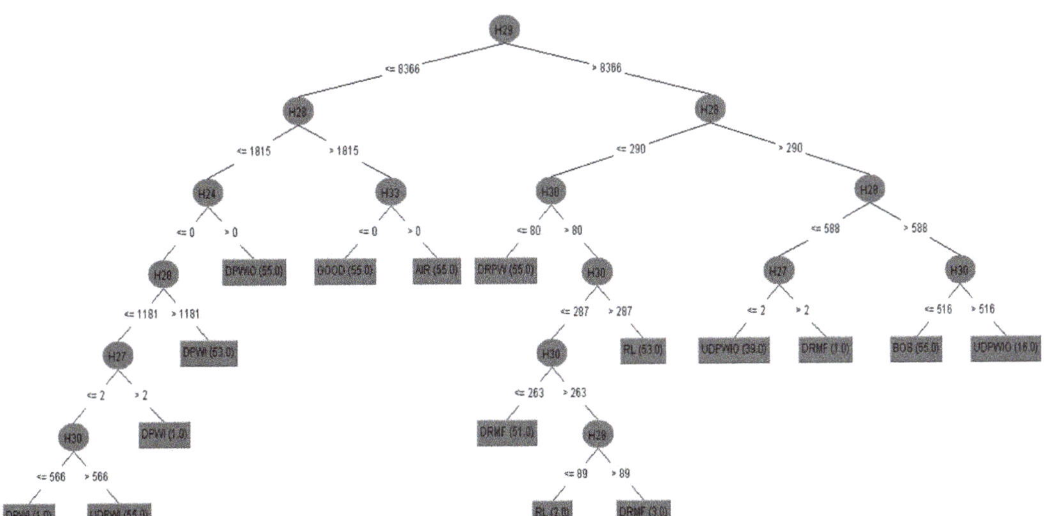

Figure 6. Graphical DT with histogram features.

The categorization accuracy is displayed in Figure 7. In Figure 7, the first cell signifies the number of data conforming to the "GOOD" class. The sum of data points in the first row is 55. All are correctly categorized and yield a classification accuracy of 100%. Likewise, the other element in the first row represents the false categorization. There are no data points on the first row beside the first column. This means that there is no false classification. Comparably, the sixth row sixth column signifies the DBPI condition. Among the 55 datapoints, 53 were categorized accurately, and two were wrongly classified under the DBP(UE)I condition. The classification accuracy was estimated as 96.4%. Among the 550 datapoints, 538 were accurately classified, with an overall classification accuracy of 97.5%.

Category	GOOD	BOS	DBM	DBP(UE)I	DBP(UE)IO	DBPI	DBPIO	DBPW	ABF	RL	Precision	False discovery rate
GOOD	55	0	0	0	0	0	0	0	0	0	100	0.00
BOS	0	55	0	0	0	0	0	0	0	0	100	0.00
DBM	0	0	53	0	2	0	0	0	0	0	96	3.64
DBP(UE)I	0	0	0	54	0	1	0	0	0	0	98	1.82
DBP(UE)IO	0	0	0	0	54	0	0	1	0	0	98.2	1.82
DBPI	0	0	0	2	0	53	0	0	0	0	96	3.64
DBPIO	1	0	0	0	0	0	54	0	0	0	98	1.82
DBPW	0	0	0	0	1	0	0	54	0	0	98.2	1.82
ABF	0	0	0	0	0	0	0	3	52	0	95	5.45
RL	0	0	2	0	1	0	0	0	0	52	94.5	5.45
TPR	98.2	100.0	96.4	96.4	93.1	98.1	100.0	93.1	100.0	100.0	97.5	97%
FNR	1.8	0.0	3.6	3.6	6.9	1.9	0.0	6.9	0.0	0.0	2.4	

Figure 7. Confusion matrix for ND with histogram features.

The rows and columns represent the anticipated and actual conditions, respectively. The diagonal cells correspond to data points that are appropriately categorized. The off-diagonal cells represent wrongly categorized data. The no. of data is shown in each cell, and the exact prediction accuracy is shown in the last cell. The last two columns on the right show the % of all the data predicted to belong to each appropriate (precision) and wrongly classified (false discovery rate) condition. The last two rows at the bottom show the % of the data predicted appropriately (True Positive Rate (TPR)) and wrongly (False Negative Rate (FNR)) classified condition. The cell in the bottom right shows the overall accuracy.

Similarly, the selected features were classified using CBND and DNBND classifiers, and the corresponding classification accuracies were found. The overall accuracy for CBND and DNBND was 97% and 99%, respectively. Figures 8 and 9 represent the confusion matrix obtained for CBND and DNBND with histogram features.

Category	GOOD	BOS	DBM	DBP(UE)I	DBP(UE)IO	DBPI	DBPIO	DBPW	ABF	RL	Precision	False discovery rate
GOOD	55	0	0	0	0	0	0	0	0	0	100	0.00
BOS	0	54	0	0	0	0	0	0	1	0	98	1.82
DBM	0	0	52	0	1	0	0	0	0	2	95	5.45
DBP(UE)I	0	0	0	54	0	0	1	0	0	0	98	1.82
DBP(UE)IO	0	0	0	0	55	0	0	0	0	0	100.0	0.00
DBPI	0	0	0	1	0	54	0	0	0	0	98	1.82
DBPIO	1	0	0	0	0	0	54	0	0	0	98	1.82
DBPW	0	0	0	0	1	0	0	54	0	0	98.2	1.82
ABF	0	5	0	0	0	0	0	3	47	0	85	14.55
RL	0	0	1	0	0	0	0	0	0	54	98.2	1.82
TPR	98.2	91.5	98.1	98.2	96.5	100.0	98.2	94.7	97.9	96.4	97.5	97%
FNR	1.8	8.5	1.9	1.8	3.5	0.0	1.8	5.3	2.1	3.6	2.4	

Figure 8. Confusion matrix for CBND with histogram features.

7.2. Wavelet Feature Classification Using ND Family Classifiers

Eight families of wavelets were selected, and a denoised signal was obtained from each family. From the denoised signal, twelve statistical features were extracted. The attribute evaluator was used to select the most significant features, and nine statistical features were acknowledged and considered as input for the classifier. The nominated nine features were appraised using the influence of the no. of feature study. Table 1 indicates the influence of the no. of features studied performed on the Coiflets wavelet. The CBND produces a better accuracy (99.45%) with the top eight features.

The feature assortment process was applied for all wavelet families, and the resultant accuracy was recorded and presented in Table 2. In Table 2, the Coiflets wavelet (Coif) produced a maximum classification accuracy compared to other wavelets. Therefore, the Coiflets wavelet was selected. From the influence of the no. of feature study, the eight features that contributed to the classification were chosen for the study. They were as follows: (1) Mean, (2) Standard Error, (3) Median, (4) Standard Deviation, (5) Kurtosis, (6) Skewness, (7) Range, and (8) Minimum. The classification accuracy for the ND family (ND, CBND, and DNBND) algorithms was found. Figure 10 represents the confusion matrix for ND. Figure 10 reveals the overall classification accuracy of 99%.

Category	GOOD	BOS	DBM	DBP(UE)I	DBP(UE)IO	DBPI	DBPIO	DBPW	ABF	RL	Precision	False discovery rate
GOOD	55	0	0	0	0	0	0	0	0	0	100	0.00
BOS	0	54	0	0	1	0	0	0	0	0	98	1.82
DBM	0	0	52	0	1	0	0	0	0	2	95	5.45
DBP(UE)I	0	0	0	55	0	0	0	0	0	0	100	0.00
DBP(UE)IO	0	0	0	0	55	0	0	0	0	0	100.0	0.00
DBPI	0	0	0	1	0	54	0	0	0	0	98	1.82
DBPIO	0	0	0	0	0	0	55	0	0	0	100	0.00
DBPW	0	0	0	0	1	0	0	54	0	0	98.2	1.82
ABF	0	0	0	0	0	0	0	0	55	0	100	0.00
RL	0	0	1	0	0	0	0	0	0	54	98.2	1.82
TPR	100.0	100.0	98.1	98.2	94.8	100.0	100.0	100.0	100.0	96.4	97.5	99%
FNR	0.0	0.0	1.9	1.8	5.2	0.0	0.0	0.0	0.0	3.6	2.4	

Figure 9. Confusion matrix for DNBND with histogram features.

Table 1. Influence of no. of feature study with Coiflets wavelet.

No. of Features	Categorization Accuracy (%)		
	ND	CBND	DNBND
1	28.91	28	28.36
2	89.45	88.55	90
3	96	95.27	95.82
4	94.55	94.91	95.27
5	98	98.73	98.55
6	98	99.27	98.73
7	98.73	98.55	98.73
8	98.73	99.45	99.27
9	98.55	99.27	98.91

Table 2. Classification accuracy for various wavelet families.

Wavelet	Classification Accuracy (%) ND	CBND	DNBND
Coif1L5	98.73	99.45	99.27
Dmey	99.27	99.09	99.09
Bior2.2L5	98.18	98.18	98.36
Db5L5	99.09	98.73	98.36
Haar	98.73	98.55	98.55
Sym3L5	98.73	98.55	98.55
Rbior2.8L5	86.73	85.82	86
Fk4L5	97.64	98.18	98.36

Category	GOOD	BOS	DBM	DBP(UE)I	DBP(UE)IO	DBPI	DBPIO	DBPW	ABF	RL	Precision	False discovery rate
GOOD	55	0	0	0	0	0	0	0	0	0	100	0.00
BOS	0	55	0	0	0	0	0	0	0	0	100	0.00
DBM	0	0	53	0	1	0	0	0	0	1	96	3.64
DBP(UE)I	0	0	0	55	0	0	0	0	0	0	100	0.00
DBP(UE)IO	0	1	0	0	53	1	0	0	0	0	96.4	3.64
DBPI	0	0	0	0	0	55	0	0	0	0	100	0.00
DBPIO	0	0	0	0	0	0	55	0	0	0	100	0.00
DBPW	0	0	1	0	0	0	0	54	0	0	98.2	1.82
ABF	0	0	0	0	0	0	0	0	55	0	100	0.00
RL	0	0	2	0	0	0	0	0	0	53	96.4	3.64
TPR	100.0	98.2	94.6	100.0	98.1	98.2	100.0	100.0	100.0	98.1	97.5	99%
FNR	0.0	1.8	5.4	0.0	1.9	1.8	0.0	0.0	0.0	1.9	2.4	

Figure 10. Confusion matrix for ND with Coiflets wavelet.

Comparably, the remaining data points in the first row signify the false classification details. All other values are zero in the first row other than the first element, revealing no false classification. Similarly, the last element in the last row represents RL. Among the 55 datapoints, 53 belong to RL and are accurately categorized, and 2were wrongly categorized as the DBM condition. Among the 550 datapoints, 543 were categorized accurately, with an accuracy of 99%.

Likewise, the classification accuracy was computed for CBND and DNBND. Both the classifier produced 99.45% and 99.27% as classification accuracy, respectively. Figures 11 and 12 show the graphical representation of the confusion matrix attained for the CBND and DNBND. The overall accuracy of the ND classifier is presented in Table 3. The CBND gives an enhanced accuracy of 99.45% among the three ND family classifiers. Among the 550 datapoints, 547 were accurately classified. Denoising using wavelet significantly reduced noise and increased classification accuracy.

Category	GOOD	BOS	DBM	DBP(UE)I	DBP(UE)IO	DBPI	DBPIO	DBPW	ABF	RL	Precision	False discovery rate
GOOD	55	0	0	0	0	0	0	0	0	0	100	0.00
BOS	0	55	0	0	0	0	0	0	0	0	100	0.00
DBM	0	0	55	0	0	0	0	0	0	1	98	1.79
DBP(UE)I	0	0	0	55	0	0	0	0	0	0	100	0.00
DBP(UE)IO	0	1	0	0	54	0	0	0	0	0	98.2	1.82
DBPI	0	0	0	0	0	55	0	0	0	0	100	0.00
DBPIO	0	0	0	0	0	0	55	0	0	0	100	0.00
DBPW	0	0	1	0	0	0	0	54	0	0	98.2	1.82
ABF	0	0	0	0	0	0	0	0	55	0	100	0.00
RL	0	0	1	0	0	0	0	0	0	54	98.2	1.82
TPR	100.0	98.2	96.5	100.0	100.0	100.0	100.0	100.0	100.0	98.2	97.5	99%
FNR	0.0	1.8	3.5	0.0	0.0	0.0	0.0	0.0	0.0	1.8	2.4	

Figure 11. Confusion matrix for CBND with wavelet.

Category	GOOD	BOS	DBM	DBP(UE)I	DBP(UE)IO	DBPI	DBPIO	DBPW	ABF	RL	Precision	False discovery rate
GOOD	55	0	0	0	0	0	0	0	0	0	100	0.00
BOS	0	55	0	0	0	0	0	0	0	0	100	0.00
DBM	0	0	55	0	0	0	0	0	0	1	98	1.79
DBP(UE)I	0	0	0	55	0	0	0	0	0	0	100	0.00
DBP(UE)IO	0	2	0	0	53	0	0	0	0	0	96.4	3.64
DBPI	0	0	0	0	0	55	0	0	0	0	100	0.00
DBPIO	0	0	0	0	0	0	55	0	0	0	100	0.00
DBPW	0	0	1	0	0	0	0	54	0	0	98.2	1.82
ABF	0	0	0	0	0	0	0	0	55	0	100	0.00
RL	0	0	1	0	0	0	0	0	0	54	98.2	1.82
TPR	100.0	96.5	96.5	100.0	100.0	100.0	100.0	100.0	100.0	98.2	97.5	99%
FNR	0.0	3.5	3.5	0.0	0.0	0.0	0.0	0.0	0.0	1.8	2.4	

Figure 12. Confusion matrix for DNBND with Coiflets wavelet.

Table 3. Overall classification accuracy of ND family classifiers.

Classifier	Features	Total No. of Data	Correctly Categorized Instances		Incorrectly Categorized Instances		Kappa Statistic
ND	Histogram	550	540	98.18%	10	1.82%	0.9758
CBND	Histogram	550	542	98.55%	8	1.45%	0.9818
DNBND	Histogram	550	542	98.55%	8	1.45%	0.9818
ND	Coiflets wavelet	550	543	98.73%	7	1.27%	0.9859
CBND	Coiflets wavelet	550	547	99.45%	3	0.55%	0.9939
DNBND	Coiflets wavelet	550	546	99.27%	4	0.73%	0.9919

7.3. Comprehensive Accuracy by Condition

The specific condition implementation of the ND can be studied using comprehensive accuracy by condition study, termed as TPR, FPR, Precision, and ROC. The TPR corresponds to the accurately identified conditional class, and the FPR is the rate at which the classes are wrongly classified. For a machine learning algorithm, TPR must be 1, and FPR should be 0. This study's TPR and FPR are closer to 1 and 0, respectively. Similarly, the other parameter values are also closer. Table 4 presents the classification accuracy for CBND for wavelet feature extraction.

Table 4. Comprehensive accuracy by conditions—CBND with Coiflets wavelets.

Category	TPR	FPR	Precision	Recall	F-Measure	ROC Area
ABF	1	0	1	1	1	1
BOS	1	0.002	0.982	1	0.991	1
DBM	1	0.004	0.965	1	0.982	1
DBP(UE) inner	1	0	1	1	1	1
DB(UE) IO	0.982	0	1	0.982	0.991	1
DBPI	1	0	1	1	1	1
DBPIO	1	0	1	1	1	1
DBPW	0.982	0	1	0.982	0.991	1
GOOD	1	0	1	1	1	1
RL	0.982	0	1	0.982	0.991	1
Weighted Average	0.995	0.001	0.995	0.995	0.995	1

7.4. Comparative Analysis

The captured raw vibration signals were denoised through WT. The carefully chosen features were categorized using ND, CBND, and DNBND, and the overall accuracy is shown in Table 5. The results of wavelet features were compared with histogram features. Table 5 shows the WT of the vibration signal with CBND, revealing a maximum overall accuracy of 99.45%. The comparative results show that the wavelet features and ND outperform the statistical features and all other classifiers considered in this study. Hence, the wavelets with CBND can be considered an appropriate classifier for diagnosing brake faults.

Table 5. Classification accuracy of ND classifiers.

S. No.	Name of the Classifier	Classification Accuracy (%)	
		Wavelet Features	Histogram Features
1	ND	99.20	97.50
2	CBND	99.45	97.00
3	DNBND	99.30	99.00

8. Conclusions

The paper discussed the wavelet and histogram features for supervising the brake system using vibration signals. The present study considered nine good simulated faults of the brake system. The vibration signals were recorded under all simulated defects. The wavelet and histogram features were extracted from the vibration signal, and the most significant features were selected. Then, the preferred features were classified using ND, CBND, and DNBND. The results reveal that the CBND classifier performed better than the ND and DNBND in classification accuracy. The competency of the wavelet features was evaluated with the histogram features, and the wavelet feature yielded an overall prediction accuracy of 99.45%. Henceforth, the CBND with wavelet features can be recommended for online brake fault diagnostics. This research is protracted with actual road conditions for monitoring the hydraulic brake system.

Author Contributions: Conceptualization, S.G. and J.R.; methodology, J.R.; software, L.P.J.; validation, J.R., M.T. and Y.B.; formal analysis, J.R.; investigation, L.P.J.; resources, J.R.; data curation, J.R.; writing—original draft preparation, J.R.; writing—review and editing, M.T.; visualization, S.G.; supervision, L.P.J.; project administration, J.R. and S.G.; funding acquisition, J.R. All authors have read and agreed to the published version of the manuscript.

Funding: The authors want to acknowledge the research funding from DST—Science and Engineering Research Board (SERB Grant) File No. ECR/2016/002068.

Institutional Review Board Statement: Not applicable.

Informed Consent Statement: Not applicable.

Data Availability Statement: The data used to support the findings of this study will be provided upon request and institutional approval.

Acknowledgments: The authors thank the Management of VIT University, School of Mechanical Engineering, VIT Chennai and the Amrita School of Engineering, Coimbatore, Amrita Vishwa Vidyapeetham, India, and for their support to publish this work.

Conflicts of Interest: The authors declared no conflict of interest.

References

1. National Highway Traffic Safety Administration. *Critical Reasons for Crashes Investigated in the National Motor Vehicle Crash Causation Survey*; US Department of Transportation: Washington, DC, USA, 2015; Volume 2, pp. 1–2.
2. NickJohnson. Brake Failure Accidents. 2007. Available online: www.articles3k.com/article/144/176520 (accessed on 1 September 2023).
3. Mohapatra, A.G.; Mohanty, A.; Pradhan, N.R.; Mohanty, S.N.; Gupta, D.; Alharbi, M.; Alkhayyat, A.; Khanna, A. An Industry 4.0 implementation of a condition monitoring system and IoT-enabled predictive maintenance scheme for diesel generators. *Alex. Eng. J.* **2023**, *76*, 525–541. [CrossRef]
4. Molla Salilew, W.; Karim, Z.A.A.; Lemma, T.A. Investigation of fault detection and isolation accuracy of different Machine learning techniques with different data processing methods for gas turbine. *Alex. Eng. J.* **2022**, *61*, 12635–12651. [CrossRef]
5. Goel, S.; Ghosh, R.; Kumar, S.; Akula, A. A methodical review of condition monitoring techniques for electrical equipment. In Proceedings of the National Seminar & Exhibition on Non-Destructive Evaluation (NDE 2014), Pune, India, 4–6 December 2014.
6. Wallace, T.E.; Rink, R.J.; Chandler, M.D. Brake Monitoring System and Method. US Patent No. 8,319,623, 27 November 2012.
7. Kanda, Y.; Terashima, N.; Katou, T. Vehicle Condition Monitoring System. US Patent No. 7,321,814, 22 January 2008.
8. Fiechter, C.-N.; Göker, M.H.; Grill, D.; Kaufmann, R.; Engelhardt, T.; Bertsche, A. Method and System for Condition Monitoring of Vehicles. US Patent No. 6,609,051, 19 August 2003.
9. Haugse, E.D.; Ikegami, R.; Trego, A. Vehicle Condition Monitoring System. US Patent No. 6,691,007, 10 February 2004.
10. Miller, R.J.; Marshall, R.J.; Bailey, D.A.; Griffin, N.C. Brake Condition Monitoring. US Patent No. 7,086,503, 3 August 2001.
11. Praveenkumar, T.; Saimurugan, M.; Ramachandran, K. Intelligent Fault Diagnosis of Synchromesh Gearbox Using Fusion of Vibration and Acoustic Emission Signals for Performance Enhancement. *Int. J. Progn. Health Manag.* **2019**, *10*, 2738. [CrossRef]
12. Yuvaraju, E.; Rudresh, L.; Saimurugan, M. Vibration signals based fault severity estimation of a shaft using machine learning techniques. *Mater. Today Proc.* **2020**, *24*, 241–250. [CrossRef]
13. Kim, Y.; Lee, S.; Seo, H.; Jung, J.; Yang, H. Development of dissolved gas analysis (DGA) expert system using new diagnostic algorithm for oil-immersed transformers. In Proceedings of the 2012 IEEE International Conference on Condition Monitoring and Diagnosis, Bali, Indonesia, 23–27 September 2012; pp. 365–369. [CrossRef]
14. Sreedhar, U.; Krishnamurthy, C.; Balasubramaniam, K.; Raghupathy, V.; Ravisankar, S. Automatic defect identification using thermal image analysis for online weld quality monitoring. *J. Mater. Process. Technol.* **2012**, *212*, 1557–1566. [CrossRef]
15. Sachin Krishnan, P.; Rameshkumar, K.; Krishnakumar, P. Hidden Markov modelling of high-speed milling (HSM) process using acoustic emission (AE) signature for predicting tool conditions. In *Advances in Materials and Manufacturing Engineering*; Springer: Singapore, 2020; pp. 573–580.
16. Elasha, F.; Greaves, M.; Mba, D.; Addali, A. Application of acoustic emission in diagnostic of bearing faults within a helicopter gearbox. *Procedia Cirp* **2015**, *38*, 30–36. [CrossRef]
17. Tan, C.K.; Irving, P.; Mba, D. A comparative experimental study on the diagnostic and prognostic capabilities of acoustics emission, vibration and spectrometric oil analysis for spur gears. *Mech. Syst. Signal Process.* **2007**, *21*, 208–233. [CrossRef]
18. Altaf, M.; Akram, T.; Khan, M.A.; Iqbal, M.; Ch, M.M.I.; Hsu, C.-H. A New Statistical Features Based Approach for Bearing Fault Diagnosis Using Vibration Signals. *Sensors* **2022**, *22*, 2012. [CrossRef] [PubMed]
19. Gnanasekaran, S.; Jakkamputi, L.; Thangamuthu, M.; Marikkannan, S.K.; Rakkiyannan, J.; Thangavelu, K.; Kotha, G. Condition Monitoring of an All-Terrain Vehicle Gear Train Assembly Using Deep Learning Algorithms with Vibration Signals. *Appl. Sci.* **2022**, *12*, 10917. [CrossRef]
20. Mohanraj, T.; Yerchuru, J.; Krishnan, H.; Aravind, R.N.; Yameni, R. Development of tool condition monitoring system in end milling process using wavelet features and Hoelder's exponent with machine learning algorithms. *Measurement* **2021**, *173*, 108671. [CrossRef]
21. Steffan, J.J.; Jebadurai, I.J.; Asirvatham, L.G.; Manova, S.; Larkins, J.P. Prediction of Brake Pad Wear Using Various Machine Learning Algorithms. In *Recent Trends in Design, Materials and Manufacturing*; Springer: Singapore, 2022; pp. 529–543.
22. Liu, J.; Li, Y.-F.; Zio, E. A SVM framework for fault detection of the braking system in a high speed train. *Mech. Syst. Signal Process.* **2017**, *87*, 401–409. [CrossRef]
23. Joshuva, A.; Anaimuthu, S.; Selvaraju, N.; Muthiya, S.J.; Subramaniam, M. *A Machine Learning Approach for Vibration Signal Based Fault Classification on Hydraulic Braking System through C4. 5 Decision Tree Classifier and Logistic Model Tree Classifier*; SAE Technical Paper; SAE International: Warrendale, PA, USA, 2020. [CrossRef]

24. Satapathy, S.; Loganathan, D.; Kondaveeti, H.K.; Rath, R. Performance analysis of machine learning algorithms on automated sleep staging feature sets. *CAAI Trans. Intell. Technol.* **2021**, *6*, 155–174. [CrossRef]
25. Mazzoleni, M.; Sarda, K.; Acernese, A.; Russo, L.; Manfredi, L.; Glielmo, L.; Del Vecchio, C. A fuzzy logic-based approach for fault diagnosis and condition monitoring of industry 4.0 manufacturing processes. *Eng. Appl. Artif. Intell.* **2022**, *115*, 105317. [CrossRef]
26. Bajaj, N.S.; Patange, A.D.; Jegadeeshwaran, R.; Kulkarni, K.A.; Ghatpande, R.S.; Kapadnis, A.M. A Bayesian optimized discriminant analysis model for condition monitoring of face milling cutter using vibration datasets. *J. Nondestruct. Eval. Diagn. Progn. Eng. Syst.* **2022**, *5*, 021002. [CrossRef]
27. Vos, K.; Peng, Z.; Jenkins, C.; Shahriar, M.R.; Borghesani, P.; Wang, W. Vibration-based anomaly detection using LSTM/SVM approaches. *Mech. Syst. Signal Process.* **2022**, *169*, 108752. [CrossRef]
28. Chao, Q.; Wei, X.; Tao, J.; Liu, C.; Wang, Y. Cavitation recognition of axial piston pumps in noisy environment based on Grad-CAM visualization technique. *CAAI Trans. Intell. Technol.* **2023**, *8*, 206–218. [CrossRef]
29. Jia, Y.; Xu, M.; Wang, R. Symbolic important point perceptually and hidden Markov model based hydraulic pump fault diagnosis method. *Sensors* **2018**, *18*, 4460. [CrossRef] [PubMed]
30. Wang, Y.; Yan, Y.; Wang, Q. Wavelet-based feature extraction in fault diagnosis for biquad high-pass filter circuit. *Math. Probl. Eng.* **2016**, *2016*, 5682847. [CrossRef]
31. Yao, Y.; Ma, J.; Ye, Y. Regularizing Autoencoders with Wavelet Transform for Sequence Anomaly Detection. *Pattern Recognit.* **2022**, *134*, 109084. [CrossRef]
32. Leathart, T.; Frank, E.; Pfahringer, B.; Holmes, G. Ensembles of nested dichotomies with multiple subset evaluation. In Proceedings of the Pacific-Asia Conference on Knowledge Discovery and Data Mining, Macau, China, 14–17 April 2019.

Disclaimer/Publisher's Note: The statements, opinions and data contained in all publications are solely those of the individual author(s) and contributor(s) and not of MDPI and/or the editor(s). MDPI and/or the editor(s) disclaim responsibility for any injury to people or property resulting from any ideas, methods, instructions or products referred to in the content.

sensors

MDPI

Article

Regional Thermal Radiation Characteristics of FY Satellite Remote Sensing Based on Big Data Analysis

Tao Wen [1], Congxin Wei [2,*], Zhenyi Wang [2], Linzhu Wang [1], Zihan Yang [1] and Tingting Gu [3]

1 Gansu Computing Center, Lanzhou 730030, China; went@cc.gs.cn (T.W.); wanglz@cc.gs.cn (L.W.)
2 School of Computer and Communication, Lanzhou University of Technology, Lanzhou 730050, China
3 Gansu Zijin Cloud Big Data Development Co., Ltd., Jinchang 737100, China; gutt@gepic.cn
* Correspondence: weicx@lut.edu.cn

Abstract: It is of great significance to study the thermal radiation anomalies of earthquake swarms in the same area in terms of selecting abnormal characteristic determination parameters, optimizing and determining the processing model, and understanding the abnormal machine. In this paper, we investigated short-term and long-term thermal radiation anomalies induced by earthquake swarms in Iran and Pakistan between 2007 and 2016. The anomalies were extracted from infrared remote sensing black body temperature data from the China Geostationary Meteorological Satellites (FY-2C/2E/2F/2G) using the multiscale time-frequency relative power spectrum (MS T-FRPS) method. By analyzing and summarizing the thermal radiation anomalies of series earthquake groups with consistency law through a stable and reliable MS T-FRPS method, we first obtained the relationship between anomalies and ShakeMaps from USGS and proposed the anomaly regional indicator (ARI) to determine seismic anomalies and the magnitude decision factor (MDF) to determine seismic magnitude. In addition, we explored the following discussions: earthquake impact on regional thermal radiation background and the relationship between thermal anomalies and earthquake magnitude and the like. Future research directions using the MS T-FRPS method to characterize regional thermal radiation anomalies induced by strong earthquakes could help improve the accuracy of earthquake magnitude determination.

Keywords: satellite thermal radiation remote sensing; multiscale time-frequency relative power spectrum analysis; thermal radiation background; thermal anomaly

1. Introduction

The application of satellite remote sensing thermal radiation data to study earthquake anomalies and mechanisms is one of the research directions of geophysics, which has obtained many research results. One of the earliest studies on the relationship between satellite radiation remote sensing and earthquakes dates back to the 1980s [1,2]. With the development of satellite remote sensing technology, many efforts have been devoted to studying the earthquake thermal radiation characteristics. It has been reported that significant thermal anomalies can be identified prior to earthquakes in seismically active areas [2–11]. Such examples include the following: Bhuj (Gujarat), Mw = 7.7, 26 January 2001, India [6,11]; Colima, Mw = 7.8, 21 January 2003, Mexico [11]; Kashmir, Mw = 7.6, 8 October 2005, [3]; Bhuj (Gujarat), Mw = 7.7, 26 January 2001, India [6,10]; Colima, Mw = 7.8, 21 January 2003, Mexico [11]; Yutian, Mw = 7.3, 21 March 2008, China [12]; Wenchuan, Mw = 8.0, 12 May 2008, China [13]; the Andaman Islands, Mw = 7.5, 11 August 2009 [14]; Honshu, Mw = 9.1, 11 March 2011, Japan [15]; and many others.

However, in seismic thermal radiation research, anomaly extraction is difficult because of the weak signals and complicated background noise. First of all, satellite remote sensing images were used in structural geological and geomorphologic research. Active faults and structures were mapped on the base of satellite images. This method is very limited in time series analysis. There was no possibility to measure short-term processes before and after the earthquake. It was as simple as an extrapolation of air photo geological interpretation methods applied to space. The modern version of this method is active tectonic analysis with the application of alignment analysis. Case studies of various thermal remote sensing application methods for earthquakes have been recently reported. The use of MISR (Multi-angle Imaging Spectra Radiometer) data revealed a significant increase in surface moisture after the Bhuj (Gujarat) earthquake on 26 January 2001. This indicates that changes in surface moisture after an earthquake can be detected through thermal remote sensing technology [16]. The discovery that anomalous areas are consistent with seismic zones or active tectonics, with epicenters often located at the intersections of stress tropics, suggests a certain correlation between seismic activity and thermal remote sensing anomalies, possibly related to tectonic activity [17–21]. The analysis of satellite thermal infrared brightness temperature anomaly images reveals that short-term anomalies exhibit clearly identifiable radiation characteristics. This suggests that by analyzing radiation anomalies at short-term time scales, specific feature information can be obtained [22]. The study of the temporal-spatial variations of associated faulting inferred from satellite infrared information reveals a close correlation between the Mani earthquake radiation anomaly and the activity of the Altyn Tagh fault, illustrating the interactions between faults and defection among the deformation exception, current fault activities, and earthquakes [23]. Secondly, time series of alignment distributions on the Earth's surface are investigated before and after an earthquake. Tramutoli and colleagues have developed a robust satellite data analysis technique for space–time thermal anomalies on the Earth's surface recorded by satellites months to weeks before the occurrence of earthquakes. Lastly, since 2000, Zhang et al. have researched anomaly extraction using case studies and have investigated underlying mechanisms [24]. They developed the time-frequency relative power spectrum (T-FRPS) method based on radiation brightness temperature remote sensing data from the China Geostationary Meteorological Satellites (FY-2C/2E/2F/2G) and proposed the 'union origin mechanism' to explain pre-earthquake thermal anomalies [25]. The thermal radiation anomaly characteristics, including characteristic period, amplitude, duration, and morphology, were clear and valuable for pre-earthquake thermal radiation research. The epicenters were located near the intersection of the anomaly migration processes, which was consistent with the remote sensing rock test mechanics [26,27].

Researchers have achieved certain results in the extraction methods of seismic anomaly information (from viewing pictures to time series and time-frequency analysis) and earthquake case studies. However, there are few studies in some directions, such as the characteristics of thermal radiation anomalies and the characteristics of swarms of earthquakes over many years; there are many studies on short-term and anomalous thermal radiation, but there are few studies on mid- and long-term background anomalies, and thermal radiation background characteristics, as well as the characteristics of thermal radiation backgrounds and the impacts of strong earthquakes on thermal radiation background field.

Therefore, this article details the multiscale time-frequency relative power spectrum (MS T-FRPS), it initially scrutinizes the results of Wavelet filtering results, and conducts an annual background anomaly analysis of Mw 7+ earthquakes. Additionally, it extracts short- and long-term thermal radiation anomalies of several strong earthquake swarms in Iran and Pakistan. Furthermore, it compares and analyzes the similarities and differences between short-term and long-term background abnormalities for the first time. The impacts of strong earthquakes on thermal radiation background field are also discussed for the first time.

2. Methodology

2.1. Multiscale Time-Frequency Relative Power Spectrum (MS T-FRPS) Analysis

Time-frequency analysis (TFA) is a powerful tool for analyzing time-varying non-stationary signals and has received increasing attention in recent years. TFA provides joint distribution information of the time domain and the frequency domain, which clearly describes the signal frequency over time. The basic idea of TFA is to design a joint function of time and frequency that simultaneously describes the energy density or intensity of the signal at different times and frequencies. This joint function of time and frequency is referred to simply as the time-frequency distribution, by which we can draw the instantaneous frequency and amplitude at each moment and conduct time-frequency filtering and time-varying analysis of the signal.

Firstly, Daubechies wavelet transform (2nd-order scale and 7th-order scale filtering) is used to process the multi-year data of satellite thermal radiation daily values to separate the temperature changes caused by the basic temperature field, the annual temperature field, and other factors. In addition, the 6th- and 7th- order scales are respectively decomposed to obtain the annual and approximate annual background information. Secondly, the Welch's FFT power spectrum is applied to process the above wavelet filtering and decomposition results, in order to obtain the dominant frequency and amplitude in the time domain. Therefore, the method of successive wavelet processing and power spectrum processing is called time-frequency power spectrum analysis. Thirdly, in order to compare and analyze the power spectrum of the thermal radiation change before and after the earthquake, the power spectrum values of all the frequencies of each pixel are relatively processed to generate spatial-temporal data of the relative time-frequency power spectrum. The frequency analysis method through wavelet transform at different scales (6th-order scale, 7th-order scale, and 2nd- and 7th-order bandpass filtering) and relative Fourier power spectrum calculation for different window lengths (64 days, 128 days, 256 days, and 512 days) is called multiscale time-frequency relative power spectrum (MS T-FRPS).

2.1.1. Wavelet and Fourier Transform

Wavelet theory is a relatively new digital signal processing technology that originated a few decades ago. Due to its high resolution in both the time and frequency domains, it is called the 'mathematical microscope', which has been widely used and is developing rapidly in the field of digital signal processing. All wavelet transforms may be considered as forms of time-frequency representation for continuous-time signals and thus are related to harmonic analysis. Almost all practically useful discrete wavelet transforms use discrete-time filter banks, which are called wavelet and scaling coefficients in the wavelet nomenclature. The most commonly used set of discrete wavelet transforms was formulated by Ingrid Daubechies in 1988 [28]. The Daubechies wavelets are a family of orthogonal wavelets, with D2-D20 and db1-db10 commonly used. In this paper, the db8 wavelet basis function of the Daubechies (dbN) wavelet system is used as the basic transform window function to process daily value brightness temperature data.

Spectral analysis is an important method for studying signal characteristics. For deterministic signals, Fourier transform (FT) can be used to examine its spectral properties. Any signal that can be represented by amplitude that varies in time has a corresponding frequency spectrum, which contains all the information of the original signal. This means that the original function can be completely reconstructed by an inverse FT. In classical spectral estimation using a periodical diagram, the variance of spectral estimate at a specific frequency does not decrease with the increasing number of samples. Although it is successful in examining the frequency characteristics of noise-free functions, it cannot process noise-like signals, e.g., sinusoids with low signal-to-noise ratios. To mitigate this issue, Welch modified the periodic diagram method by taking average over time, which becomes popular for estimating the power of a signal at different frequencies. Due to its ease of understanding and calculation, it has been widely used and adopted in our analysis.

2.1.2. Relative Processing

In order to obtain the relative time-frequency power spectrum of each pixel, we calculate it by Equation (1),

$$R_{PS} = \frac{R_{i,j}}{R_{i,Mean}} \tag{1}$$

where R_{PS} is the calculated relative power spectrum value, $R_{i,j}$ is the power spectrum value of a certain pixel at any time, and $R_{i,Mean}$ is the ten-year average of the corresponding pixel.

2.2. Data Acquisition

Geostationary meteorological satellites are preferable to polar orbit satellites for earthquake research as they consistently observe the same location. We used radiation brightness temperature remote sensing data acquired from the China Geostationary Meteorological Satellites FY-2C/2E/2F/2G, which were launched during 2004–2014. The scan period was 30 min or 1 h, and the spatial resolution was 5 km. The data were collected from 1 January 2007 to 31 December 2016. Five midnight observations were selected from each day because of the lower solar radiation at night.

2.3. Data Processing

In this paper, the wavelet transform basis function and the relative FT are designed as a joint function of time and frequency, which describes the energy density or intensity of the signal at different times and frequencies. The TFA function was used to analyze the similarities and differences between time-frequency analyses at each scale.

Data processing and extraction were conducted in four steps, as shown in Figure 1:

1. The original database was formed after pre-processing the radiation remote sensing black body temperature data from the Chinese geostationary meteorological satellite. This process included data format conversion, data space range interception, removal of clouds, and selection of nighttime data for daily value calculations. The influence of partial cloud cover was removed by making up the window. Then, we calculated the mean value of five data points to obtain the daily value.
2. Various scale-factor transformation results were obtained by applying discrete Daubechies wavelet transformations. And the dominant frequency and the peak-to-peak amplitude values were obtained using the FFT power spectral method.
3. The decade (2007–2016) average value of each pixel was used as the denominator to obtain the relative time-frequency PS value of the corresponding pixel.
4. Time-frequency mapping was used to extract earthquake thermal radiation anomaly data. The non-seismic factors were effectively removed and earthquake thermal radiation anomalies were highlighted by scanning the time-frequency space data over the full spatial-temporal and band range.

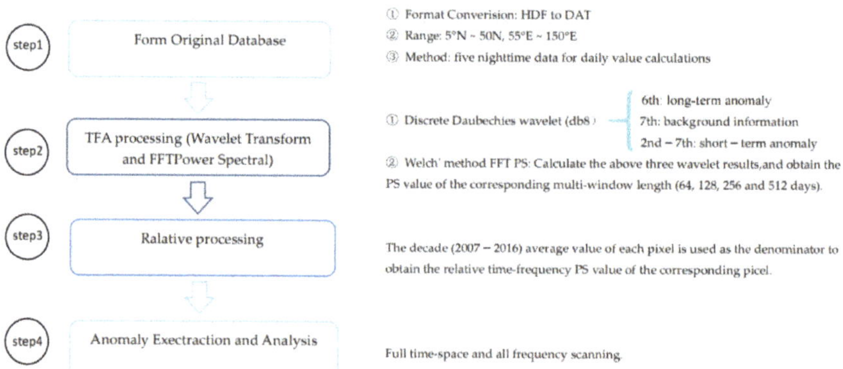

Figure 1. Data processing flowchart.

3. Study Area

3.1. Regional Tectonic Background

The tectonics of this investigated region is dominated by the collisions of the Arabian and Indian plates with Eurasia. The Arabian plate lithosphere subducts beneath the Eurasian plate at the Makran coast of Pakistan and Iran, and extends progressively deeper to the north [29]. Four major tectonic plates (Arabia, Eurasia, India, and Africa) and one smaller tectonic block (Anatolia) are responsible for seismicity in the Middle East and surrounding region. The geologic evolution of this region is a consequence of several first-order plates' tectonic processes including seductions, large-scale transform faulting, compression mountain building, and crustal extension. Continental thickening of the northern and western edge of the India subcontinent has produced the highest mountains in the world, including the Himalayan, Karakoram, Pamir, and Hindu Kush ranges. Earthquake activity and faulting found in this region as well as adjacent parts of Afghanistan and India are due to collisional plate tectonics. Shallower crustal earthquakes in the Pamir-Hindu Mountains occur primarily along the Main Pamir Thrust and other active Quaternary faults, which accommodate much of the region's crustal shortening. From 2007 to 2016, 4 earthquakes with Mw 6.0+ occurred in this region; the ShakeMaps of the main shocks are shown in Figure 2 and the source is the following website: (a) https://earthquake.usgs.gov/earthquakes/eventpage/usp000gma2#shakemap, (accessed on 1 July 2023); (b) https://earthquake.usgs.gov/earthquakes/eventpage/usp000htb4#shakemap, (accessed on 1 July 2023); (c) https://earthquake.usgs.gov/earthquakes/eventpage/usb000g7x7#shakemap, (accessed on 1 July 2023); (d) https://earthquake.usgs.gov/earthquakes/eventpage/usb000jyiv#shakemap, (accessed on 1 July 2023). In Figure 2, the star marks the epicenter, and the lines around the star represent the rupture direction of the earthquake.

3.2. Regional Earthquake Activity History

This region had strong seismic activity from 1900 to 2016, including 130 earthquakes (100 shallow earthquakes) of magnitude 6+, as shown in Figure 3, and the source is the following website: https://earthquake.usgs.gov/earthquakes/search/, accessed on 10 July 2023, Date: 1 January 1990–31 December 2016. The number of earthquakes decreased as magnitude increased. Previous studies have found that the thermal radiation anomalies of shallow earthquakes are more obvious and easier to extract. Therefore, we focused on analyzing shallow-source seismic activity.

Since 1900, the frequency of earthquakes per decade has gradually risen, peaking between 1997 and 2006, as shown in Figure 4, and the source is the following website: https://earthquake.usgs.gov/earthquakes/search/, accessed on 10 July 2023, Date: 1 January 1990–31 December 2016. Earthquakes occurred less frequently before 1947, which may be associated with poor seismic 87c monitoring capacity and record keeping. Since 1947, the number of shallow earthquakes per decade has continuously increased. The last decade generally reflects the region's average seismic activity; therefore, its thermal radiation background information should be representative as well.

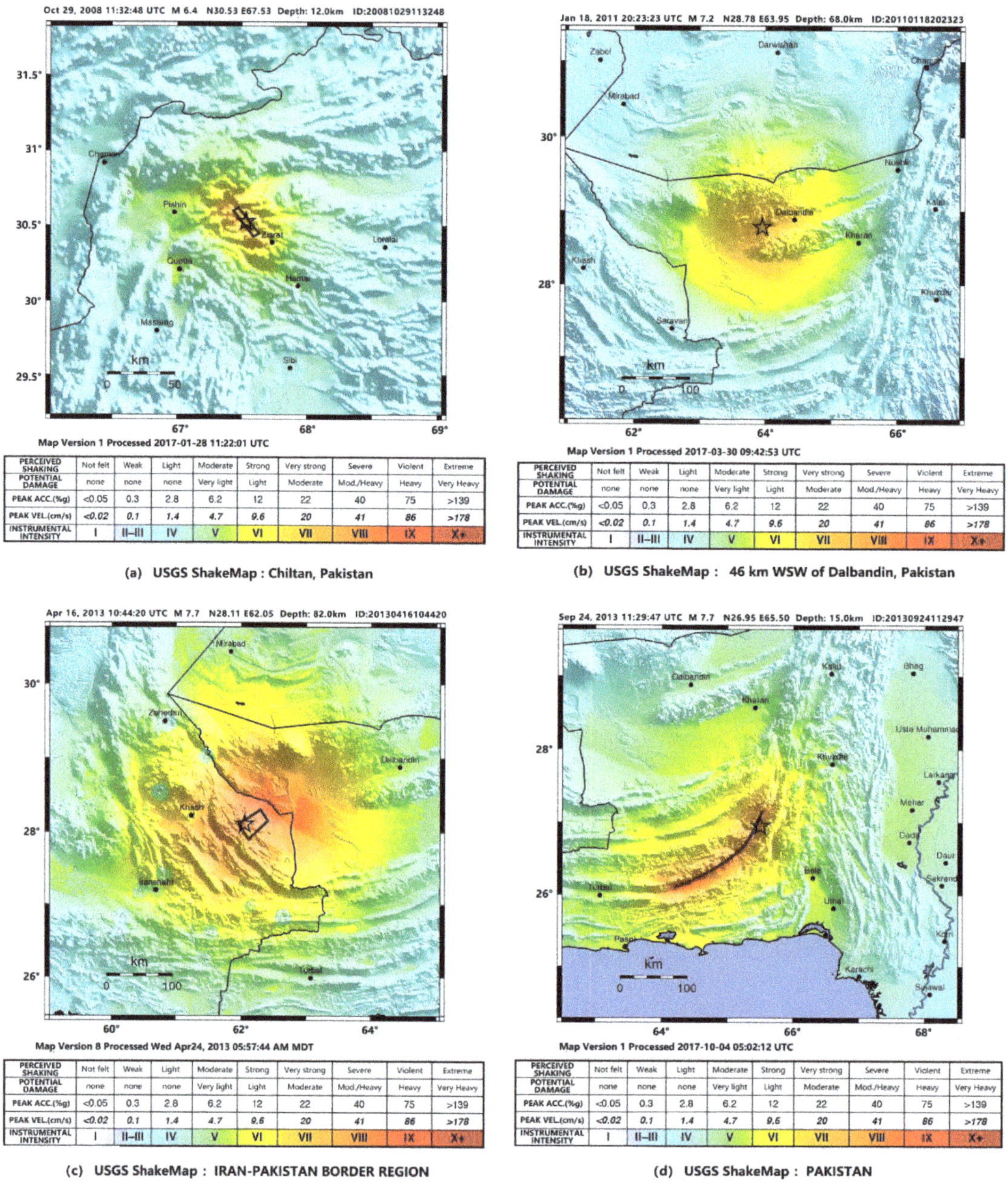

Figure 2. ShakeMap of four earthquakes. Source: United States Geological Survey.

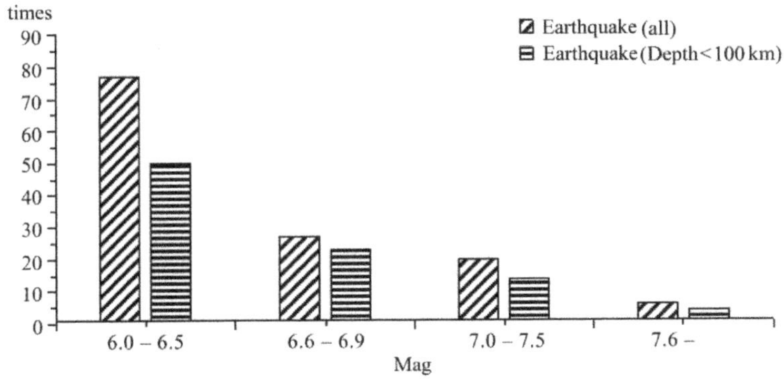

Figure 3. Regional MAG time map from 1900 to 2016. Region: 22–38° N, 55–71° E, earthquake magnitudes: Mw 6+; source: United States Geological Survey.

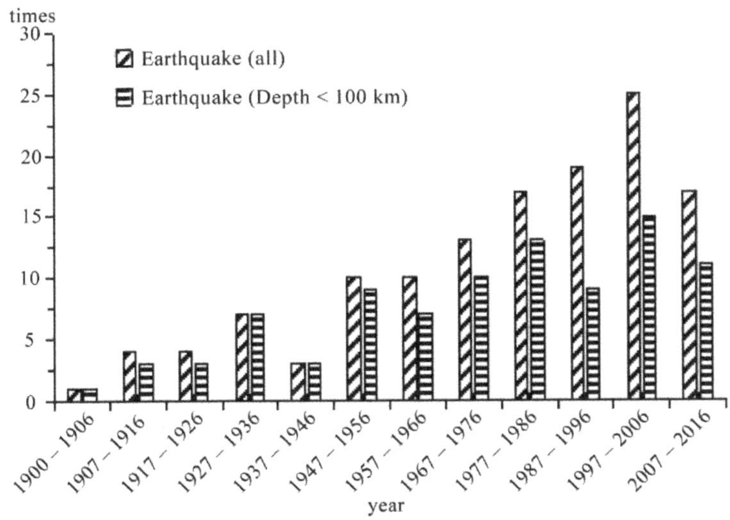

Figure 4. Regional decade interval earthquake frequency chart from 1900 to 2016. Region: 22–38° N, 55–71° E; earthquake magnitude: Mw 6+; source: United States Geological Survey.

3.3. Earthquake Catalogue and Grouping

Table 1 lists 10 shallow earthquakes that occurred over the last decade, which can be divided into four chronological swarms. The first earthquake group has three earthquakes in 2008, including an Mw 6.1 single earthquake and an Mw 6.4 double shock. The second group has three earthquakes in 2010 and 2011, including an Mw 7.2 single earthquake and an Mw 6.7 main earthquake in 2010 and its Mw 6.2 aftershock in 2011. The third group has two earthquakes of Mw 7.7 and Mw 6.1 in the first half of 2013. The last earthquake group has two earthquakes in the second half of 2013, including an Mw 7.7 main earthquake and its Mw 6.8 aftershock.

Table 1. Earthquake catalogue and grouping.

Date	Latitude (°)	Longitude (°)	Depth (km)	Magnitude and Type	Place	Group
10 September 2008	26.74	55.83	12	6.1 mwc	Southern Iran	G1
28 October 2008	30.64	67.35	15	6.4 mwc	Pakistan	
29 October 2008	30.60	67.46	14	6.4 mwc	Pakistan	
20 December 2010	28.41	59.18	12	6.7 mwc	Southeastern Iran	G2
18 January 2011	28.78	63.95	68	7.2 mww	Southwestern Pakistan	
27 January 2011	28.20	59.06	10	6.2 mww	Southeastern Iran	
16 April 2013	28.03	62.00	80	7.7 mww	83 km E of Khash, Iran	G3
11 May 2013	26.56	57.77	15	6.1 mww	94 km SE of Minab, Iran	
24 September 2013	26.95	65.50	15	7.7 mww	61 km NNE of Awaran, Pakistan	G4
28 September 2013	27.18	65.51	12	6.8 mww	85 km NNE of Awaran, Pakistan	

4. Results

In this section, the wavelet filtering results and wavelet annual background anomaly analysis of Mw 7+ earthquake are first analyzed. Secondly, the short-term seismic thermal radiation anomalies of the four earthquake swarms are analyzed. Then, the long-term background anomaly characteristics of regional thermal radiation are examined. Finally, the effects of strong earthquakes on the regional thermal radiation environment are investigated.

4.1. Wavelet Filtering Results and Wavelet Annual Background Anomaly Analysis of Mw 7+ Earthquake

The positive–negative time domain waveform processed by the second- to seventh-order band-pass filtering can be separated from the low-frequency annual temperature field and the high-frequency temperature field caused by other short time factors (cloud, short time heating up or cooling down), as shown in Figure 5.

In Figure 5a, the raw data exhibit distinguishable annual variations caused by season alternation. Due to the short-term weather changes, the fluctuations of the data are relatively large, which can be handled by wavelet analysis. By applying the sixth and seventh order wavelet transform, it is possible to separate the approximate annual period (sixth scale) and annual period (seventh scale) change information, respectively. The seventh scale has a longer period, which is closer to one year. Figure 5b,c show that Mw 7+ can cause changes in the annual period (seventh scale) and approximate annual period (6th scale) background curves. Also, three earthquakes occurred in the background transition.

In conclusion, the second- to seventh-order band-pass Daubechies wavelet filtering is used to reduce noise of the low-frequency annual temperature field and the high-frequency temperature field caused by other short time factors. The sixth- and seventh-order scales indicate the annual background anomaly and correspond to the Mw 7+ earthquake time.

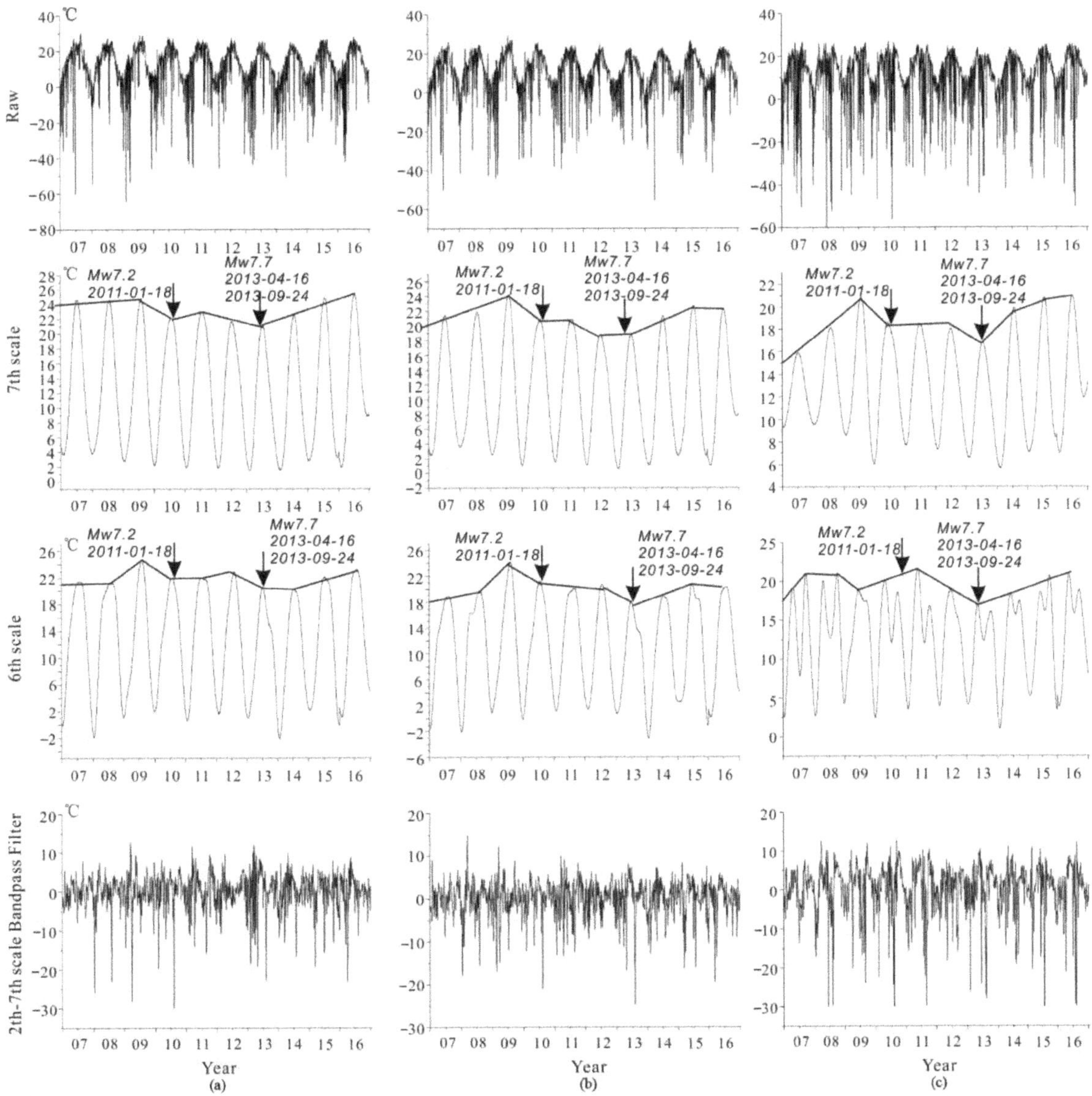

Figure 5. Different scale factor and band-pass filtering wavelet transform results for three Mw 7+ earthquakes. (**a**) Earthquake 1: 28.78° N, 63.95° E, Mw 7.2, 18 January 2011; (**b**) Earthquake 2: 28.03° N, 62.00° E, Mw 7.7, 16 April 2013; (**c**) Earthquake 3: 26.95° N, 65.50° E, Mw 7.7, 24 September 2013. Raw: raw data per day; 6th: sixth order; 7th: seventh order; 2nd–7th: second to seventh band-pass filter.

4.2. TFA of Earthquake Swarm Short-Term Thermal Radiation Anomalies

The TFRPS method was used to extract and identify the characteristic information of earthquake swarm thermal radiation anomalies. The analysis included the anomaly characteristic period, amplitude, duration, form, and the relationship among the anomaly range and earthquake epicenter and evolution characteristics. In the following section, the TFAs of the thermal radiation anomalies of the four earthquake swarms (G1–G4) are described in chronological order.

1. Short-term thermal radiation anomalies of G1

As shown in Figure 6, the anomaly lasted about 70 days from 21 July 2008 to 29 September 2008. The maximum amplitude of the anomaly occurred on 5 August 2008, reaching more than 10 times the average value. The area of the anomaly was close to the double-shock epicenter, with the expansion direction nearly parallel to the line connection of these two epicenters (E(a), E(b)). The temporal and spatial evolution of the anomaly was clear and easy to extract and identify. The anomaly evolution presents a three-stage process of the abnormal occurrence–maximum–disappearance, in line with the earthquake propagation process.

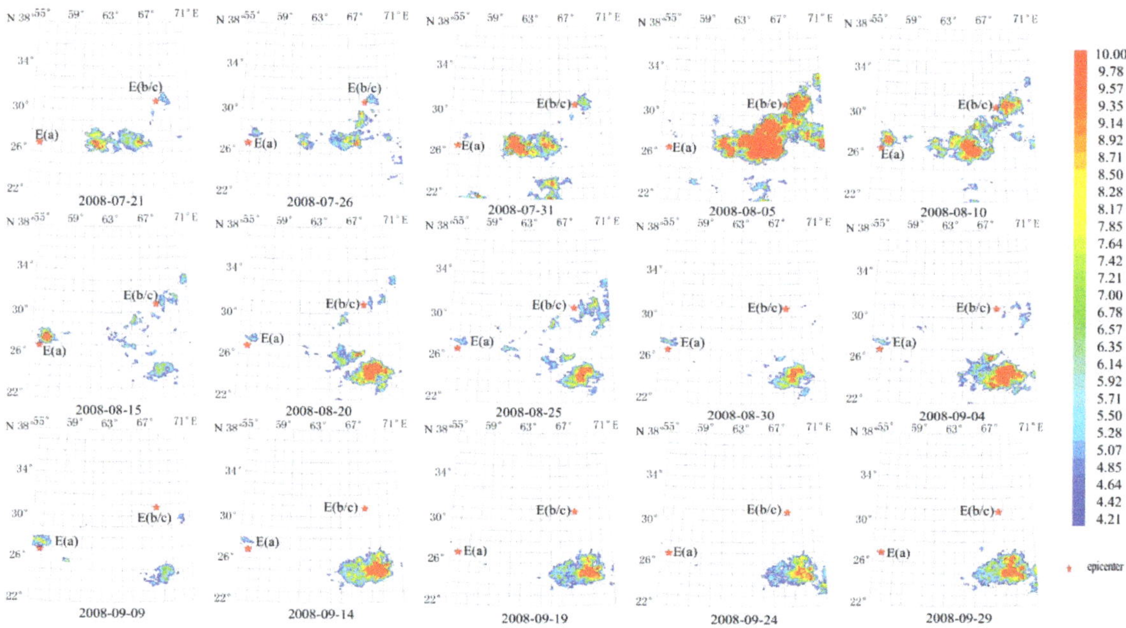

Figure 6. Temporal and spatial evolution of thermal radiation anomalies of earthquake group G1. Single earthquake: E(a) 26.74° N, 55.83° E, Mw 6.1, 10 September 2008. Double earthquake: E(b and c) 30.64° N, 67.35° E and 30.60° N, 67.40° E, Mw 6.4 and Mw 6.4 on 28 October 2008 and 29, respectively. Wavelet transform: second- to seventh-order band-pass filter using Daubechies wavelets. Window length of FT power spectrum: 64 days using Welch's method. Characteristic period: 13 days.

2. Short-term thermal radiation anomalies of G2

The anomaly duration of these earthquakes was also about 70 days from 20 August 2010 to 29 September 2010, as shown in Figure 7. The maximum amplitude of the anomaly reached more than 10 times the average value on 14 September 2010. The anomaly center was close to the Mw 7.2 epicenter. The anomalous shape is consistent with the ShakeMap of Figure 2b, which is mainly located in the southern epicenter.

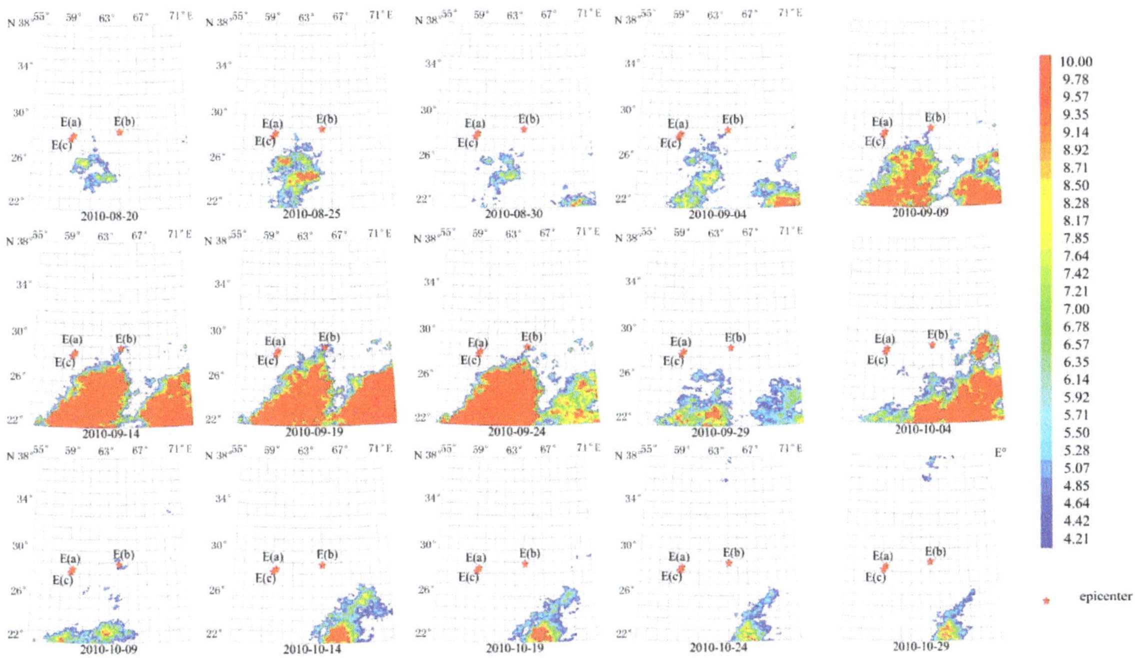

Figure 7. Temporal and spatial evolution of thermal radiation anomalies of earthquake group G2. Single earthquake: E(b) 28.78° N, 63.95° E, Mw 7.2, 18 January 2011. Main shock: E(a) 28.41° N, 59.18° E, Mw 6.7, 20 December 2010. Aftershock: E(c) 28.20° N, 59.02° E, Mw 6.2, 27 January 2011. Wavelet transform: second- to seventh-order scale band-pass filter using Daubechies wavelets. Window length of FT power spectrum: 64 days using Welch's method. Characteristic period: 13 days.

Using a longer characteristic period of 64 days, temporal and spatial evolution of thermal radiation anomalies of earthquakes in group G2 demonstrates a pattern that lasted for about 60 days, as shown in Figure 8. The maximum amplitude of the anomaly occurred on 14 September 2010. The circular anomaly appears to the southeast of the epicenter and agrees with the ShakeMap in Figure 2b.

3. Short-term thermal radiation anomalies of G3

The anomaly, lasting about 60 days from 7 March 2013 to 1 May 2013, is elliptical with the major axis extending diagonally. The anomaly center lies on top of the E(a) epicenter, and its maximum amplitude occurred from 27 March 2013 to 16 April 2013, rising to about 10 times the average value. The area of the anomaly overlapped with the Mw 7.7 epicenter.

4. Short-term thermal radiation anomalies of G4

The anomaly slightly extends from the epicenter to the southwest direction in an elliptical shape, which is consistent with Figure 2d. Its width along the south–north direction is shorter than the anomaly in Figure 9. The duration of this anomaly was about 55 days from 9 September 2013 to 3 October 2013. The maximum amplitude of the anomaly appeared during 14 August 2013 to 24 August 2013, reaching more than 10 times the average value.

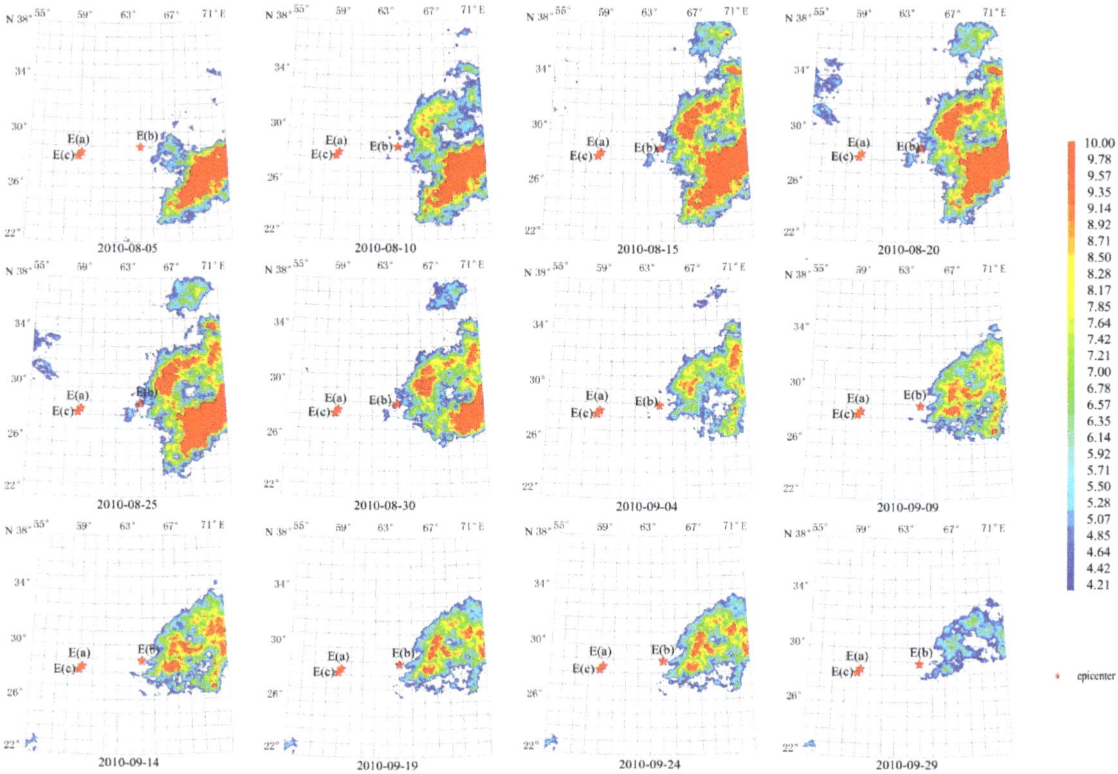

Figure 8. Temporal and spatial evolution of thermal radiation anomalies of earthquake group G2. Single earthquake: E(b) 28.78° N, 63.95° E, Mw 7.2, 18 January 2011. Main shock: E(a) 28.41° N, 59.18° E, Mw 6.7, 20 December 2010. Aftershock: E(c) 28.20° N, 59.02° E, Mw 6.2, 27 January 2011. Wavelet transform: second- to seventh-order scale band-pass filter using Daubechies wavelets. Window length of FT power spectrum: 64 days using Welch's method. Characteristic period: 64 days.

Compared to Figure 10, the anomaly in Figure 9, with a short characteristic period of 13 days, appeared 40 days pre-earthquake, reached the maximum when the earthquake occurred, and disappeared in 15 days after the earthquake. In addition, the anomaly area is much larger with the epicenter in the center, while in Figure 10, the anomaly appeared 45 days before the earthquake, and already started waning at the earthquake occurrence, and completely disappeared in the 10 days post-earthquake. The anomaly is more concentrated with the epicenter lying at its right edge.

As summarized in Table 2, short-term thermal radiation anomalies are irrelevant to the earthquake magnitude in terms of characteristic period, duration, and earthquake time and epicenter location. All anomalies lasted about 2 months and the earthquake occurred after they reached the maximum. Anomalies either overlap with or extend away from the epicenters. A positive relationship is identified between the anomaly morphology and the ShakeMap. That is, the higher the magnitude, the better they match each other.

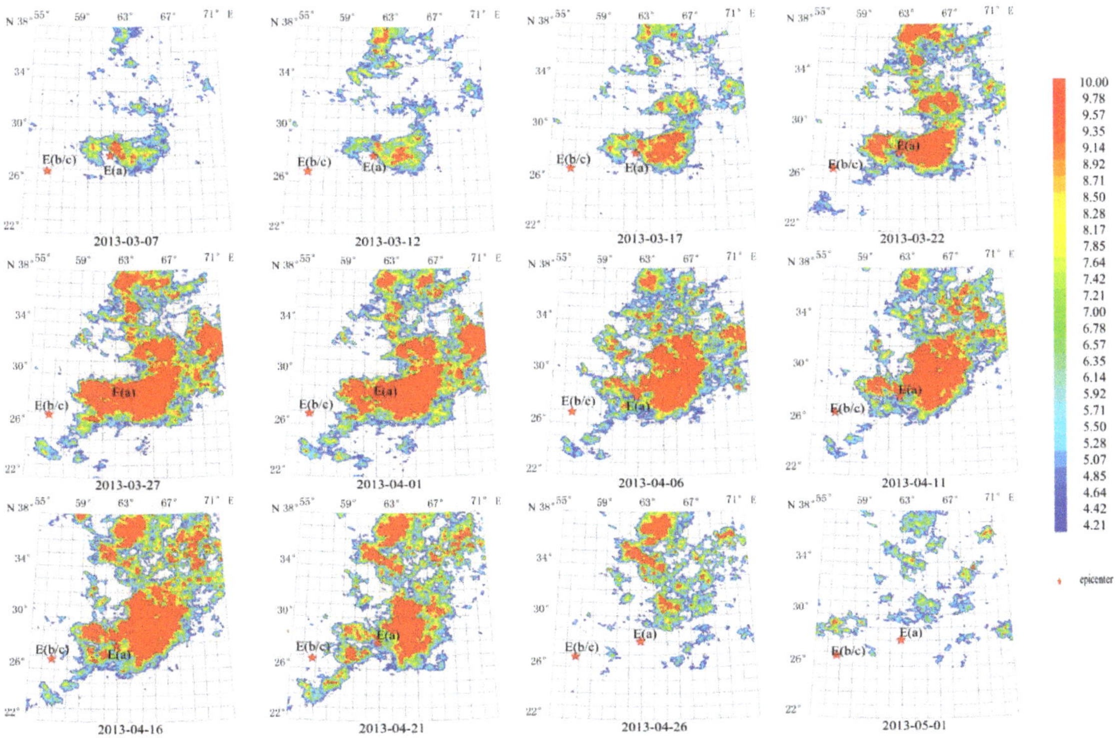

Figure 9. Temporal and spatial evolution of thermal radiation anomalies of earthquake group G3. Single earthquake: E(a) 28.03° N,62.00° E, Mw 7.7, 16 April 2013; E(b) 26.56° N, 57.77° E, Mw 6.1, 11 May 2013. Wavelet transform: second- and seventh-order band-pass filter using Daubechies wavelets. Window length of FT power spectrum: 64 days using Welch's method. Characteristic period: 13 days.

In conclusion, the temporal and spatial evolution of the anomaly is clearly evident and can be easily extracted and identified. The anomaly evolution process presents a process of the appearance reaching the maximum disappearance, in line with the earthquake propagation process. The processes of the anomaly were not related to changes in weather but related to ShakeMap. Short-term seismic thermal radiation anomalies showed no obvious trends with earthquake magnitude, i.e., characteristic period, duration, earthquake time, and epicenter location are not positively related to earthquake magnitude. Strong earthquakes can affect regional thermal radiation background in that both the characteristic periods and the duration of impact became longer with increasing magnitude. By analyzing and summarizing the earthquake case, we found the following rules. The characteristic period of all earthquakes is 13 days if there is no earthquake of magnitude 7 within one year. If a magnitude 7 earthquake occurs within a year, the characteristic period of the earthquake in the region will increase from 13 days to 64 days. Therefore, we can use the short-term characteristic periods as an anomaly area indicator (AAI) to determine seismic anomalies.

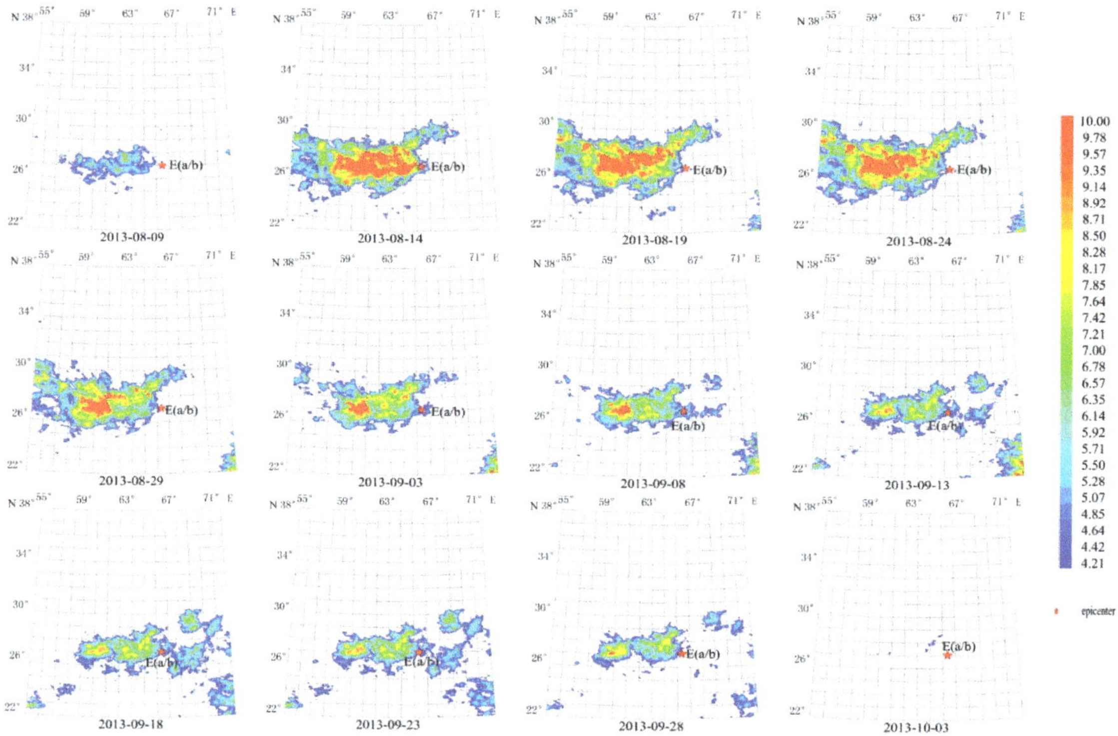

Figure 10. Temporal and spatial evolution of thermal radiation anomalies of earthquake group G4. Main shock: E(a) 26.95° N, 65.50° E, Mw 7.7, 24 September 2013. Aftershock: E(b) 27.18° N, 65.51° E, Mw 6.8, 28 September 2013. Wavelet transform: second- to seventh-order band-pass filter using Daubechies wavelets. Window length of FT power spectrum: 64 days using Welch's method. Characteristic period: 64 days.

Table 2. A summary of short-term thermal radiation anomalies of earthquakes.

Magnitude	Characteristic Period	Relationship of Abnormal Morphology and ShakeMap	Epicenter Location	Duration	Earthquake Time
Double Mw 6.4 28&29 October 2008	13 Days	Inconsistency	Abnormal edge	about 2 months	2 months after the maximum anomaly
Mw 6.7 20 December 2010	13 Days	Consistency	Abnormal edge	about 2 months	1 month after the maximum anomaly
Mw 7.2 18 January 2011	64 Days	Consistency	Abnormal edge	about 2 months	2 months after the maximum anomaly
Mw 7.7 16 April 2013	13 Days	Very consistent	Abnormal center	about 2 months	20 days after the maximum anomaly
Mw 7.7 24 September 2013	64 Days	Very consistent	Abnormal edge	about 2 months	1 month after the maximum anomaly

4.3. Multiscale TFA of the Long-Term Thermal Radiation Anomalies of Mw 6+ Earthquake Swarms

In the following section, a multiscale TFA of the thermal radiation anomaly caused by large earthquakes is described in chronological order.

1. TF characteristics of long-term thermal radiation anomaly of Mw 6+ earthquakes in 2008

To extract the thermal radiation anomaly, the sixth-order transform using Daubechies wavelets and FT using Welch's method with the window length of 256 days and the characteristic period of 128 days were conducted. As shown in Figure 11, the anomaly that arose 3 months pre-earthquake lasted about 70 days from 10 July 2008 to 20 September 2008. The maximum amplitude of the anomaly occurred on 31 August 2008, reaching more than eight times the average value. The anomaly was far from the epicenter of the first shock but close to the double shock epicenter. The temporal and spatial evolution of the anomaly was clear and matches the ShakeMap in Figure 2a. The anomaly evolution process presents a process of the abnormal occurrence–maximum–disappearance, in line with the earthquake propagation process. The processes of the anomaly were not related to changes in weather, indicating that it was caused by the earthquakes.

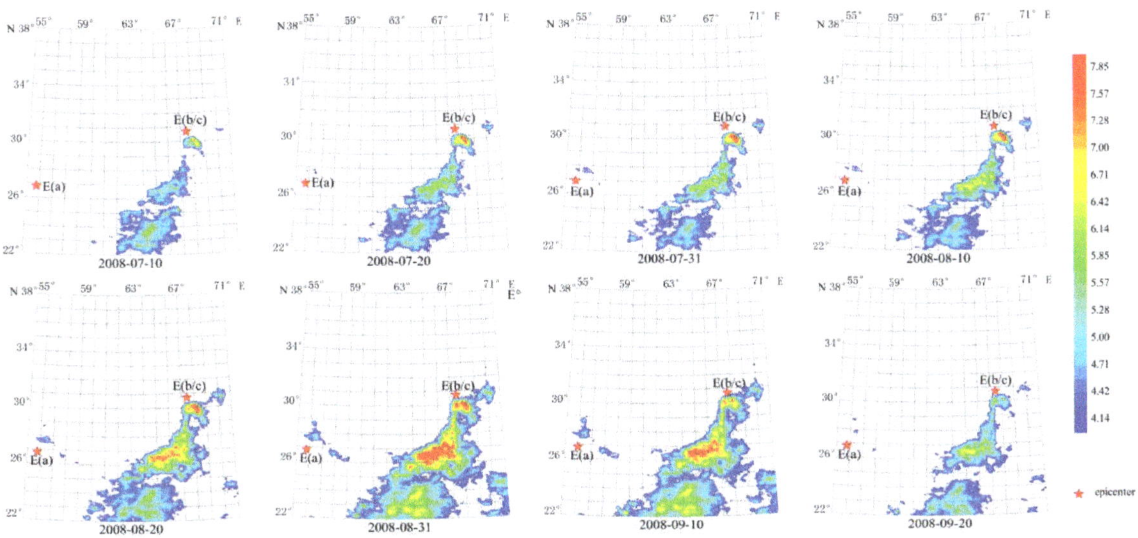

Figure 11. Temporal and spatial evolution of long-term thermal radiation anomaly of strong earthquakes. Single earthquake: E(a) 26.74° N, 55.83° E, Mw 6.1, 10 September 2008. Double earthquakes: E(b and c) 30.64° N, 67.35° E and 30.60° N, 67.40° E, Mw 6.4 and Mw 6.4, 28 October 2008 and 29, respectively. Wavelet transform: sixth order using Daubechies wavelets. Window length of FT power spectrum: 256 days using Welch's method. Characteristic period: 128 days.

2. TF characteristics of long-term thermal radiation anomalies of Mw7+ earthquakes in 2010 and 2011.

As shown in Figure 12, this anomaly showed up about 5 months pre-earthquake, lasting from 31 August 2010 to 31 January 2011. It reached the maximum amplitude from 31 October 2010 to 30 November 2011, which is more than eight times the average value. The area of the anomaly was close to the Mw 7.2 epicenter, and its shape is similar to that of the ShakeMap in Figure 1b. Different from the short-term anomaly, this long-term one disappeared with the earthquake occurrence. Thus, it can indicate the earthquake in advance. In this case, both the sixth- and seventh-order wavelet transforms showed anomalies, and FT used the window length of 512 days and the characteristic period of 256 days.

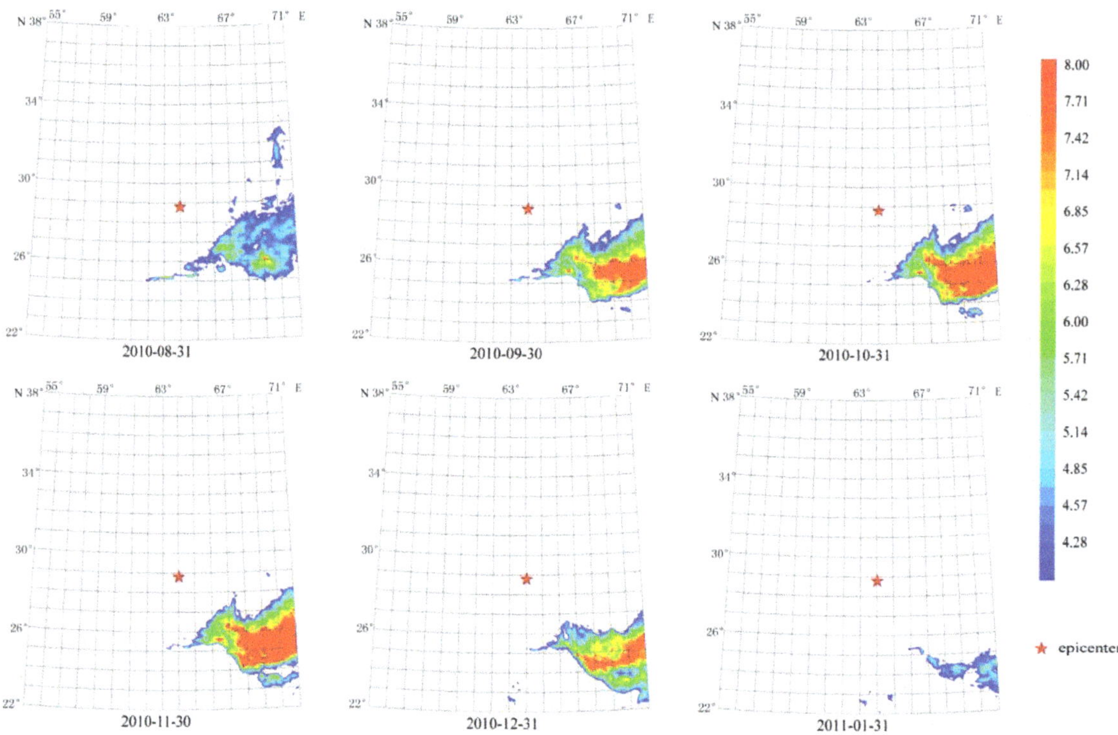

Figure 12. Temporal and spatial evolution of long-term thermal radiation anomaly of strong earthquakes. Single earthquake: E, 28.78° N, 63.95° E, Mw 7.2, 2011-01-18. Wavelet transform: sixth order using Daubechies wavelets. Window length of FT power spectrum using Welch's method: 512 days. Characteristic period: 256 days.

Figures 12 and 13 are similar in anomaly morphology, amplitude, characteristic period, and evolution processes. Nonetheless, the anomaly from the seventh-order transform appeared 3 months earlier and lasted 3 months longer than that from the sixth order transform. In addition, the sixth order anomaly disappeared with the earthquake occurrence, while the seventh order one faded in 2 months post-earthquake.

3. TF characteristics of long-term thermal radiation anomalies of Mw 7+ earthquakes in 2013.

Two Mw 7.7 earthquakes occurred separately in the first and second half of 2013 in this region. From Figures 9 and 10, the characteristic period of the short-term anomaly elongated from 13 days to 64 days, indicating that the first Mw 7.7 earthquake changed the regional thermal radiation. The second earthquake, along with the first one, increased the characteristic period of the long-term anomaly to 512 days (Figures 14 and 15).

The duration of the first anomaly was about 6 months from 31 July 2013 to 31 January 2014 (Figure 14). The maximum amplitude occurred from 30 September 2013 to 31 December 2013, which is eight times the average value. The second anomaly lasted about 9 months from 30 June 2013 to 31 March 2014. The maximum amplitude occurred from 30 September 2013 to 31 November 2013, which is also about eight times the average value. The shape of these two anomalies is similar to that of the ShakeMap provided by the USGS for the Mw 7.7 earthquakes (Figure 2d). These two anomalies showed similar characteristics, but their appearance and lasting time are obviously different. Specifically, the seventh-order anomaly appeared 1 month earlier and lasted 3 months longer than the sixth order one.

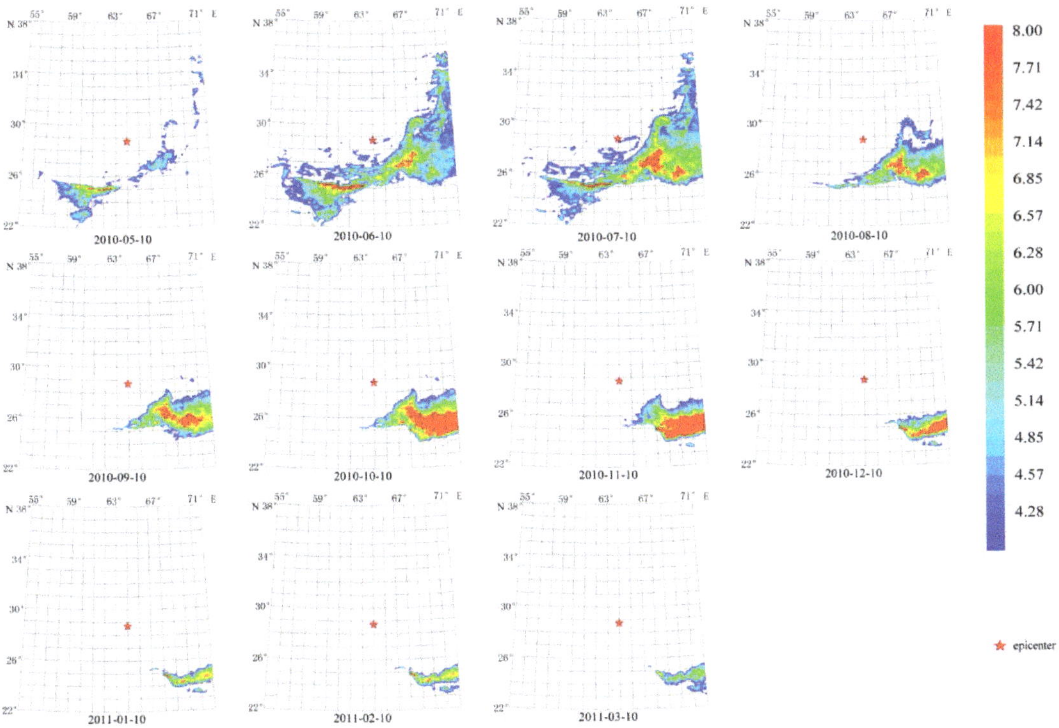

Figure 13. Temporal and spatial evolution of long-term thermal radiation anomaly of strong earthquakes. Single earthquake: E 28.78° N, 63.95° E, Mw 7.2, 18 January 2011. Wavelet transform: seventh order using Daubechies wavelets. Window length of FT power spectrum using Welch's method: 512 days. Characteristic period: 256 days.

Long-term anomalies are positively correlated to the earth magnitude in that the higher the magnitude the longer the window length and characteristic period (Table 3). For earthquakes < Mw 7, the window length of relative power spectrum is 256 days; for Mw 7.0+ earthquakes, the window length is 512 days. Moreover, only the sixth-order transform shows an anomaly when the earthquake Mw is smaller than 7.0; while both the sixth- and seventh-order transforms show anomalies for Mw 7.0+ earthquakes. In addition, the anomalies of Mw 7.0+ earthquakes lasted 6 months or more and the anomaly duration from seventh-order transforms is longer than that from sixth-order transforms, but those of earthquakes < Mw 7.0 only lasted 2 months. Last, all anomalies reached their maximum before or at the earthquake occurrence.

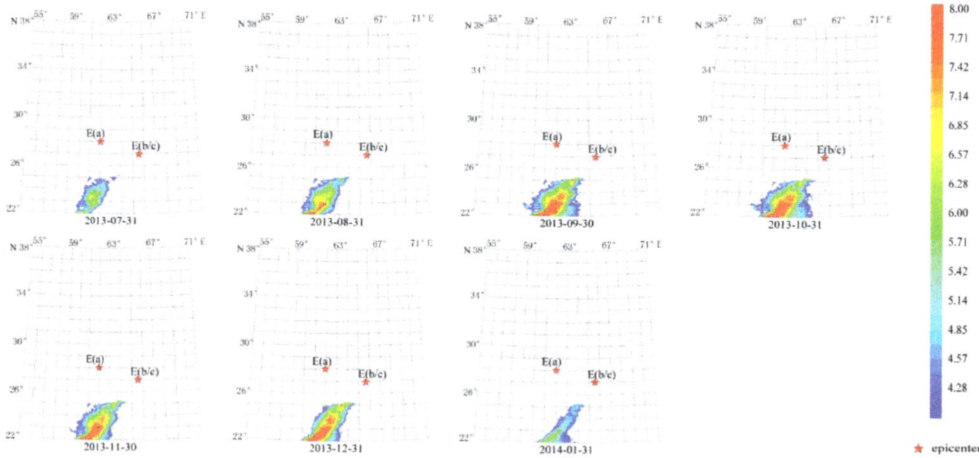

Figure 14. Temporal and spatial evolution of long-term thermal radiation anomaly of strong earthquakes. Single earthquake: E(a) 28.03° N, 62.00° E, Mw 7.7, 16 April 2013; E(b) 26.95° N, 65.50° E, Mw 7.7, 24 September 2013. Wavelet transform: sixth order using Daubechies wavelets. Window length of FT power spectrum using Welch's method: 512 days. Characteristic period: 512 days.

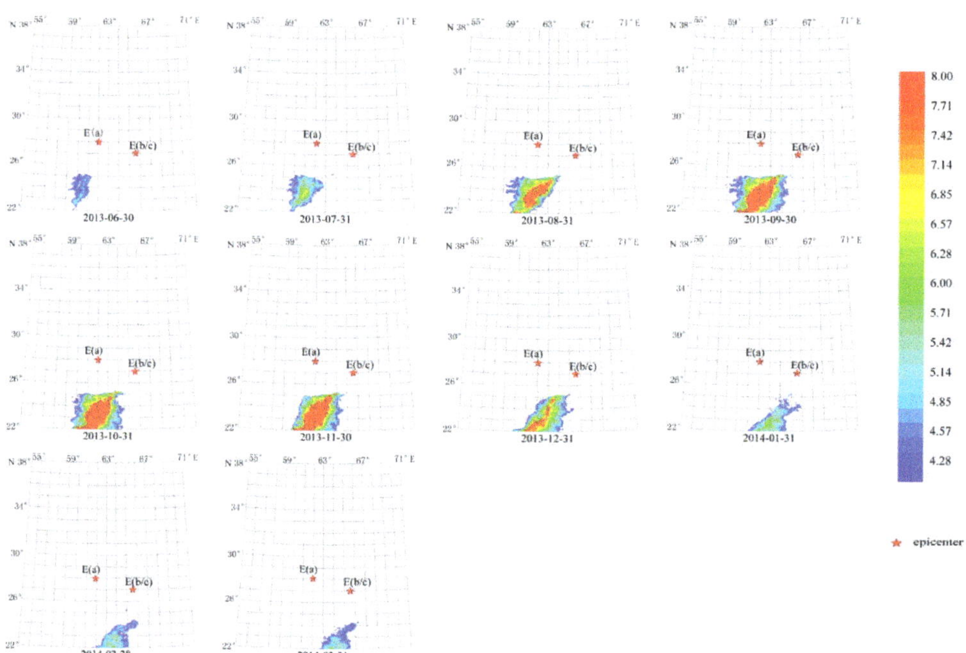

Figure 15. Temporal and spatial evolution of long-term thermal radiation anomalies of strong earthquakes. Single earthquake: E(a) 28.03° N, 62.00° E, Mw 7.7, 16 April 2013; E(b) 26.95° N, 65.50° E, Mw 7.7, 24 September 2013. Wavelet transform: seventh order using Daubechies wavelets. Window length of FT power spectrum using Welch's method: 512 days. Characteristic period: 512 days.

Table 3. A summary of the relationship between seismic anomaly and magnitude.

Magnitude	Wavelet Scale	Window Length of RPS	Characteristic Period	Duration	Earthquake Time
Mw 6.4	Only 6th	256 Days	128 Days	About 2 months	2 months after the maximum anomaly
Mw 6.7 Mw 7.2	6th and 7th	512 Days	256 Days	About 5 months (6th) About 10 months (7th)	2 months after the maximum anomaly 3 months after the maximum anomaly
Two Mw 7.7	6th and 7th	512 Days	512 Days	About 6 months (6th) About 9 months (7th)	When the maximum anomaly occurred When the maximum anomaly occurred

In conclusion, long-term anomalies are positively correlated to magnitude, i.e., both the window length of RPS and characteristic period increased with increasing magnitude. In addition, for Mw 7.0− earthquakes, only the sixth-order wavelet transform showed anomalies, while both the sixth- and seventh-order wavelet transforms showed anomalies. Therefore, we can use the long-term characteristic periods and window lengths as a magnitude decision factor (MDF) to determine seismic magnitude.

5. Discussion

5.1. Earthquake Impact on Regional Thermal Radiation Background

1. Characteristic period of short-term anomalies

The periods of the thermal radiation anomalies were changed by Mw 7+ earthquakes. Earthquake group G2 consists of foreshock, mainshock, and aftershock. From August to October 2010, short-term thermal radiation anomalies with characteristic periods of 13 and 64 days appeared (Figures 7 and 8). Later from September to November, another anomaly with the period of 64 appeared (Figure 16). Similarly, for G3 and G4, the characteristic period of the 16 April 2013 Mw 7.7 earthquake was 13 days but that of the 24 September 2013 Mw 7.7 earthquake increased to 64 days. However, Mw 6+ earthquakes from G1 do not show such an increasing trend in characteristic period. Therefore, the relative power spectrum of the 64-day characteristic period can be used to analyze the regional thermal radiation anomalies caused by strong earthquakes.

2. Duration of the impact and earthquake magnitude

Thermal radiation anomalies could be useful for studying strong earthquake processes and stress adjustment mechanisms. The 18 January 2011 Mw 7.2 earthquake affected the regional thermal radiation background for about 2 years (2010–2011), as shown in Figures 8 and 16, while the 16 April 2013 Mw 7.7 and 24 September 2013 Mw 7.7 earthquakes had a 4-year effect (2013–2016), as shown in Figures 10, 17 and 18. The duration of the impact increases with the magnitude of the earthquake, which may be further related to the regional stress adjustments and recovery post-earthquakes. Thus, longer duration of the impact could probably indicate slower regional stress adjustment and recovery.

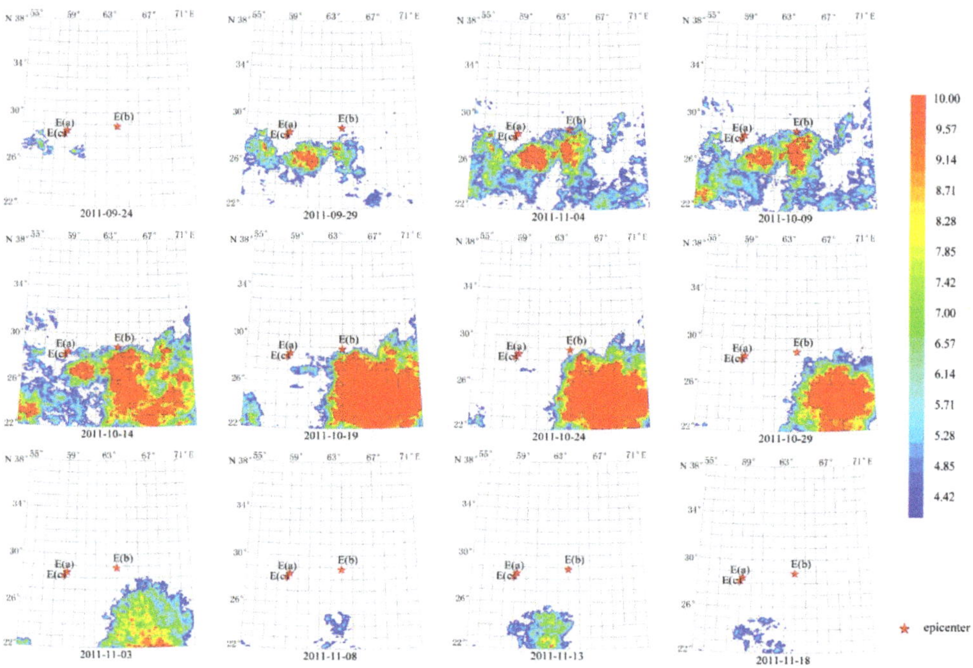

Figure 16. Spatial-temporal evolution of the thermal radiation background in 2011 caused by the Mw 7.2 earthquake. Single earthquake: E 28.78° N, 63.95° E, Mw 7.2, 18 January 2011. Wavelet transform: second- and seventh-order band-pass filter using Daubechies wavelets. Window length of FT power spectrum using Welch's method: 64 days. Characteristic period: 64 days.

By carrying out the TFA for the past 10 years' data with the characteristic period of 64 days, we found the thermal radiation anomaly caused by the Mw 7.2 earthquake impact lasted 2 years from 2010 to 2011. The duration of the anomaly was about 55 days from 24 September 2011 to 18 November 2011. The maximum amplitude of the anomaly occurred from 14 October 2011 to 29 October 2011, reaching more than 10 times the average value. The anomalous area was closer to the Mw 7.2 epicenter. The long-term and concentrated thermal radiation anomalies were similar to those of the Mw 7.2 earthquake group in terms of characteristic period, amplitude, form, and evolution.

From the data from 2013 to 2016, we found the thermal radiation anomaly caused by the double Mw 7.7 earthquakes. Their impact lasted for 4 years from 2013 to 2016. The anomaly continued for about 70 days from 17 March 2015 to 26 May 2015, which is consistent with the anomaly of the 16 April 2011 Mw 7.7 earthquake. The maximum amplitude of the anomaly that is more than 10 times the average value occurred from 26 April 2015 to 16 May 2015. The anomaly was close to the double Mw 7.7 epicenter. In addition, the long-term and concentrated thermal radiation anomalies are similar to those of the 16 April 2013 Mw 7.7 earthquake in terms of amplitude, morphology, and evolution (Figure 17).

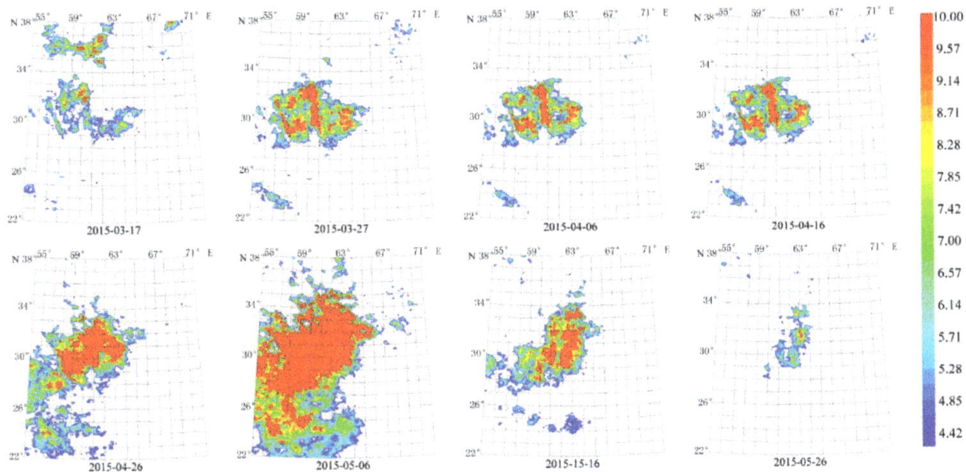

Figure 17. Spatial-temporal evolution of the thermal radiation background in 2015 caused by the Mw 7.7 earthquake. Earthquakes: E 28.03° N, 62.00° E, Mw 7.7, 16 April 2013; E 26.95° N, 65.50° E, Mw 7.7, 24 September 2013. Wavelet transform: second- and seventh-order band-pass filter using Daubechies wavelets. Window length of FT power spectrum: 64 days using Welch's method. Characteristic period: 64 days.

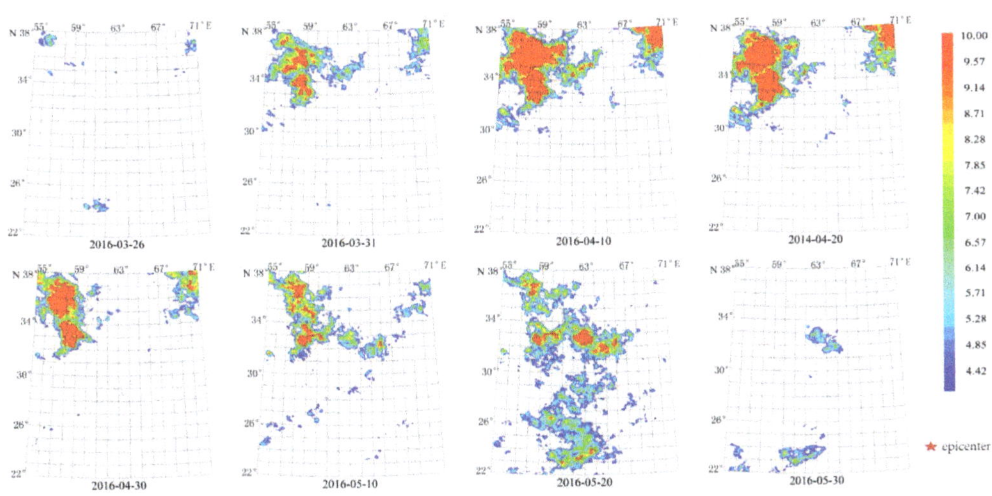

Figure 18. Spatial-temporal evolution of the thermal radiation background from 2016 caused by Mw 7.7 earthquakes. Earthquakes: E 28.03° N, 62.00° E, Mw 7.7, 16 April 2013; E 26.95° N, 65.50° E, Mw 7.7, 24 September 2013. Wavelet transform: second- and seventh-order band-pass filter using Daubechies wavelets. Window length of FT power spectrum: 64 days using Welch's method. Characteristic period: 64 days.

The anomaly from 26 March 2016 to 30 May 2016 with the maximum amplitude occurring from 10 April 2016 to 30 April 2016 was also closer to the double Mw 7.7 epicenter (Figure 18) and similar to that of the 16 April 2013 Mw 7.7 earthquake. The thermal radiation anomalies of the 24 September 2013 Mw 7.7 earthquake are different from that of the 2015–2016 in morphology. Specifically, the thermal radiation anomalies from the 24 September 2013 Mw 7.7 earthquake were concentrated at the epicenter and nearby areas, and the area of the anomaly was smaller (Figure 10) than that of the anomaly from 2015 (Figure 17), and less dispersed than that of the anomaly from 2016 (Figure 18). The absence of thermal radiation in 2014 might be a consequence of the Mw 7.7 earthquake in September 2013.

3. Superposition of short-term anomalies induced by multiple earthquakes

If more than one Mw 6.0+ earthquake occurred over a 1-year time scale, the characteristics of the seismic thermal radiation anomalies became complex and were difficult to extract and identify, which is due to the superposition of the regional thermal radiation caused by the mutual coupling between strong earthquakes. The thermal radiation anomalies of the 2010–2011 earthquake swarms were complex, and the characteristic periods of 13 and 64 days appeared simultaneously, possibly because the earlier Mw 7.2 earthquake changed the latter regional thermal radiation. In addition, the characteristic period of the 16 April 2013 Mw 7.7 earthquake was 13 days, whereas that of the 24 September 2013 Mw 7.7 earthquake was 64 days, which may also have been affected by the earlier 16 April 2013 Mw 7.7 earthquake. This is also supported by the fact that there were no thermal radiation anomalies in March–May 2014.

5.2. Relationship between Thermal Anomalies and Earthquake Magnitude

From Table 2, short-term seismic thermal radiation anomalies showed no obvious change with earthquake magnitude, indicating that this method is suitable for extracting anomaly characteristics. In addition, characteristic period, duration, earthquake time, and epicenter location are not positively related to earthquake magnitude. All anomalies endured about 2 months, with the maximum amplitude appearing before the earthquake. The positive correlation identified is that the higher the magnitude, the better correspondence between the anomaly morphology and the ShakeMap.

Table 3 suggests strong relationships between long-term anomaly characteristics and earthquake magnitude. Since generally the longer the energy accumulates, the higher the earthquake magnitude, a longer characteristic period indicates longer energy accumulation time. This finding could serve as a criterion for determining the earthquake magnitude. In terms of wavelet scale, if only the sixth-order transform shows anomaly, the earthquake magnitude can be determined to be Mw 7.0−; if both the sixth- and seventh-order transforms show anomalies, the magnitude can be determined to be Mw 7.0+. Moreover, the longer the characteristic period, the higher the earthquake magnitude. Specifically, a period of 256 days corresponds to Mw 7.5−, and a period of 512 days indicates Mw 7.5+ earthquakes.

6. Conclusions

In this paper, we analyzed the characteristics of seismic thermal radiation anomalies in the same area. The results of this study have found the regular pattern and mechanism of anomaly seismic thermal radiation in the same region, and also improved the accuracy of the analytical processing model. The MS T-FRPS method was used to identify and determine the anomalous information of series earthquake groups with consistency law, which proves the stability and reliability of the method. By analyzing and summarizing the earthquake case, we propose the anomaly area indicator (AAI) to determine seismic anomalies using short-term characteristic periods and the magnitude decision factor (MDF) to determine seismic magnitude using the long-term characteristic periods and window lengths. In addition, we have explored the following: earthquake impact on regional thermal radiation background and relationship between thermal anomalies and earthquake magnitude and the like. Future research directions using the MS T-FRPS method to characterize regional thermal radiation anomalies induced by strong earthquakes could be (1) carrying out more case studies with accurate parameter quantification; (2) studying the relationship between thermal radiation anomalies and the earthquake magnitude; (3) employing methodologies where the correlation and causality between spatial points is examined in other phenomena; (4) utilizing machine learning algorithms to analyze earthquake precursors, crustal movements, and other data to enhance the accuracy of earthquake predictions. Meaningful findings could help improve the accuracy of earthquake magnitude determination.

Author Contributions: Conceptualization, T.W. and C.W.; Validation, T.W.; Formal analysis, C.W.; Investigation, T.W. and Z.W.; Data curation, Z.W.; Visualization, L.W. and Z.Y.; Writing—review & editing, T.G. All authors have read and agreed to the published version of the manuscript.

Funding: This research was funded by the grant of National Natural Science Foundation of China, grant number 42161056, by Natural Science Foundation of Gansu Province, China, grant number 20JR10RA501, by National Special Projects of China, grant number Y612ZX1701, by the Central Funds Guiding the Local Science and Technology Development of Gansu Province, China, grant number 22ZY1QA006 and 22ZY2QC001.

Institutional Review Board Statement: Not applicable.

Informed Consent Statement: Not applicable.

Data Availability Statement: Data sharing is not applicable to this article as no datasets were generated or analyzed during the current study.

Acknowledgments: The authors gratefully acknowledge the anonymous reviewers for their helpful comments and suggestions. A portion of the computational resources was also provided by Gansu Computing Center.

Conflicts of Interest: The authors declare no conflict of interest.

References

1. Gorny, V.I.; Salman, A.G.; Tronin, A.A. The earth outgoing IR radiation as an indicator of seismic activity. *Proc. Acad. Sci. USSR* **1988**, *301*, 67–69.
2. Tronin, A.A.; Hayakawa, M.; Molchanov, O.A. Thermal IR satellite data application for earthquake research in Japan and China. *J. Geodyn.* **2002**, *33*, 519–534. [CrossRef]
3. Panda, S.K.; Choudhury, S.; Saraf, A.K.; Das, J.D. MODIS land surface temperature data detects thermal anomaly preceding 08 October 2005 Kashmir earthquake. *Int. J. Remote Sens.* **2007**, *28*, 4587–4596. [CrossRef]
4. Qiang, Z.J.; Xu, X.D.; Dian, C.G. Thermal infrared anomaly precursor of short-term earthquakes. *Chin. Sci. Bull.* **1991**, *36*, 319–323.
5. Qiang, Z.J.; Xu, X.D.; Dian, C.G. Short-term-earthquake satellite thermal infrared and ground temperature increase anomalies. *Chin. Sci. Bull.* **1991**, *36*, 1894–1900.
6. Saraf, A.K.; Choudhury, S. Satellite detects surface thermal anomalies associated with the Algerian earthquakes of May, 2003. *Int. J. Remote Sens.* **2005**, *26*, 2705–2713. [CrossRef]
7. Singh, R.P.; Ouzounov, D. Earth processes in wake of Gujarat earthquake reviewed from space. *EOS Trans. Am. Geophys. Union* **2003**, *84*, 244. [CrossRef]
8. Ouzounov, D.; Freund, F. Mid-infrared emission prior to strong earthquakes analyzed by remote sensing data. *Adv. Space Res.* **2003**, *33*, 268–273. [CrossRef]
9. Tramutoli, V.; Bello, G.D.; Pergola, N.; Piscitelli, S. Robust satellite techniques for remote sensing of seismically active areas. *Ann. Geofis.* **2001**, *44*, 295–312. [CrossRef]
10. CORDIS RTD-PROJECTS. *Regular Update of Seismic Hazard Maps through Thermal Space Observations*; Project Reference: ENV4980741; European Communities: Madrid, Spain, 2000; Volume 10.
11. Ouzounov, D.; Bryant, N.; Logan, T.; Pulinets, S.; Taylor, P. Satellite thermal IR phenomena associated with some of the major earthquakes in 1999–2003. *Phys. Chem. Earth* **2006**, *31*, 154–163. [CrossRef]
12. Guo, X.; Zhang, Y.S.; Wei, C.X.; Zhong, M.J. Anomalies of middle infrared brightness before 2008 Yutian Ms7.3 and 2010 Yushu Ms7.1 earthquakes. *Acta Seismol. Sin.* **2014**, *36*, 175–183, (In Chinese with English Abstract).
13. Zhang, Y.S.; Guo, X.; Zhong, M.J.; Shen, W.; Li, W.; He, B. Wenchuan earthquake: Brightness temperature changes from satellite infrared information. *Chin. Sci. Bull.* **2010**, *55*, 1917–1924. [CrossRef]
14. Wei, C.; Zhang, Y.; Guo, X.; Hui, S.; Qin, M.; Zhang, Y. Thermal Infrared Anomalies of Several Strong Earthquakes. *Sci. World J.* **2013**, *2013*, 208407. [CrossRef] [PubMed]
15. Zhang, Y.S.; Guo, X.; Wei, C.; Shen, W.R.; Hui, S.X. The characteristics of seismic thermal radiation about Japan Ms9.0 and Myanmar Ms7.2 earthquake. *Chin. J. Geophys.* **2011**, *34*, 153–157, (In Chinese with English Abstract).
16. Pinty, B.; Gobron, N.; Verstraete, M.M.; Mélin, F.; Widlowski, J.-L.; Govaerts, Y.; Diner, D.J.; Fielding, E.; Nelson, D.L.; Madariaga, R.; et al. Observing earthquake-related dewatering using MISR/Terra satellite Data. *EOS Trans. Am. Geophys. Union* **2003**, *84*, 37–48. [CrossRef]
17. Qiang, Z.J.; Xu, X.D.; Dian, C.G. Thermal infrared anomalies under the non-temperature-increment background and the mechanism discussion. *Chin. Sci. Bull.* **1991**, *36*, 1901–1906.

18. Qiang, Z.J.; Dian, C.G.; Wang, X.J.; Hu, S.Y. Satellite thermal infrared anomalous temperature increase and short-term earthquake precursor. *China Sci. Bull.* **1992**, *37*, 1642–1646.
19. Qiang, Z.J.; Kong, L.C.; Guo, M.H.; Wang, G.P.; Zhao, Y. An experimental study on temperature increasing mechanism of satellitic thermal-infrared. *Acta Seismol. Sin.* **1997**, *10*, 247–252. [CrossRef]
20. Qiang, Z.J.; Dian, C.G. Satellite infrared brightness temperature anomaly earthquake imminent precursor. *Sci. China Ser. D* **1998**, *28*, 564–573, (In Chinese with English Abstract).
21. Qiang, Z.J.; Dian, C.G.; Li, L.Z.; Xu, M.; Ge, F.; Liu, T.; Zhao, Y.; Guo, M. Satellitic thermal infrared brightness temperature anomaly image—Short-term and short-term earthquake precursors. *Sci. China Ser. D-Earth* **1999**, *42*, 313–324. [CrossRef]
22. Xu, X.D.; Xu, X.M.; Ma, S.D.; Luo, Y. Preliminary inquiry into the causes of anomaly increase of air temperature by an short-term earthquake. *Acta Seismol. Sin.* **1995**, *8*, 149–154. [CrossRef]
23. Ma, J.; Chen, S.-Y.; Liu, P.-X.; Wang, Y.-P.; Liu, L.-Q. Temporal-spatial variations of associated faulting inferred from satellite infrared information: A case study of N-S seismo-tectonic zone in China. *Chin. J. Geophys.* **2006**, *49*, 707–716, (In Chinese with English Abstract). [CrossRef]
24. Zhang, L.F. The Applied Study in Predicting Earthquake Using Satellite Remote Sensing Data of Medium Wave Infrared. Master's Thesis, Lanzhou Institute of Seismology, Lanzhou, China, 2016.
25. Wei, C.X.; Zhang, Y.S.; Wang, Y. Time-frequency analysis of the influence of Japan Mw9.1 earthquake on regional thermal radiation background field. *Acta Seismol. Sin.* **2018**, *40*, 205–214, (In Chinese with English Abstract). [CrossRef]
26. Wang, Q.L.; Wang, W.P.; Liang, W.F.; Chen, B. Stress-dissipative heat geotemperature precursor mechanism of earthquakes. *Acta Seismol. Sin.* **1998**, *11*, 615–621, (In Chinese with English Abstract). [CrossRef]
27. Wu, L.X.; Liu, S.J.; Wu, Y.H.; Li, Y.Q. Remote-sensing rock mechanics (IV)—laws of thermal infrared radiation from compressively sheared fracturing of rock and its meanings for earthquake omens. *Chin. J. Rock Mech. Eng.* **2004**, *23*, 539–544, (In Chinese with English Abstract).
28. Akansu, A.N.; Haddad, R.A. *Multiresolution Signal Decomposition: Transforms, Subbands, and Wavelets*; Academic Press: Boston, MA, USA, 1992; ISBN 978-0-12-047141-6.
29. Hessami, K.; Jamali, F. Explanatory notes to the map of major active faults of Iran. *J. Seismol. Earthq. Eng.* **2006**, *8*, 1–11.

Disclaimer/Publisher's Note: The statements, opinions and data contained in all publications are solely those of the individual author(s) and contributor(s) and not of MDPI and/or the editor(s). MDPI and/or the editor(s) disclaim responsibility for any injury to people or property resulting from any ideas, methods, instructions or products referred to in the content.

sensors

MDPI

Article

A Novel Deep Convolutional Neural Network Combining Global Feature Extraction and Detailed Feature Extraction for Bearing Compound Fault Diagnosis

Shuzhen Han [1,2], Pingjuan Niu [1,3,*], Shijie Luo [2], Yitong Li [2], Dong Zhen [4], Guojin Feng [4] and Shengke Sun [5]

1 School of Mechanical Engineering, Tiangong University, Tianjin 300387, China; hanshuzhen@tiangong.edu.cn
2 Office of the Cyberspace Affairs, Tiangong University, Tianjin 300387, China
3 School of Electronics and Information Engineering, Tiangong University, Tianjin 300387, China
4 School of Mechanical Engineering, Hebei University of Technology, Tianjin 300401, China; G.Feng@hebut.edu.cn (G.F.)
5 School of Software, Tiangong University, Tianjin 300387, China; sunshengke@tiangong.edu.cn
* Correspondence: niupingjuan@tiangong.edu.cn; Tel.: +86-13821536307

Abstract: This study researched the application of a convolutional neural network (CNN) to a bearing compound fault diagnosis. The proposed idea lies in the ability of CNN to automatically extract fault features from complex raw signals. In our approach, to extract more effective features from a raw signal, a novel deep convolutional neural network combining global feature extraction with detailed feature extraction (GDDCNN) is proposed. First, wide and small kernel sizes are separately adopted in shallow and deep convolutional layers to extract global and detailed features. Then, the modified activation layer with a concatenated rectified linear unit (CReLU) is added following the shallow convolution layer to improve the utilization of shallow global features of the network. Finally, to acquire more robust features, another strategy involving the GMP layer is utilized, which replaces the traditional fully connected layer. The performance of the obtained diagnosis was validated on two bearing datasets. The results show that the accuracy of the compound fault diagnosis is over 98%. Compared with three other CNN-based methods, the proposed model demonstrates better stability.

Keywords: fault diagnosis; bearing compound fault; GDDCNN; CReLU; GMP

1. Introduction

With the development of modern industry, machine health monitoring becomes more and more important to maintain the safe operation of modern mechanical equipment [1,2]. Bearings are an indispensable part of modern machinery, especially in rotating machinery. Rolling bearing faults account for a large proportion of mechanical equipment faults, so it is necessary to ensure the normal operation of bearings [3–5]. In practical engineering, because of the complex and changeable operating conditions, once a fault occurs, it often affects other parts, leading to a compound fault [6–8]. Compared with single faults, the vibration signals are affected by mutual coupling and interference from various fault features, so the difficulty of feature extraction and fault diagnosis is greatly increased [9]. An effective and reliable compound fault diagnosis of rolling bearings is of great significance in guaranteeing safe operation.

In recent years, the compound fault diagnosis of rolling bearings has received increasing interest and attention from researchers. The research methods for compound fault diagnosis mainly include compound fault mechanism research [10,11], the blind source separation algorithm [12–15], the signal decomposition algorithm [16,17], and the artificial intelligence algorithm [18,19]. The method based on the compound fault mechanism, taking specific machineries as research objects, limits the application and portability of the model. Methods based on blind source separation have high requirements for the number of channels of a raw signal. The number of sensors must meet the requirements of the algorithm, which increases the cost of a fault diagnosis. The modal decomposition algorithm is a typical signal decomposition method that is suitable for non-stationary signal processing.

However, there are some issues, such as mode aliasing and the end effect, that can directly affect the accuracy of a compound fault diagnosis. With the rapid development of artificial intelligence techniques, machine learning methods, including support vector machines (SVMs), Bayesian classifiers, artificial neural networks (ANNs), and convolutional neural networks (CNNs), have already been applied as potential tools to fault diagnosis, especially deep learning approaches. Ref. [20] proposed a deep inception net with atrous convolution and applied this model to a bearing fault diagnosis. This model overcomes the problem caused by the different feature distributions of characteristics between two data sets and achieves high accuracy. Ref. [21] proposed a model based on CNN, named WDCNN, for a bearing fault diagnosis that achieves high accuracy on raw vibration signals directly in the Case Western Reserve University (CWRU) bearing data set. Ref. [22] utilized an improved CNN for a fault diagnosis. In that network, a multiscale cascaded layer is added to CNN, which can enhance the classification information of the input. Ref. [23] constructed a CNN with feature alignment that addressed the finite-shift-invariance problem, which can extract a robust fault feature. Ref. [24] adopted a multi-task CNN through information fusion for a fault diagnosis on two bearing data sets. The experiment results showed that the proposed model improved the accuracy of the fault diagnosis. Ref. [25] implemented a lightweight CNN that combined transfer learning and self-attention, which can achieve higher diagnosis accuracy than traditional CNN models. Ref. [26] proposed a novel multiscale CNN model that incorporates multiscale learning during the feature extraction process to diagnose the fault of a wind turbine gearbox. A novel multiscale residual attention CNN model was proposed in [27], which utilizes multiscale features, an attention mechanism, and residual learning to enhance feature extraction ability. Experimental validation on two bearing datasets demonstrated that the algorithm achieved higher accuracy. Based on a CNN model, [28] designed a multi-task CNN model that utilizes a speed identification task and a load identification task as auxiliary tasks to improve the performance of a fault diagnosis task. The experimental results showed that multi-task learning can enhance the fault diagnosis performance of the model. In [29], an end-to-end fault diagnosis model combining CNN with LSTM is designed, which can realize a bearing fault diagnosis in a short time. Ref. [30] presented an improved one-dimensional multiscale model that combined different extended convolutional kernels with varying dilation rates. The superiority of this approach was validated on the CWRU and PU datasets. Ref. [31] integrated vibration signals and sound signals through a one-dimensional CNN for fusion and validated that it has a higher diagnostic accuracy. Ref. [32] used short-time Fourier transform to convert vibration signals into spectrograms and adopted a model based on CNN for feature extraction and health status classification. Ref. [33] proposed a lightweight CNN combined with data augmentation technology for a bearing fault diagnosis. A novel hybrid CNN-MLP model for diagnoses was proposed in [34], which combined mixed input to achieve a rolling bearing diagnosis. In [35], a lightweight CNN model with fixed feature graph dimensions is constructed with down-sampling vibration signals to construct spectral graphs, which can achieve high classification accuracy on low-dimensional input data. Ref. [36] put forward a model based on optimized-parameter maximum-correlated kurtosis deconvolution and CNN for a bearing compound fault diagnosis and verified its effectiveness.

The above models demonstrate that deep learning approaches have significantly improved fault diagnosis effectiveness. However, the relationship between compound fault and single fault is one-to-many or many-to-many, which is not a simple superposition of a single fault signal. Compared with signal faults, feature extraction and location of compound faults are more difficult and challenging, which brings great difficulties to compound fault diagnosis. Currently, intelligent diagnosis methods based on deep learning mainly focus on single faults. Some models are suitable for single fault diagnosis; however, they are not effective for compound fault diagnosis. The research and application of deep learning approaches to bearing compound fault diagnosis are still in their infancy. Thus, in this study, a novel deep convolutional neural network combining global feature extraction with detailed feature extraction (GDDCNN) was proposed.

The proposed GDDCNN model is a deep convolutional neural network that incorporates both global feature extraction and detailed feature extraction, where G means global feature extraction, D means detailed feature extraction, and DCNN represents a deep convolutional neural network. DCNN is a feature-progressive learning algorithm in which the deep network can continue to learn more advanced fault features based on shallow features. By designing two feature extraction modules, G and D, better learning ability can be achieved for DCNN. Therefore, more abundant fault features can be extracted through GDCNN, thus improving the performance of fault diagnosis. The contributions of this study are summarized as follows:

1. A novel deep convolutional neural network combining global feature extraction with detailed feature extraction (GDDCNN) is proposed to extract features adaptively from a raw signal.
2. The modified activation concatenated ReLU (CReLU) is applied in the shallow layer of GDDCNN. It can improve the performance of global feature extraction.
3. The global max pooling (GMP) strategy is designed to replace the traditional fully connected layer, which can extract shift-invariance features and reduce model parameters. It overcomes the overfitting problem in the training process.

The rest of this study is described as follows. Section 2 reviews the basic theory of CNN. Section 3 expresses the proposed model in detail. Section 4 presents the experimental validation through two bearing datasets. Finally, Section 5 draws the conclusion.

2. Theoretical Background

CNN can be built with multiple layers, comprising a convolution layer, a pooling layer, an activation layer, and a classification layer. The convolution layer, pooling layer, and activation layer are used to extract features from input signals. One-dimensional CNN has been widely applied to 1-D vibration signal processing due to its powerful feature extraction ability. The classification layer applies the extracted features to classify.

The convolution layer is the core layer of the CNN structure, which convolves the input data with filter kernels. The network makes the filter learn to activate when it extracts certain features, then realizes feature extraction. The mathematical form can be described as follows:

$$y_j^l = K_i^l * x_i^l = \sum_{i \in M_j} x_i^l * w_{ij}^l + b_j^i \tag{1}$$

where the y_j^l denotes the output of the l-th layer; K_i^l is the i-th convolution kernel of the l-th layer; x_i^l denotes the input of the l-th layer; the notation $*$ represents the convolution operation; w_{ij}^l denotes the weights the convolution kernel; and b_j^i is the offset.

The activation layer is usually followed by the convolution layer, which is an essential layer. The activation function defines the output and input connections of a neuron, which is usually nonlinear. It makes the network learn nonlinear features from an input vibration signal to improve the capability of feature extraction. Rectified linear unit (ReLU) [37] is commonly used as an activation function in CNN, which is defined as follows:

$$\mathrm{Re}LU(y_j^l) = \begin{cases} 0 \; y_j^l < 0 \\ y_j^l \; y_j^l \geq 0 \end{cases} \tag{2}$$

which denotes the activation value is 0 on the negative half-axis.

Batch normalization (BN) [38] is applied to deep neural networks, which can reduce the shift of internal covariance and improve the accuracy of the training model. In addition,

BN compels the learned options into a standard distribution with a mean value of 0 and a variance of 1, which can accelerate the training speed of the model. The transform of BN is described as follows:

$$\begin{cases} \mu_B = \frac{1}{m}\sum_{i=1}^{m} x_i \\ \delta_B^2 = \frac{1}{m}\sum_{i=1}^{m} (x_i - \mu_B)^2 \\ \hat{x}_i = \frac{x_i - \mu_B}{\sqrt{\delta_B^2 + \varepsilon}} \\ y_i = \gamma \hat{x}_i + \beta \end{cases} \tag{3}$$

where m represents the mini-batch size, μ_B expresses the mini-batch mean, and δ_B^2 represents the mini-batch variance. γ and β are learnable parameters of the network.

The pooling layer performs a down-sampling operation, which removes redundant features and extracts deeper features. Max pooling and average pooling are the most common pooling operations. Max pooling generally outperforms average pooling for time series classification tasks, which is expressed as follows:

$$p_i^l = \max_{(j-1)s+1 \leq t \leq js} \left\{ a_i^l(t) \right\} \tag{4}$$

where the p_i^l represents the output features of the l-th layer; max is max pooling, $a_i^l(t)$ means the output value of the t-th neuron in the i-th channel of the l-th layer, s denotes the stride of the pooling.

In the classification layer, the Softmax function is applied to normalize the probability of each category in the output. Softmax in the neural network is defined as

$$\text{Softmax}(x_i) = \frac{e^{x_i}}{\sum_{j=1}^{K} e^{x_j}} \tag{5}$$

where K is the number of categories and x_i represents the logits of the j-th output neuron.

3. Proposed GDDCNN

To overcome the problems of compound fault feature extraction, a novel deep CNN is proposed. The structure of the proposed model, GDDCNN, is shown in Figure 1, composed of an input module, a feature extractor module, a GMP layer, and a Softmax classifier. Global feature extraction and detailed feature extraction constitute the feature extractor module. Besides, this study used two strategies: modified activation operation and GMP strategy during the feature extraction process to enhance the feature extraction ability of the proposed model, GDDCNN. Finally, a compound fault diagnosis is implemented. More details are illustrated in subsequent parts.

3.1. GDDCNN Architecture Design

Different kernel sizes for convolution: In CNN, convolutional kernel size plays an important role in the convolutional layer because kernels of different sizes can obtain different features. Generally, wider kernels pay more attention to global information during convolution operations, thereby extracting more global features, while smaller kernels can capture more detailed features. To obtain more robust features from a raw vibration signal, global feature extraction and detailed feature extraction are combined and applied to the feature extractor module. This study designed different kernel sizes for convolution; wide kernel sizes are applied in the shallow (first and second) convolution layers to extract global features, while the deep convolutional kernels are small, which help to obtain detailed features. Multi-convolution layers that adopt small convolutional kernels make the CNN networks deeper, which can improve the performance of compound fault feature extraction. Finally, the size of the convolution kernel for different layers is set to be [64, 16, 3].

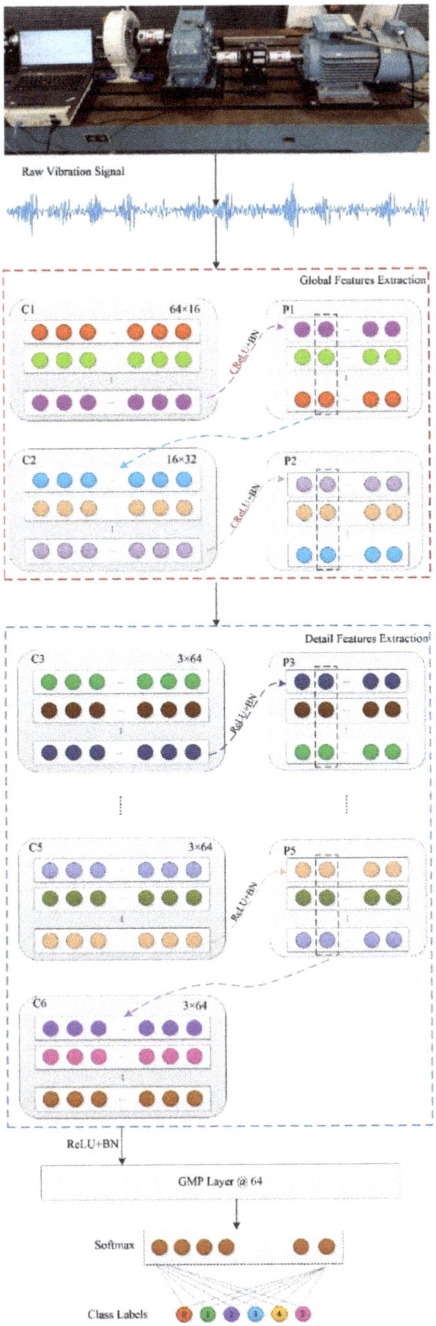

Figure 1. Structure of GDDCNN.

Modified activation operation: In addition to the convolution, the activation operation also affects the performance of CNN. Due to the advantages of simple calculation and no vanishing gradient problem, ReLU is widely applied in CNN as an activation unit.

However, when the inputs are negative, the neurons are always inactive. These dead neurons in a network may never activate, which stops learning and thus affects the learning ability of the network. In [39], it was found that in the shallow layer of CNNs, the parameter distribution of the network exhibits a stronger negative correlation, while with the deepening of the network, this negative correlation gradually becomes weak. CReLU is a concatenated ReLU, which, through inverting the feature map to activate the negative inputs, helps features transmit better backward. CReLU is defined as follows:

$$\mathrm{CRe}LU(x) = [\mathrm{Re}LU(x), \mathrm{Re}LU(-x)] \tag{6}$$

where ReLU represents the ReLU activation function. The modified activation CReLU is used in the shallow (first and second) convolution layers, which can improve the performance of global feature extraction.

GMP strategy: The fully connected layer is generally applied after the last convolutional or pooling layer, which can integrate the class-differentiated local features extracted by CNN. Each neuron in the fully connected layer is fully connected to all the neurons in the previous layer. Due to its fully connected characteristics, the parameters of the fully connected layer are numerous, and the calculation is extremely complicated. The right part of Figure 2 shows the GMP progress, that is, the max value of each channel is used as the new feature vector. The GMP strategy clearly reduces the dimension of the feature vector, which can avoid overfitting. Another advantage is that it is more native to the convolution structure by enforcing correspondences between feature maps and categories. Furthermore, this strategy can retain the spatial information that keeps shift-invariance, resulting in robust extracted features.

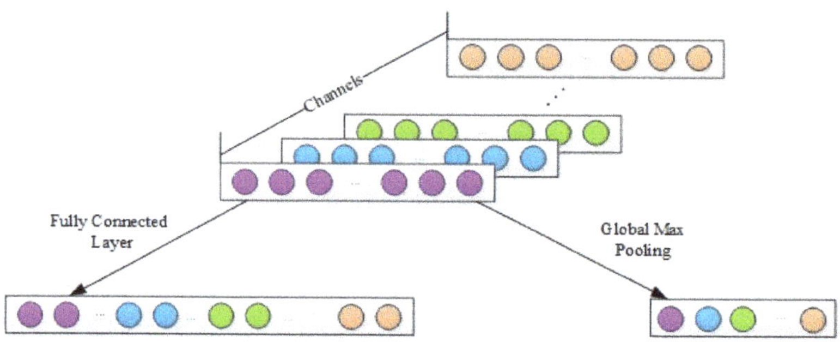

Figure 2. Global max pooling.

3.2. *Training of GDDCNN*

The architecture of GDDCNN is designed to take advantage of one-dimensional CNN. The cross-entropy loss function is used to estimate the consistency between the Softmax output probability distribution and the target class probability distribution. Suppose that $p(x)$ and $q(x)$ represent the target distribution and estimated distribution, respectively. The loss function can be expressed as follows:

$$Loss(L) = H(p(x), q(x)) = -\sum_{x} p(x) log q(x) \tag{7}$$

The gradient of loss L and the gradient of parameters related to BN should be back-propagated in GDDCNN during training as per the following rules:

$$\frac{\partial L}{\partial \overline{x}_i} = \frac{\partial L}{\partial y_i} \cdot \gamma \tag{8}$$

$$\frac{\partial L}{\partial \delta_B^2} = \sum_{i=1}^{m} \frac{\partial L}{\partial \overline{x}_i} \cdot (x_i - \mu_B) \cdot \frac{-1}{2} (\delta_B^2 + \varepsilon)^{-3/2} \tag{9}$$

$$\frac{\partial L}{\partial \mu_B} = (\sum_{i=1}^{m} \frac{\partial L}{\partial \overline{x}_i} \cdot \frac{-1}{\sqrt{\delta_B^2 + \varepsilon}}) + \frac{\partial L}{\partial \delta_B^2} \cdot \frac{\sum_{i=1}^{m} -2(x_i - \mu_B)}{m} \tag{10}$$

$$\frac{\partial L}{\partial x_i} = \frac{\partial L}{\partial \overline{x}_i} \cdot \frac{-1}{\sqrt{\delta_B^2 + \varepsilon}} + \frac{\partial L}{\partial \delta_B^2} \cdot \frac{2(x_i - \mu_B)}{m} + \frac{\partial L}{\partial \mu_B} \cdot \frac{1}{m} \tag{11}$$

$$\frac{\partial L}{\partial \gamma} = \sum_{i=1}^{m} \frac{\partial L}{\partial y_i} \cdot \overline{x}_i \tag{12}$$

$$\frac{\partial L}{\partial \beta} = \sum_{i=1}^{m} \frac{\partial L}{\partial y_i} \tag{13}$$

Adam is a learning rate adaptive optimization algorithm that assembles the Adagrad algorithm and the RMSProp algorithm, which allows a model to allocate more updates to rarely occurring features, thereby aiding in the convergence of the optimization process. By bringing together the adaptive learning rates of Adagrad and the stability of RMSProp, the Adam algorithm provides an effective solution to optimize the proposed model, enabling it to converge quickly and efficiently while avoiding common optimization pitfalls. The choice of optimizer plays a crucial role in the success of the proposed model's training process.

In order to minimize the value of the loss function, the Adam optimization algorithm is applied to update the weights and obtain the optimal weights. The Adam optimizer is chosen for the following reasons: Bearing vibration signals typically contain a large number of data points due to their time-series nature. The efficient convergence property of the Adam algorithm allows the model to learn from a large volume of data more quickly. This, in turn, accelerates the assessment of the bearing's health status, saving training time. Features in bearing vibration signals may vary in importance over different time intervals. Adam's adaptive learning rate automatically adjusts the learning rate based on the gradient of each parameter. This capability helps the model better adapt to dynamic variations within a signal. When the Adam optimizer is initialed, it is necessary to set the learning rate ε, exponential decay rate of moment estimation ρ_1, ρ_2, and constant δ. In this experiment, we set ε to 0.001, ρ_1, ρ_2 to default 0.9, 0.999, and the constant was set to prevent the numerical mutation from being set to 10^{-8} during the dividing operation. The detailed process of the Adam algorithm is shown in Table 1. More details on the Adam algorithm can be found in [40].

Table 1. Detailed flow of the Adam algorithm.

Adam Algorithm

Initialize variables $\varepsilon, \rho_1, \rho_2, \delta, \theta, s = 0, r = 0, t = 0$
While stopping criterion not met **do**
Sample a minibatch of m examples from the training set
$\left\{x^{(1)}, \ldots, x^{(\mathrm{m})}\right\}$ with corresponding targets $y(i)$
Computing gradient:
$g \leftarrow \frac{1}{m} \nabla_\theta \sum_i L(f(x^{(i)};\theta), y^{(i)})$
$t \leftarrow t + 1$
Update biased first-order moment estimation: $s \leftarrow \rho_1 s + (1 - \rho_1) g$
Update biased second-order moment estimation: $r \leftarrow \rho_2 r + (1 - \rho_2) g \odot g$
Correcting the deviation of the first moment: $\hat{s} \leftarrow \frac{s}{1-\rho_1^t}$
Correcting the deviation of the second moment: $\hat{r} \leftarrow \frac{r}{1-\rho_2^t}$
Calculation update: $\Delta\theta = -\varepsilon \frac{\hat{s}}{\sqrt{\hat{r}} + \delta}$
Apply update: $\theta \leftarrow \theta + \Delta\theta$
end while

3.3. *Diagnosis Procedure*

1. The raw vibration signals of rolling bearings are collected by a data acquisition device.
2. The obtained vibration signals are sliced into samples of length 2048 for standardization processing and then used as network input.
3. Extracting fault features by combining global feature extraction with detailed feature extraction. Utilizing the GMP layer to integrate features. Then, Softmax is employed as a classifier to classify fault features.
4. Testing samples are entered into the network to realize fault diagnosis and validate the performance of the proposed model, GDDCNN.

4. Experiments and Results

4.1. *Data Acquisition*

In this study, the two-stage gear drive test bench shown in Figure 3, is composed of a drive motor, a two-stage helical gear box, a powder brake, a torque speed sensor, a coupling, tested bearings, and a data acquisition device. The experimental data was obtained by the intermediate shaft end bearing of the two-stage helical gearbox. In this task group, three rotating speeds (1200 rmp, 1500 rmp, and 1800 rmp) and three electrical machinery loads (0 A, 0.5 A, and 1.0 A) were set up in 9 working conditions in the experiment, respectively. In addition, the sampling frequency is 96 kHz, and the sampling time is 10 s. Collecting 9 raw data samples from each category, with a sampling length of 96,000 for each raw data sample. There are six categories of bearing faults: one normal, three single faults, and two compound faults. These six health conditions of bearings are designated as normal (NO), inner fault (IF), outer fault (OF), roll fault (RF), inner and roll compound fault (IRF), and outer and roll fault (ORF). All bearing faults simulated by electrical discharge machining (EDM) technology are shown in Figure 4.

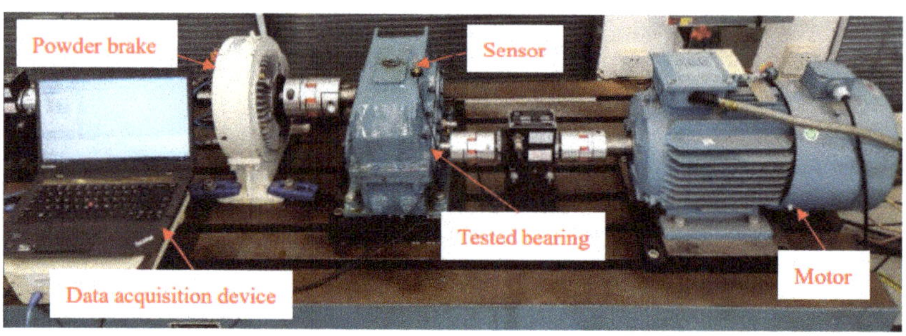

Figure 3. Two-stage gear drive test bench.

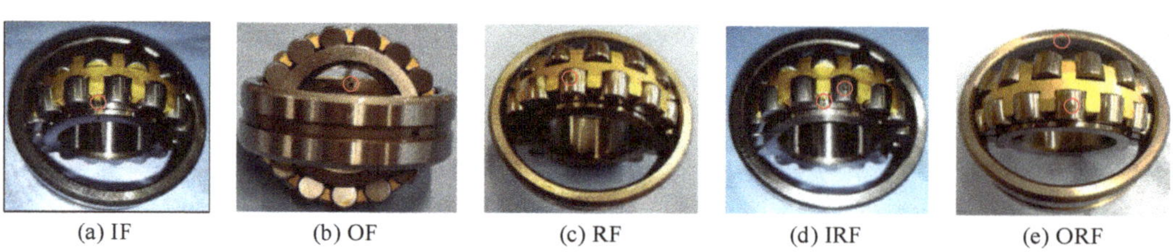

Figure 4. Bearing fault simulation: (**a**) inner fault; (**b**) outer fault; (**c**) roll fault; (**d**) inner and roll compound fault; and (**e**) outer and roll fault.

In addition to the data obtained from the above two-stage gear drive test bench, this study also utilized the public Case Western Reserve University (CWRU) dataset [41] to simulate compound fault signals. The raw vibration signals under a load of 1 hp with a rotating speed of 1772 r/min and a sampling frequency of 12kHz are selected. Four compound fault categories are simulated by selected signals, which are IF + RF, OF + RF, IF + OF, and IF + OF + RF.

4.2. Data Preprocessing

(1) Sample Segmentation: After data acquisition, an easy data augmentation technique of slicing the raw vibration signals with overlap is applied to process training samples. In order to reflect the periodicity of raw signals, this study set the sampling size to 2048 and the overlapping window size to 512. The data segmentation process is shown in Figure 5. After data segmentation, 16,848 samples of each category were obtained, totaling 101,088 samples.

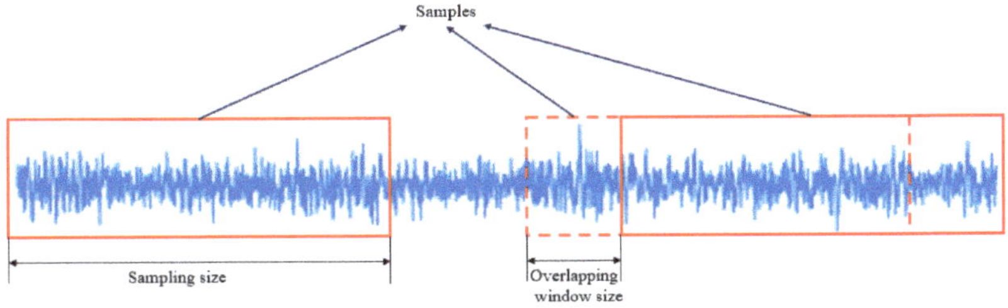

Figure 5. Data segmentation process.

(2) Standardization: After sample segmentation, in order to improve the convergence speed and accuracy of the model, the samples need to be standardized. The samples are standardized by the z-score method and follow the distribution of the mean value 0 and variance 1. The standardization process for the raw data $(x_{1,}x_{2,}x_3 \ldots x_n)$ is as follows:

$$\overline{x} = \frac{1}{n}\sum_{i=1}^{n} x_i \tag{14}$$

$$\delta = \sqrt{\frac{1}{n-1}\sum_{i=1}^{n} (x_i - \overline{x})^2} \tag{15}$$

$$\text{Standard}(x_i) = \frac{x_i - \overline{x}}{\delta} \tag{16}$$

4.3. Parameters of the Proposed Network

The experiments were performed using the above data to verify the performance of the proposed method. The architecture of the proposed method involves stacking 2 convolutional layers with a wide kernel size and pooling layers, 4 convolutional layers with a small kernel size and pooling layers, followed by GMP and a Softmax layer to implement fault diagnosis. The first convolutional kernel size and channel depth are, respectively, set to 64 and 16; the second convolutional kernel size and channel size are set to be 16 and 32, respectively; and the rest of the convolutional kernel size and channel are set to be 3 and 64, respectively. All five categories of pooling are max pooling, with a size of 2. The first two activation functions are the improved activation connection ReLU (CReLU), and the remaining activation functions are ReLU. To improve the performance of the proposed model, GDDCNN, BN is applied after each activation layer. Each convolution layer and pooling layer utilizes a zero-padding of causal and same type to prevent the loss of edge information. The detailed parameters of the proposed network are listed in Table 2.

Table 2. Detailed parameters of the proposed network.

Layer Type	Kernel Size/Stride	Kernel Channel Size	Padding	Output Size
Input	/	/	/	2048
C1	64/8	16	causal	256*16
P1	2/2	/	same	128*32
C2	16/2	32	causal	64*32
P2	2/2	/	same	32*64
C3	3/1	64	causal	32*64
P3	2/2	/	same	16*64
C4	3/1	64	causal	16*64
P4	2/2	/	same	8*64
C5	3/1	64	causal	8*64
P5	2/2	/	same	4*64
C6	3/1	64	same	4*64
GMP	/	/	/	64*1
Softmax	/	/	/	6*1

4.4. Comparative Experiment

In order to validate the feasibility of the proposed method, which is compared with several currently popular models, including classical CNN, CNN-SVM, and WDCNN [21]. This study selected 1500 samples from each category under all working conditions randomly for experiments: 1000 samples as training data, 300 samples as verification data, and 200 samples as testing data. The details of the experimental dataset are shown in Table 3. All experiments were run on a server equipped with an Intel Core i7-6800k CPU, an Nvidia GeForce GTX 1080 Ti GPU, and 32G of RAM.

Table 3. Description of samples distribution.

Fault Location	Sample Number	Training/Validation/Test	Label
Normal	1500	1000/300/200	0
IF	1500	1000/300/200	1
OF	1500	1000/300/200	2
RF	1500	1000/300/200	3
IRF	1500	1000/300/200	4
ORF	1500	1000/300/200	5

Accuracy is the most common metric used to evaluate the performance of a classifier. It calculates the proportion of correctly classified samples out of the total number of samples. Specifically, accuracy can be calculated using the following formula:

$$Accuray = \frac{TP + TN}{TP + TN + FP + FN} \tag{17}$$

where TP represents the number of true positive samples, i.e., the number of samples correctly predicted as positive by the classifier; TN represents the number of true negative samples, i.e., the number of samples correctly predicted as negative by the classifier; FP represents the number of false positive samples, that is, the number of samples incorrectly predicted as positive by the classifier; and FN represents the number of false negative samples, that is, the number of samples incorrectly predicted as negative by the classifier.

The results of the diagnosis accuracy for each category using four different methods are summarized in Table 4. According to the statistical results, it can be observed that the single fault diagnosis accuracy of the proposed method can reach over 98% or even approach 100%, and the compound fault diagnosis accuracy is also over 95%. The accuracy of other methods in category 4 (IRF) of the compound fault diagnosis is lower than 90%.

Table 4. Classification accuracy results of the four methods.

Method	Class Label					
	0	1	2	3	4	5
CNN	92.9%	92.9%	95.7%	92.9%	84.6%	87.6%
CNN-SVM	92.8%	93.9%	94.6%	93.1%	84.3%	90.1%
WDCNN	95%	88.2%	99%	95.7%	88.6%	94.7%
GDDCNN	100%	99%	99%	98%	96.9%	95.1%

In order to demonstrate the advantages of the proposed method more visually and evidently, a line chart is used to compare the accuracy of the four methods, as shown in Figure 6. Where fault category 0–5 indicates a fault label, the details about fault labels correspond to the fault category, as shown in Table 3. It can be clearly observed that the proposed method has better performance than the other three methods. The diagnosis accuracy of each category is higher than that of other methods. Moreover, the diagnosis accuracy of each category is relatively balanced, with none of the six categories having particularly low accuracy. Although the WDCNN model has higher accuracy than CNN and CNN-SVM, its performance and stability are still not comparable to the proposed method.

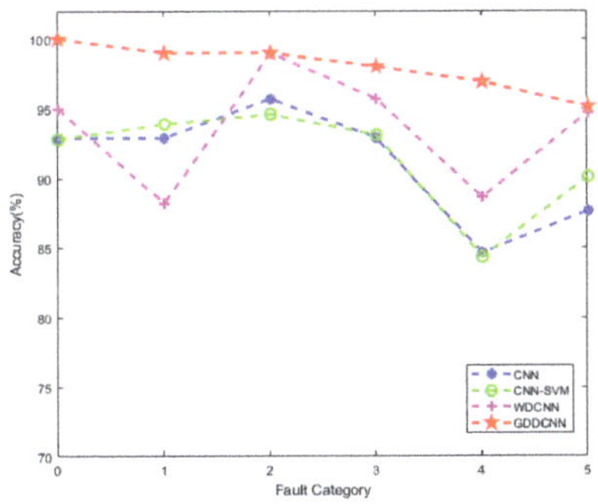

Figure 6. Accuracy comparison of four methods.

Precision is similar to but different from accuracy, which is the proportion of correctly predicted positive samples among all samples identified as positive. Precision represents the prediction accuracy of the positive sample results. Recall refers to the proportion of correctly predicted positive samples among all true positive samples, which reflects the proportion of positive samples correctly predicted. F1-Score is the harmonic mean of precision and recall, which combines information from these two metrics. The calculation of precision, recall, and F1-score is as follows:

$$Precision = \frac{TP}{TP + FP} \tag{18}$$

$$Recall = \frac{TP}{TP + FN} \tag{19}$$

$$F1 - Score = \frac{2 \times Precision \times Recall}{Precision + Recall} \tag{20}$$

in which TP, FP, and FN have the same meanings as those in Formula 17. The precision, recall, and F1-score of the proposed method are shown in Figure 7. Where Class 0 represents normal, Class 1, Class 2, and Class 3 represent inner fault, outer fault, and roll fault. Both Class 4 and Class 5 are compound faults; Class 4 expresses inner and roll compound faults, while Class 5 represents outer and roll faults.

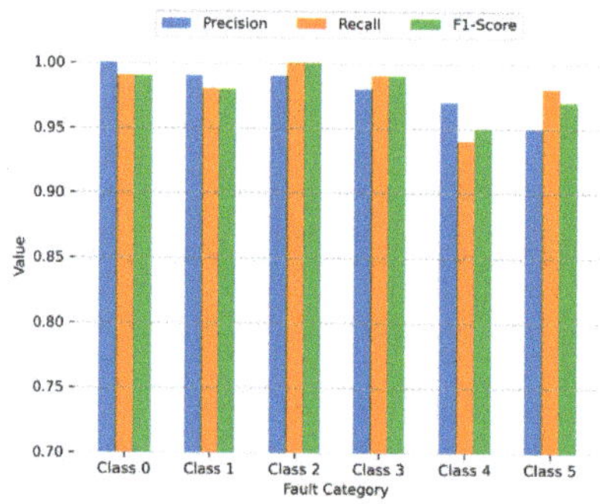

Figure 7. Precision, recall, and F1-score of the GDDCNN.

Precision, recall, and F1-score are also crucial performance metrics for a classifier; a higher precision, recall, and F1-score indicates a better performance of the classifier. It can be observed from Figure 7 that all the values of the proposed model for all conditions are high.

In order to further validate the effectiveness of the fault diagnosis model, three error metrics—maximum average error (MAE), mean squared error (MSE), and root mean squared error (RMSE)—are selected to evaluate the fitting effect of the fault diagnosis model. The formulas for each metric are defined as follows:

$$MAE = \frac{1}{N}\sum_{\text{i}=1}^{N}\left|Y_{i,true} - Y_{i,pred}\right| \tag{21}$$

$$MSE = \frac{1}{N}\sum_{\text{i}=1}^{N}\left(Y_{i,true} - Y_{i,pred}\right)^2 \tag{22}$$

$$RMSE = \sqrt{\frac{1}{N}\sum_{\text{i}=1}^{N}\left(Y_{i,true} - Y_{i,pred}\right)^2} \tag{23}$$

where N is the number of testing samples, the value of $Y_{i,true}$ is 1, and $Y_{i,pred}$ is the probability value of the category to which it belongs. The value of $Y_{i,pred}$ calculated through the Softmax classifier is closer to 1, indicating a better fitting effect of the model. Therefore, smaller values of the three metrics represent better stability for the model. The metrics results of the four methods are listed in Table 5. From the statistical results, it can be seen that the three metrics of the GDDCNN are all smaller than those of other methods, indicating that the GDCNN proposed in this study has good stability.

Table 5. Metrics results of the four methods.

Method	MAE	MSE	RMSE
CNN	0.265	0.861	0.928
CNN-SVM	0.270	0.860	0.927
WDCNN	0.108	0.275	0.524
GDDCNN	0.032	0.065	0.255

Finally, the feature distribution in different layers of the proposed model, GDDCNN, can be visualized through the t-SNE technique. t-SNE is a nonlinear dimensionality reduction algorithm that calculates the positions of sample points based on the similarity between sample points in high-dimensional space and the distances between sample points in low-dimensional space. The experiment results, i.e., the position information after dimension reduction of the raw signal, six convolutional layers, and the GMP layer, are shown in Figure 8. It can be obviously seen that the raw signal and the features in the shallow convolution layers are inseparable. With the deepening of the convolution layer, the feature distribution can be improved significantly. From the visualization of the GMP layer, we can observe the feature distribution, which suggests that the proposed model cannot discriminate between category 1 and category 4 very well. It may be due to the fact that category 4 is a compound fault that contains both category 1 and category 3. The samples of other categories can cluster in their own area, and the inter-category differences are good. The experiment results indicate that the proposed model, GDDCNN, can extract robust and effective features from a raw signal and achieve efficient fault diagnosis.

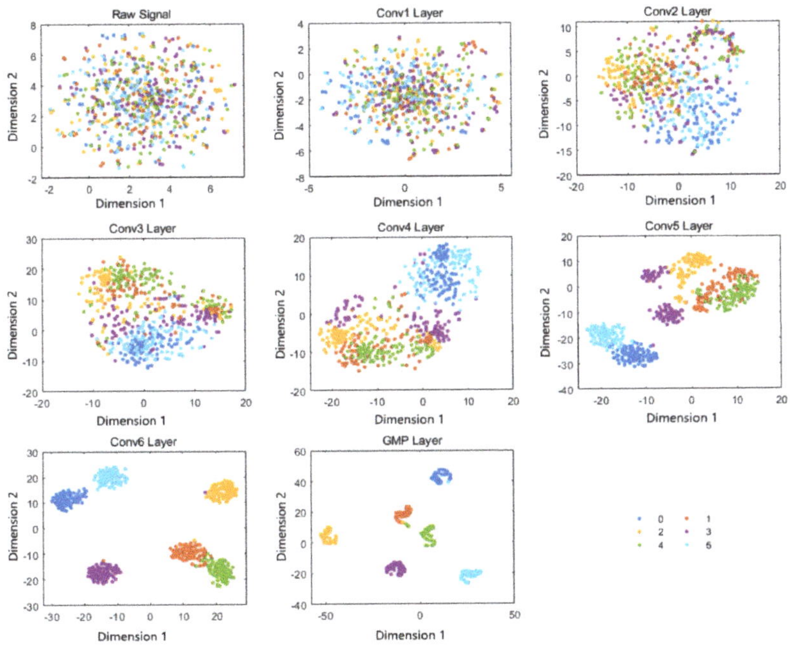

Figure 8. Feature visualization via t-SNE.

In order to further verify the effectiveness and performance of the proposed model, GDDCNN, fault diagnosis experiments are carried out with the simulated data from the CWRU dataset. Data is processed by overlapping sampling during simulation. The distribution of samples is listed in Table 6.

Table 6. Distribution of samples.

Fault Location	Sample Number	Training/Validation/Test	Label
Normal	1000	700/200/100	0
IF	1000	700/200/100	1
OF	1000	700/200/100	2
RF	1000	700/200/100	3
IF + RF	1000	700/200/100	4
OF + RF	1000	700/200/100	5
IF + OF	1000	700/200/100	6
IF + OF + RF	1000	700/200/100	7

The average accuracy of the four models is shown in Table 7. It is easy to see from the results that all models achieve high diagnostic accuracy. The accuracy of the proposed model, GDDCNN, reaches 100%, which indicates it has good performance.

Table 7. Average accuracy of the four models.

Model	CNN	CNN-SVM	WDCNN	GDDCNN
Accuracy (%)	98	98	98.75	100

5. Conclusions

This study proposed a novel deep CNN named GDDCNN for the diagnosis of practical engineering compound faults. GDDCNN extracted robust features adaptively from raw vibration signals and realized compound fault diagnosis without any manual processing. The datasets acquired by a two-stage gear drive test bench and simulated with the published dataset of CWRU were applied to experiments to verify the performance of the proposed model. The experimental results demonstrate that GDDCNN can achieve more than 98% accuracy on both datasets. Compared with three other existing CNN-based models, this model not only achieves satisfactory diagnostic accuracy but also exhibits strong stability. By combining different convolutional kernel sizes, improved activation CReLU, and GMP strategy, the performance of GDDCNN has been improved, and compound fault diagnosis has been achieved more effectively. In further work, we will continue to explore how to adaptively select hyperparameters to enhance the accuracy and stability of the diagnosis model. We also aim to research the application of transfer learning to achieve compound fault diagnosis across different operating conditions.

Author Contributions: Data curation, Y.L.; investigation, S.H., G.F. and S.S.; methodology, S.H.; resources, D.Z.; software, S.L.; validation, P.N.; writing—original draft, S.H.; writing—review and editing, P.N. and Y.L.; funding acquisition, P.N. All authors have read and agreed to the published version of the manuscript.

Funding: This research is financially supported by the Tianjin Education Commission (2019ZD08).

Institutional Review Board Statement: Not applicable.

Informed Consent Statement: Not applicable.

Data Availability Statement: Not applicable.

Conflicts of Interest: The authors declare no conflict of interest.

References

1. Wei, S.; Wang, D.; Wang, H.; Peng, Z. Time-varying envelope filtering for exhibiting space bearing cage fault features. *IEEE Trans. Instrum. Meas.* **2021**, *70*, 3504313. [CrossRef]
2. Wang, L.; Zhang, X.; Liu, Z.; Wang, J. Sparsity-based fractional spline wavelet denoising via overlapping group shrinkage with non-convex regularization and convex optimization for bearing fault diagnosis. *Meas. Sci. Technol.* **2020**, *31*, 055003. [CrossRef]
3. Gao, S.; Xu, L.; Zhang, Y.; Pei, Z. Rolling bearing fault diagnosis based on intelligent optimized self-adaptive deep belief network. *Meas. Sci. Technol.* **2020**, *31*, 055009. [CrossRef]
4. Rai, A.; Upadpyay, S.H. A review on signal processing techniques utilized in the fault diagnosis of rolling element bearings. *Tribol. Int.* **2016**, *96*, 289–306. [CrossRef]
5. Qin, Y.; Jin, L.; Zhang, A.; He, B. Rolling bearing fault diagnosis with adaptive harmonic kurtosis and improved bat algorithm. *IEEE Trans. Instrum. Meas.* **2021**, *70*, 3508112. [CrossRef]
6. Tang, G.; Wang, Y.; Huang, Y.; Liu, N.; He, J. Compound bearing fault detection under varying speed conditions with virtual multichannel signals in angle domain. *IEEE Trans. Instrum. Meas.* **2020**, *69*, 5535–5545. [CrossRef]
7. Zhang, J.; Zhang, Q.; He, X.; Sun, G.; Zhou, D. Compound-Fault diagnosis of rotating machinery: A fused imbalance learning method. *IEEE Trans. Control Syst. Technol.* **2020**, *29*, 1462–1474. [CrossRef]
8. Ma, X.; Hu, Y.; Wang, M.; Li, F.; Wang, Y. Degradation state partition and compound fault diagnosis of rolling bearing based on personalized multilabel learning. *IEEE Trans. Instrum. Meas.* **2021**, *70*, 3520711. [CrossRef]
9. Jin, Y.; Qin, C.; Zhang, Z.; Tao, J.; Liu, C. A multi-scale convolutional neural network for bearing compound fault diagnosis under various noise conditions. *Sci. China Tech. Sci.* **2022**, *65*, 2551–2563. [CrossRef]
10. Zhang, X.; Yan, C.; Liu, Y.; Yan, P.; Wang, Y.; Wu, L. Dynamic modeling and analysis of rolling bearing with compound fault on raceway and rolling element. *Shock Vib.* **2020**, *5*, 8861899. [CrossRef]
11. Huang, C.; Zhang, B.; Yi, C.; Jin, H. Vibration model for axle box bearings with multiple defective rolling elements for high-speed trains and the defects localization method. *Shock Vib.* **2020**, *39*, 34–43.

12. Tang, G.; Luo, G.; Zhang, W.; Yang, C.; Wang, H. Underdetermined blind source separation with variational mode decomposition for compound roller bearing fault signals. *Sensors* **2016**, *16*, 897. [CrossRef]
13. Huang, D.; Chen, D.; Ke, L.; Zhao, L.; Sun, G. A blind source separation method for multi-fault modal characteristic signals of rolling bearings with error influences. *Acta Armamentarii* **2018**, *39*, 1419–1428.
14. Xi, J.; Cui, J.; Jiang, L. JADE-ICA-based blind source separation of multi-fault signals of rolling bearings. *Shock Vib.* **2017**, *5*, 231–237.
15. Mika, D.; Budzik, G.; Józwik, J. Single Channel Source Separation with ICA-Based Time-Frequency Decomposition. *Sensors* **2020**, *20*, 2019. [CrossRef] [PubMed]
16. Pan, H.; Yang, Y.; Li, X.; Zheng, J.; Cheng, J. Symplectic geometry mode decomposition and its application to rotating machinery compound fault diagnosis. *Mech. Syst. Signal Process.* **2019**, *114*, 189–211. [CrossRef]
17. Miao, Y.; Zhao, M.; Lin, J. Identification of mechanical compound-fault based on the improved parameter-adaptive variational mode decomposition. *ISA Trans.* **2018**, *84*, 82–95. [CrossRef] [PubMed]
18. Islam, M.M.; Manjurul, J.M. Kim Reliable multiple combined fault diagnosis of bearings using heterogeneous feature models and multiclass support vector Machines. *Reliab. Eng. Syst. Safty* **2019**, *184*, 55–56. [CrossRef]
19. Jin, Y.; Qin, C.; Huang, Y.; Liu, C. Actual bearing compound fault diagnosis based on active learning and decoupling attentional residual network. *Measurement* **2021**, *173*, 108500. [CrossRef]
20. Chen, Y.; Peng, G.; Xie, C.; Zhang, W.; Li, C.; Liu, S. ACDIN: Bridging the gap between artificial and real bearing damages for bearing fault diagnosis. *Neurocomputing* **2018**, *294*, 61–71. [CrossRef]
21. Zhang, W.; Peng, G.; Li, C.; Chen, Y.; Zhang, Z. A new deep learning model for fault diagnosis with good anti-noise and domain adaptation ability on raw vibration signals. *Sensors* **2017**, *17*, 425. [CrossRef] [PubMed]
22. Huang, W.; Cheng, J.; Yang, Y.; Guo, G. An improved deep convolutional neural network with multi-scale information for bearing fault diagnosis. *Neurocomputing* **2019**, *359*, 77–92. [CrossRef]
23. Chen, J.; Huang, R.; Zhao, K.; Wang, W.; Liu, L.; Li, W. Multiscale convolutional neural network with feature alignment for bearing fault diagnosis. *IEEE Trans. Instrum. Meas.* **2021**, *70*, 3517010. [CrossRef]
24. Guo, S.; Zhang, B.; Yang, T.; Lyu, D.; Gao, W. Multitask convolutional neural network with information fusion for bearing fault diagnosis and localization. *IEEE Trans. Ind. Electron.* **2020**, *67*, 8005–8016. [CrossRef]
25. Zhong, H.; Lv, Y.; Yuan, R.; Yang, D. Bearing fault diagnosis using transfer learning and self-attention ensemble lightweight convolutional neural network. *Neurocomputing* **2022**, *501*, 765–777. [CrossRef]
26. Jiang, G.; He, H.; Yan, J.; Xie, P. Multiscale Convolutional Neural Networks for Fault Diagnosis of Wind Turbine Gearbox. *IEEE Trans. Ind. Electron.* **2019**, *66*, 3196–3207. [CrossRef]
27. Jia, L.; Chow, T.W.S.; Wang, Y.; Yuan, Y. Multiscale Residual Attention Convolutional Neural Network for Bearing Fault Diagnosis. *IEEE Trans. Instrum. Meas.* **2022**, *71*, 3519413. [CrossRef]
28. Liu, Z.; Wang, H.; Liu, J.; Qin, Y.; Peng, D. Multitask Learning Based on Lightweight 1DCNN for Fault Diagnosis of Wheelset Bearings. *IEEE Trans. Instrum. Meas.* **2021**, *70*, 3501711. [CrossRef]
29. Khorram, A.; Khalooei, M.; Rezghi, M. End-to-End CNN + LSTM Deep Learning Approach for Bearing Fault Diagnosis. *Appl. Intell.* **2021**, *51*, 736–751. [CrossRef]
30. He, J.; Wu, P.; Tong, Y.; Zhang, X.; Lei, M.; Gao, J. Bearing Fault Diagnosis via Improved One-Dimensional Multi-Scale Dilated CNN. *Sensors* **2021**, *21*, 7319. [CrossRef]
31. Wang, X.; Mao, D.; Li, X. Bearing Fault Diagnosis Based on Vibro-Acoustic Data Fusion and 1D-CNN Network. *Measurement* **2021**, *173*, 108518. [CrossRef]
32. Pham, M.T.; Kim, J.-M.; Kim, C.H. Accurate Bearing Fault Diagnosis under Variable Shaft Speed Using Convolutional Neural Networks and Vibration Spectrogram. *Appl. Sci.* **2020**, *10*, 6385. [CrossRef]
33. Oh, J.W.; Jeong, J. Data Augmentation for Bearing Fault Detection with a Light Weight CNN. *Procedia Comput. Sci.* **2020**, *175*, 72–79. [CrossRef]
34. Sinitsin, V.; Ibryaeva, O.; Sakovskaya, V.; Eremeeva, V. Intelligent Bearing Fault Diagnosis Method Combining Mixed Input and Hybrid CNN-MLP Model. *Mech. Syst. Signal Process.* **2022**, *180*, 109454. [CrossRef]
35. Yoo, Y.; Jo, H.; Ban, S.-W. Lite and Efficient Deep Learning Model for Bearing Fault Diagnosis Using the CWRU Dataset. *Sensors* **2023**, *23*, 3157. [CrossRef]
36. Gao, S.; Shi, S.; Zhang, Y. Rolling bearing compound fault diagnosis based on parameter optimization MCKD and convolutional neural network. *IEEE Trans. Instrum. Meas.* **2022**, *71*, 3508108. [CrossRef]
37. Nair, V.; Hinton, G.E. Rectified linear units improve restricted boltzmann machines. In Proceedings of the 27th International Conference on International Conference on Machine Learning, Haifa, Israel, 21–24 June 2010; pp. 807–814.
38. Ioffe, S.; Szegedy, C. Batch normalization: Accelerating deep network training by reducing internal covariate shift. *arXiv* **2015**, arXiv:1502.03167.
39. Shang, W.; Sohn, K.; Almeida, D.; Lee, H. Understanding and improving convolutional neural networks via concatenated rectified linear units. In Proceedings of the 33rd International Conference on Machine Learning, New York, NY, USA, 19–24 June 2016; pp. 2217–2225.

40. Kingma, D.; Ba, J. Adam: A Method for Stochastic Optimization. *arXiv* **2014**, arXiv:1412.6980.
41. Lou, X.; Loparo, K. A Bearing fault diagnosis based on wavelet transform and fuzzy inference. *Mech. Syst. Signal Process.* **2004**, *18*, 1077–1095. [CrossRef]

Disclaimer/Publisher's Note: The statements, opinions and data contained in all publications are solely those of the individual author(s) and contributor(s) and not of MDPI and/or the editor(s). MDPI and/or the editor(s) disclaim responsibility for any injury to people or property resulting from any ideas, methods, instructions or products referred to in the content.

sensors

MDPI

Article

Controlled Symmetry with Woods-Saxon Stochastic Resonance Enabled Weak Fault Detection

Jian Liu [1], Jiaqi Guo [1], Bing Hu [2,*], Qiqing Zhai [3], Can Tang [1] and Wanjia Zhang [1]

1 College of Information Engineering, Nanjing University of Finance and Economics, Nanjing 210023, China; liujian@nufe.edu.cn (J.L.); guojiaqi.orange@foxmail.com (J.G.)

2 College of Modern Posts, Nanjing University of Posts and Telecommunications, Nanjing 210003, China

3 College of Science, Nanjing University of Posts and Telecommunications, Nanjing 210003, China

* Correspondence: hubing@njupt.edu.cn

Abstract: Weak fault detection with stochastic resonance (SR) is distinct from conventional approaches in that it is a nonlinear optimal signal processing to transfer noise into the signal, resulting in a higher output SNR. Owing to this special characteristic of SR, this study develops a controlled symmetry with Woods-Saxon stochastic resonance (CSwWSSR) model based on the Woods-Saxon stochastic resonance (WSSR), where each parameter of the model may be modified to vary the potential structure. Then, the potential structure of the model is investigated in this paper, along with the mathematical analysis and experimental comparison to clarify the effect of each parameter on it. The CSwWSSR is a tri-stable stochastic resonance, but differs from others in that each of its three potential wells is controlled by different parameters. Moreover, the particle swarm optimization (PSO), which can quickly find the ideal parameter matching, is introduced to attain the optimal parameters of the CSwWSSR model. Fault diagnosis of simulation signals and bearings was carried out to confirm the viability of the proposed CSwWSSR model, and the results revealed that the CSwWSSR model is superior to its constituent models.

Keywords: weak fault detection; controlled symmetry with Woods-Saxon stochastic resonance (CSwWSSR); particle swarm optimization (PSO); signal-to-noise ratio (SNR)

1. Introduction

As the industrialization has found its way into society, the safety and reliability of engineering systems and mechanical equipment play a critical role in the production process, which is an essential subject involving technology and people's livelihoods. It is common knowledge that mechanical failures occur from time to time with the increase in the age of the equipment or the imperfection of the technology in the use of mechanical equipment. Therefore, it is advantageous to find equipment problems as soon as possible because doing so can minimize financial damage and potentially save lives. Yet, in real-world applications, mechanical equipment operates in the presence of very strong background noise, which is manifested as a high frequency of noise overwhelming the presence of the effective signal in the time and frequency domain diagrams. Therefore, the development of novel methods that can identify and extract useful signals under the interference of substantial background noise has become an increasingly important topic in the field of weak fault detection [1,2].

In general, it is believed that the presence of noise will obstruct the detection of effective signals. The previous approaches to weak fault detection, such as filter technique, empirical mode decomposition, singular value decomposition, and wavelet transform, all use a noise suppression viewpoint [3–5]. These signal processing methods based on noise cancellation, however, are inevitably interfering with useful signals that may even lead to compromised detection performance. In their investigation of paleometeorological variability, Benzi et al. initially put forth the stochastic resonance (SR) theory [6–8]. Researchers began to notice the phenomenon of noise-enhanced weak periodic excitation in ring lasers in the following years, and as a result, SR started to spark a research boom in a wide range of fields. For the first time, Hu et al. utilized SR to mechanical problem diagnostics in [9]. The traditional view of noise suppression has been challenged by the development of SR theory, and the notion of turning noise into valuable signal has captured the interest of many academics and researchers [10].

Over the past two decades, SR has been used in many different fields, including meteorology [11], pharmacology [12], nanomechanics [13–15], and medicine [16]. SR-aided signal processing is particularly useful for wireless communication [17], mechanical fault diagnosis [18], underwater target identification [19], image enhancement [20], and other applications. SR was found in [21] to improve the performance of the system, especially for the case of neural networks whose hidden layer consists of a threshold function. Nishiguchi and Fujiwara demonstrated that weak fault characteristic signals drowned in noise can be detected by a classical bistable stochastic resonance (CBSR) method performed by nano field effect transistors [22]. Li et al. proposed a noise-controlled second-order CBSR system method based on wavelet transform, and successfully implemented this method in wind turbine vibration signals [23]. Liu et al. applied SR to a lake eutrophic ecosystem and discussed in detail the effect of noise, time delay on the state transition of the lake in [24]. To overcome the restriction that CBSR can only be applied to periodic signals, Liu et al. concentrated on applying CBSR to binary pulse amplitude modulation (BPAM) signals [25]. He et al. proposed a detection method for non-collaborative SR for cognitive radio networks, which alleviates the signal-to-noise ratio (SNR) wall and the corresponding noise uncertainty problem of conventional energy detectors, especially in the case of low SNRs [26].

For the sake of practical engineering needs, time-delayed SR systems have attracted the attention of many scholars. In [27], Shi et al. proposed a time-delayed feedback tristable system and analyzed the effects of the time-delayed terms on the signal-to-noise ratio (SNR) and the probability density function (PDF), where they concluded that the time-delay can influence the SR phenomenon. Lu et al. [28] proposed a nonstationary weak signal detection strategy based on a time-delayed feedback SR model and demonstrated that it is applicable to the detection of strong nonlinear nonstationary signals. Mei et al. studied SR with time delay, correlated noise, and periodic signals using SNR theory to determine how time delay affected SR [29]. The SR problem in time-delayed bistable systems subject to Gaussian white noise was covered by He et al. in [30], along with an analysis of the impact of each parameter on the average first crossing time, the Shannon entropy, and the SNR.

The CBSR model has received the most attention among the several theoretical representations of SR. However, CBSR is only effective for small-amplitude signal identification due to the limitations of its own potential function. The frequent output saturation that CBSR encounters, which limits its ability to enhance performance in weak signal detection, is clearly described in [31–34]. In an effort to address this weakness of CBSR, Qiao et al. presented an adaptive unsaturated bistable stochastic resonance approach that, by adopting segmented potential walls with a fixed potential structure in [35] rather than the steep potential walls.

With the innovative development of SR, the model is not only limited to the CBSR, but also the multistable and monostable SR is gradually researched by scholars. Lu et al. investigated a tri-stable stochastic resonance (TSR) system and adaptively adjusted the parameters using particle swarm optimization (PSO) to obtain the best SNR and spectral output in [36]. Through simulation and actual fault signal detection, Lu et al. in [37] demonstrated the superiority of the proposed Woods-Saxon stochastic resonance (WSSR) method over the CBSR. Zhang et al. proposed a combined hybrid TSR system by combining WSSR and CBSR with simulations to demonstrate its effectiveness [38]. However, the TSR suggested in [38] is unable to controllably change the potential structure since the potential well depth and width of the CBSR change simultaneously with a single parameter. The two TSR models mentioned above have a significant impact on the potential structure when a parameter is changed [36,38]. Hence, building a novel controlled potential structure with an optimal control strategy to catch weak signals remains a difficult problem.

In this paper, we propose a novel SR model called controlled symmetry with Woods-Saxon stochastic resonance (CSwWSSR). Since the parameters of the WSSR model control the potential wells singly and explicitly, we construct a controlled TSR based on the WSSR with three potential wells controlled by different parameters. In order to establish the optimal control strategy with the highest SNR, we then confirm the ideal parameters for the CSwWSSR model in various noise environments by utilizing PSO in the scenarios of controlled potential well depth and controlled potential well width. The feasibility and superiority of CSwWSSR for practical engineering applications are further verified by comparing the results with the original signal and its constituent model through the detection experiments of simulation signal and bearing.

The rest of this essay is structured as follows. The CSwWSSR model and its components are introduced in Section 2, which is then followed by a brief explanation of the procedure for optimizing the CSwWSSR parameters. Section 3 confirms the usefulness of the proposed CSwWSSR-based weak fault detection in simulation signals, and presents an analysis of its performance under various noise intensities and various fault characteristic frequencies. In Section 4, We demonstrate the efficiency of the CSwWSSR model by applying it to the bearing experiments. Finally, we conclude this paper in Section 5.

2. The Proposed CSwWSSR

2.1. Introduction of CSwWSSR

Most of the studies on SR primarily investigated the classical bistable SR model based on the four-time reflection-symmetric potentials. In these models, a change in a single parameter may induce a large change in the shape of the potential function. However, the shape of Woods-Saxon potential illustrated in Equation (1) can be adjusted separately or combinedly with the parameters H, W, and a. The potential well of WSSR is depicted in Figure 1 with steep potential well walls and a smooth bottom. Comparing the four curves in Figure 1, H is to control the depth of the potential well, W is to control the width of the potential well, and a is to determine the steepness of the potential well.

$$U_1(x) = -\frac{H}{1 + exp(\frac{|x| - W}{a})} \tag{1}$$

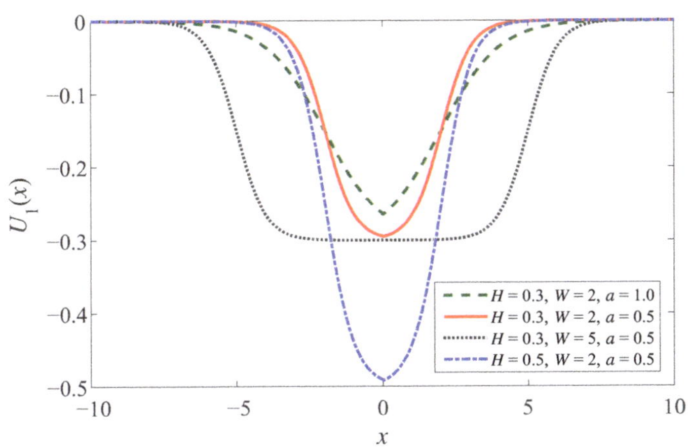

Figure 1. The potential function of WSSR.

Based on the previous analysis of WSSR, we propose a novel stochastic resonance model for weak fault detection, named CSwWSSR, whose potential function is expressed as

$$U(x) = U_1(x) + U_2(x) = -\frac{H}{1 + exp\left(\frac{|x|-W}{a}\right)} - \frac{1}{2}A\omega_0^2 x^2 + \frac{1}{4}B\beta x^4 \quad (2)$$

where $U_1(x)$ indicates the potential function of WSSR as Equation (1), and $U_2(x)$ is the potential function of controlled symmetric stochastic resonance (CSSR), which may be represented as Equation (3). The parameters of $U(x)$ are (H, W, a, A, B, ω_0^2, β), where (H, W, a) are the parameters of the WSSR and (A, B, ω_0^2, β) are the parameters of the CSSR, all of which have real values higher than 0.

$$U_2(x) = -\frac{1}{2}A\omega_0^2 x^2 + \frac{1}{4}B\beta x^4 \quad (3)$$

Further, CSSR is defined as the controlled potential well depth and the controlled potential well width. In the case of the controlled well depth, let $A = B = k$. Then, through analyzing the potential function Equation (3), two minima, the potential wells, are located at $x_m = \pm\sqrt{\omega_0^2/\beta}$. The potential barrier is located at $x = 0$, with a height of $k\omega_0^4/4\beta$. So the parameter k is independent of the positions of the potential wells and the potential barrier, which are only related to ω_0^2 and β. For the ease of analysis, we assume that $\omega_0^2 = \beta = 1$, Figure 2 shows the effect of different k on the potential well depth, where k is taken as 0.5, 1, and 1.5, respectively. It is not tough to find that the depth of the two potential wells gradually increases and the walls of the potential wells gradually steepen as k increases [39–41].

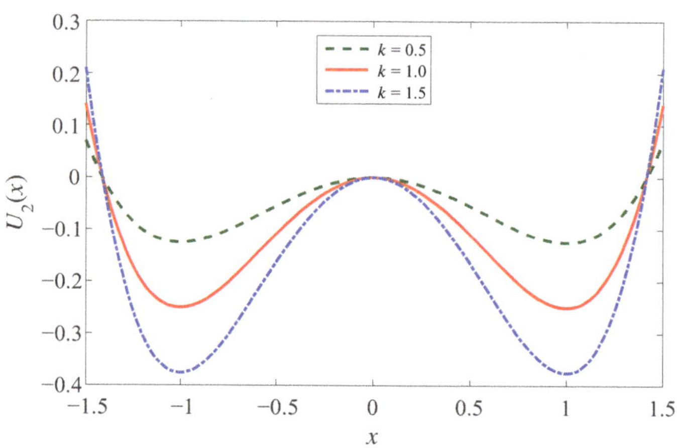

Figure 2. The potential function of CSSR in controlled depth case.

Under the circumstance of the controlled well width, let $A = 1/k^2$, and $B = 1/k^4$. Then, analyzing the potential function Equation (3), the potential barrier of Equation (3) exists at $x = 0$, which is similar to the controlled well depth. The difference is that the potential wells are at $x_m = \pm k\sqrt{\omega_0^2/\beta}$, where the depth of potential well is $\omega_0^4/4\beta$ and the width of the potential well is $k\sqrt{\omega_0^2/\beta}$. It is uncomplicated to note that the depth of the potential well is only related to ω_0^2 and β, and k controls the width of the potential well. Consistent with the above, we assume that $\omega_0^2 = \beta = 1$, Figure 3 shows the effect of different k on the potential well width, where k is taken respectively as 0.5, 1, and 1.5. With the increase of k, the depth of both potential wells remains constant, the width of the potential well increases, and the bottom of the potential well becomes smoother [42–45].

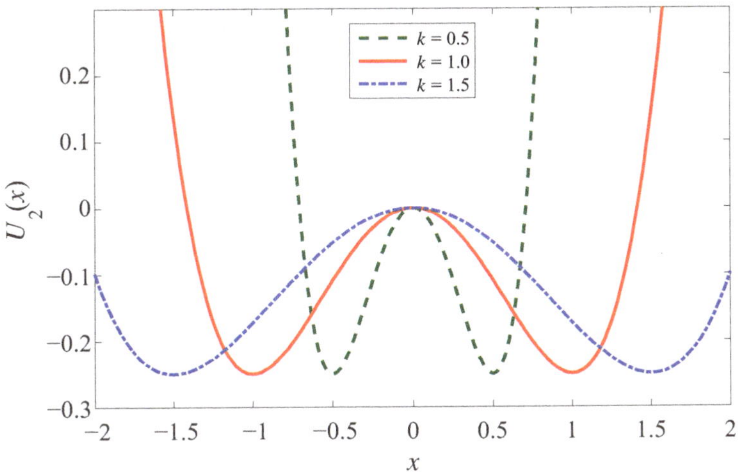

Figure 3. The potential function of CSSR in controlled width case.

Since the potential function of CSwWSSR is generated in the combination of WSSR and CSSR where CSSR has two cases, i.e., the controlled potential well depth and the controlled potential well width, CSwWSSR should also be discussed in two cases. On the basis of the aforementioned analysis of CSSR, we find that ω_0^2 and β do not affect the potential function either in the controlled well depth or the controlled well width, so we assume that $\omega_0^2 = \beta = 1$. Accordingly, the total four parameters controlling CSwWSSR are (H, W, a, k) in any case.

The parameters (H, W, a) of the WSSR control the intermediate potential well of the CSwWSSR, which have the same impact on the potential function as for the WSSR, as shown in Figures 4 and 5. In the case of the controlled potential well depth, the depth of the intermediate potential well gradually increases with increasing H, the width of the well gradually increases with increasing W, and the steepness of the potential well wall gradually becomes slower with increasing a in Figure 4a–c. The potential wells on the left and right sides depend on parameter k of the CSSR, which can be simply realized that the well depth increases with the increase of k in Figure 4d.

For the example of the controlled well width, Figure 5 shows the variation of the CSwWSSR potential function with each parameter. Similar to the controlled well depth, in Figure 5a–c, the depth and width of the intermediate potential well are proportional to the parameters H and W, as well as the steepness of the potential well walls is inversely proportional to the parameter a. Differently, k controls the width of the potential wells on both sides, which is positive to the width in Figure 5d. Comparing Figures 4a and 5a, the potential function is exactly the same for both cases when $k = 1$. Surprisingly, it is noteworthy that the CSwWSSR potential function will resemble that of the bistable state. After the analysis of the CSwWSSR, it contains both the characteristic of steep potential well walls in WSSR that allow for stable particle motion and the characteristic of controlled adjustment of the potential wells on both sides in CSSR. More importantly, the CSwWSSR is strongly adjustable, thus it has a significant impact on the SR efficacy and helps the system produce a strong output.

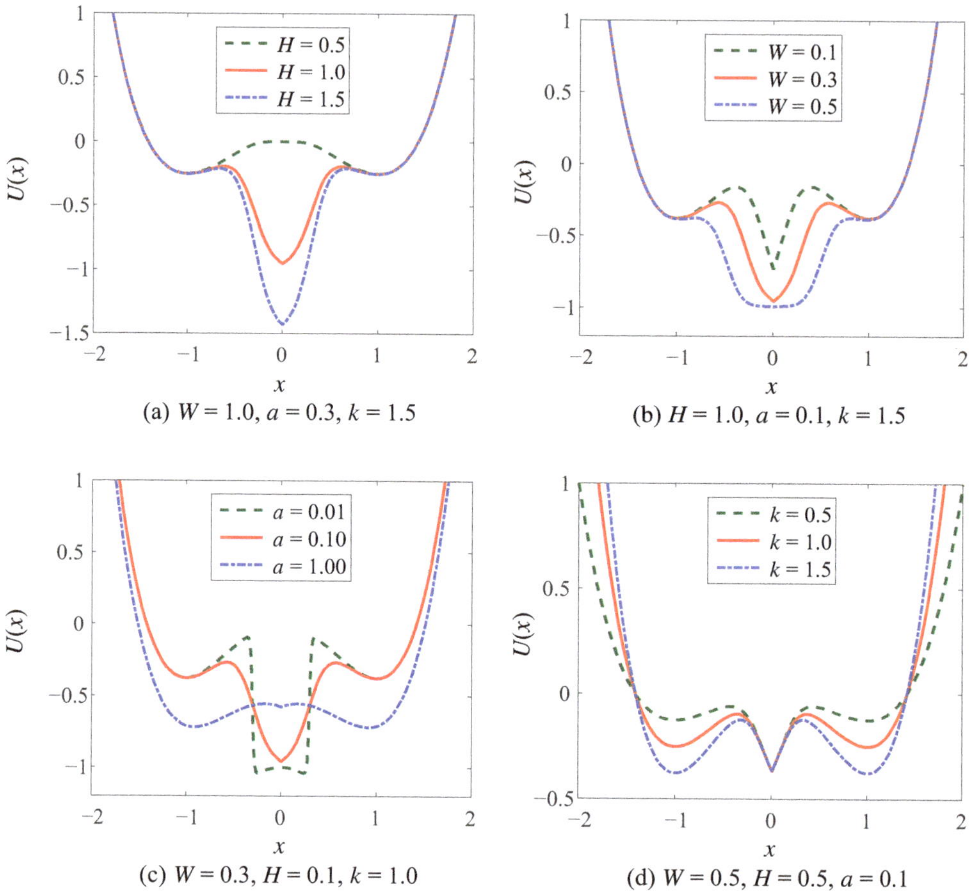

Figure 4. The potential function of CSwWSSR in controlled depth case.

2.2. Optimization of CSwWSSR Parameters

In light of the analysis of the CSwWSSR in Section 2.1, We draw the conclusion that the CSwWSSR depends on four parameters (H, W, a, k). In order to optimize the output, the particle swarm optimization (PSO) [46–48] is used to make the best match between the system parameters and the input signal. Figure 6 provides an illustration of the CSwWSSR model for weak fault detection using PSO. Here, the objective function of the PSO is output SNR from [35], which is defined as follows

$$SNR = 10\log_{10}\left(\frac{A_d}{\sum_{i=1}^{N/2} A_i - A_d}\right) \tag{4}$$

where A_d denotes the amplitude of the fault characteristic frequency in the output power spectrum, $\sum_{i=1}^{N/2} A_i - A_d$ represents the power amplitude sum of noise in the output power spectrum, and N stands for the length of the signal. A larger output SNR indicates superior performance in weak signal detection.

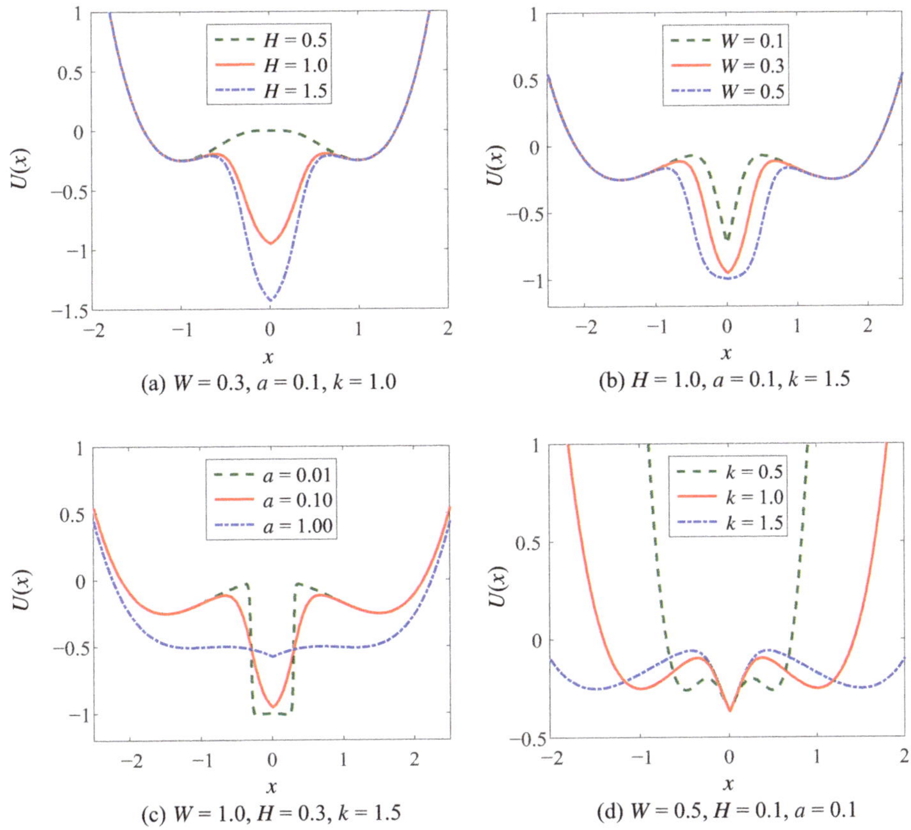

Figure 5. The potential function of CSwWSSR in controlled width case.

Step 1. Preprocess input signal. The adiabatic approximation theory, which stipulates that the driving frequency, signal amplitude, and noise intensity should all be considerably less than 1, was the basis for the initial proposal of SR. However, the majority of real signals are high frequency signals in the presence of severe noise, which does not agree with the theory of adiabatic approximation. A rescaling method was used in [27] to circumvent the restriction that SR is only applied to small parameter signals in an effort to more effectively apply it to numerous fields. The specific method first scales down the high frequency signal to get it into the adiabatic approximation theory's narrow parameter region, where SR is possible. The compressed signal is then transformed using SR, and finally the output signal is converted back to a high frequency signal with the same scale.

Step 2. Optimize parameters. The PSO process is shown in the dotted box in Figure 6. First, we need to initialize the PSO, including the population size, the maximum generations of iterations, the learning factor, the initial position of particles, the individual optimal position, the population optimal position, etc. Then, the fitness corresponding to the current position is calculated, i.e., the output SNR. In order to prevent the algorithm from entering a local optimum, the step to update the particle positions involves the use of a dynamic inertia factor whose value is correlated with the number of current iterations [49]. Afterwards, the fitness of the updated position is evaluated against the population and individual optimums. The updated position will take the place of the optimal position when the updated fitness is outperforming the individual optimal fitness and the population optimal fitness. The population optimal position is produced after multiple iterations.

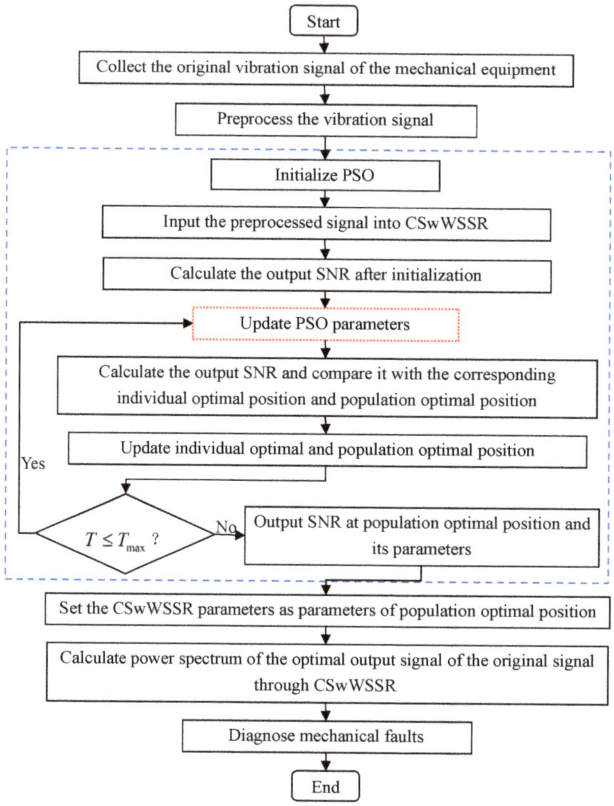

Figure 6. Process flow of CSwWSSR for weak fault detection through PSO.

Step 3. Calculate the CSwWSSR. The optimal combination of parameters output in the previous step is substituted into CSwWSSR to calculate the output signal. Due to the fact that the SR is based on a phenomenon generated by a nonlinear system, the analytical and approximate analytical methods are no longer applicable. As a consequence, a numerical method is usually used to solve the problem, i.e., fourth-order Runge-Kutta [36]. The solution method is as follows

$$\begin{cases} k_1 = \Delta * func\_CSwWSSR(x_{in}(i), x_{out}(i), H, W, a, k) \\ k_2 = \Delta * func\_CSwWSSR(x_{in}(i), x_{out}(i) + k_1/2, H, W, a, k) \\ k_3 = \Delta * func\_CSwWSSR(x_{in}(i+1), x_{out}(i) + k_2/2, H, W, a, k) \\ k_4 = \Delta * func\_CSwWSSR(x_{in}(i+1), x_{out}(i) + k_3, H, W, a, k) \\ x_{out}(i+1) = x_{in}(i) + \frac{1}{6}(k_1 + 2k_2 + 2k_3 + k_4) \end{cases} \quad (5)$$

in which $x_{in}(i)$ is the i-th sample value of the input signal after summing the vibration signal and the noise signal, and $x_{out}(i)$ is the i-th sample value of the output signal. Δ is employed to represent the step size, which is scaled from $1/f_s$ to $1/f_{s\partial} = 1/(f_s/\partial)$ when using the rescaling method (∂ is scale compression ratio, f_s is the frequency of sample, $f_{s\partial}$ is two sampling frequency). $func\_CSwWSSR()$ stands for the Langevin equation for CSwWSSR, expressed as follows

$$\begin{cases} func\_CSwWSSR = \frac{dx}{dt} = -\frac{dU(x)}{dt} + s(t) + \xi(t) \\ \frac{dU(x)}{dt} = \frac{H}{a}\mathrm{sgn}(x)exp\left(\frac{|x|-W}{a}\right)\left(1 + exp\left(\frac{|x|-W}{a}\right)\right)^{-2} - A\omega_0^2 x + B\beta x^3 \end{cases} \quad (6)$$

where $s(t)$ and $\xi(t)$ are vibration signal and noise respectively, $(H, W, a, A, B, \omega_0^2, \beta)$ are the parameters of CSwWSSR. We analyzed in Section 2.1 to know that ω_0^2 and β have no effect on the potential structure and set $\omega_0^2 = \beta = 1$. Both A and B can be represented by k in different cases, so the $func\_CSwWSSR()$ only contains four parameters (H, W, a, k).

Step 4. Diagnose faults. A power spectrum is created from the output signal via the Fourier transform to detect the weak fault and diagnose mechanical faults. In this work, the output SNR and the amplitude of the power spectrum at the fault characteristic frequency are observed and compared to assess the model's outstanding performance.

3. Simulation

3.1. Output Analysis of Simulation Signal

To illustrate the effectiveness of the proposed CSwWSSR model, the simulation signal is imported into the CSwWSSR model's weak signal detection for fault diagnosis. The simulation signal can be represented as

$$x(t) = A\sin(2\pi f_0 t) + \xi(t) \tag{7}$$

where $\xi(t)$ is additive Gaussian white noise with $\langle \xi(t)\xi(t+\tau)\rangle = 2D\xi(\tau)$ where noise intensity D = 5 and $\langle\bullet\rangle$ represent statistical average operator. A denotes the amplitude of the simulation signal with A = 0.25. To be specific, the fault characteristic frequency f_0 = 140 Hz, the sampling frequency f_s = 10,000 Hz, and the data length N = 4000. To satisfy the requirement of adiabatic approximation theory to achieve small parameters of SR, the scale compression ratio ∂ = 2000, then the step size $\Delta = 1/(f_s/\partial) = 0.2$. Figure 7 displays the simulation signal's time domain waveform and power spectrum. It is obvious that the input signal without SR processing is completely drowned in the surrounding noise at the fault characteristic frequency f_0 = 140 Hz in the frequency domain diagram, so detecting the fault of simulation signal is challenging. Moreover, the waveform of the signal is not visible in Figure 7a.

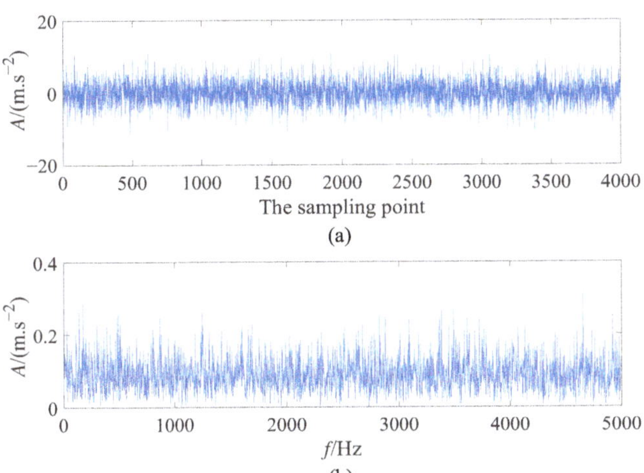

Figure 7. The input signal: (**a**) time domain diagram and (**b**) frequency domain diagram.

Aiming to demonstrate the fault characteristic detection capability of CSwWSSR, three different models are used to process the above simulation signal, that is WSSR, CSSR, and CSwWSSR. In addition, since CSSR and CSwWSSR are divided into controlled well depth and controlled well width, the input signal is subjected to five experiments. For the sake of fairness, the PSO is used for all five experiments to optimize parameters, where the population size and the number of iterations are both set to 50. After the PSO, Table 1 lists the optimized parameter combinations and output SNR, and Figure 8 provides their corresponding output signal time domain diagram and frequency domain diagram.

Table 1. Parameter combination and output SNR of each model after optimization of simulation signal.

Model	*H*	*W*	*a*	*k*	SNR/dB
WSSR	8.5624	6.2607	3.5634	-	−10.16
CSSR-depth	-	-	-	0.0417	−10.85
CSSR-width	-	-	-	1.6886	−11.63
CSwWSSR-depth	5.5987	5.1328	0.6898	0.0015	−9.863
CSwWSSR-width	1.6607	2.3355	0.1399	3.3592	−9.777

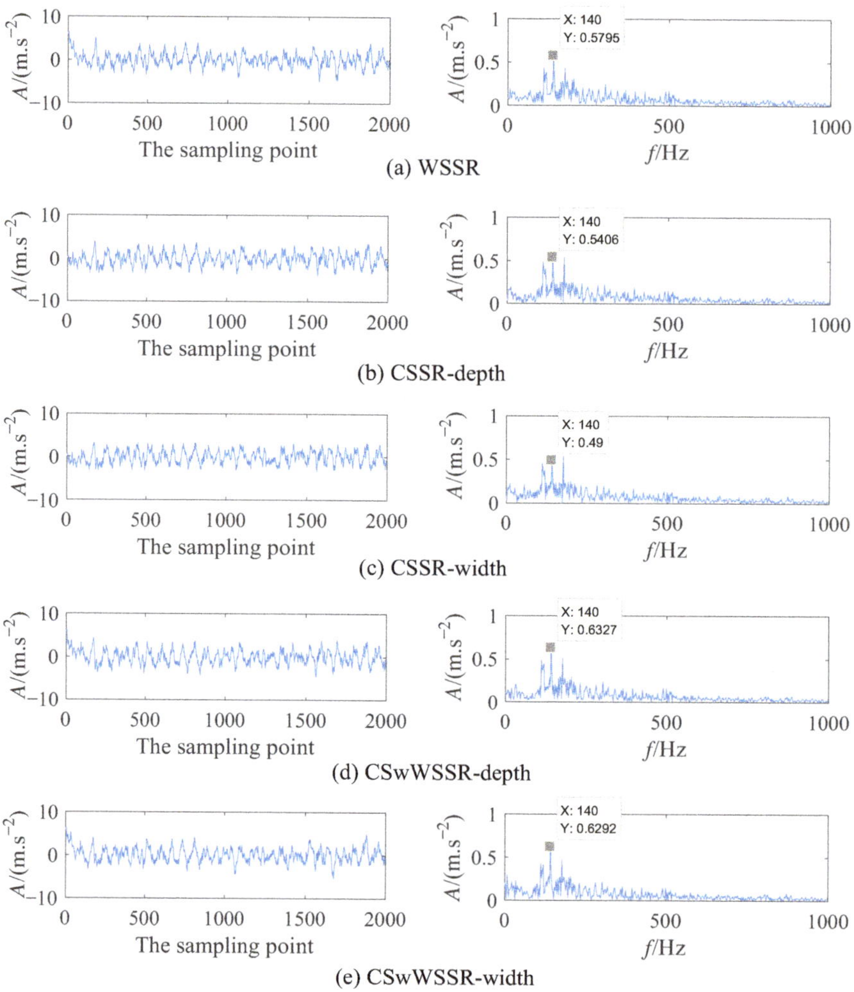

Figure 8. The output signal time-frequency diagram of each model in simulation signal. (**Left** column: Time domain diagram; **Right** column: Frequency domain diagram).

From Table 1, in view of the output SNR of the above simulation signal, it is seen clearly that the output SNR is: −9.777 dB (CSwWSSR-width) > −9.863 dB (CSwWSSR-depth) >

−10.16 dB (WSSR) > −10.85 dB (CSSR-depth) > −11.63 dB (CSSR-width). In Figure 8a, the amplitude of the output signal after WSSR processing at the fault characteristic frequency of 140 Hz is 0.5795. From Figure 8b,c, the amplitude of the output signal after CSSR processing at the fault characteristic frequency is 0.5406 and 0.49. It can be viewed that the amplitude of the output signal at the fault characteristic frequency after CSwWSSR processing is 0.6327 and 0.6292 in Figure 8d,e. The amplitude of fault characteristic frequency is 0.6327 (CSwWSSR-depth) > 0.6292 (CSwWSSR-width) > 0.5795 (WSSR) > 0.5406 (CSSR-depth) > 0.49 (CSSR-width). So the fault characteristic extraction capability is CSwWSSR > WSSR > CSSR. Furthermore, Figure 8b,c show that the fault characteristic frequency is more significantly influenced by the neighboring frequencies, which is evident from the fact that the amplitude of the non-fault characteristic frequency is nearly similar to the fault characteristic frequency after CSSR processing. Based on the analysis hereinbefore, the method of using the CSwWSSR model to implement weak fault detection for simulation signal appeared more suitable.

3.2. Capability of Detecting Different Simulation Signals

To evaluate the capability of the proposed CSwWSSR model in detecting different fault signals, it is demonstrated by a series of simulation signals with different fault characteristic frequencies and noise intensities. Consistent with the above, Equation (7) is used as the input signal by simply changing its fault characteristic frequency f_0 and noise intensity D, in which the fault characteristic frequency f_0 is from 0.05 to 1500 Hz and the noise intensity D is from 1 to 6. In Figure 9a, let the sampling points $N = 4000$, the signal amplitude $A = 0.25$, the sampling frequency $f_s = 10{,}000$ Hz, the fault characteristic frequency $f_0 = 140$ Hz, and the rescaling ratio $\partial = 1000$. In Figure 9b, let the sampling points $N = 4000$, the signal amplitude $A = 0.2$, the noise intensity $D = 2$, and the sampling frequency $f_s = f_0 * 100$. To meet the adiabatic approximation theory, the rescaling ratio $\partial = f_0 * 8$. As before, PSO was used to find the best parameter match with a population size and iteration number of 50 and 10, respectively. The average of the five final output SNRs was taken as the output SNR of the current signal. Figure 9 shows the variation of SNR with D and f_0 under three different models.

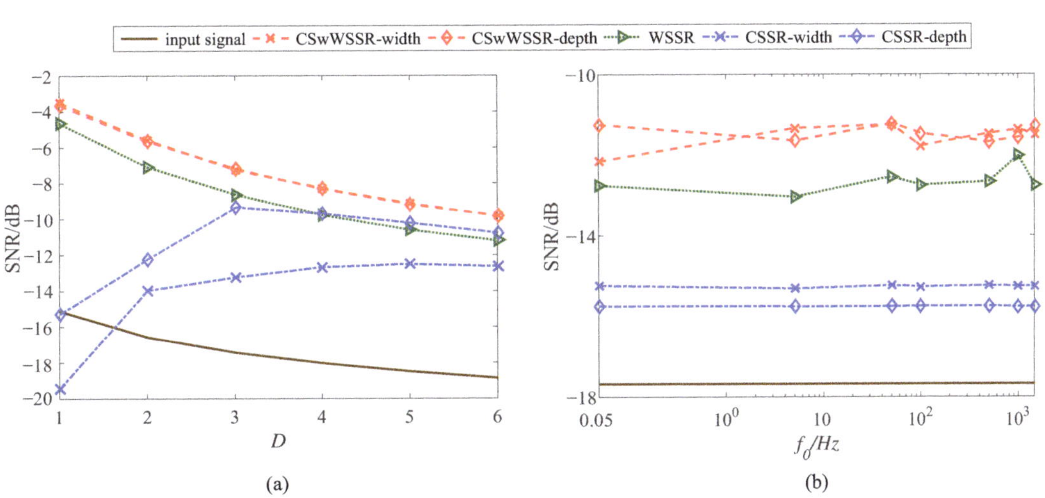

Figure 9. The variation of SNR with (**a**) D and (**b**) f_0.

We can apparently conclude CSwWSSR is preferable to the other two models in terms of its ability to detect fault signals, whether anti-noise or enhancement of high frequency signals in Figure 9. On the one hand, the improvement of CSSR by CSwWSSR is particularly significant, as can be seen that CSSR has no improvement effect on the input signal at low noise intensities, which is manifested by its output SNR being lower than the input SNR at low noise intensities. Although the enhanced performance of WSSR at high noise intensities is noticeably poorer and even worse than the output SNR of CSSR, it still outperforms input SNR in general. With the adequacies of CSSR and WSSR combined, CSwWSSR has an excellent ability to improve for both high and low noise intensities. On the other hand, the curves of the CSwWSSR are always higher than those of the other two models in Figure 9b for signals of various frequencies. Focusing on the above comparison, the suggested CSwWSSR shows more enhancement capability than CSSR and WSSR for the same input signal, different fault characteristic frequencies and noise intensities.

4. Experimental Demonstration of Bearing

In the previous section, we demonstrated the effectiveness of the CSwWSSR model for simulation signals. Furthermore, it is applied to practical engineering applications in this section to demonstrate effectiveness of the CSwWSSR model. Here, we adopt the data of rolling bearing provided by Case Western Reserve University, whose sampling frequency is 12 kHz, and speed is 1796 r/min. Taking the outer loop fault as an example, the theoretical value of the outer loop fault frequency is calculated by Equation (8) which is 107.30 Hz. The parameters needed to calculate the fault frequency are shown in Table 2.

$$f_i = \frac{n_1 N_1}{120}\left(1 - \frac{d_1 \cos\beta_1}{D_1}\right) \tag{8}$$

Table 2. The main structure parameters of the tested bearing 6205-2RSJEM.

Inner Diameter/mm	Outer Diameter/mm	D_1/mm	n_1	d_1/mm	Contact Angle/(°)
25.001	51.999	39.040	9.000	7.940	0

In Equation (8), n_1 represents the number of rolling elements, N_1 stands for the rotation rate, d_1 is employed to denote the rolling element diameter, β_1 is the bearing contact angle, and D_1 represents the pitch diameter. In this way, set the sampling point N = 3000 and the noise intensity D = 4. Since the fault characteristic frequency does not fulfill the adiabatic approximation theory, using a rescaling method is necessary, with two sampling frequency $f_{s\partial}$ = 6 Hz.

Figure 10 plots the time domain waveform and spectrum of the input signal. It is shown that, with the absence of noise, the spectrum of the input signal is concentrated at 2000–4000 Hz, which resulted in the failure to detect the fault characteristic frequency. Furthermore, the time domain diagram contains waveforms of periodic shocks and vibrations. To accomplish weak fault detection, additive noise is added to the input signal with the purpose of generating the SR phenomenon. As in Section 3.1, we use PSO to achieve the best match between the signal and the parameters. The best matches for the five cases and their corresponding output SNRs are given in Table 3. Comparing the output SNR for the five cases in Table 3, it is clear that output SNR: −8.996 dB (CSwWSSR-width) > −9.702 dB (CSwWSSR-depth) > −9.813 dB (CSSR-depth) > −9.911 dB (CSSR-width) > −11.21 dB (WSSR).

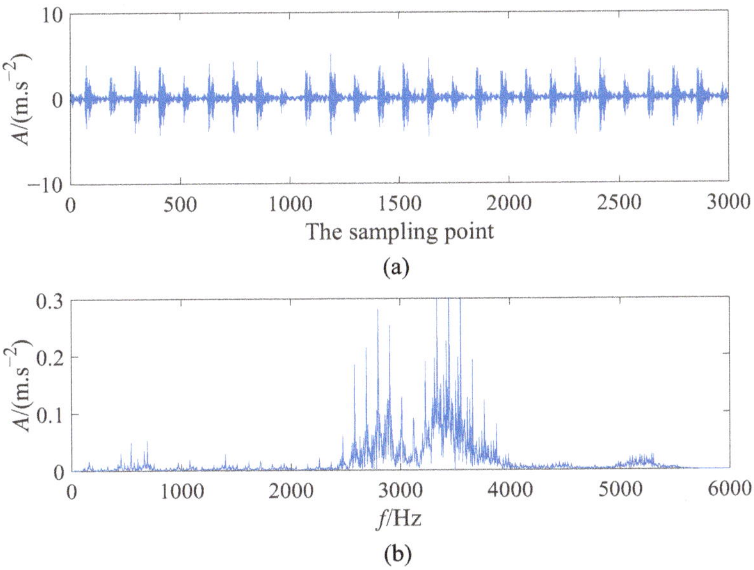

Figure 10. The input signal: (**a**) time domain diagram and (**b**) frequency domain diagram.

Table 3. Parameter combination and output SNR of each model after optimization of bearing.

Model	H	W	a	k	SNR/dB
WSSR	8.2734	3.8796	0.7568	-	−11.21
CSSR-depth	-	-	-	0.2897	−9.813
CSSR-width	-	-	-	1.1389	−9.911
CSwWSSR-depth	3.4026	3.4026	0.0367	0.1631	−9.702
CSwWSSR-width	3.2067	2.2868	0.0164	1.7698	−8.996

Figure 11 illustrates the time domain and frequency domain diagrams of the output signals for the five cases at the best match. 108 Hz is not hard to obtain from Figure 11, which is close to the fault characteristic frequency 107.30 Hz. In Figure 11a, the amplitude of the output signal after WSSR processing at the fault characteristic frequency of 108 Hz is 0.5423. From Figure 11b,c, the amplitude of the output signal after CSSR processing at the fault characteristic frequency is 0.4721 and 0.4431. It can be viewed that the amplitude of the output signal at the fault characteristic frequency after CSwWSSR processing is 0.5092 and 0.6009 in Figure 11d,e. The relationship between the amplitude at fault characteristic frequency in the five cases is clear: 0.6009 (CSwWSSR-width) > 0.5421 (WSSR) > 0.5092 (CSwWSSR-depth) > 0.4721 (CSSR-depth) > 0.4431 (CSSR-width). Although the fault characteristic frequency amplitude of WSSR is higher than this of CSwWSSR-depth, it is influenced by the low frequency, whose amplitude is almost equal to that of the fault characteristic frequency in Figure 11a. Taking into account, the fault characteristic extraction capability: CSwWSSR > CSSR > WSSR. Unlike the result in Section 3.1, CSSR outperforms WSSR in practical engineering applications. Consistently, CSwWSSR exhibits the excellent capability of weak fault detection. On account of the above analysis. CSwWSSR is also preferable to CSSR and WSSR in terms of effectiveness in practical engineering applications.

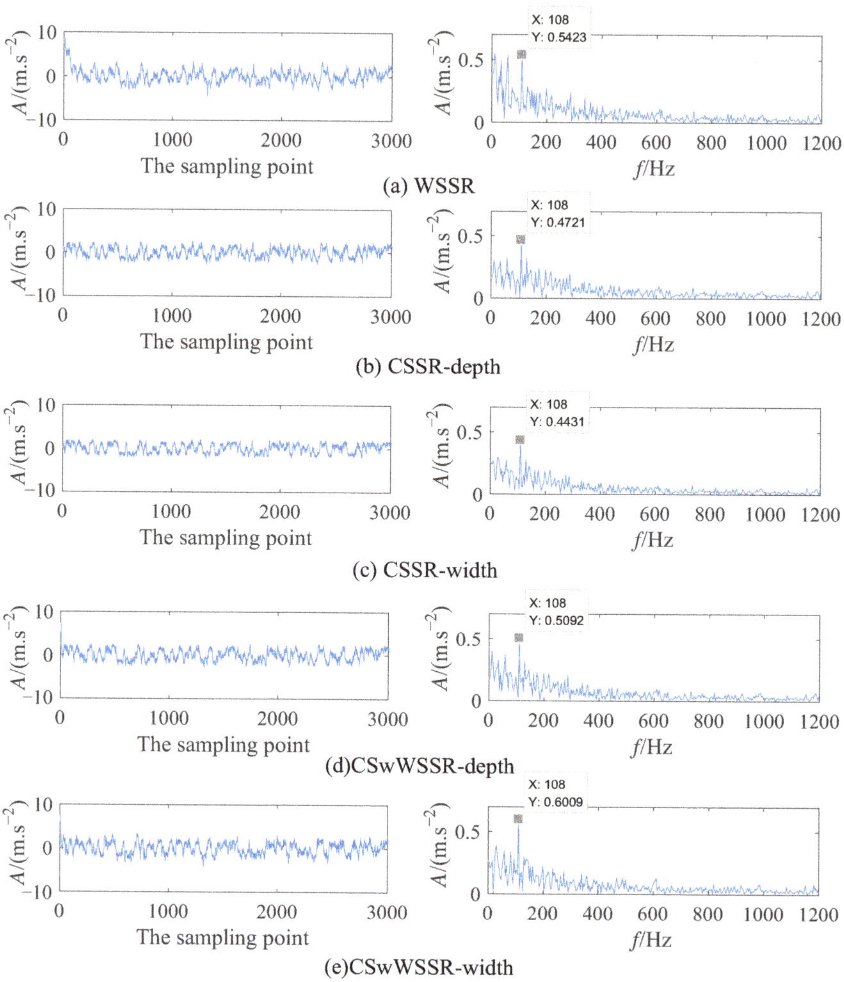

Figure 11. The output signal time-frequency diagram of each model in simulation signal. (**Left** column: Time domain diagram; **Right** column: Frequency domain diagram).

5. Conclusions

Linear optimal signal processing is typically less effective with a higher SNR in weak signal detection. However, because weak fault detection is regularly hindered by strong noise, the CSwWSSR model we proposed in this paper exploits noise to enhance weak signal detection. In addition to excellent noise utilization, it also exhibits notable features at the fault characteristic frequency, which is one of the benefits of combining WSSR and CSSR. What's more, by modifying the parameters, it may visibly alter the potential function's shape directly. The model parameters with the best SNR may be found for different input signals and noise thanks to an optimization algorithm. Then, comparative experiments are conducted for the controlled potential well depth and controlled potential well width to demonstrate the viability and superiority of the CSwWSSR model in fault characteristic detection. The following conclusions are drawn:

1. We analyze the potential functions of WSSR, CSSR, and CSwWSSR to investigate the effects of parameters on the potential structure and dynamical properties. The

intermediate potential well of the CSwWSSR is discovered to be controlled by the parameters (H, W, a), which have the same impact on the potential structure as the WSSR, and the potential wells on both sides depend on the parameter k. The CSwWSSR parameters are transparent to the control of the potential structure, which means that each parameter controls a different aspect of the potential structure.
2. The output SNR curves of WSSR, CSSR, and CSwWSSR were examined for various noise intensities and various fault characteristic frequencies, demonstrating how well CSwWSSR combines the benefits of WSSR and CSSR both in terms of anti-noise and enhancement of high frequency signals. It has both the features of stable particle motion in WSSR and the controlled adjustment of potential wells on both sides in CSSR.
3. We compare the output signal time domain and frequency domain diagrams of the WSSR, CSSR, and CSwWSSR after PSO optimized parameters in the simulation signal and bearing experiment. In terms of output SNR and amplitude of fault characteristic frequency, it is apparent that CSwWSSR outperforms CSSR and WSSR, which indicates that CSwWSSR has robust engineering applicability in the future.

In future work, we will consider the optimization problem of two-dimensional signals based on multistable stochastic resonance. In the low SNR environment, it will be a new challenge to make full use of SR techniques to overcome the spatial correlation of two-dimensional signals and recover two-dimensional signals disturbed by environmental noise. It is obvious that the results obtained in this study make a good foundation for the discussed future work.

Author Contributions: Methodology, J.L.; Validation, J.G.; Resources, B.H.; Data curation, C.T.; Writing—original draft, J.G.; Writing—review & editing, J.L.; Visualization, J.G. and W.Z.; Project administration, Q.Z. All authors have read and agreed to the published version of the manuscript.

Funding: This research is financially supported National Natural Science Foundation of China (62001210, 62201284), Natural Science Foundation of Jiangsu (BK20190789).

Institutional Review Board Statement: Not applicable.

Informed Consent Statement: Not applicable.

Data Availability Statement: Not applicable.

Conflicts of Interest: The authors declare no conflict of interest.

References

1. Gupta, P.; Pradhan, M.K. Fault detection analysis in rolling element bearing: A review. *Mater. Today Proc.* **2017**, *4*, 2085–2094. [CrossRef]
2. Wei, Y.; Li, Y.; Xu, M.; Huang, W. A review of early fault diagnosis approaches and their applications in rotating machinery. *Entropy* **2019**, *21*, 409. [CrossRef] [PubMed]
3. Duan, F.; Pan, Y.; Chapeau-Blondeau, F.; Abbott, D. Noise benefits in combined nonlinear bayesian estimators. *IEEE Trans. Signal Process.* **2019**, *67*, 4611–4623. [CrossRef]
4. Liu, J.; Hu, B.; Yang, F.; Zang, C.; Ding, X. Stochastic Resonance in a delay-controlled dissipative bistable potential for weak signal enhancement. *Commun. Nonlinear Sci. Numer. Simul.* **2020**, *85*, 105245. [CrossRef]
5. Qiao, Z.; Pan, Z. SVD principle analysis and fault diagnosis for bearings based on the correlation coefficient. *Meas. Sci. Technol.* **2015**, *26*, 085014. [CrossRef]
6. Xu, M.; Zheng, C.; Sun, K.; Xu, L.; Qiao, Z.; Lai, Z. Stochastic resonance with parameter estimation for enhancing unknown compound fault detection of bearings. *Sensors* **2023**, *23*, 3860. [CrossRef] [PubMed]
7. Benzi, R.; Sutera, A.; Vulpiani, A. The mechanism of stochastic resonance. *J. Phys. Math. Gen.* **1981**, *14*, L453–L457. [CrossRef]
8. Lu, S.; He, Q.; Wang, J. A review of stochastic resonance in rotating machine fault detection. *Sensors* **2019**, *116*, 230–260. [CrossRef]
9. Hu, N.; Chen, M.; Wen, X. The application of stochastic resonance theory for early detecting rub-impact fault of rotor system. *Mech. Syst. Signal Process.* **2003**, *17*, 883–895. [CrossRef]
10. Gammaitoni, L.; Hänggi, P.; Jung, P.; Marchesoni, F. Stochastic resonance: A remarkable idea that changed our perception of noise. *Eur. Phys. J. B* **2009**, *69*, 1–3. [CrossRef]
11. Caccamo, M.T.; Magazù, S. A physical–mathematical approach to climate change effects through stochastic resonance. *Climate* **2019**, *7*, 21. [CrossRef]

12. Wang, W.; Xiang, S.; Xie, S.; Xiang, B. An adaptive single-well stochastic resonance algorithm applied to trace analysis of Clenbuterol in human urine. *Molecules* **2012**, *17*, 1929–1938. [CrossRef] [PubMed]
13. Markina, A.; Muratov, A.; Petrovskyy, V.; Avetisov, V. Detection of single molecules using stochastic resonance of bistable oligomers. *Nanomaterials* **2020**, *10*, 2519. [CrossRef] [PubMed]
14. Zhang, Y.; Wang, Z.; Tazeddinova, D.; Ebrahimi, F.; Habibi, M.; Safarpour, H. Enhancing active vibration control performances in a smart rotary sandwich thick nanostructure conveying viscous fluid flow by a PD controller. *Waves Random Complex Media* **2021**, *33*, 1–24. [CrossRef]
15. Zare, R.; Najaafi, N.; Habibi, M.; Ebrahimi, F.; Safarpour, H. Influence of imperfection on the smart control frequency characteristics of a cylindrical sensor-actuator GPLRC cylindrical shell using a proportional-derivative smart controller. *Smart Struct. Syst.* **2020**, *26*, 469–480. [CrossRef]
16. Ko, L.-W.; Chikara, R.K.; Chen, P.-Y.; Jheng, Y.-C.; Wang, C.-C.; Yang, Y.-C.; Li, L.P.-H.; Liao, K.-K.; Chou, L.-W.; Kao, C.-L. Noisy galvanic vestibular stimulation (stochastic resonance) changes electroencephalography activities and postural control in patients with bilateral vestibular hypofunction. *Brain Sci.* **2020**, *10*, 740. [CrossRef] [PubMed]
17. Li, Q.; Li, Z. A novel sequential spectrum sensing method in cognitive radio using suprathreshold stochastic resonance. *IEEE Trans. Veh. Technol.* **2013**, *63*, 1717–1725. [CrossRef]
18. Qiao, Z.; Lei, Y.; Li, N. Applications of stochastic resonance to machinery fault detection: A review and tutorial. *Mech. Syst. Signal Process.* **2019**, *122*, 502–536. [CrossRef]
19. Dong, H.; He, K.; Shen, X.; Ma, S.; Wang, H.; Qiao, C. Adaptive intrawell matched stochastic resonance with a potential constraint aided line enhancer for passive sonars. *Sensors* **2020**, *20*, 3269. [CrossRef]
20. Liu, J.; Hu, B.; Wang, Y. Optimum adaptive array stochastic resonance in noisy grayscale image restoration. *Phys. Lett. A* **2019**, *383*, 1457–1465. [CrossRef]
21. Ikemoto, S.; DallaLibera, F.; Hosoda, K. Noise-modulated neural networks as an application of stochastic resonance. *Neurocomputing* **2018**, *277*, 29–37. [CrossRef]
22. Nishiguchi, K.; Fujiwara, A. Detecting signals buried in noise via nanowire transistors using stochastic resonance. *Appl. Phys. Lett.* **2012**, *101*, 193108. [CrossRef]
23. Li, J.; Chen, X.; Du, Z.; Fang, Z.; He, Z. A new noise-controlled second-order enhanced stochastic resonance method with its application in wind turbine drivetrain fault diagnosis. *Renew. Energy* **2013**, *60*, 7–19. [CrossRef]
24. Liu, J.; Guo, J.; Ding, X.; Qiao, Z.; Zang, C. Colored correlated noises induced regime shifts in a time-delayed lake eutrophication ecosystem. *Front. Sustain. Dev.* **2022**, *2*, 24–37. [CrossRef]
25. Liu, J.; Li, Z.; Guan, L.; Pan, L. A novel parameter-tuned stochastic resonator for binary PAM signal processing at low SNR. *IEEE Commun. Lett.* **2014**, *18*, 427–430. [CrossRef]
26. He, D.; Chen, X.; Pei, L.; Jiang, L.; Yu, W. Improvement of noise uncertainty and signal-to-noise ratio wall in spectrum sensing based on optimal stochastic resonance. *Sensors* **2019** , *19*, 841. [CrossRef]
27. Shi, P.; Yuan, D.; Han, D.; Zhang, Y.; Fu, R. Stochastic resonance in a time-delayed feedback tristable system and its application in fault diagnosis. *J. Sound Vib.* **2018**, *424*, 1–14. [CrossRef]
28. Lu, S.; He, Q.; Zhang, H.; Kong, F. Enhanced rotating machine fault diagnosis based on time-delayed feedback stochastic resonance. *J. Vib. Acoust.* **2015**, *137*, 051008. [CrossRef]
29. Mei, D.C.; Du, L.C.; Wang, C.J. The effects of time delay on stochastic resonance in a bistable system with correlated noises. *J. Stat. Phys.* **2009**, *137*, 625–638. [CrossRef]
30. He, M.; Xu, W.; Sun, Z. Dynamical complexity and stochastic resonance in a bistable system with time delay. *Nonlinear Dyn.* **2015**, *79*, 1787–1795. [CrossRef]
31. Liao, Z.; Wang, Z.; Yamahara, H.; Tabata, H. Echo state network activation function based on bistable stochastic resonance. *Chaos Solitons Fractals* **2021**, *153*, 111503. [CrossRef]
32. Lai, Z.H.; Wang, S.B.; Zhang, G.Q.; Zhang, C.L.; Zhang, J.W. Rolling bearing fault diagnosis based on adaptive multiparameter-adjusting bistable stochastic resonance. *Shock Vib.* **2020**, *2020*, 1–15. [CrossRef]
33. Liao, Z.; Wang, Z.; Yamahara, H.; Tabata, H. Low-power-consumption physical reservoir computing model based on overdamped bistable stochastic resonance system. *Neurocomputing* **2022**, *468*, 137–147. [CrossRef]
34. Zhang, X.; Wang, J.; Liu, Z.; Wang, J. Weak feature enhancement in machinery fault diagnosis using empirical wavelet transform and an improved adaptive bistable stochastic resonance. *ISA Trans.* **2019**, *84*, 283–295. [CrossRef] [PubMed]
35. Qiao, Z.; Lei, Y.; Lin, J.; Jia, F. An adaptive unsaturated bistable stochastic resonance method and its application in mechanical fault diagnosis. *Mech. Syst. Signal Process.* **2017**, *84*, 731–746. [CrossRef]
36. Lu, J.; Huang, M.; Yang, J.-J. A novel spectrum sensing method based on tri-stable stochastic resonance and quantum particle swarm optimization. *Wirel. Pers. Commun.* **2017**, *95*, 2635–2647. [CrossRef]
37. Lu, S.; He, Q.; Kong, F. Stochastic resonance with Woods-Saxon potential for rolling element bearing fault diagnosis. *Mech. Syst. Signal Process.* **2014**, *45*, 488–503. [CrossRef]
38. Zhang, G.; Li, H. Hybrid tri-stable stochastic resonance system used for fault signal detection. *J. Vib. Shock* **2019**, *38*, 9–17.
39. Khan, M.B.; Santos-García, G.; Noor, M.A.; Soliman, M.S. Some new concepts related to fuzzy fractional calculus for up and down convex fuzzy-number valued functions and inequalities. *Chaos Solitons Fractals* **2022**, *164*, 112692. [CrossRef]

40. Khan, M.B.; Santos-García, G.; Noor, M.A.; Soliman, M.S. New Hermite–Hadamard inequalities for convex fuzzy-number-valued mappings via fuzzy riemann integrals. *Mathematics* **2022**, *10*, 3251. [CrossRef]
41. Khan, M.B.; Othman, H.A.; Santos-García, G.; Saeed, T.; Soliman, M.S. On fuzzy fractional integral operators having exponential kernels and related certain inequalities for exponential trigonometric convex fuzzy-number valued mappings. *Chaos Solitons Fractals* **2023**, *169*, 113274. [CrossRef]
42. Macías-Díaz, J.E.; Khan, M.B.; Noor, M.A.; Abd Allah, A.M.; Alghamdi, S.M. Hermite-Hadamard inequalities for generalized convex functions in interval-valued calculus. *AIMS Math.* **2022**, *7*, 4266–4292. [CrossRef]
43. Khan, M.B.; Treanta, S.; Soliman, M.S. Generalized preinvex interval-valued functions and related Hermite–Hadamard type inequalities. *Symmetry* **2022**, *14*, 1901. [CrossRef]
44. Khan, M.B.; Noor, M.A.; Noor, K.I.; Chu, Y.-M. Higher-order strongly preinvex fuzzy mappings and fuzzy mixed variational-like inequalities. *Int. J. Comput. Intell. Syst.* **2021**, *14*, 1856. [CrossRef]
45. Khan, M.B.; Noor, M.A.; Noor, K.I.; Chu, Y.-M. New Hermite–Hadamard-type inequalities for -convex fuzzy-interval-valued functions. *Adv. Differ. Equ.* **2021**, *149*, 1-20 . [CrossRef]
46. Wang, D.; Tan, D.; Liu, L. Particle swarm optimization algorithm: An overview. *Soft Comput.* **2018**, *22*, 387–408. [CrossRef]
47. Xia, X.; Gui, L.; Yu, F.; Wu, H.; Wei, B.; Zhang, Y.; Zhan, Z. Triple archives particle swarm optimization. *IEEE Trans. Cybern.* **2019**, *50*, 4862–4875. [CrossRef]
48. Shami, T.M.; El-Saleh, A.A.; Alswaitti, M.; Al-Tashi, Q.; Summakieh, M.A.; Mirjalili, S. Particle swarm optimization: A comprehensive survey. *IEEE Access.* **2022**, *10*, 10031–10061. [CrossRef]
49. Tian, D.; Shi, Z. MPSO: Modified particle swarm optimization and its applications. *Swarm Evol. Comput.* **2018**, *41*, 49–68. [CrossRef]

Disclaimer/Publisher's Note: The statements, opinions and data contained in all publications are solely those of the individual author(s) and contributor(s) and not of MDPI and/or the editor(s). MDPI and/or the editor(s) disclaim responsibility for any injury to people or property resulting from any ideas, methods, instructions or products referred to in the content.

sensors

MDPI

Article

High-Performance Adaptive Weak Fault Diagnosis Based on the Global Parameter Optimization Model of a Cascaded Stochastic Resonance System

Zhihui Lai [1,2], Zhangjun Huang [1,2], Min Xu [3], Chen Wang [1,2], Junchen Xu [1,2], Cailiang Zhang [4], Ronghua Zhu [4,*] and Zijian Qiao [5,*]

1 Shenzhen Key Laboratory of High Performance Nontraditional Manufacturing, College of Mechatronics and Control Engineering, Shenzhen University, Shenzhen 518060, China; laizh@szu.edu.cn (Z.L.)
2 Guangdong Key Laboratory of Electromagnetic Control and Intelligent Robots, College of Mechatronics and Control Engineering, Shenzhen University, Shenzhen 518060, China
3 Ningbo Cigarette Factory, China Tobacco Zhejiang Industry Co., Ltd., Ningbo 315040, China
4 Ocean College, Zhejiang University, Zhoushan 316021, China
5 School of Mechanical Engineering and Mechanics, Ningbo University, Ningbo 315211, China
* Correspondence: zhu.richard@zju.edu.cn (R.Z.); qiaozijian@nbu.edu.cn (Z.Q.)

Abstract: Stochastic resonance (SR), as a type of noise-assisted signal processing method, has been widely applied in weak signal detection and mechanical weak fault diagnosis. In order to further improve the weak signal detection performance of SR-based approaches and realize high-performance weak fault diagnosis, a global parameter optimization (GPO) model of a cascaded SR system is proposed in this work. The cascaded SR systems, which involve multiple multi-parameter-adjusting SR systems with both bistable and tri-stable potential functions, are first introduced. The fixed-parameter optimization (FPO) model and the GPO models of the cascaded systems to achieve optimal SR outputs are proposed based on the particle swarm optimization (PSO) algorithm. Simulated results show that the GPO model is capable of achieving a better SR output compared to the FPO model with rather good robustness and stability in detecting low signal-to-noise ratio (SNR) weak signals, and the tri-stable cascaded SR system has a better weak signal detection performance compared to the bistable cascaded SR system. Furthermore, the weak fault diagnosis approach based on the GPO model of the tri-stable cascaded system is proposed, and two rolling bearing weak fault diagnosis experiments are performed, thus verifying the effectiveness of the proposed approach in high-performance adaptive weak fault diagnosis.

Keywords: weak fault diagnosis; stochastic resonance; cascaded system; global parameter optimization; particle swarm optimization algorithm

1. Introduction

Condition monitoring and fault diagnosis of rotating mechanical components, such as bearings, gears, and rotors, are important to ensure the safe operation of equipment [1]. Vibration analysis and magnetic flux leakage detection [2] are common and effective technologies in the condition monitoring and fault diagnosis of machinery. In many cases, especially when the machine operates in poor working conditions or the sensors have to be installed far away from the fault source, the fault characteristics are always submerged in strong background noise and other interference components, making them difficult to be identified. Therefore, many weak signal detection methods, such as spectral kurtosis (SK) [3], empirical mode decomposition (EMD) [4,5], digital filter [6], wavelet transform (WT) [7], etc., have been widely investigated to extract weak fault signals from noisy backgrounds. However, these weak signal detection methods inevitably weaken the fault characteristic when filtering out the noise; hence, the weak signal detection performance is limited. Different from the conventional weak signal detection methods, stochastic resonance (SR)-based methods can utilize the noise (rather than eliminate the noise) to achieve weak signal detection, thus showing potential in weak fault diagnosis.

The concept of SR was first proposed by Benzi et al. [8,9] in 1981 when they studied the paleo-meteorological problem of alternating glacial and warm climate cycles. When SR occurs, a part of the noise energy is transferred to the low-frequency signal through a nonlinear system. Thus, the low-frequency weak signal can be significantly enhanced while filtering out the high-frequency noise, and the output signal-to-noise ratio (SNR) is significantly improved [10,11]. Due to the advantages of SR in weak signal detection, SR has attracted continuous attention in the field of weak fault diagnosis in the past 40 years [12–17].

In order to make SR-based methods suitable for practical engineering applications of mechanical fault diagnosis, many efforts have been made by researchers. To overcome the limitation of conventional SR that it can only process a signal with small parameters (e.g., a signal frequency that is much less than 1 Hz) due to the limitation of the adiabatic approximation theory [18], researchers have proposed many solutions to make SR capable of processing large-parameter signals, such as frequency re-scaling SR (FRSR) [19], frequency-shifted and re-scaling SR (FSRSR) [20], parameter-normalized SR (PNSR) [21], modulated SR (MSR) [22], and scale-transformation SR [23], among others. To appropriately select the system parameters to generate the optimal SR output for given signals, many adaptive parameter-tuning SR methods have been investigated [10,18,24], which are further applied in the fault diagnosis of planetary gearboxes [24], rolling bearing fault diagnosis [25,26], and rotor misalignment fault diagnosis [27].

The most important objective of SR-based fault diagnosis methods is to achieve a high-performance weak signal detection ability. The weak signal detection ability of an SR-based method is related to the potential function of the system [12]. It was reported that SR can be produced in nonlinear systems with monostable, bistable, tri-stable, and multi-stable potential functions [23,28–31]. Moreover, many novel forms of potential functions, such as the Woods–Saxon potential function proposed by Gao et al. [32], the pining potential function employed by Zhang et al. [33], the piecewise linear potential function employed by Li et al. [34], and the compound tri-stable potential function proposed by Tang et al. [1], have been proposed recently for the realization of SR and the successful application of weak signal detection. However, how to quantitively and reasonably evaluate the detection performance of an SR system with different potential functions is still an issue that needs to be solved.

Moreover, the weak signal detection ability of a single SR system is still limited [35], especially for weak signals with extremely low SNRs. Hence, some enhancement models based on the SR systems, such as the cascaded SR systems [36–38], coupled SR systems [29,39], and parallel SR systems [40,41], have been constructed to further enhance the weak signal detection ability of SR-based methods. Among these enhancement models, the cascaded SR models, which consist of several single SR systems connected in series, are of most interest. It is known that a single SR system acts as a low-pass filter that can transfer the energy from a high-frequency region to a low-frequency region, thus amplifying the low-frequency weak signal [42]. Hence, multiple filtering processes can be realized through a cascaded SR system, thus making the output signal smoother and making the characteristic signal more obvious. Therefore, the cascaded SR system may have a better weak signal detection performance compared to a single SR system.

Compared to a single SR system, the number of adjustable parameters of a cascaded system is significantly enhanced, and different parameter-adjusting strategies have been proposed for a cascaded system to realize SR output. Up to now, the so-called "fixed-parameter adjusting strategy" and "stage-by-stage parameter adjusting strategy" have been the most adopted. The fixed-parameter strategy means that the corresponding parameters in each subsystem are kept the same. Using this parameter-adjusting strategy, Hairan et al. [43] proposed a signal detection method under a dual-mode non-Gaussian noise background based on a cascaded SR. This parameter-adjusting strategy can be further combined with a multi-parameter adaptive optimization algorithm to find out the optimal SR output adaptively [44]. The stage-by-stage parameter-adjusting strategy means the optimal SR output of each subsystem is obtained by adjusting the parameters of this subsystem, and its SR output is input into the next subsystem (if necessary). Therefore, the parameters of each subsystem may be different. Using this strategy, Xiao et al. [45] proposed an adaptive cascaded vibration resonance method and applied it to rotating machinery fault diagnosis; Liu et al. [46] proposed an adaptive cascaded piecewise linear SR system for bearing fault diagnosis; Xu et al. [37] proposed an adaptive cascaded SR method using the quantum particle swarm optimization (QPSO) algorithm; and Zhang et al. [47] proposed a cascaded three-steady-state SR and applied it to rolling bearing fault diagnosis. However, it is noted that in most of the literature, this strategy is adopted for a cascaded system with only two subsystems, as the output SNR cannot continue to increase with more subsystems. Hence, the fixed-parameter-adjusting strategy can produce a better SR output compared to the stage-by-stage parameter-adjusting strategy. Furthermore, it can be easily understood that as the cascaded system is not treated as a system as a whole, the optimal SR output of the cascaded system cannot be found by either the "fixed-parameter adjusting strategy" or the "stage-by-stage parameter adjusting strategy". In contrast, a novel "global-parameter adjusting strategy", which means all parameters of all subsystems are adjusted jointly, has the potential to realize the optimal SR output for the whole cascaded system. This parameter-adjusting strategy, which has not been fully studied yet, is worthy to be further investigated.

Therefore, in this paper, the global parameter optimization model of a cascaded SR system is proposed for high-performance adaptive weak signal detection and weak fault diagnosis. As it can achieve the theoretically optimal SR output of a cascaded system, the output SNR of this model can be also used to quantitively evaluate the detection performance of SR systems with different potential functions. In this paper, the classical bistable and tri-stable potential functions are considered for demonstration. The rest of the paper is organized as follows. In Section 2, the background knowledge of the multi-parameter-adjusting SR is introduced, and both the fixed-parameter optimization (FPO) model and the global parameter optimization (GPO) model of cascaded SR systems are constructed with the help of the particle swarm optimization (PSO) algorithm. The weak signal detection performance of the single SR system and the FPO model and the GPO model of the cascaded systems with either bistable or tri-stable potential functions is fully studied and compared through numerical simulations in Section 3. Two rolling bearing weak fault diagnosis experiments are performed in Section 4 to demonstrate the effectiveness of the proposed GPO model of a cascaded system in weak signal detection and weak fault diagnosis. Conclusions are drawn in Section 5.

2. Weak Signal Detection Approaches Based on Adaptive Cascaded SR Systems

2.1. Multi-Parameter-Adjusting SR

SR is a nonlinear physical phenomenon reflecting the synergistic effect of the nonlinear system, the weak periodic signal, and the noise. When SR occurs, the weak periodic signal is enhanced by gaining energy from high-frequency noise. The mathematical model of an underdamped SR system, which describes the motion of a unit-mass Brownian particle in a potential field in the presence of periodic force and noise, can be written as:

$$\frac{dx^2(t)}{dt^2}+k\frac{dx(t)}{dt}=-\frac{dU(x)}{dx}+s(t)+n(t) \tag{1}$$

Here, k denotes the damping ratio; $U(x)$ denotes the potential field of the system, and $-dU(x)/dx$ is the potential field force applied to the particle; $s(t) = A\cos(2\pi f_0 t)$ represents a harmonic signal with amplitude A and frequency f_0; and $n(t) = \sqrt{2D}\xi(t)$ represents noise with an intensity of D, where $\xi(t)$ is zero-mean and unit-variance Gaussian white noise. In this system, $sn(t) = s(t) + n(t)$ is defined as the input signal, and $x(t)$ denotes the output signal, which can be obtained by numerically solving the equation.

The nonlinear system shown in Equation (1) can generate SR with many different potential fields. Among these, the bistable potential field and the tri-stable potential field are two classical potential fields, whose generalized potential functions can be written as follows [28,48]:

$$U_B(x) = -\frac{1}{2}a_1x^2 + \frac{1}{4}b_1x^4 \tag{2}$$

$$U_T(x) = \frac{1}{2}a_2x^2 - \frac{1}{4}b_2x^4 + \frac{1}{6}c_2x^6 \tag{3}$$

where the potential field parameters a_1, b_1, a_2, b_2, and c_2 all have positive values.

Examples of these two potential functions are plotted in Figure 1. It can be seen that the bistable potential field is symmetrical with two potential wells, and the tri-stable one is symmetrical with three potential wells.

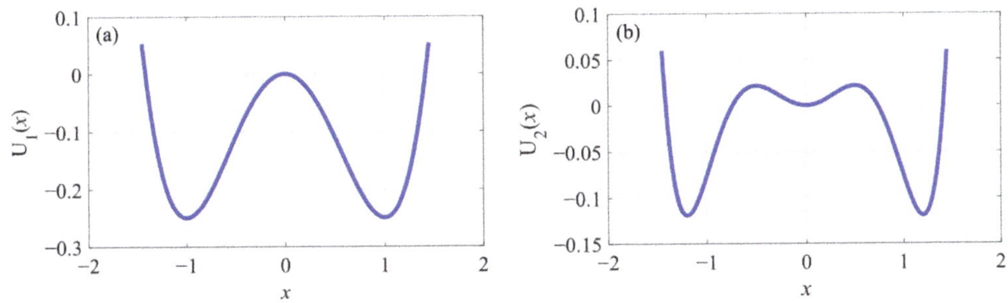

Figure 1. Potential functions of (**a**) the bistable potential field with $a_1 = 1$ and $b_1 = 1$ and (**b**) the tri-stable potential field with $a_2 = 0.36$, $b_2 = 1.69$, and $c_2 = 1.69$.

The output signal $x(t)$ of the SR system can be understood as the displacement of the Brownian particle moving in the potential field driven by the periodic signal and the random noise. When the signal is weak and noise is absent, i.e., $A < A_C$ and $D = 0$ (A_C indicates the critical amplitude for the particle to pass over the potential barrier to another potential well [48]), the Brownian particle can only oscillate within one of the potential wells. When noise is present ($D > 0$), the particle may pass over the barrier even though the signal is weak, as the particle can accumulate energy from high-frequency noise with the help of the nonlinear system. When the transition rate of the particle between potential wells accords with the signal period, the particle can realize regular inter-well oscillation, and SR occurs. Hence, the weak features of the input signal can be enhanced through SR and can be identified from the output signal $x(t)$. This advantage makes SR an advanced weak signal detection model, which has been applied in many fields including weak fault diagnosis.

Due to the limitation of the adiabatic approximation theory [23], the classical SR system shown in Equation (1) is only applicable to small-parameter conditions ($A < A_C$, $f_0 \ll 1$, and $D \ll 1$), which limits its applications in practical engineering. To solve this issue, a multi-parameter-adjusting SR system is further proposed by introducing an amplitude-transformation coefficient ε and a scale-transformation coefficient R to Equation (1), which can be written as [23]:

$$\frac{dx^2(t')}{dt'^2} + k\frac{dx(t')}{dt'} = -\frac{dU(x)}{dx} + \varepsilon{\cdot}sn(t') \tag{4}$$

Here, ε is used to linearly transform the amplitude of the input signal to be in an appropriate small-parameter range, and $t' = Rt$ is the times series after scale transformation, where R is used to transform the frequency/time scale of the input signal to meet the small-parameter conditions. The scale transformation can be simply realized by adopting a time step of $h' = R/F_s$ instead of $h = 1/F_s$ in the calculation when numerically solving the mathematic models of the SR systems [23], where F_s is the sampling frequency of the input signal. Thus, the frequency of the characteristic signal can be regarded as $f_0{}' = f_0/R$ in the calculations, and the small-frequency condition of $f_0{}' \ll 1$ can be realized by appropriately adjusting the value of R. Hence, SR can be achieved under large-parameter conditions using the multi-parameter-adjusting SR system.

2.2. Models of Cascaded SR Systems

The SR system has shown its potential in weak signal detection, but its detection ability for a weak signal with an extremely low SNR is still limited, which should be further improved. One possible solution is to construct a cascaded SR system, which represents multiple SR systems connected in series, as shown in Figure 2. Note that the number of connected subsystems is defined as N in this work. Previous research shows that as the low-frequency weak signal can continuously accumulate energy from high-frequency signal and noise through multiple SR systems, the cascaded SR system has better output performance compared to the single SR system, such as a higher output SNR, a more obvious signal spectral peak, and a smoother output signal resulting from multiple high-frequency filterings [35].

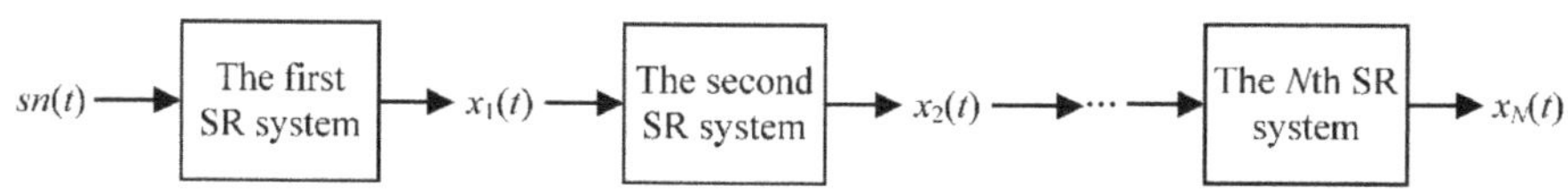

Figure 2. Framework of a cascaded SR system.

In order to utilize the cascade SR system to detect a large-parameter weak signal, the multi-parameters (k, ε, R, and the potential field parameters) of each subsystem should be optimized to achieve an optimal SR output. For this purpose, the multi-parameters can be optimized through different optimization models of the cascaded SR system. The first one is termed the fixed-parameter optimization (FPO) model of the cascade SR system, as shown in Equation (5), which has been studied by many scholars previously [44,49]. One can see that the multi-parameters in all subsystems stay the same. Thus, the cascaded bistable SR system only has five parameters (k, a_1, b_1, ε, and R) to be optimized, and the cascaded tri-stable SR system only has six parameters (k, a_2, b_2, c_2, ε, and R) to be optimized. In the optimization process, the objective of the FPO model is to realize the optimal SR output of the last SR system by optimizing these limited parameters. Therefore, the optimization process based on the FPO model is relatively simple due to the small quantity of the parameters to be optimized.

$$\begin{cases} \frac{dx_1{}^2(t')}{dt'^2} + k\frac{dx_1(t')}{dt'} = -\frac{dU(x_1)}{dx_1} + \varepsilon \cdot sn(t') \\ \frac{dx_2{}^2(t')}{dt'^2} + k\frac{dx_2(t')}{dt'} = -\frac{dU(x_2)}{dx_2} + x_1(t') \\ \cdots \\ \frac{dx_N{}^2(t')}{dt'^2} + k\frac{dx_N(t')}{dt'} = -\frac{dU(x_N)}{dx_N} + x_{N-1}(t') \end{cases} \tag{5}$$

In order to further improve the SR output of the cascaded SR system, a novel optimization model of the cascaded SR system is proposed in this work, which is termed the global parameter optimization (GPO) model, as shown in Equation (6). It can be seen that in this model, the multi-parameters of each subsystem are independent. In the optimization process, the objective is to realize the optimal SR output of the last subsystem by optimizing all multi-parameters jointly, and the number of parameters to be optimized is decided by the number of connected subsystems (N). It is easy to understand that the optimized result obtained from the FPO model is one of the special cases of the results obtained from the GPO model. Therefore, the proposed GPO model has the potential to achieve a better SR output compared to the FPO model.

$$\begin{cases} \frac{dx_1^2(t')}{dt'^2} + k_1 \frac{dx_1(t')}{dt'} = -\frac{dU_1(x_1)}{dx_1} + \varepsilon \cdot sn(t') \\ \frac{dx_2^2(t')}{dt'^2} + k_2 \frac{dx_2(t')}{dt'} = -\frac{dU_2(x_2)}{dx_2} + x_1(t') \\ \cdots \\ \frac{dx_N^2(t')}{dt'^2} + k_N \frac{dx_N(t')}{dt'} = -\frac{dU_N(x_N)}{dx_N} + x_{N-1}(t') \end{cases} \tag{6}$$

2.3. Optimization of the Cascaded SR System Based on the PSO Algorithm

To achieve the optimal SR output of both the FPO model and the GPO model of the cascaded system containing many adjustable parameters, all these parameters should be optimized through a multi-parameter optimization algorithm, which should be carefully selected to match the optimization objective. The particle swarm optimization (PSO) algorithm is a global algorithm based on the foraging behavior of birds. This method requires fewer parameters to adjust compared with other algorithms such as the genetic algorithm and the mosquito swarm, making it easier to implement and simplify the algorithm [28]. Therefore, the PSO algorithm is adopted in this work to find out the optimal SR output of both the FPO model and the GPO model of the cascaded system. The optimization process is briefly introduced in this subsection.

In order to evaluate the output performance of the cascaded SR system appropriately, the output SNR ($\mathrm{SNR_{out}}$) of the system is first defined as the evaluation index [28]:

$$\mathrm{SNR_{out}} = 10\log_{10}\left(\frac{P_0{}^2}{\sum[X(f)]^2 - P_0{}^2}\right) \tag{7}$$

where $X(f)$ represents the single-side spectrum of the output signal of the single or cascade SR system and P_0 is the amplitude value of the output signal at the characteristic frequency. It is assumed that the optimal SR output of the single or cascaded system is achieved when $\mathrm{SNR_{out}}$ reaches its maximum value. Hence, $\mathrm{SNR_{out}}$ can be also used as the objective function of the PSO algorithm.

The PSO algorithm can be described as follows: assuming a particle swarm is a population of n particles in a Z-dimensional search space, The position and velocity of the ith particle are defined as $X_i = (x_{i1}, x_{i2}, x_{i3}, \ldots, x_{iZ})^T$ and $V_i = (V_{i1}, V_{i2}, V_{i3}, \ldots, V_{iZ})^T$, respectively. Initially, random position and velocity populations are created. In each iteration, the particles update their own by tracking two optimal solutions: one is the individual optimal solution of each particle $p_i = (p_{i1}, p_{i2}, p_{i3}, \ldots, p_{iZ})$, and the other one is the population optimal solution of all particles $p_g = (p_{g1}, p_{g2}, p_{g3}, \ldots, p_{gZ})$ [10]. Then, the positions and velocities are updated according to the iterative formulas; thus, the individual and population optimal solution can be updated accordingly at each iteration. The optimization process is finished when the number of iterations reaches the pre-set value $I_{\max}$, and the final population extreme value and the corresponding position of the particle can be obtained. In optimizing the cascaded system, the final population extreme value indicates the optimal $\mathrm{SNR_{out}}$ of the system, and the optimal parameter indicates the population optimal solution. Moreover, the iterative formulas of the velocity and position are given as:

$$V_{i,j}(t+1) = \omega V_{i,j}(t) + c_1 r_1 \left[p_{i,j} - x_{i,j}(t)\right] + c_2 r_2 \left[p_{g,j} - x_{i,j}(t)\right] \tag{8}$$

$$x_{i,j}(t+1) = x_{i,j}(t) + V_{i,j}(t+1),\ j = 1, \ldots, Z \tag{9}$$

where c_1 and c_2 are positive learning factors or acceleration constants and r_1 and r_2 are random numbers uniformly distributed within 0~1. In order to effectively control the flight velocity of the particle, an inertia weight ω is also introduced. In addition, by setting the velocity range $[V_{\min}, V_{\max}]$ and position range $[x_{\min}, x_{\max}]$ of the particle, the movement of the particle can be appropriately limited. In this paper, c_1 and c_2 are set to 1.718, and ω is set to 0.8.

In this paper, the optimal $\mathrm{SNR}_{\mathrm{out}}$ of the signal SR system and the FPO model and the GPO model of a cascaded SR system are obtained by using the PSO algorithm. The only difference in the optimization of these three systems is their optimized parameters.

3. Numerical Simulations

3.1. Weak Signal Detection Based on the FPO Model and the GPO Model

In this subsection, the weak signal detection results of the FPO model and GPO model of the cascaded bistable system are studied and compared by processing a simulated weak signal. The cascaded tri-stable SR system has similar results, which are not presented in this paper. It is noted that the fourth-order Runge–Kutta algorithm is adopted in this work to solve the equations of the SR systems.

A weak signal $sn(t)$ with $A = 0.1$, $f_0 = 0.01$ Hz, and $D = 0.4$ is first generated. The sampling frequency of the signal is set to $F_s = 5$ Hz, and the signal contains 4096 points. The waveform $sn(t)$ and its spectrum $sn(f)$ are shown in Figure 3a,b. One can see that the characteristic signal is almost submerged in the strong background noise due to a relatively low $\mathrm{SNR}_{\mathrm{in}}$ of -22.90 dB, which can be calculated similarly to Equation (7). This weak signal is further processed by the single SR system Equation (4), the FPO model Equation (5), and the GPO model Equation (6) of a cascaded SR system ($N = 4$) with the bistable potential function Equation (2). It is noted that in the optimization processes, the values of ε and R are both set to 1, and the optimization ranges of other parameters are all set to [0.01, 6]. Moreover, the number of iterations $I_{\max}$ is set to 200, and the population n is set to 30 in the PSO algorithm.

First, the weak signal $sn(t)$ is processed by the single bistable SR system. In the optimization process, the values of k, a_1, and b_1 are optimized to achieve an optimal SR output. The optimization result shows that with $sn(t)$ input, the single bistable system achieves optimal SR when $[k, a_1, b_1] = [1.58, 0.08, 0.20]$, and the optimal $\mathrm{SNR}_{\mathrm{out}}$ is -3.43 dB. The output waveform $x(t)$ and its spectrum $x(f)$ are shown in Figure 3c,d. It can be seen that the characteristic of the weak signal is significantly enhanced after being processed by the single bistable system.

Next, the FPO model of the cascaded bistable system ($N = 4$) is used to process the weak signal $sn(t)$. The optimized parameters contain k, a_1, and b_1. An optimal $\mathrm{SNR}_{\mathrm{out}}$ of -1.16 dB can be obtained from the FPO model when $[k, a_1, b_1] = [5.24, 0.01, 5.67]$. The output waveform $x_F(t)$ of the FPO model and its spectrum $x_F(f)$ are shown in Figure 3e,f. One can see that the characteristic of the weak signal is further enhanced through the FPO model by filtering other interference signals.

Lastly, the weak signal $sn(t)$ is processed by the GPO model of the cascaded bistable system ($N = 4$). In this case, there exist 12 optimized parameters including k_1, a_{11}, b_{11}, k_2, a_{12}, b_{12}, ..., k_4, a_{14}, and b_{14}. When their values are 0.54, 0.01, 0.01, 4.74, 0.01, 0.03, 5.91, 0.95, 3.06, 2.25, 0.03, and 0.01, respectively, the GPO model achieves its optimal $\mathrm{SNR}_{\mathrm{out}}$ of -0.72 dB. The output waveform $x_G(t)$ of the GPO model and its spectrum $x_G(f)$ are shown in Figure 3g,h. It can be seen that the spectral peak at the frequency of $f_0 = 0.01$ Hz is more obvious compared to previous results.

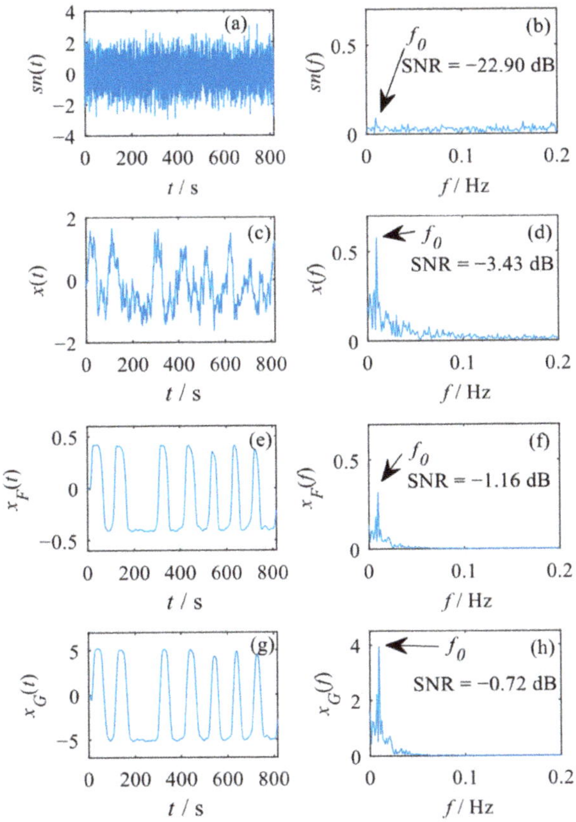

Figure 3. The waveform and spectrum of (**a**,**b**) the weak signal, (**c**,**d**) the output signal of the single bistable system, (**e**,**f**) the output signal of the FPO model of the cascaded bistable system ($N = 4$), and (**g**,**h**) the output signal of the GPO model of the cascaded bistable system ($N = 4$).

The simulated results show that by improving the signal spectral peak and filtering the interference signals/noise, the characteristic of the weak signal can be significantly enhanced through the single bistable system and the FPO model and the GPO model of the cascaded bistable system. The FPO model and the GPO model can achieve better SR outputs compared to the single bistable system, demonstrating the priority of the cascaded SR system in weak signal detection. Moreover, the GPO model is capable of achieving a better SR output compared to the FPO model, which accords with our previous analysis in Section 3.1. However, these conclusions are obtained from a special simulation case, which should be further demonstrated through statistical analysis. Moreover, it should be noted that the single bistable system and the FPO model and the GPO model of the cascaded bistable system all can detect the frequencies of multiple frequency components by setting the corresponding detected frequencies, and they will not misidentify non-existent frequency components. This can be demonstrated through further simulations, which are not presented in this paper.

3.2. Statistical Analysis of the Weak Signal Detection Performances of Different Models

In this subsection, the weak signal detection performances of different models are fully investigated through statistical analysis. In this paper, six models are studied, including the single SR model, the FPO model, and the GPO model of the cascaded SR system with both bistable and tri-stable potential functions, as shown in Table 1.

Table 1. The SR models investigated in this work.

	Bistable Potential Field	Tri-Stable Potential Field
Single SR system	Model 1	Model 4
FPO model of the cascaded SR system	Model 2	Model 5
GPO model of the cascaded SR system	Model 3	Model 6

In the simulations, a weak signal with $A = 0.1$ and $f_0 = 0.01$ Hz is generated, and noises with intensities of 0.4, 1.6, 2.8, and 4.0 are added. Thus, four weak signals with an SNR_{in} of −22.90 dB, −28.65 dB, −30.61 dB, and −31.66 dB are obtained. The noisy signals are processed by the presented 6 models to obtain their optimal SR results, and the searching ranges of the optimization parameters are all set to $[0.01, 6]$ except $\varepsilon = 1$ and $R = 1$. Every signal is processed by each model 100 times, and the results are analyzed and compared statistically.

The statistical graph of the optimization results based on Models 1–3 is shown in Figure 4, in which the mean value of the output SNRs of each model for each input signal is plotted using a histogram and the maximal and minimum SNR_{out} for each case are plotted as well. One can see from Figure 4 that all the SNR_{out} are much larger than their corresponding SNR_{in}, indicating that all presented SR systems can significantly enhance the characteristic of the weak signal using the PSO algorithm. Moreover, for each input signal, the highest average SNR_{out} can be achieved from Model 3, and the second highest average SNR_{out} can be achieved from Model 2, indicating that in general, the GPO model of the cascaded bistable system has the best weak signal detection performance compared to the FPO model of the cascaded bistable system (second best) and the single bistable system. This conclusion can be also drawn from the fact that Model 3 can produce the highest maximum SNR_{out} for each input signal and also the highest minimum SNR_{out} in most cases. Another interesting finding is that as the SNR_{in} decreases from −28.65 dB to −31.66 dB, the average SNR_{out} for each case maintains a relatively stable value, demonstrating the robustness and stability of the SR systems in detecting low-SNR weak signals.

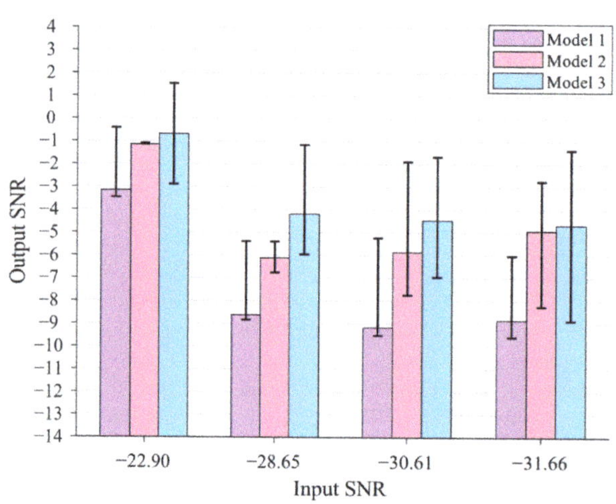

Figure 4. The statistical graph of the optimization results based on Models 1–3.

The statistical graph of the optimization results based on Models 4–6 is further shown in Figure 5. It can be seen that the statistical graph shows a similar pattern to Figure 4, indicating that the conclusions of the weak signal detection performance based on the bistable systems are applicable to those based on the tri-stable systems.

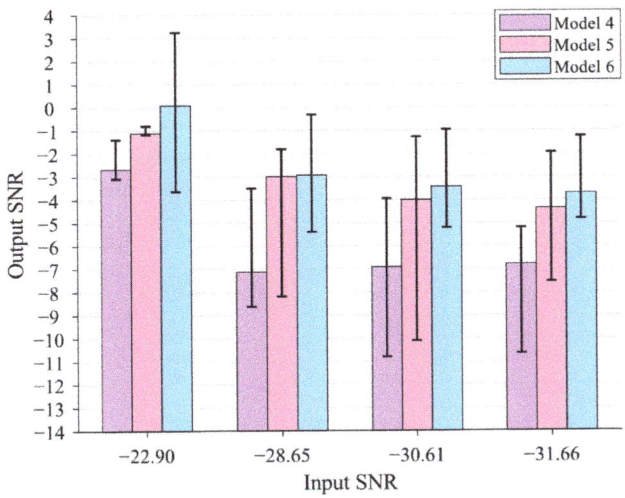

Figure 5. The statistical graph of the optimization results based on Models 4–6.

Previous analyses show the superiority of the proposed GPO model of the cascaded SR system compared to the conventional FPO model of the cascaded SR system and the single SR system. Furthermore, the weak signal detection performance of the cascaded systems with a bistable potential function and tri-stable potential function is compared through statistical analysis. In Figure 6, the mean values, maximum values, and minimum values of the output SNRs in 100 simulations based on the GPO models of the cascaded bistable systems and tri-stable systems are plotted for comparison. It can be seen that in most cases with different weak signals, especially the low-SNR_{in} cases, the GPO model of the cascaded tri-stable system has a higher average SNR_{out}, higher maximum SNR_{out}, and higher minimum SNR_{out} (except for one case) compared to those of the cascaded bistable system. This result indicates that the tri-stable SR system has a better weak signal detection performance compared to the bistable SR system, which has been reported in Refs. [28,50] for single SR systems.

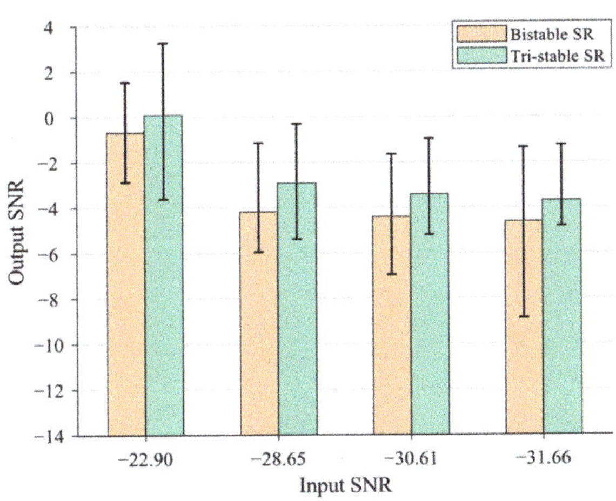

Figure 6. The statistical graph of the optimization results based on Model 3 and Model 6.

3.3. *Influence of the Number of Subsystems*

In previous simulations, the cascaded SR systems were pre-set as four-stage cascaded systems. Actually, the number of subsystems (N) has an influence on the weak signal detection performance and optimization time of the cascaded system, which are fully investigated in this subsection.

The GPO model of the cascaded tri-stable system is used for investigation. A total of 5 cascaded tri-stable systems with $N = 2, 3, 4, 5, 6, 7$, are adopted to analyze the same noisy signal with $A = 0.1$, $f_0 = 0.01$ Hz, and $D = 0.4$. The searching ranges of all adjustable parameters are set to $[0.01, 6]$ except $\varepsilon = 1$ and $R = 1$; the number of iterations $I_{\max}$ is set to 200, and the population n is set to 30 in the PSO algorithm. The input signal is processed by each cascaded system 100 times, and the average SNR_{out} and running times of each case are presented in Figure 7.

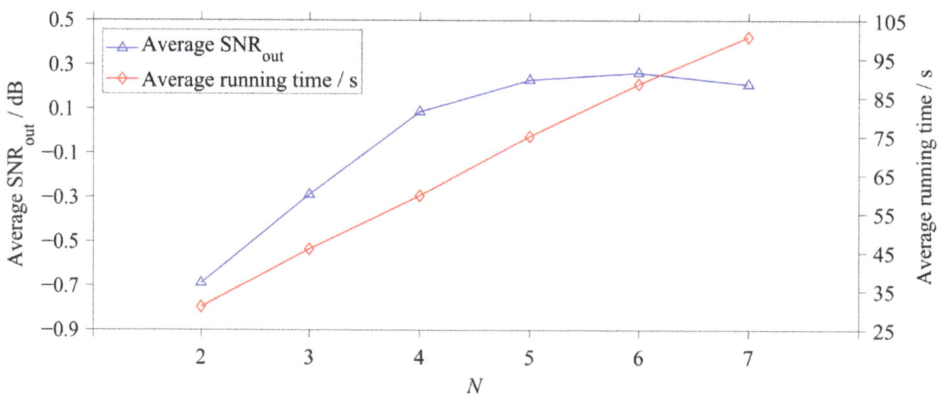

Figure 7. The statistical graph of the optimization results based on Model 6 with different numbers of subsystems.

It can be seen from Figure 7 that as N increases from 2 to 4, the average SNR_{out} gradually increases, which stays almost constant when the number of subsystems continues to increase from 4 to 6. The average SNR_{out} value starts to decrease when N increases to 7. The reason is that when the number of subsystems is large, too many subsystem parameters make it easy for the whole system to achieve a local optimum instead of the global optimum. On the other hand, with N increasing, the number of optimized parameters increases accordingly, and the average running time increases linearly, which can be seen in Figure 7. Therefore, the four-stage cascaded system is a high-cost-performance cascaded system that has a relatively high output performance and an acceptable optimization time. In the following sections of this paper, four-stage cascaded systems are used for weak signal detection and fault diagnosis.

4. Experimental Verifications

The simulation results of Section 3 show that the GPO model of the cascaded SR system has a rather good weak signal detection performance. Hence, it has great potential in mechanical weak fault diagnosis for faulty signals with extremely low SNR. In this section, two examples of practical mechanical weak fault diagnosis are presented to verify the effectiveness of the proposed GPO model of the cascaded SR system. It is noted that the SR system with tri-stable potential functions is used due to its better weak signal detection performance compared to that with a bistable potential function. The fault diagnosis process is shown in Figure 8.

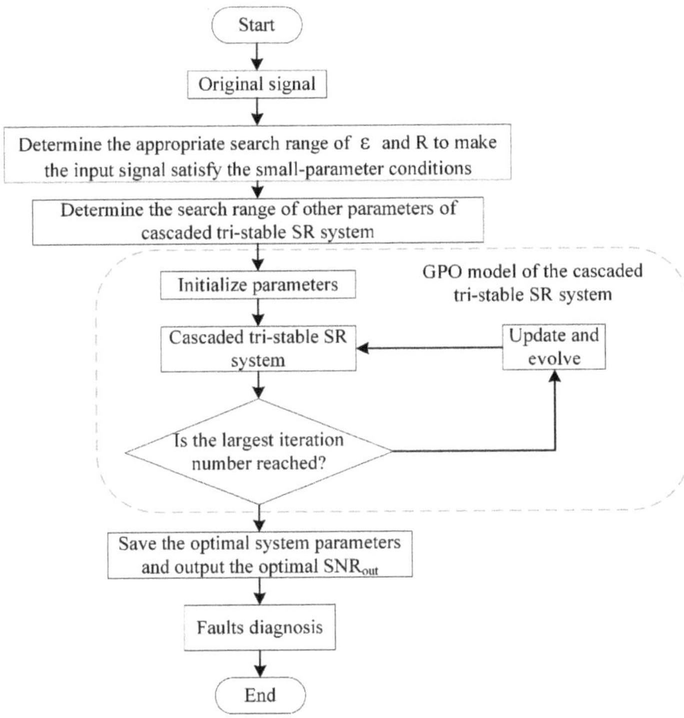

Figure 8. Flowchart of the fault diagnosis based on the GPO model of the cascaded tri-stable system.

4.1. Case 1: Diagnosis of Rolling Bearing Inner Raceway Fault

The data of the rolling bearing inner ring fault are obtained from the bearing data center website of the Case Western Reserve University (CWRU). The associated test rig consists of a 2-horsepower reliance electric motor, a torque transducer/encoder, a dynamometer, and control electronics. In the tests, a groove of width of 0.021 inches was cut in the face of the inner raceway of the drive-end bearing by using electrical discharge machining (EDM), thus introducing a single-point inner ring fault. The faulty bearing was a deep groove ball bearing (6205-2RS JEM SKF), whose parameters are shown in Table 2. According to the fault mechanism of the inner raceway fault, a characteristic frequency of f_{i} is contained in the vibration signal of the vibrating system. The value of f_{i} can be calculated according to [51]:

$$f_{\mathrm{i}} = \frac{z}{2}\left(1 + \frac{d_r}{D_p}\cos\beta\right)f_R \tag{10}$$

where z is the number of rolling elements; d_r and D_p are the rolling elements' diameter and the pitch diameter, respectively; β is the contact angle; and f_R is the rotation frequency of the bearing. During the test, the shaft speed was 1774 rpm, i.e., $f_R = 29.57$ Hz; thus, the characteristic frequency of the inner raceway fault can be calculated as $f_{\mathrm{i}} = 160.1$ Hz.

Table 2. The parameters of the tested rolling bearing.

Model of Bearing	Inside Diameter d_r (Inches)	Outside Diameter (Inches)	Thickness (Inches)	Ball Diameter (Inches)	No. of Rolling Elements Z	Contact Angle β
SKF 6205-2RS JEM	0.9843	2.0472	0.5906	0.3126	9	0°

The bearing vibration signals were acquired using an accelerometer placed at the motor housing, and the sampling frequency was 48,000 Hz. The length of the original signal used in this paper was $M = 4096$, and its input SNR was calculated as $\text{SNR}_{\text{in}} = -39.45$ dB. The waveform, global spectrum, and low-frequency spectrum of the original signal are shown in Figure 9. It can be seen that the fault characteristic frequency of $f_i = 160.1$ Hz cannot be identified from either the time domain or the frequency domain due to the strong noise interference. Therefore, the existing inner raceway fault of the bearing cannot be diagnosed directly from the original signal.

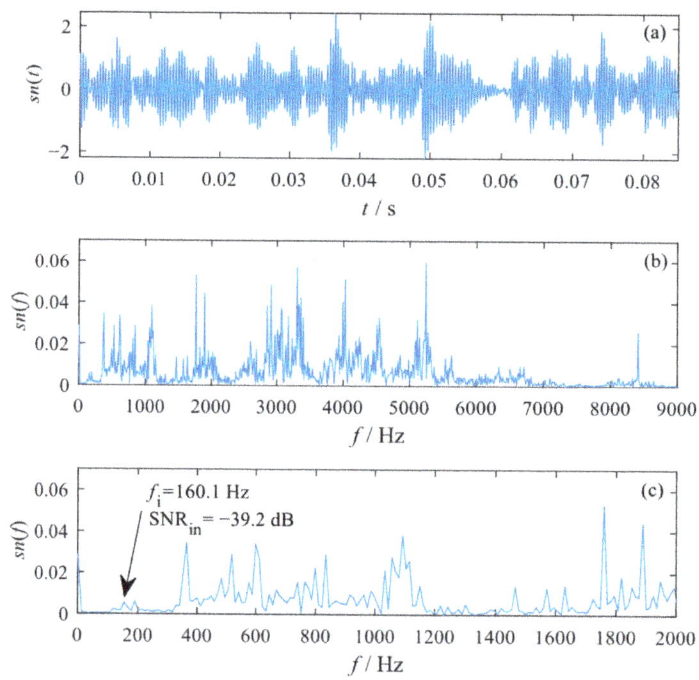

Figure 9. Original vibration signal with bearing inner raceway fault. (**a**) Waveform; (**b**) global spectrum; (**c**) low-frequency spectrum.

In order to diagnose the potential inner raceway fault of the rolling bearing, the proposed GPO model of a four-stage cascaded tri-stable system was adopted to further analyze the original signal. In the optimization of the PSO algorithm, the SNR_{out} at the frequency closest to the characteristic frequency f_i was selected as the objective function. The searching ranges of the amplitude-transformation coefficient and the scale-transformation coefficient were set to $\varepsilon \in [0, 30]$ and $R \in [2000, 3000]$, respectively, and the searching ranges of other optimization parameters $(a_1, b_1, c_1, k_1, a_2, b_2, c_2, k_2, \ldots, a_4, b_4, c_4, k_4)$ were all set to $[0.01, 6]$. The values of the optimal parameters of ε, R, a_1, b_1, c_1, k_1, a_2, b_2, c_2, k_2, …, a_4, b_4, c_4, and k_4 obtained from the GPO model were 6.13, 2000.60, 5.95, 1.10, 1.07, 4.48, 5.39, 4.63, 2.77, 3.72, 3.79, 5.08, 0.93, 4.77, 0.27, 2.38, 1.62, and 0.01, respectively, which enabled the cascaded tri-stable system to produce an optimal SR output with an optimal SNR_{out} of +3.31 dB. The corresponding waveform, global spectrum, and low-frequency spectrum of the output signal were presented in Figure 10. One can see from the waveform that the system output presents a regular inter-well oscillation, and a spectral peak is prominent at $f = 164.1$ Hz with other interference components greatly reduced and nearly eliminated. It can be calculated that the frequency resolution of the spectrum is $\Delta f = F_s/N = 11.7$ Hz; thus, the frequency of the prominent spectral peak ($f = 164.1$ Hz) indicates the characteristic frequency of an inner raceway fault ($f_i = 160.1$ Hz). Hence, an inner raceway fault in the experimental bearing can be identified from the system SR output, which accords with the practical truth. Thus, the effectiveness of the proposed GPO model of the cascaded SR system in mechanical fault diagnosis can be verified.

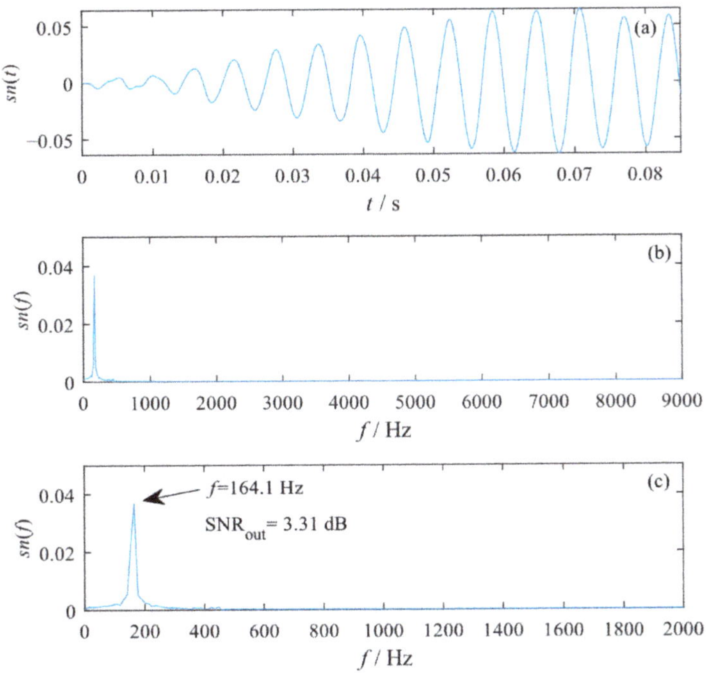

Figure 10. The output signal obtained from the 4-stage cascaded tri-stable system optimized by the GPO model with the rolling bearing inner raceway faulty signal input. (**a**) Waveform; (**b**) global spectrum; (**c**) low-frequency spectrum.

4.2. Case 2: Diagnosis of Rolling Bearing Outer Raceway Fault

The bearing outer raceway failure test rig is shown in Figure 11, which mainly contains a rotor shaft and two rolling bearings driven by an electric motor. In the tests, the outer raceway of the left bearing (cylindrical roller bearing, NU205) had a groove 0.2 mm in width and 0.1 mm in depth, which simulated a bearing with an outer raceway fault. According to the fault mechanism of the outer raceway fault, a characteristic frequency of f_o is contained in the vibration signal of the vibrating system, whose value can be calculated according to [51]:

$$f_o = \frac{z}{2}\left(1 - \frac{d_r}{D_p}\cos\beta\right)f_R \tag{11}$$

In the tests, the motor rotated at a speed of 950 rpm, i.e., $f_R = 15.83$ Hz. Hence, $f_o = 82.95$ Hz can be obtained with basic parameters of the faulty bearing of $z = 13$, $\beta = 0°$, $d_r = 7.5$ mm, and $D_p = 39$ mm.

An accelerometer was installed on the test rig to acquire the vibration signal of the system using a signal-acquiring device (NI PXI-1033). In the tests, the sampling frequency was $F_s = 15{,}000$ Hz. An original signal with length $M = 4096$ was used for analysis. The waveform, global spectrum, and low-frequency spectrum of the original signal are shown in Figure 12. It can be seen that as the SNR of the original signal was as low as $\text{SNR}_{\text{in}} = -39.45$ dB, the waveform is disordered, and the fault characteristic is also submerged in the spectrum. Therefore, the outer raceway fault of the bearing cannot be identified from the original signal.

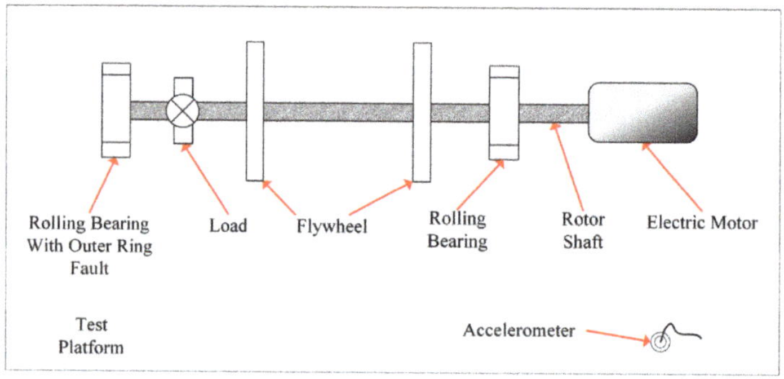

Figure 11. Bearing outer raceway fault test rig.

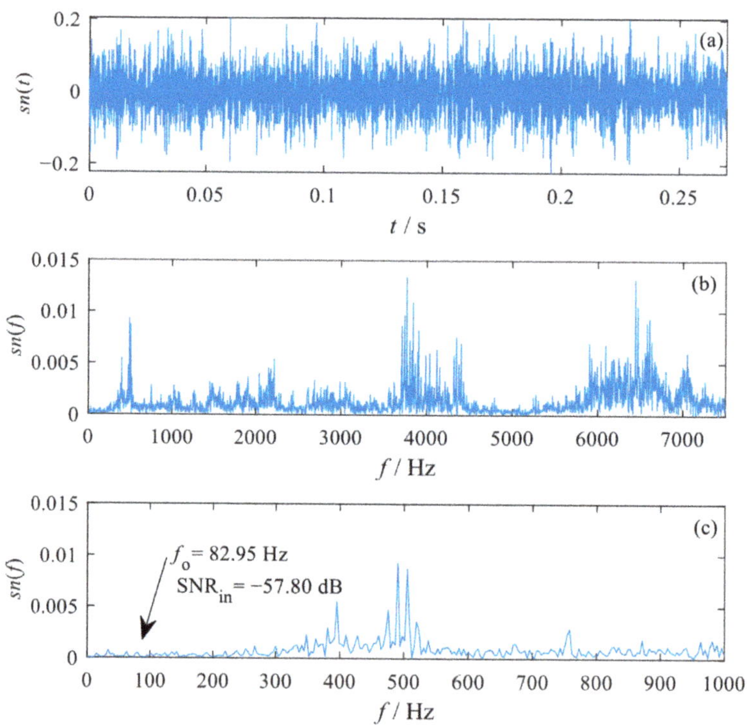

Figure 12. Original vibration signal with bearing outer raceway fault. (**a**) Waveform; (**b**) global spectrum; (**c**) low-frequency spectrum.

Again, the proposed GPO model of the four-stage cascaded tri-stable system was used to further process the original signal. The searching ranges of the amplitude-transformation coefficient and the scale-transformation coefficient were set to $\varepsilon \in [0, 100]$ and $R \in [1000, 2000]$, respectively, and the searching ranges of a_1, b_1, c_1, k_1, a_2, b_2, c_2, k_2, … , a_4, b_4, c_4, k_4 were all set to $[0.01, 6]$. The optimal SR output ($SNR_{out} = -11.33$ dB) was achieved when the values of ε, R, a_1, b_1, c_1, k_1, a_2, b_2, c_2, k_2, … , a_4, b_4, c_4, and k_4 were 74.06, 1002.90, 3.40, 2.81, 3.48, 3.88, 0.27, 0.97, 2.41, 0.01, 4.19, 3.44, 2.59, 3.49, 3.00, 5.16,

1.56, and 3.90, respectively. The corresponding waveform and spectrum of the output signal are presented in Figure 13. It can be seen from the waveform that the output signal presents a regular inter-well oscillation. The most prominent component has a frequency of 80.57 Hz, which can be clearly observed from the spectrum. Considering the frequency resolution of $\Delta f = F_s/N = 3.66$ Hz, this characteristic frequency indicates an outer raceway fault frequency of 82.95 Hz. Thus, an outer raceway fault in the rolling bearing can be identified, and the effectiveness of the proposed methods in mechanical fault diagnosis is verified again.

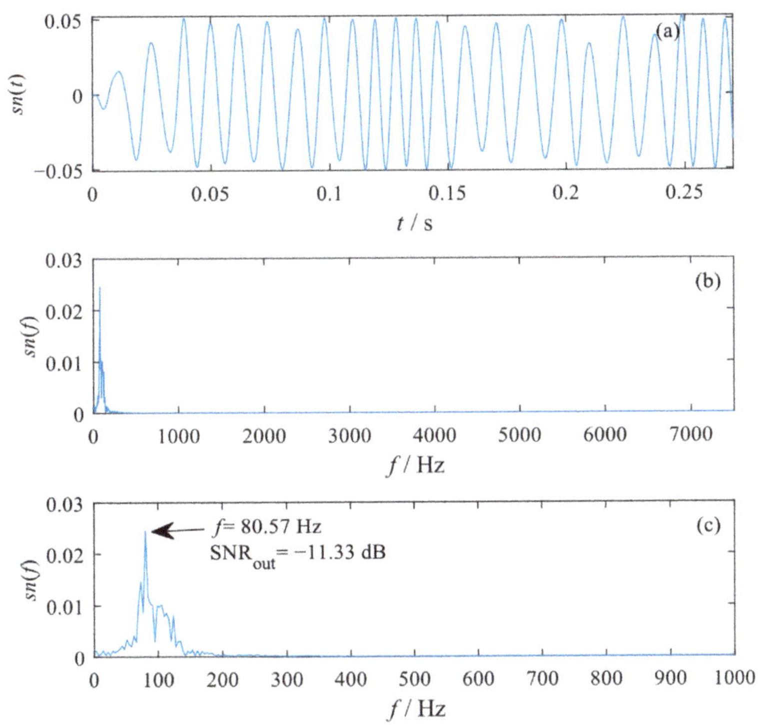

Figure 13. The output signal obtained from the 4-stage cascaded tri-stable system optimized by the GPO model with the rolling bearing outer raceway faulty signal input. (**a**) Waveform; (**b**) global spectrum; (**c**) low-frequency spectrum.

5. Conclusions

In this paper, by combining the multi-parameter-adjusting cascaded stochastic resonance (SR) system and the particle swarm optimization (PSO) algorithm, a global parameter optimization (GPO) model of a cascaded SR system is proposed for high-performance adaptive weak fault diagnosis. The cascaded SR system may involve different potential functions and contain multiple SR systems connected in series. Simulated results show that the GPO model of a cascaded SR system can realize adaptive weak signal detection even under large-parameter conditions with a better performance compared to the single SR system and the pre-proposed fixed-parameter optimization (FPO) model of the cascaded SR system, and the robustness and stability of the system in detecting low-SNR weak signals are rather good. Moreover, the cascaded SR system with a tri-stable potential function has a higher probability of producing a better SR output compared to that with a bistable potential function. The influence of the number of subsystems on the system output is also investigated. It is found that a four-stage cascaded system is a high-cost-performance cascaded system that has a relatively high output performance and also an acceptable optimization time. The weak signal detection and weak fault diagnosis approach is further proposed based on the GPO model of a tri-stable cascaded SR system, which is applied to diagnose the inner raceway fault and the outer raceway fault of a rolling bearing. Experimental work shows that the proposed approach can be utilized to identify the weak characteristic frequency component submerged in a noisy background by significantly enhancing the signal-to-noise ratio (SNR) of the characteristic signal, thus demonstrating its effectiveness in high-performance adaptive mechanical weak fault diagnosis.

Author Contributions: Conceptualization, Z.L. and Z.Q.; Data curation, Z.H., C.W. and J.X.; Formal analysis, Z.H.; Funding acquisition, Z.L. and Z.Q.; Methodology, Z.L. and Z.H.; Project administration, R.Z. and Z.Q.; Resources, M.X.; Software, Z.H.; Supervision, Z.L. and R.Z.; Validation, C.Z.; Visualization, M.X.; Writing—original draft, Z.H.; Writing—review & editing, Z.L. and Z.Q. All authors have read and agreed to the published version of the manuscript.

Funding: This work was funded by the National Natural Science Foundation of China (grant nos. 51905349, U2013603, and 52205569), Guangdong Basic and Applied Basic Research Foundation (grant nos. 2022A1515010126 and 2020A1515011509), Zhejiang Provincial Natural Science Foundation of China (grant no. LQ22E050003), Ningbo Science and Technology Major Project (grant nos. 2020Z110, 2022Z057, and 2022Z002), Chuying Planning Project of Zhejiang Provincial Administration for Market Regulation (grant no. CY2023328), and Shandong Provincial Innovation Ability Improvement Project of Middle and Small-sized High-tech Enterprises (grant no. 2022TSGC2364).

Institutional Review Board Statement: Not applicable.

Informed Consent Statement: Not applicable.

Data Availability Statement: Data are available on request from the authors.

Acknowledgments: We sincerely thank the support by the National Natural Science Foundation of China (grant nos. 51905349, U2013603, and 52205569), Guangdong Basic and Applied Basic Research Foundation (grant nos. 2022A1515010126 and 2020A1515011509), Zhejiang Provincial Natural Science Foundation of China (grant no. LQ22E050003), Ningbo Science and Technology Major Project (grant nos. 2020Z110, 2022Z057, and 2022Z002), Chuying Planning Project of Zhejiang Provincial Administration for Market Regulation (grant no. CY2023328), and Shandong Provincial Innovation Ability Improvement Project of Middle and Small-sized High-tech Enterprises (grant no. 2022TSGC2364).

Conflicts of Interest: The authors declare no conflict of interest.

Abbreviations

The following abbreviations are used in this manuscript:

CWRU	Case Western Reserve University
EDM	Electrical Discharge Machining
EMD	Empirical Mode Decomposition
FPO	Fixed-parameter Optimization
FRSR	Frequency Re-scaling Stochastic Resonance
FSRSR	Frequency-Shifted and Re-scaling Stochastic Resonance
GPO	Global parameter Optimization
MSR	Modulated Stochastic Resonance
PNSR	Parameter-Normalized Stochastic Resonance
PSO	Particle Swarm Optimization
QPSO	Quantum Particle Swarm Optimization
SK	Spectral Kurtosis
SNR	Signal-to-noise Ration
SR	Stochastic Resonance
WT	Wavelet Transform

References

1. Tang, J.; Shi, B.; Li, Z.; Li, Y. Weak fault feature extraction method based on compound tri-stable stochastic resonance. *Chin. J. Phys.* **2020**, *66*, 50–59. [CrossRef]
2. Liu, J.; Shen, X.; Wang, J.; Jiang, L.; Zhang, H. An Intelligent Defect Detection Approach Based on Cascade Attention Network Under Complex Magnetic Flux Leakage Signals. *IEEE Trans. Ind. Electron.* **2023**, *70*, 7417–7427. [CrossRef]
3. Kolbe, S.; Gelman, L.; Ball, A. Novel Prediction of Diagnosis Effectiveness for Adaptation of the Spectral Kurtosis Technology to Varying Operating Conditions. *Sensors* **2021**, *21*, 6913. [CrossRef] [PubMed]
4. Yang, J.; Han, S.; Zhang, S.; Liu, H.; Tang, C. Empirical mode decomposition of weak fault characteristic signal of rolling bearing under strong noise background. *J. Vib. Eng.* **2020**, *33*, 582–589.
5. Zheng, J.D.; Su, M.X.; Ying, W.M.; Tong, J.Y.; Pan, Z.W. Improved uniform phase empirical mode decomposition and its application in machinery fault diagnosis. *Measurement* **2021**, *179*, 10. [CrossRef]
6. Wang, L.M.; He, J.F.; Xiao, H.F.; Zeng, Q.; Ding, X.X.; Shao, Y.M. Joint suppression of normal gear mesh component and background noise for early local fault detection based on dynamic evolutionary digital filter. *Measurement* **2022**, *201*, 16. [CrossRef]
7. Han, B.; Li, C.S.; Zhou, Y.Q.; Yu, G.; Wei, C.L. Second-Order Multisynchrosqueezing Wavelet Transform for Bearing Fault Detection. *J. Vib. Eng. Technol.* **2022**, *10*, 1541–1559. [CrossRef]
8. Benzi, R.; Parisi, G.; Sutera, A.; Vulpiani, A. Stochastic resonance in climatic change. *Tellus* **1982**, *34*, 10–16. [CrossRef]
9. Benzi, R.; Sutera, A.; Vulpiani, A. The mechanism of stochastic resonance. *J. Phys. A Math. Gen.* **1981**, *14*, L453. [CrossRef]
10. Shi, P.M.; Su, C.J.; Han, D.Y. Fault diagnosis of rotating machinery based on adaptive stochastic resonance and AMD-EEMD. *Shock. Vib.* **2016**, *2016*, 9278581. [CrossRef]
11. Tang, J.C.; Shi, B.Q.; Bao, H.R.; Li, Z.X. A new method for weak fault feature extraction based on piecewise mixed stochastic resonance. *Chin. J. Phys.* **2020**, *68*, 87–99. [CrossRef]
12. Lu, S.; He, Q.; Wang, J. A review of stochastic resonance in rotating machine fault detection. *Mech. Syst. Signal Process.* **2019**, *116*, 230–260. [CrossRef]
13. Qiao, Z.J.; Elhattab, A.; Shu, X.D.; He, C.B. A second-order stochastic resonance method enhanced by fractional-order derivative for mechanical fault detection. *Nonlinear Dyn.* **2021**, *106*, 707–723. [CrossRef]
14. Qiao, Z.; Shu, X. Coupled neurons with multi-objective optimization benefit incipient fault identification of machinery. *Chaos Solitons Fractals* **2021**, *145*, 110813. [CrossRef]
15. Zhang, J.J.; Zhang, T. Parameter-induced stochastic resonance based on spectral entropy and its application to weak signal detection. *Rev. Sci. Instrum.* **2015**, *86*, 6. [CrossRef]
16. Zhang, G.; Liu, Y.L.; He, L.F. Research on fault detection of asymmetric piecewise well-posed stochastic resonance system. *Rev. Sci. Instrum.* **2021**, *92*, 11. [CrossRef]
17. Qiao, Z.; Lei, Y.; Li, N. Applications of stochastic resonance to machinery fault detection: A review and tutorial. *Mech. Syst. Signal Process.* **2019**, *122*, 502–536. [CrossRef]
18. Yang, Z.; Li, Z.Q.; Zhou, F.X.; Ma, Y.J.; Yan, B.K. Weak fault feature extraction method based on improved stochastic resonance. *Sensors* **2022**, *22*, 6644. [CrossRef]
19. Leng, Y.G.; Leng, Y.S.; Wang, T.Y.; Guo, Y. Numerical analysis and engineering application of large parameter stochastic resonance. *J. Sound Vib.* **2006**, *292*, 788–801. [CrossRef]
20. Tan, J.; Chen, X.; Wang, J.; Chen, H.; Cao, H.; Zi, Y.; He, Z. Study of frequency-shifted and re-scaling stochastic resonance and its application to fault diagnosis. *Mech. Syst. Signal Process.* **2009**, *23*, 811–822. [CrossRef]
21. Yang, D.X.; Hu, Z.; Yang, Y.M. The analysis of stochastic resonance of periodic signal with large parameters. *Acta Phys. Sin.* **2012**, *61*, 10. (In Chinese)
22. Lin, M.; Huang, Y.M. Modulation and demodulation for detecting weak periodic signal of stochastic resonance. *Acta Phys. Sin.* **2006**, *55*, 3277–3282. (In Chinese) [CrossRef]
23. Lai, Z.-H.; Leng, Y.-G. Weak-signal detection based on the stochastic resonance of bistable Duffing oscillator and its application in incipient fault diagnosis. *Mech. Syst. Signal Process.* **2016**, *81*, 60–74. [CrossRef]
24. Xu, B.M.; Shi, J.C.; Zhong, M.; Zhang, J. Incipient fault diagnosis of planetary gearboxes based on an adaptive parameter-induced stochastic resonance method. *Appl. Acoust.* **2022**, *188*, 108587. [CrossRef]
25. Hu, B.B.; Guo, C.; Wu, J.M.; Tang, J.H.; Zhang, J.L.; Wang, Y. An adaptive periodical stochastic resonance method based on the grey wolf optimizer algorithm and its application in rolling bearing fault diagnosis. *J. Vib. Acoust. Trans. Asme* **2019**, *141*, 9. [CrossRef]
26. Cheng, W.; Xu, X.M.; Ding, Y.P.; Sun, K.H. Stochastic resonance in a single-well potential and its application in rolling bearing fault diagnosis. *Rev. Sci. Instrum.* **2020**, *91*, 16. [CrossRef]
27. Tong, L.; Li, X.G.; Hu, J.H.; Ren, L.T. A PSO optimization scale-transformation stochastic-resonance algorithm with stability mutation operator. *IEEE Access* **2018**, *6*, 1167–1176. [CrossRef]
28. Lai, Z.H.; Liu, J.S.; Zhang, H.T.; Zhang, C.L.; Zhang, J.W.; Duan, D.Z. Multi-parameter-adjusting stochastic resonance in a standard tri-stable system and its application in incipient fault diagnosis. *Nonlinear Dyn.* **2019**, *96*, 2069–2085. [CrossRef]
29. Cui, H.J.; Guan, Y.; Chen, H.Y.; Deng, W. A novel advancing signal processing method based on coupled multi-stable stochastic resonance for fault detection. *Appl. Sci.* **2021**, *11*, 5385. [CrossRef]

30. Qiao, Z.; Liu, J.; Ma, X.; Liu, J. Double stochastic resonance induced by varying potential-well depth and width. *J. Frankl. Inst.* **2021**, *358*, 2194–2211. [CrossRef]
31. Li, J.; Wang, X.; Li, Z.; Zhang, Y. Stochastic resonance in cascaded monostable systems with double feedback and its application in rolling bearing fault feature extraction. *Nonlinear Dyn.* **2021**, *104*, 971–988. [CrossRef]
32. Gao, R.; Jiao, S.; Wang, Y.; Li, Y. Weak signal detection of composite multistable stochastic resonance with Woods-Saxon potential. *Z. Fur Nat. Sect. A J. Phys. Sci.* **2022**, *77*, 963–976. [CrossRef]
33. Zhang, H.; He, Q.; Kong, F. Stochastic resonance in an underdamped system with pinning potential for weak signal detection. *Sensors* **2015**, *15*, 21169–21195. [CrossRef]
34. Li, Z.X.; Liu, X.D.; Han, S.J.; Wang, J.G.; Ren, X.P. Fault diagnosis method and application based on unsaturated piecewise linear stochastic resonance. *Rev. Sci. Instrum.* **2019**, *90*, 10. [CrossRef]
35. Li, J.M.; Zhang, Y.G.; Xie, P. A new adaptive cascaded stochastic resonance method for impact features extraction in gear fault diagnosis. *Measurement* **2016**, *91*, 499–508. [CrossRef]
36. Guo, W.; Zhou, Z.; Chen, C.; Li, X. Multi-frequency weak signal detection based on multi-segment cascaded stochastic resonance for rolling bearings. *Microelectron. Reliab.* **2017**, *75*, 239–252. [CrossRef]
37. Xu, D.; Ge, J.; Wang, Y.; Shao, J. Multi-Frequency Weak Signal Decomposition and Reconstruction of Rolling Bearing Based on Adaptive Cascaded Stochastic Resonance. *Machines* **2021**, *9*, 275. [CrossRef]
38. Zhang, G.; Wang, H.; Zhang, T.Q. Cascaded Nonlinear Mass Fluctuation Stochastic Resonance System and Its Application in Bearing Fault Diagnosis. *Fluct. Noise Lett.* **2021**, *20*, 2150045. [CrossRef]
39. Li, J.M.; Zhang, J.F.; Li, M.; Zhang, Y.G. A novel adaptive stochastic resonance method based on coupled bistable systems and its application in rolling bearing fault diagnosis. *Mech. Syst. Signal Process.* **2019**, *114*, 128–145. [CrossRef]
40. Kim, J.; Harne, R.L.; Wang, K.W. Online signal denoising using adaptive stochastic resonance in parallel array and its application to acoustic emission signals. *J. Vib. Acoust. Trans. Asme* **2022**, *144*, 12. [CrossRef]
41. Liu, W.; Liu, Z.; Zhang, Q.; Xu, Y.; Liu, S.; Chen, Z.; Zhu, C.; Wang, Z.; Pan, M.; Hu, J.; et al. Magnetic Anomaly Signal Detection Using Parallel Monostable Stochastic Resonance System. *IEEE Access* **2020**, *8*, 162230–162237. [CrossRef]
42. He, H.L.; Wang, T.Y.; Leng, Y.G.; Zhang, Y.; Li, Q. Study on non-linear filter characteristic and engineering application of cascaded bistable stochastic resonance system. *Mech. Syst. Signal Process.* **2007**, *21*, 2740–2749. [CrossRef]
43. Hairan, R.; Jalimhan, E.; Dalabaev, S. Cascaded Bistable Stochastic Resonance Signal Detection under Dual-mode Non-Gaussian Noise Background. *Telecommun. Eng.* **2014**, *54*, 1260–1264. (In Chinese)
44. Li, B.; Li, J.; He, Z. Fault feature enhancement of gearbox in combined machining center by using adaptive cascade stochastic resonance. *Sci. China Technol. Sci.* **2011**, *54*, 3203–3210. [CrossRef]
45. Xiao, L.; Bajric, R.; Zhao, J.; Tang, J.; Zhang, X. An adaptive vibrational resonance method based on cascaded varying stable-state nonlinear systems and its application in rotating machine fault detection. *Nonlinear Dyn.* **2021**, *103*, 715–739. [CrossRef]
46. Liu, H.; Han, S.; Yang, J.; Liu, S. Improving the weak feature extraction by adaptive stochastic resonance in cascaded piecewise-linear system and its application in bearing fault detection. *J. Vibroeng.* **2017**, *19*, 2506–2520. [CrossRef]
47. Zhang, G.; Gao, J.P.; Hong-Wei, L.I. Research on stochastic resonance characteristics of cascaded three-steady-state and its application. *Comput. Sci.* **2018**, *45*, 146–151.
48. Lai, Z.H.; Leng, Y.G. Generalized parameter-adjusted stochastic resonance of Duffing oscillator and its application to weak-signal detection. *Sensors* **2015**, *15*, 21327–21349. [CrossRef]
49. Sun, H.; Wang, Z. Weak signal detection based on cascaded piecewise-linear stochastic resonance. *China Mech. Eng.* **2014**, *25*, 3343. (In Chinese)
50. Lai, Z.H.; Leng, Y.G. Dynamic response and stochastic resonance of a tri-stable system. *Acta Phys. Sin.* **2015**, *64*, 12. (In Chinese)
51. Zhang, G.; Yi, T.; Zhang, T.; Cao, L. A multiscale noise tuning stochastic resonance for fault diagnosis in rolling element bearings. *Chin. J. Phys.* **2018**, *56*, 145–157. [CrossRef]

Disclaimer/Publisher's Note: The statements, opinions and data contained in all publications are solely those of the individual author(s) and contributor(s) and not of MDPI and/or the editor(s). MDPI and/or the editor(s) disclaim responsibility for any injury to people or property resulting from any ideas, methods, instructions or products referred to in the content.

sensors

MDPI

Communication

Stochastic Resonance with Parameter Estimation for Enhancing Unknown Compound Fault Detection of Bearings

Min Xu [1], Chao Zheng [1], Kelei Sun [1], Li Xu [1], Zijian Qiao [2,3,4,5,*] and Zhihui Lai [6]

1 Ningbo Cigarette Factory, China Tobacco Zhejiang Industry Co., Ltd., Ningbo 315040, China
2 School of Mechanical Engineering and Mechanics, Ningbo University, Ningbo 315211, China
3 Yangjiang Offshore Wind Power Laboratory, Yangjiang 529500, China
4 State Key Laboratory of Performance Monitoring and Protecting of Rail Transit Infrastructure, East China Jiaotong University, Nanchang 330013, China
5 Zhejiang Provincial Key Laboratory of Part Rolling Technology, Ningbo 315211, China
6 Shenzhen Key Laboratory of High Performance Nontraditional Manufacturing, College of Mechatronics and Control Engineering, Shenzhen University, Shenzhen 518060, China
* Correspondence: qiaozijian@nbu.edu.cn

Abstract: Although stochastic resonance (SR) has been widely used to enhance weak fault signatures in machinery and has obtained remarkable achievements in engineering application, the parameter optimization of the existing SR-based methods requires the quantification indicators dependent on prior knowledge of the defects to be detected; for example, the widely used signal-to-noise ratio easily results in a false SR and decreases the detection performance of SR further. These indicators dependent on prior knowledge would not be suitable for real-world fault diagnosis of machinery where their structure parameters are unknown or are not able to be obtained. Therefore, it is necessary for us to design a type of SR method with parameter estimation, and such a method can estimate these parameters of SR adaptively by virtue of the signals to be processed or detected in place of the prior knowledge of the machinery. In this method, the triggered SR condition in second-order nonlinear systems and the synergic relationship among weak periodic signals, background noise and nonlinear systems can be considered to decide parameter estimation for enhancing unknown weak fault characteristics of machinery. Bearing fault experiments were performed to demonstrate the feasibility of the proposed method. The experimental results indicate that the proposed method is able to enhance weak fault characteristics and diagnose weak compound faults of bearings at an early stage without prior knowledge and any quantification indicators, and it presents the same detection performance as the SR methods based on prior knowledge. Furthermore, the proposed method is more simple and less time-consuming than other SR methods based on prior knowledge where a large number of parameters need to be optimized. Moreover, the proposed method is superior to the fast kurtogram method for early fault detection of bearings.

Keywords: stochastic resonance; weak fault detection; mechanical fault diagnosis

1. Introduction

Rolling bearings are key components of rotating machinery and can provide reliable and stable support. However, during the operating process, rolling bearings are inevitable; their defects include wear, cracking and pitting due to insufficient lubrication, contact fatigue, etc. Therefore, it is very important for us to achieve early fault detection and fault diagnosis of bearings to avoid a serious accident. Up until now, there were two well-known categories of bearing fault diagnosis, including model updating [1,2] and direct feature extraction and analysis using machine learning [3,4] or signal processing [5,6]. In addition, some scholars focus on the prediction of remaining useful life [7,8].

Among them, because direct feature extraction and analysis based on signal processing does not require building complex mathematical models and complete fault samples, they have attracted sustained attention in fault detection. However, most signal processing methods attempt to cancel the noise embedded in signals to detect weak fault features, while stochastic resonance is able to harvest the energy of noise to enhance weak useful signals and therefore has been widely used in weak fault detection and fault diagnosis of machinery. In 2019, Lu et al. [9] and Qiao et al. [10] reviewed these articles on SR-based fault diagnosis of machinery and pointed out the potential development directions, respectively. Two review articles are inspiring researchers to explore the potential of SR as well as to develop advanced research in this field.

Up until now, there were many scholars who investigated the application of SR in mechanical fault diagnosis. Among them, some paid attention to designing novel indicators to quantify SR. For example, López et al. [11] proposed hidden Markov models and a Box–Cox sparse measures-based SR method for bearing fault diagnosis in which the designed indicator does not depend on prior knowledge of the fault signature to be detected, but the parameter selection of hidden Markov models and Box–Cox sparse measures would complicate the proposed method. Lin et al. [12] designed a novel indicator to quantify the SR for bearing fault diagnosis, where the cross-correlation coefficient, impulse index and signal-to-noise ratio (SNR) are fused. Li et al. [13] presented a multi-parameter constrained potential underdamped SR method and applied it to weak fault diagnosis, where SNR is used to evaluate SR. Lai et al. [14] defined SNR_{input} and SNR_{output} as the signal-to-noise ratios at SR system input and output to quantify SR for mechanical fault diagnosis. Zhang et al. [15] used the opposite of SNR as a fitness function of salp swarm algorithms to optimize the parameters of the proposed SR method. Wang et al. [16] applied order tracking to address the time-varying signals and then designed a tristable SR method to enhance weak fault characteristics with the fitness function of SNR. Zhou et al. [17] used generative adversarial networks to fuse 10 statistical parameters as the fitness function for optimizing the proposed SR method. Shi et al. [18] proposed a novel adaptive multi-parameter unsaturation bistable SR method in which the output SNR is selected as the objective function. Qiao et al. [19] proposed a second-order SR method enhanced by a fractional-order derivative for mechanical fault detection where SNR is used to optimize its parameters, even in coupled neurons [20] or the proposed nonlinear resonant decomposition [21]. Li et al. [22] presented a frequency-shift multiscale noise tuning SR method for fault diagnosis of generator bearings in wind turbines and even presented the coupled bistable SR method [23] in which the parameters of the proposed method are optimized through modified SNR and genetic algorithms. Fu et al. [24] studied the SR in Duffing oscillators and proposed a moment-method-based bearing fault diagnosis algorithm where the output SNR is seen as the objective function of optimization algorithms. Xu et al. [25] studied the SR behaviors in a high-order-degradation bistable system, and the ratio of the amplitudes around the target orders to those of the interference orders was calculated as the objective function of optimization algorithms. In summary, most of the proposed SR methods' objective functions depend on prior knowledge of the fault characteristics to be detected, which would result in false SR and weaken the detection performance of SR further. It is improper to enhance and detect unknown weak fault characteristics, especially for real-world equipment. In addition, a large number of tuning parameters in the existing SR methods need to be optimized by using artificial intelligence, which is very time-consuming and complex and is not suitable to engineering application.

Triggering SR has clear mathematical conditions, but among these variables in the conditions, the intensity of noise is indefinite. If we could estimate the intensity of noise in a signal, a parameter-matching equation could be built. Motivated by such an idea, this paper will attempt to design a parameter-matched second-order SR method for enhancing unknown weak fault characteristics of bearings due to second-order SR that has a bandpass filtering property. The remainder of this paper is organized as follows. Section 2 gives clear mathematical conditions of parameter-matched second-order SR and then proposes the corresponding SR method for bearing fault enhancement. In Section 3, a bearing fault experiment was performed to demonstrate the effectiveness of the proposed method, and a comparison is made. Finally, conclusions are drawn in Section 4.

2. A Parameter-Matched Second-Order SR Method

A second-order SR can be described as below [26],

$$\frac{\mathrm{d}^2x}{\mathrm{d}t^2}+\gamma\frac{\mathrm{d}x}{\mathrm{d}t}=-\frac{\mathrm{d}V(x)}{\mathrm{d}x}+s(t)+n(t) \quad (1)$$

where γ is the damping factor and $0<\gamma\le 2\sqrt{2a}$ [27], $s(t)$ is a periodic signal to be detected and its expression can be written as $A\cos(2\pi f_0 t+\varphi)$ with the amplitude A, driving frequency f_0 and initial phase φ. $n(t)$ is noise with $\langle n(t),n(t+\tau)\rangle=2D\delta(t)$ and D as the noise intensity. $V(x)$ is the bistable potential as below,

$$V(x)=-\frac{a}{2}x^2+\frac{b}{4}x^4, a,b>0 \quad (2)$$

where two stable states and one unstable state are located at $\pm x_m=\pm\sqrt{a/b}$ and $x_0=0$, respectively. Moreover, the barrier height is $\Delta V=\frac{a^2}{4b}$. According to two-state theory, the matched conditions for triggering the SR induced by a periodic signal is [28]

$$r_K=2f_0 \quad (3)$$

where r_K is the Kramers' rate given by [29]

$$r_K=\frac{\omega_b\omega_0}{2\pi\gamma}\exp\left(-\frac{\Delta V}{D}\right) \quad (4)$$

in which $\omega_b=\sqrt{2a}$ and $\omega_0=\sqrt{a}$.

For convenience, we rewrite the matched condition in Equation (3) as [30]

$$F(a,b,D,\gamma,f_0)=\frac{a}{2\sqrt{2}\pi\gamma f_0}\exp\left(-\frac{a^2}{4bD}\right) \quad (5)$$

It can be noticed from Equation (5) that SR can be induced when $F=1$. In general, the input and output SNR in Equation (1) for a periodic signal plus additive noise can be written as [31]

$$SNR_{\text{input}}=\frac{A^2}{4D} \quad (6)$$

$$SNR_{\text{output}}\approx\frac{aA^2\sqrt{2a}}{4bD^2}\exp\left(-\frac{a^2}{4bD}\right) \quad (7)$$

Then, the SNR gain (Signal-to-noise ratio gain, SNRG) can be calculated as

$$SNRG=\frac{SNR_{\text{output}}}{SNR_{\text{input}}}\approx\frac{a\sqrt{2a}}{bD}\exp\left(-\frac{a^2}{4bD}\right) \quad (8)$$

It can be seen from Equation (8) that the SNRG would be controlled by bistable potential parameters and the noise intensity. Thus, the objective function of the signal to be detected is given as the following optimization problem:

$$D_{opt}=\text{argmax } SNRG \quad (9)$$

By solving Equation (8) with respect to D, we can obtain the optimal condition as below:

$$D_{opt}=\frac{a^2}{4b}=\Delta V \quad (10)$$

Therefore, the optimal parameter-matched condition for weak fault enhancement can be given as below:

$$\begin{cases} F = 1 \\ D = \frac{a^2}{4b} \end{cases} \tag{11}$$

However, Equation (1) is suitable to process weak signals subject to small-parameter limitation. In the real world, the mechanical signals are generally large parameter. For solving this issue, a normalized scale transformation is performed to Equation (1). Let $z = x\sqrt{b/a}$ and $\tau = at$, Equation (1) could be rewritten as

$$\frac{d^2z}{d\tau^2} + \gamma_h \frac{dz}{d\tau} = a_h z - b_h z^3 + A_h \cos(2\pi f_{0h}\tau + \varphi) + D_h \xi\left(\frac{\tau}{a}\right) \tag{12}$$

where $\gamma_h = \gamma/a$, $a_h = b_h = 1/a$, $f_{0h} = f_0/a$, $A_h = \sqrt{b/a^5}A$, and $D_h = b/a^5 D$. By comparison, it is found that large-parameter signals can be transferred into small-parameter ones in which the amplitude is reduced by the time of $\sqrt{b/a^5}$ and the frequency is reduced by the time of $1/a$. By substituting the above parameters in Equation (12) into Equation (11), the parameter-matched conditions for triggering SR for weak large-parameter signal detection can be achieved as below:

$$\begin{cases} F(a_h, b_h, D_h, f_{0h}) = \frac{a_h}{2\sqrt{2}\pi\gamma_h f_{0h}} \exp\left(-\frac{a_h^2/4b_h}{D_h}\right) = 1 \\ D_h = a_h^2/4b_h \end{cases} \tag{13}$$

which is calculated further as

$$\begin{cases} a = 2\sqrt{2}\pi f_0 \gamma e \\ b = a^4/4D \end{cases} \tag{14}$$

It can be seen from Equation (14) that the damping factor γ can be adjusted to enhance the weak fault characteristics of bearings. According to the condition $0 < \gamma_h \leq 2\sqrt{2a_h}$, we can obtain the limitation.

$$0 < \gamma < 16\sqrt{2}\pi f_0 e \tag{15}$$

Therefore, the parameter-matched conditions can be given by using Equations (14) and (15) mathematically for weak large-parameter signal detection. According to the above parameter-matched conditions, we can design an SR method with parameter estimation to enhance the weak large-parameter fault signature of bearings. The proposed method is shown in Figure 1, and its detailed steps are given as below.

1. Signal pre-processing. In general, the bearing vibration signal $v(t)$ is a large-parameter signal where the bearing fault characteristics are modulated by natural vibration from the machinery itself. Hence, some signal demodulations are performed to the raw signals. Here, Hilbert envelope demodulation is used to pre-process the raw vibration signal of the tested bearing, and the corresponding envelope signal is denoted as $\widetilde{v}(t)$.
2. Noise intensity estimation. In Equation (14), an important parameter D, namely, noise intensity, needs to be estimated from the envelope signal $\widetilde{v}(t)$ of the raw signal of the tested bearing. Here, we use the principle of maximum likelihood estimation (MLE) [32] to achieve this, and we can obtain $D = MLE(\widetilde{v}(t))$. The MLE algorithm can be downloaded by using the following website: http://www.biomecardio.com/matlab/evar_doc.html (accessed on 5 December 2022).
3. Damping factor initialization. The damping factor needs to be tuned to obtain the optimal detected result. Here, according to its range in Equation (15), we initialize the damping factor γ.

4. Output signal calculation and evaluation. By substituting $a_{\text{match}} = 2\sqrt{2}\pi f_0 \gamma e$, $b_{\text{match}} = a^4/4D$ and the corresponding damping factor into Equation (16), we can obtain the output signal $x(t)$. Then, we can calculate the corresponding $SNRG$ as the objective function $\gamma_{\text{opt}} = \text{argmax } SNRG$ for optimizing the damping factor. Finally, the optimal γ_{opt} is substituted into Equation (17) to solve the optimal $x(t)$.

$$\frac{\mathrm{d}^2 x}{\mathrm{d}t^2} + \gamma \frac{\mathrm{d}x}{\mathrm{d}t} = a_{\text{match}} x - b_{\text{match}} x^3 + \widetilde{v}(t) \tag{16}$$

$$\frac{\mathrm{d}^2 x}{\mathrm{d}t^2} + \gamma_{\text{opt}} \frac{\mathrm{d}x}{\mathrm{d}t} = a_{\text{match}} x - b_{\text{match}} x^3 + \widetilde{v}(t) \tag{17}$$

5. Signal post-processing. Output the optimal $x(t)$ corresponding to the maximum of SNRG as the detected signal. Here, the frequency spectral analysis is used to process the optimal $x(t)$ for observing the spectral peaks at the fault characteristic frequencies of bearings.

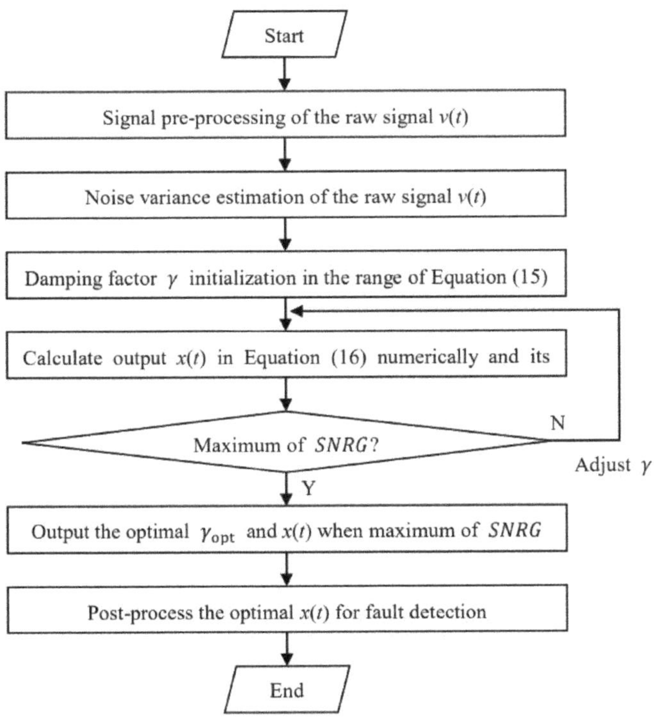

Figure 1. The flowchart of the proposed method.

3. Bearing Fault Experimental Verification

The bearing fault experimental setup is shown in Figure 2 where a bearing accelerated degradation test was performed throughout the whole operating life of the bearing, and the corresponding vibration signals were acquired by using two sensors [33]. The two sensors were placed on the housing of the tested bearings and positioned at 90 degrees to each other, i.e., one was placed on the vertical axis and the other one was placed on the horizontal axis. The tested bearing's parameters are given in Table 1. The sampling frequency was 25.6 kHz and 32,768 samples (i.e., 1.28 s) were recorded every 1 min. In addition, the rotating speed was 2100 rpm, and the radial force was 12 kN.

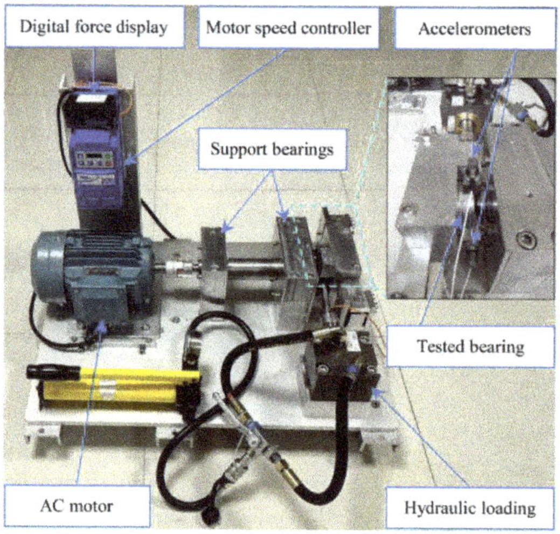

Figure 2. Bearing fault experimental setup.

Table 1. The tested bearing's parameters.

Parameters	Values	Parameters	Values
Outer race diameter	39.80 mm	Inner race diameter	29.30 mm
Mean diameter	34.55 mm	Ball diameter	7.92 mm
Number of balls	8	Contact angle	0°

In the process of the bearing accelerated degradation test, inner race and outer race wear occurred in the tested bearing, as shown in Figure 3. Figure 4 depicts the horizontal and vertical root mean square (RMS) indicators of the tested bearing throughout its whole operating life. It can be seen from Figure 3 that horizontal and vertical RMS indicators were kept unchanged initially and then rose rapidly after 30 min, suggesting that early wear may have occurred in the tested bearing. According to the horizontal and vertical RMS indicators, we can deduce that early compound wear may happen in the inner race and outer race of the tested bearing at the 24th min. Therefore, the horizontal and vertical raw vibration signals and their spectrum are shown in Figures 5 and 6, respectively. In the figures, we mark the inner race, outer race, roller and cage fault characteristic frequencies in the envelope spectrum by using different colors. According to the tested bearing's parameters, we can calculate the theoretical inner race, outer race, roller and cage fault characteristic frequencies as 172.09, 107.91, 72.33 and 13.49 Hz, respectively, by using the following equations:

$$f_{inner} = \frac{N}{2} f_r \left(1 + \frac{d}{D}\cos\theta\right) \tag{18}$$

$$f_{outer} = \frac{N}{2} f_r \left(1 - \frac{d}{D}\cos\theta\right) \tag{19}$$

$$f_{cage} = \frac{1}{2} f_r \left(1 - \frac{d}{D}\cos\theta\right) \tag{20}$$

$$f_{roller} = \frac{D}{d} f_r \left(1 - \left(\frac{d}{D}\cos\theta\right)^2\right) \tag{21}$$

where N is the number of balls, f_r is the rotating frequency, d is the ball diameter, D is the mean diameter and θ is the contact angle.

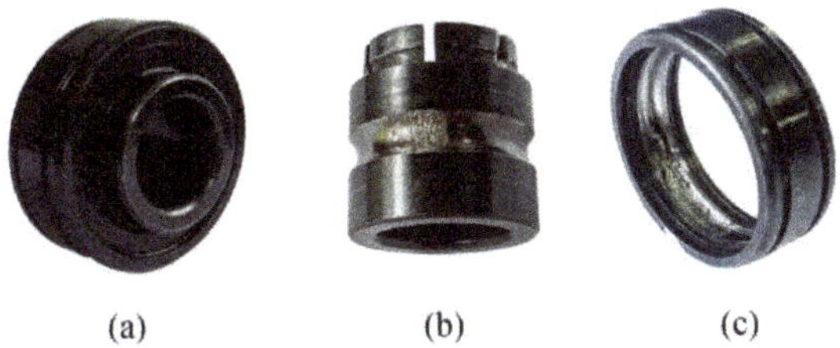

Figure 3. Photos of tested bearings: (**a**) Normal bearing, (**b**) Inner race wear and (**c**) Outer race wear.

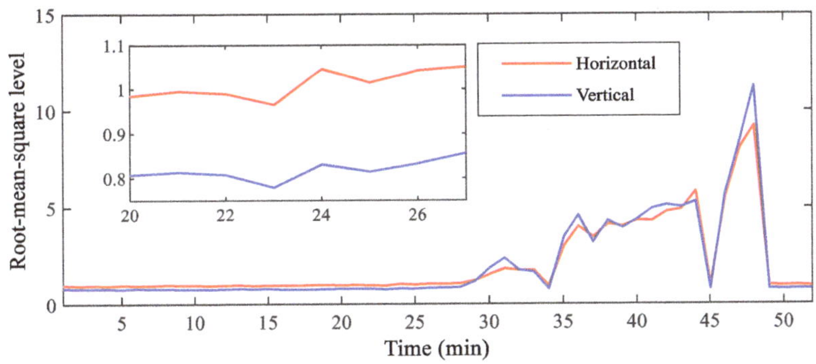

Figure 4. Horizontal and vertical RMS indicators of the tested bearing throughout its whole operating life.

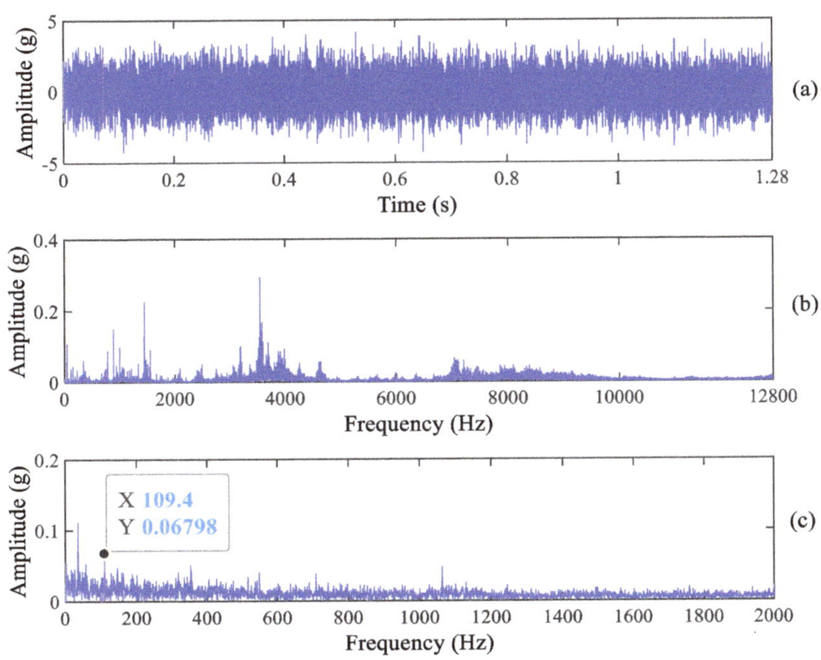

Figure 5. Horizontal raw vibration signal and its spectrum. (**a**) Raw signal, (**b**) Frequency spectrum and (**c**) Zoomed envelope spectrum.

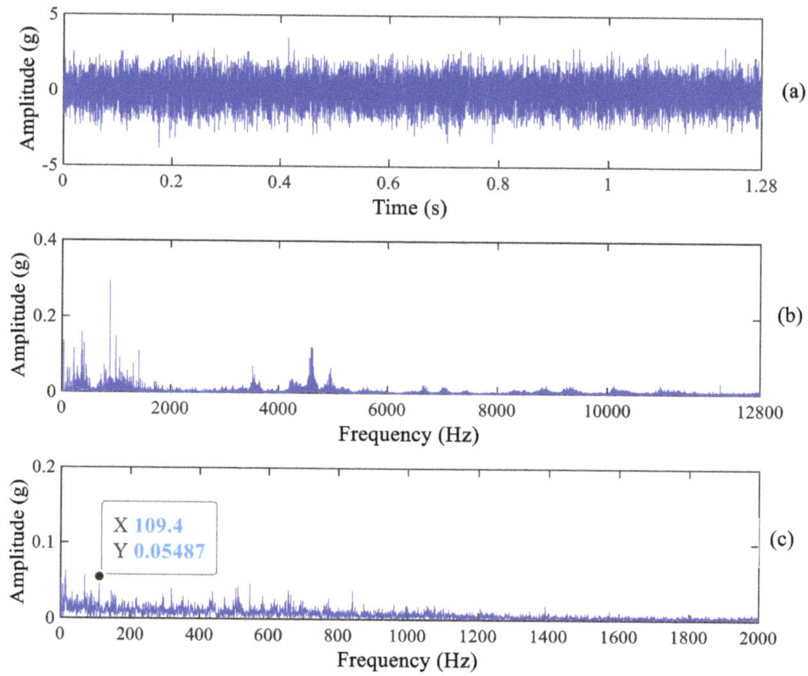

Figure 6. Vertical raw vibration signal and its spectrum: (**a**) Raw signal, (**b**) Frequency spectrum and (**c**) Zoomed envelope spectrum.

Compared with the information in the envelope spectrum in Figures 5 and 6, it is found that there is a clear spectrum peak at 109.4 Hz which is close to the theoretical value of the outer race fault characteristic frequency 107.91 Hz, but the spectrum peaks of its harmonics cannot be recognized. Obviously, the above signature demonstrates that early outer race wear occurs, which is consistent with the real experimental results. In addition, we can see from Figure 5 that there is a clear spectrum peak at 68.75 Hz, which is not a roller fault characteristic frequency. Other information cannot be observed from Figures 5 and 6. As a result, we can deduce that early outer race wear occurs in the tested bearing from the raw signal and its spectrum.

Although we can observe the weak spectrum peak at the outer race fault characteristic frequency, it cannot recognize the fault information excited by inner race wear. Therefore, the proposed method is used to process the raw vibration signal of the tested bearing. Inner race and outer race fault information can be enhanced by using the proposed method, as shown in Figures 7 and 8. It is found from Figure 7b that there is a clear spectrum peak at 171.9 Hz which is close to the theoretical value of the inner race fault characteristic frequency 172.09 Hz, suggesting that early inner race wear happens in the tested bearing. In fact, an inner race wear has been formed as shown in Figure 3c. Meanwhile, we can also observe the clear spectrum peak at 109.4 Hz from Figure 8b which is close to the theoretical value of the outer race fault characteristic frequency 107.91 Hz, indicating that early outer race wear happens. The period of enhanced signal in Figures 7a and 8a is equal to an inner race or outer race fault period, respectively.

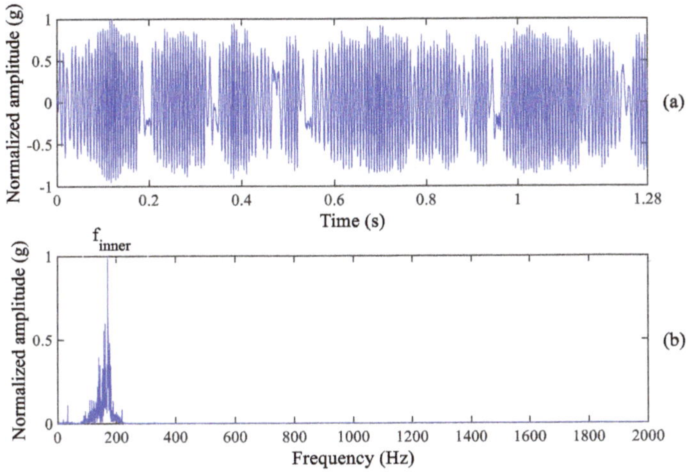

Figure 7. Inner race wear diagnosis result: (**a**) Enhanced signal (**b**) Zoomed frequency spectrum.

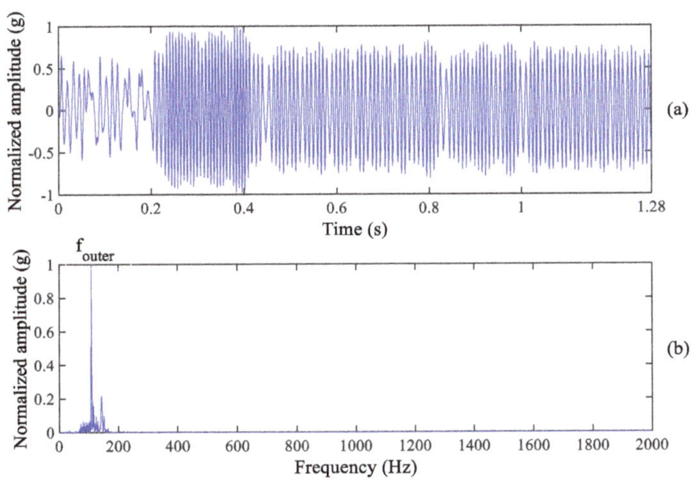

Figure 8. Outer race wear diagnosis result: (**a**) Enhanced signal (**b**) Zoomed frequency spectrum.

For comparison, the fast kurtogram [34,35], as a widely used method in mechanical fault diagnosis, is applied to process the raw vibration signal of the tested bearing. The kurtogram as shown in Figure 9 selects the optimal filtering frequency band as the optimal carrier frequency 3900 Hz with bandwidth 200 Hz to filter out the weak fault signature excited by the outer race and inner race wear in the tested bearing. The corresponding filtered result is shown in Figure 10. It can be seen from the amplitude spectrum of the filtered signal squared envelope in Figure 10 that there are clear spectrum peaks at 34.68 Hz and even at 75.81 Hz. Obviously, these are not fault signatures excited by inner race and outer race wear of the tested bearing. Therefore, we deduce that the optimal filtering frequency band may be the optimal carrier frequency 7866 Hz, as shown in Figure 9. The corresponding filtered signal is depicted in Figure 11. We cannot also observe clear spectrum peaks at the inner race and outer race fault characteristic frequencies from the amplitude spectrum of the filtered signal squared envelope in Figure 11. As a result, the fast kurtogram fails to detect the weak fault characteristics of the tested bearing in this experiment. The comparison demonstrates the feasibility and superiority of the proposed method.

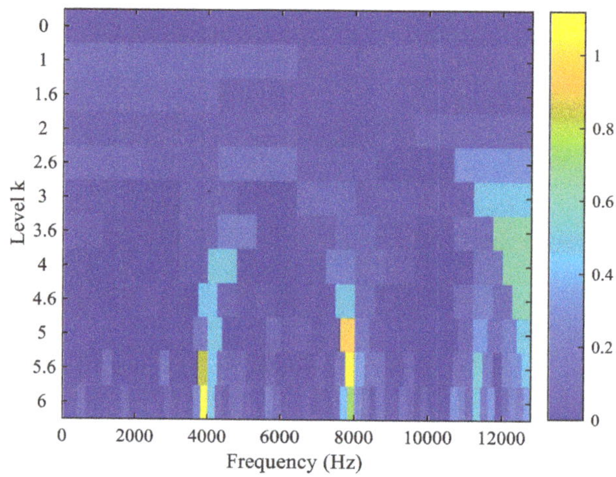

Figure 9. The kurtogram using the fast kurtogram method.

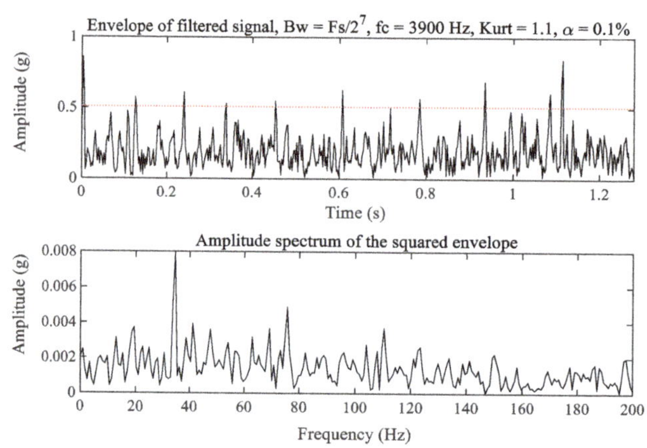

Figure 10. The filtered signal with the optimal carrier frequency 3900 Hz.

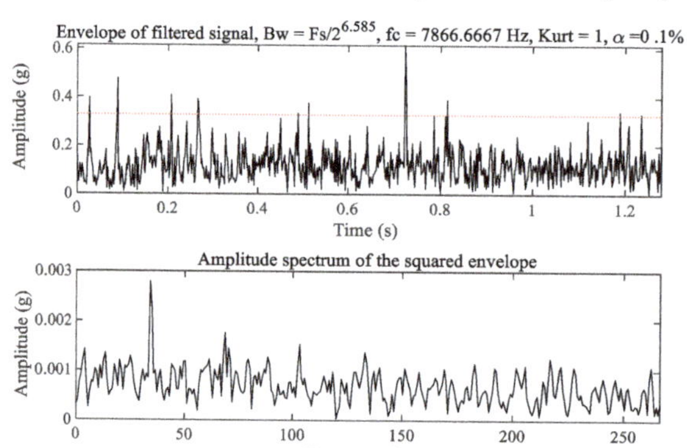

Figure 11. The filtered signal with the optimal carrier frequency 7866 Hz.

4. Conclusions

Stochastic resonances (SR) have been widely used in mechanical fault diagnosis, but most of them require designing the quantification indicators by virtue of prior knowledge of the fault signature to be detected, resulting in false SR and further impaired detection performance of SR. Therefore, it is very necessary for us to study the parameter-matched SR method based on the mathematical condition of triggering SR. For this purpose, we propose an SR method with parameter estimation for bearing fault diagnosis. Compared with other SR methods, the tuning parameters of the proposed method can be estimated by virtue of the raw signal of bearings and do not depend on prior knowledge of the defects to be detected. Moreover, the proposed method considers the conditions of triggering SR to avoid false SR and improve the detection performance of SR further. Such an advantage makes the proposed method more suitable in engineering application. In addition, the proposed method has a tuning parameter and a damping factor with clear mathematical limitation, making it more simple and less time-consuming than other SR methods with a large amount of parameter optimization. A bearing fault experiment was performed to verify this. The experimental results indicate that the proposed method can enhance the weak compound fault diagnosis of bearings at an early stage and is superior to the widely used fast kurtogram method.

Author Contributions: Conceptualization, M.X. and C.Z.; methodology, Z.Q. and K.S.; software, Z.L. and K.S.; validation, K.S., L.X. and Z.Q.; formal analysis, Z.L.; investigation, M.X.; resources, M.X.; data curation, M.X. and L.X.; writing—original draft preparation, M.X. and L.X.; writing—review and editing, K.S. and Z.Q.; visualization, C.Z. and L.X.; supervision, Z.Q.; project administration, Z.Q. and Z.L.; funding acquisition, Z.Q. and Z.L. All authors have read and agreed to the published version of the manuscript.

Funding: This research was supported by Zhejiang Provincial Natural Science Foundation of China (LQ22E050003), Foundation of the State Key Laboratory of Performance Monitoring and Protecting of Rail Transit Infrastructure of East China Jiaotong University (HJGZ2021114), Laboratory of Yangjiang Offshore Wind Power (YJOFWD-OF-2022A08), National Natural Science Foundation of China (52205569, 51905349, U2013603), Chuying Planning Project of Zhejiang Provincial Administration for Market Regulation (CY2023328), Shandong Provincial Innovation Ability Improvement Project of Middle and Small-sized High-tech Enterprises (2022TSGC2364), Ningbo Science and Technology Major Project (2020Z110, 2022Z057, 2022Z002), Ningbo Natural Science Foundation (2022J132), Guangdong Basic and Applied Basic Research Foundation (Grant Nos. 2022A1515010126, 2020A1515011509) and was also sponsored by K.C. Wong Magna Fund in Ningbo University.

Institutional Review Board Statement: Not applicable.

Informed Consent Statement: Not applicable.

Data Availability Statement: The datasets generated during and/or analyzed during the current study are available from the corresponding author on reasonable request.

Conflicts of Interest: The authors declare no conflict of interest.

References

1. Kim, W.; Kim, S.; Jeong, J.; Kim, H.; Lee, H.Y.; Youn, B.D. Digital twin approach for on-load tap changers using data-driven dynamic model updating and optimization-based operating condition estimation. *Mech. Syst. Signal Process.* **2022**, *181*, 109471. [CrossRef]
2. Jerez, D.J.; Jensen, H.A.; Beer, M. An effective implementation of reliability methods for Bayesian model updating of structural dynamic models with multiple uncertain parameters. *Reliab. Eng. Syst. Saf.* **2022**, *225*, 108634.
3. Liu, Y.; Jiang, H.; Liu, C.; Yang, W.; Sun, W. Data-augmented wavelet capsule generative adversarial network for rolling bearing fault diagnosis. *Knowledge-Based Syst.* **2022**, *252*, 109439. [CrossRef]
4. Lei, Y.; Yang, B.; Jiang, X.; Jia, F.; Li, N.; Nandi, A.K. Applications of machine learning to machine fault diagnosis: A review and roadmap. *Mech. Syst. Signal Process.* **2020**, *138*, 106587. [CrossRef]
5. Wang, Z.; Zhou, J.; Du, W.; Lei, Y.; Wang, J. Bearing fault diagnosis method based on adaptive maximum cyclostationarity blind deconvolution. *Mech. Syst. Signal Process.* **2022**, *162*, 108018. [CrossRef]
6. Liang, M.; Zhou, K. Probabilistic bearing fault diagnosis using Gaussian process with tailored feature extraction. *Int. J. Adv. Manuf. Technol.* **2021**, *119*, 2059–2076. [CrossRef]

7. Song, W.; Liu, H.; Zio, E. Long-range dependence and heavy tail characteristics for remaining useful life prediction in rolling bearing degradation. *Appl. Math. Model.* **2022**, *102*, 268–284. [CrossRef]
8. Liu, H.; Song, W.; Zhang, Y.; Kudreyko, A. Generalized Cauchy Degradation Model With Long-Range Dependence and Maximum Lyapunov Exponent for Remaining Useful Life. *IEEE Trans. Instrum. Meas.* **2021**, *70*, 1–12. [CrossRef]
9. Lu, S.; He, Q.; Wang, J. A review of stochastic resonance in rotating machine fault detection. *Mech. Syst. Signal Process.* **2018**, *116*, 230–260. [CrossRef]
10. Qiao, Z.; Lei, Y.; Li, N. Applications of stochastic resonance to machinery fault detection: A review and tutorial. *Mech. Syst. Signal Process.* **2019**, *122*, 502–536. [CrossRef]
11. López, C.; Naranjo, A.; Lu, S.; Moore, K.J. Hidden Markov Model based Stochastic Resonance and its Application to Bearing Fault Diagnosis. *J. Sound Vib.* **2022**, *528*, 116890. [CrossRef]
12. Lin, Y.; Xu, X.; Ye, C. Adaptive stochastic resonance quantified by a novel evaluation index for rotating machinery fault diagnosis. *Measurement* **2021**, *184*, 109920. [CrossRef]
13. Li, Z.; Liu, X.; Wang, X.; He, T.; Shan, Y. A multi-parameter constrained potential underdamped stochastic resonance method and its application for weak fault diagnosis. *J. Sound Vib.* **2019**, *459*, 114862. [CrossRef]
14. Lai, Z.H.; Liu, J.S.; Zhang, H.T.; Zhang, C.L.; Zhang, J.W.; Duan, D.Z. Multi-parameter-adjusting stochastic resonance in a standard tri-stable system and its application in incipient fault diagnosis. *Nonlinear Dyn.* **2019**, *96*, 2069–2085. [CrossRef]
15. Zhang, X.; Wang, J.; Liu, Z.; Wang, J. Weak feature enhancement in machinery fault diagnosis using empirical wavelet transform and an improved adaptive bistable stochastic resonance. *ISA Trans.* **2018**, *84*, 283–295. [CrossRef]
16. Wang, Z.; Yang, J.; Guo, Y.; Gong, T.; Shan, Z. Positive role of bifurcation on stochastic resonance and its application in fault diagnosis under time-varying rotational speed. *J. Sound Vib.* **2022**, *537*, 117210. [CrossRef]
17. Zhou, D.; Huang, D.; Hao, J.; Ren, Y.; Jiang, P.; Jia, X. Vibration-based fault diagnosis of the natural gas compressor using adaptive stochastic resonance realized by Generative Adversarial Networks. *Eng. Fail. Analy.* **2020**, *116*, 104759. [CrossRef]
18. Shi, P.; Li, M.; Zhang, W.; Han, D. Weak signal enhancement for machinery fault diagnosis based on a novel adaptive multi-parameter unsaturated stochastic resonance. *Appl. Acoust.* **2022**, *189*, 108609. [CrossRef]
19. Qiao, Z.; Elhattab, A.; Shu, X.; He, C. A second-order stochastic resonance method enhanced by fractional-order derivative for mechanical fault detection. *Nonlinear Dyn.* **2021**, *106*, 707–723. [CrossRef]
20. Qiao, Z.; Liu, J.; Xu, X.; Yin, A.; Shu, X. Nonlinear resonance decomposition for weak signal detection. *Rev. Sci. Instruments* **2021**, *92*, 105102. [CrossRef]
21. Qiao, Z.; Shu, X. Coupled neurons with multi-objective optimization benefit incipient fault identification of machinery. *Chaos Solitons Fractals* **2021**, *145*, 110813. [CrossRef]
22. Li, J.; Li, M.; Zhang, J.; Jiang, G. Frequency-shift multiscale noise tuning stochastic resonance method for fault diagnosis of generator bearing in wind turbine. *Measurement* **2018**, *133*, 421–432. [CrossRef]
23. Li, J.; Zhang, J.; Li, M.; Zhang, Y. A novel adaptive stochastic resonance method based on coupled bistable systems and its application in rolling bearing fault diagnosis. *Mech. Syst. Signal Process.* **2019**, *114*, 128–145. [CrossRef]
24. Fu, Y.; Kang, Y.; Liu, R. Novel Bearing Fault Diagnosis Algorithm Based on the Method of Moments for Stochastic Resonant Systems. *IEEE Trans. Instrum. Meas.* **2020**, *70*, 1–10. [CrossRef]
25. Xu, H.; Zhou, S.; Yang, T. Stochastic resonance of a high-order-degradation bistable system and its application in fault diagnosis with variable speed condition. *Mech. Syst. Signal Process.* **2023**, *186*. [CrossRef]
26. Qiao, Z.; Chen, S.; Lai, Z.; Zhou, S.; Sanjuán, M.A.F. Harmonic-Gaussian double-well potential stochastic resonance with its application to enhance weak fault characteristics of machinery. *Nonlinear Dyn.* **2023**, 1–15. [CrossRef]
27. Dong, H.; Wang, H.; Shen, X.; Jiang, Z. Effects of Second-Order Matched Stochastic Resonance for Weak Signal Detection. *IEEE Access* **2018**, *6*, 46505–46515. [CrossRef]
28. Gammaitoni, L.; Hänggi, P.; Jung, P.; Marchesoni, F. Stochastic resonance. *Rev. Mod. Phy.* **1998**, *70*, 223. [CrossRef]
29. Li, F.; Duan, F.; Chapeau-Blondeau, F.; Abbott, D. Signal estimation and filtering from quantized observations via adaptive stochastic resonance. *Phys. Rev. E* **2021**, *103*(5), 052108. [CrossRef]
30. Dong, H.; Wang, H.; Shen, X.; He, K. Parameter matched stochastic resonance with damping for passive sonar detection. *J. Sound Vib.* **2019**, *458*, 479–496. [CrossRef]
31. Lu, S.; He, Q.; Kong, F. Effects of underdamped step-varying second-order stochastic resonance for weak signal detection. *Digit. Signal Process.* **2015**, *36*, 93–103. [CrossRef]
32. Kay, S.M. *Fundamentals of statistical signal processing: Estimation theory*; Prentice-Hall, Inc.: New York, NY, USA, 1993; pp. 65–75.
33. Wang, B.; Lei, Y.; Li, N.; Li, N. A Hybrid Prognostics Approach for Estimating Remaining Useful Life of Rolling Element Bearings. *IEEE Trans. Reliab.* **2018**, *69*, 401–412. [CrossRef]
34. Antoni, J. Fast computation of the kurtogram for the detection of transient faults. *Mech. Syst. Signal Process.* **2007**, *21*, 108–124. [CrossRef]
35. Lei, Y.; Lin, J.; He, Z.; Zi, Y. Application of an improved kurtogram method for fault diagnosis of rolling element bearings. *Mech. Syst. Signal Process.* **2011**, *25*, 1738–1749. [CrossRef]

Disclaimer/Publisher's Note: The statements, opinions and data contained in all publications are solely those of the individual author(s) and contributor(s) and not of MDPI and/or the editor(s). MDPI and/or the editor(s) disclaim responsibility for any injury to people or property resulting from any ideas, methods, instructions or products referred to in the content.

MDPI AG
Grosspeteranlage 5
4052 Basel
Switzerland
Tel.: +41 61 683 77 34

Sensors Editorial Office
E-mail: sensors@mdpi.com
www.mdpi.com/journal/sensors

Disclaimer/Publisher's Note: The title and front matter of this reprint are at the discretion of the Guest Editors. The publisher is not responsible for their content or any associated concerns. The statements, opinions and data contained in all individual articles are solely those of the individual Editors and contributors and not of MDPI. MDPI disclaims responsibility for any injury to people or property resulting from any ideas, methods, instructions or products referred to in the content.

www.ingramcontent.com/pod-product-compliance
Lightning Source LLC
Chambersburg PA
CBHW062043220926
51850CB00010B/2648

Genetics and Molecular Breeding of Fruit Tree Species

Guest Editors

Stefano La Malfa
Stefania Bennici

MDPI

Basel • Beijing • Wuhan • Barcelona • Belgrade • Novi Sad • Cluj • Manchester

Guest Editors
Stefano La Malfa
Department of Agriculture,
Food and Environment
University of Catania
Catania
Italy

Stefania Bennici
Department of Agriculture,
Food and Environment
University of Catania
Catania
Italy

Editorial Office
MDPI AG
Grosspeteranlage 5
4052 Basel, Switzerland

This is a reprint of the Special Issue, published open access by the journal *Horticulturae* (ISSN 2311-7524), freely accessible at: https://www.mdpi.com/journal/horticulturae/special_issues/5JNS597927.

For citation purposes, cite each article independently as indicated on the article page online and as indicated below:

Lastname, A.A.; Lastname, B.B. Article Title. *Journal Name* **Year**, *Volume Number*, Page Range.

ISBN 978-3-7258-4785-3 (Hbk)
ISBN 978-3-7258-4786-0 (PDF)
https://doi.org/10.3390/books978-3-7258-4786-0

© 2025 by the authors. Articles in this book are Open Access and distributed under the Creative Commons Attribution (CC BY) license. The book as a whole is distributed by MDPI under the terms and conditions of the Creative Commons Attribution-NonCommercial-NoDerivs (CC BY-NC-ND) license (https://creativecommons.org/licenses/by-nc-nd/4.0/).

Contents

About the Editors

Stefano La Malfa

Stefano La Malfa is a Full Professor of Arboriculture and Fruitculture at the Department of Agriculture, Food and Environment (Di3A) of Catania University. His research activities mainly deal with propagation, breeding, genetic resources characterization and post-harvest of fruit tree species. His current fields of research include the use of biotechnological approaches for propagation and breeding of fruit tree species (mainly citrus); the molecular characterization of species, varieties, and clones of several fruit tree crops; and germplasm collection and exploitation. He has been and is involved as a participant and/or scientific responsible in several research projects concerning different aspects of fruit tree crops and supported by several Italian and foreign institutions. He has been a member of the organizing committees of several national and international congresses. He is a member of the Accademia dei Georgofili, Florence, Italy, and is currently president of the fruticulture section of the Italian Society of Horticulture. He has published more than 100 items indexed on SCOPUS. He has also published some scientific books as editor or coauthor.

Stefania Bennici

Stefania Bennici is Fixed-term Assistant Professor of Arboriculture and Fruitculture at the Department of Agriculture, Food and Environment (Di3A) of Catania University. Her scientific activity concerns research topics related to biology, physiology, molecular characterization, and the genetic improvement of fruit tree species. Her main research topics include the study of reproductive biology in citrus fruits through histological and molecular analysis; marker-assisted selection and molecular characterization in collection of plant genetic resources of different fruit trees; genomic and transcriptomic analysis for the selection of genes related to agronomic traits of interest; and the use of biotechnological approaches, including genetic transformation and genome editing for the propagation and breeding of fruit tree species and mainly citrus. She has been and is involved as a participant and/or scientific responsible in several national and international research projects concerning different aspects of fruit tree crops. She is a member of international and national societies of horticulture and fruitculture.

horticulturae

Editorial

Genetics and Molecular Breeding of Fruit Tree Species

Stefano La Malfa * and Stefania Bennici *

Department of Agriculture, Food and Environment, University of Catania, Via Santa Sofia 100, 95123 Catania, Italy
* Correspondence: stefano.lamalfa@unict.it (S.L.M.); stefania.bennici@unict.it (S.B.)

Fruit tree species contribute to human nutrition and health security by providing important beneficial compounds (e.g., micronutrients, antioxidants) and by playing a key role in the economies of many countries [1]. Global demand for varieties of fruit tree varieties with improved traits—such as enhanced fruit quality, higher yield, and increased resistance to biotic and abiotic stress—is steadily rising. In addition, climate change presents new challenges for perennial crops in many regions, necessitating the development of new varieties capable of adapting to changing environmental conditions, producing high-quality products under stress, and reducing environmental impact [2].

In this context, a thorough understanding of the genetic basis of agronomic traits is essential for both unraveling their regulatory mechanisms and supporting efficient breeding strategies.

However, genetic improvement in fruit trees through conventional methods—such as sexual hybridization and selection—is significantly constrained by the complex genetic and reproductive biology of woody plants. These challenges include a long juvenile period, large plant size, high levels of heterozygosity, and the presence of reproductive barriers (e.g., male or female sterility, incompatibility).

Recent advances in biotechnology and the emergence of genomics offer new opportunities for genetic studies and molecular breeding in fruit crops, helping to overcome the limitations of conventional strategies and enabling the exploitation of novel genetic resources.

The advent of high-throughput methods—such as next-generation sequencing (NGS) and genotyping-by-sequencing (GBS)—alongside genome-wide association studies, advanced phenotyping approaches, and integrated omics technologies (e.g., transcriptomics, proteomics, metabolomics, and hormonomics), has significantly enhanced our understanding of the genetic basis of agronomically important traits [3].

The use of these technologies facilitates the characterization of genetic variability within germplasm collections—an important source of traits for breeding—and contributes to the development of molecular markers. These markers can be used for fingerprinting (e.g., varietal identification, plant-derived products traceability) or for marker-assisted selection (MAS), which enables the early identification of new genotypes with superior traits in conventional breeding programs, thereby reducing time, space, and costs [4]. Additionally, these approaches support more efficient conservation of genetic resources, particularly in neglected species.

Last but not least, genes associated with desirable traits can be introduced or modified in elite cultivars using new plant breeding techniques (e.g., cisgenesis and genome editing), which allow point-specific mutations without introducing foreign genes, thereby preserving the genetic background [5,6]. However, the successful application of these technologies

 https://doi.org/10.3390/horticulturae11070756

depends on the availability of efficient regeneration protocols, which remain a major bottleneck, as many cultivars within a species may be recalcitrant to regeneration [7].

This editorial provides an overview of the Special Issue "Genetics and Molecular Breeding of Fruit Tree Species", which aims to highlight recent advances in the genetic and molecular breeding of fruit tree species for the selection or development of new genotypes with superior characteristics.

This Issue features nine original papers contributed by groups working on various tree species.

Vodiasova et al. (Contribution 1) analyzed the genetic diversity and population structure of 161 peach cultivars from a Russian collection using the GBS approach. They identified a total of 7803 single-nucleotide polymorphism (SNP) markers, which can be used to explore associations with agronomic traits.

Song et al. (Contribution 2) performed an in-depth comparative analysis of chloroplast genome structures across 21 *Vitis* cultivars, revealing valuable genomic resources that can support cultivar selection, breeding, and conservation efforts.

Jia et al. (Contribution 3) reported a novel and efficient pipeline for the development of multi-allelic insertion/deletion markers (InDels). Using NGS-based fingerprints, they successfully discriminated among 122 grape varieties.

Saensouk et al. (Contribution 4) carried out a cytological study on four edible and ornamental *Zingiber* species, generating insights that can inform plant breeding strategies for commercial purposes.

Simoni et al. (Contribution 5) provided a comprehensive characterization of transposable elements (TEs) in *Punica granatum* through a comparative analysis of the genome assemblies of four Tunisian pomegranate cultivars, providing information that can be leveraged for breeding and crop improvement in this species.

Wang et al. (Contribution 6) developed an effective plant regeneration protocol by inducing high-vigor somatic embryos from stem explants of pomegranate, leading to an essential advancement for genetic resource conservation and breeding efforts.

Gao et al. (Contribution 7) conducted a large-scale investigation on the callus induction and adventitious root development across peach germplasms. They identified a candidate gene potentially involved in the regulation of root formation and pinpointed cultivars more prone to regeneration protocols.

Guo et al. (Contribution 8) performed whole-genome resequencing of jujube cultivars to identify genome-wide SNP markers and constructed a high-density bin map, providing valuable tools for the selection of multiple traits in jujube breeding.

Odoi et al. (Contribution 9) carried out association mapping on a panel of 374 Shea tree (*Vitellaria paradoxa*) accessions using 7530 SNPs markers, offering a foundation for molecular breeding strategies aimed at improving oil yield in the species.

Conflicts of Interest: The authors declare no conflicts of interest.

List of Contributions:

1. Vodiasova, E.; Pronozin, A.; Rozanova, I.; Tsiupka, V.; Vasiliev, G.; Plugatar, Y.; Dolgov, S.; Smykov, A. Genetic Diversity and Population Structure of *Prunus persica* Cultivars Revealed by Genotyping-by-Sequencing (GBS). *Horticulturae* **2025**, *11*, 189. https://doi.org/10.3390/horticulturae11020189.
2. Song, Y.; Wang, L.; Zhang, L.; Li, J.; Teng, Y.; Zhang, Z.; Xu, Y.; Fan, D.; He, J.; Ma, C. Unified Assembly of Chloroplast Genomes: A Comparative Study of Grapes Representing Global Geographic Diversity. *Horticulturae* **2024**, *10*, 1218. https://doi.org/10.3390/horticulturae10111218.

3. Jia, G.; Zhang, N.; Yang, Y.; Jin, Q.; Jiang, J.; Zhang, H.; Guo, Y.; Wang, Q.; Zhang, H.; Wu, J.; et al. NGS-Based Multi-Allelic InDel Genotyping and Fingerprinting Facilitate Genetic Discrimination in Grapevine (*Vitis vinifera* L.). *Horticulturae* **2024**, *10*, 752. https://doi.org/10.3390/horticulturae10070752.
4. Saensouk, P.; Saensouk, S.; Senavongse, R.; Maensiri, D.; Souladeth, P. Cytogenetics Study of Four Edible and Ornamental Zingiber Species (Zingiberaceae) from Thailand. *Horticulturae* **2024**, *10*, 409. https://doi.org/10.3390/horticulturae10040409.
5. Simoni, S.; Usai, G.; Vangelisti, A.; Castellacci, M.; Giordani, T.; Natali, L.; Mascagni, F.; Cavallini, A. Decoding the Genomic Landscape of Pomegranate: A Genome-Wide Analysis of Transposable Elements and Their Structural Proximity to Functional Genes. *Horticulturae* **2024**, *10*, 111. https://doi.org/10.3390/horticulturae10020111.
6. Wang, J.; Xia, X.; Qin, G.; Tang, J.; Wang, J.; Zhu, W.; Qian, M.; Li, J.; Cui, G.; Yang, Y.; et al. Somatic Embryogenesis and Plant Regeneration from Stem Explants of Pomegranate. *Horticulturae* **2023**, *9*, 1038. https://doi.org/10.3390/horticulturae9091038.
7. Gao, L.; Liu, J.; Liao, L.; Gao, A.; Njuguna, B.N.; Zhao, C.; Zheng, B.; Han, Y. Callus Induction and Adventitious Root Regeneration of Cotyledon Explants in Peach Trees. *Horticulturae* **2023**, *9*, 850. https://doi.org/10.3390/horticulturae9080850.
8. Guo, T.; Qiu, Q.; Yan, F.; Wang, Z.; Bao, J.; Yang, Z.; Xia, Y.; Wang, J.; Wu, C.; Liu, M. Construction of a High-Density Genetic Linkage Map Based on Bin Markers and Mapping of QTLs Associated with Fruit Size in Jujube (*Ziziphus jujuba* Mill.). *Horticulturae* **2023**, *9*, 836. https://doi.org/10.3390/horticulturae9070836.
9. Odoi, J.B.; Adjei, E.A.; Barnor, M.T.; Edema, R.; Gwali, S.; Danquah, A.; Odong, T.L.; Hendre, P. Genome-Wide Association Mapping of Oil Content and Seed-Related Traits in Shea Tree (*Vitellaria paradoxa* subsp. *nilotica*) Populations. *Horticulturae* **2023**, *9*, 811. https://doi.org/10.3390/horticulturae9070811.

References

1. Devirgiliis, C.; Guberti, E.; Mistura, L.; Raffo, A. Effect of Fruit and Vegetable Consumption on Human Health: An Update of the Literature. *Foods* **2024**, *13*, 3149. [CrossRef] [PubMed]
2. Bhattacharjee, P.; Warang, O.; Das, S.; Das, S. Impact of Climate Change on Fruit Crops—A Review. *Curr. World Environ.* **2022**, *17*, 319–330. [CrossRef]
3. Cazzonelli, C.I.; Varkonyi-Gasic, E.; Prentis, P.J. Advancing tree genomics to future proof next generation orchard production. *Front. Plant Sci.* **2024**, *14*, 1321555. [CrossRef]
4. De Mori, G.; Cipriani, G. Marker-Assisted Selection in Breeding for Fruit Trait Improvement: A Review. *Int. J. Mol. Sci.* **2023**, *24*, 8984. [CrossRef] [PubMed]
5. Sattar, M.N.; Iqbal, Z.; Al-Khayri, J.M.; Jain, S.M. Induced Genetic Variations in Fruit Trees Using New Breeding Tools: Food Security and Climate Resilience. *Plants* **2021**, *10*, 1347. [CrossRef] [PubMed]
6. Penna, S.; Jain, S.M. Fruit Crop Improvement with Genome Editing, In Vitro and Transgenic Approaches. *Horticulturae* **2023**, *9*, 58. [CrossRef]
7. Ochatt, S.J.; Akin, M.; Chan, M.T.; Dolgov, S.V.; Eimert, K.; Flachowsky, H.; Guo, W.W.; Jiménez, V.M.; Lambardi, M.; Moncaleán, P.; et al. Research is rendering the recalcitrant woody plants amenable to biotechnological approaches. *Plant Cell Tissue Organ Cult.* **2025**, *161*, 48. [CrossRef]

Disclaimer/Publisher's Note: The statements, opinions and data contained in all publications are solely those of the individual author(s) and contributor(s) and not of MDPI and/or the editor(s). MDPI and/or the editor(s) disclaim responsibility for any injury to people or property resulting from any ideas, methods, instructions or products referred to in the content.

horticulturae

MDPI

Article

Genome-Wide Association Mapping of Oil Content and Seed-Related Traits in Shea Tree (*Vitellaria paradoxa* subsp. *nilotica*) Populations

Juventine Boaz Odoi [1,2,3,4,5,*], Emmanuel Amponsah Adjei [2,3,6], Michael Teye Barnor [4], Richard Edema [2,3], Samson Gwali [1], Agyemang Danquah [5], Thomas Lapaka Odong [2] and Prasad Hendre [7]

1 National Forestry Resources Research Institute (NaFORRI), Agricultural Research Organization (NARO), Kampala P.O. Box 1752, Uganda; s.gwali2@gmail.com
2 School of Agricultural Sciences (SAS), College of Agricultural and Environmental Sciences (CAES), Makerere University, Kampala P.O. Box 7062, Uganda; emmaadjei1@gmail.com (E.A.A.); redema14@gmail.com (R.E.); thomas.l.odong@gmail.com (T.L.O.)
3 Makerere Regional Center for Crop Improvement (MaRCCI), College of Agricultural and Environmental Sciences (CAES), Makerere University, Kampala P.O. Box 7062, Uganda
4 Cocoa Research Institute of Ghana (CRIG), Bole P.O. Box BL 41, Ghana; teye.barnor@gmail.com
5 West Africa Center for Crop Improvement (WACCI), College of Basic and Applied Sciences, University of Ghana, Accra P.O. Box LG 25, Ghana; adanquah@wacci.ug.edu.gh
6 Council for Scientific and Industrial Research, Savannah Agricultural Research Institute, Tamale P.O. Box TL 52, Ghana
7 Center for International Research in Forestry—International Center for Research in Agroforestry (CIFOR—ICRAF), United Nations Avenue, Girigiri, P.O. Box 30677, Nairobi 00100, Kenya; p.hendre@cifor-icraf.org
* Correspondence: juventineboaz@gmail.com; Tel.: +256-782-568-822

Abstract: Shea tree (*Vitellaria paradoxa*) is an important fruit tree crop because of its oil used for cooking and the industrial manufacture of cosmetics. Despite its essential benefits, quantitative trait loci linked to the economic traits have not yet been studied. In this study, we performed association mapping on a panel of 374 shea tree accessions using 7530 Single-Nucleotide Polymorphisms (SNPs) markers for oil yield and seed-related traits. Twenty-three SNP markers significantly ($-\log10$ (p) = 4.87) associated with kernel oil content, kernel length, width, and weight were identified. The kernel oil content and kernel width had the most significant marker–trait associations (MTAs) on chromosomes 1 and 8, respectively. Sixteen candidate genes identified were linked to early induction of flower buds and somatic embryos, seed growth and development, substrate binding, transport, lipid biosynthesis, metabolic processes during seed germination, and disease resistance and abiotic stress adaptation. The presence of these genes suggests their role in promoting bioactive functions that condition high oil synthesis in shea seeds. This study provides insights into the important marker-linked seed traits and the genes controlling them, useful for molecular breeding for improving oil yield in the species.

Keywords: linked; marker association; annotation; genes; SNPs; shea

1. Introduction

The shea tree (*Vitellaria paradoxa* C. F. Gaertn.) is an important economic tree crop known for its oil used to produce valuable products in the food and cosmetic industries [1]. The tree is endemic to Sudano-Sahelian Africa, covering 21 countries [2], where it adds to the sustainability of sociocultural and economic wellbeing of the communities. The shea tree: *Vitellaria paradoxa* C. F. Gaertn., has two described subspecies: *V. paradoxa* subsp. *paradoxa* and *V. paradoxa* subsp. *nilotica*. The two subspecies vary for their morphological characteristics [3]. The subspecies *nilotica* has larger flowers and a dense "woolly" appearance that remains on young leaves and persists on leaf veins and midribs. It is characterized by dense ferruginous indumentum on pedicels and outer sepals, with the constituent hairs being longer, spreading and imparting a woolly appearance to the parts during the bud stage. The hermaphroditic and actinomorphic flowers are always in dense clusters on the twigs that have not formed leaves [4]. In the subspecies *paradoxa*, the flowers have longer styles measuring 12–15 cm [5].

The tree is a diploid (2n = 24), highly outcrossing that has undergone domestication in the savannah parklands of Africa for over 1000 years [6]. A molecular marker study by Allal et al. [7] placed the centre of origin of *V. paradoxa* in West Africa, where three genetic groups corresponding to West, Central, and East African types were found. The shea genome size is up to 658.7 Mbp, consisting of 38,505 coding genes [8]. There is an observable variation in stand densities within the shea parklands, due to the differences in land use, localities, soils, rainfall, temperature, daylight length, and ecological conditions [9], forming seven different morphological and structural forms [10].

The shea tree is recognized as the second-highest oil-producing plant, after the oil palm [8]. The global market for shea products was reported to be USD 30 billion in 2020 [11]. This high demand is owed to its use in the confectionary and cosmetic industries. The demand for these natural and organic cosmetics in the European market reached EUR 3.90 billion in 2019 [12]. The cosmetic sector alone exceeded USD 530 in 2020 and is expected to rise to USD 1025 million in 2027. Among this, the US market alone is projected to rise from USD 240 million in 2020 to USD 390 in 2027 and expected to grow at a compound annual growth rate (CAGR) of 7%, due to the increasing demand in the cosmetics industry [13]. The total export of oils from different plants to Europe in 2020 was estimated at 300,000 tonnes, with the Netherlands and France being the leading importers. However, both processed and unprocessed products are sold in national and international markets, contributing to national income through foreign exchange in the shea-producing countries. The leading producers of shea in Africa are Nigeria (361,017 tons/year), Mali (49,640 tons/year), Burkina Faso (45,183 tons/year), and Ghana (33,878 tons/year) [11]. There is a huge and untenable supply deficit due to heightened international demand, necessitating breeding interventions to boost production across its range.

The first recognizable shea tree improvement efforts were through a participatory selection trial of plus trees from three countries in West Africa [14]. A larger collection (from Burkina Faso, Benin, Nigeria, Ghana, Cameroun, Niger, and Mali) of carefully selected plus shea trees raised from clonal materials was also established in Mali by the World Agroforestry (CIFOR-ICRAF) [14]. Over time, there has been increasing interest in improving shea tree productivity to meet the looming domestic and international demand for shea products [15]. Some concerted efforts have been made in shea tree breeding at the University of Peleforo Gon Coulibaly (UPGC) in Korhogo, Côte d'Ivoire, where some elite trees were selected and propagated [16] by grafting to reduce the juvenile maturity period [17]. Other innovative approaches have been used in participatory plant breeding (PPB), using local knowledge to identify and select preferred traits by the communities. Such traditional and contemporary breeding and selection processes are important for tree species like shea trees, to generate new varieties with various desired properties [18].

Recent advances in shea tree genomic studies by Hale et al. [8] and Wei et al. [1] have provided insights on the new opportunities in genome-assisted breeding. Despite these advancements, the genomic resources remain underused for boosting production and improving oil yield and quality. Genome-wide association studies (GWAS) provide opportunities to identify genomic regions of an organism that are putatively associated with the traits of interest to plant breeders [8]. With the availability of affordable and economic modifications of genome sequencing approaches like genotyping by sequencing (GBS), discovering and using SNP markers has become a preferred way of genotyping. One of these technologies is the Diversity Arrays Technology Sequencing (DArTseq), where a genome is partially sequenced using a specific combination of restriction enzymes and the restriction tags are used for assembling and discovering the SNP markers [19]. The discovered SNPs are generally spread all over the genome and can be used in GWAS for the study of a wide range of tree crop traits of economic importance [20]. This study was carried out to identify genomic loci associated with seed oil content, seed weight, seed length and width of the shea tree in Uganda. Determining the marker trait association shall enhance shea tree breeding by reducing on the time required to complete the breeding cycle.

2. Materials and Methods

2.1. Plant Materials and Leaf Sampling for DNA Extraction

A total of 374 shea genotypes from the germplasm collection (Breeding Seedling Orchard) Uganda were used in this study. A total of 3600 shea fruits/seeds were collected from 180 families (Supplementary Table S1) in the districts of Amuru, Arua, Katakwi, Moyo and Otuke. The seeds were then divided into two portions: for sowing and for oil extraction. A minimum of 10 seeds were randomly picked from each family for sowing to generate seedlings used in DNA extraction. Fifteen seeds from the remaining lot were processed and used for oil extraction.

The shea seeds were sown in a tree nursery at Ngetta Zonal Agricultural Research Development Institute (NgeZARDI), Lira—Uganda in the month of June 2018. The seedlings were managed in the tree nursery for 12 months until they developed between 4–6 leaves before sampling the leaf tissues for DNA extraction. Leaf samples of 374 seedlings were randomly picked for DNA extraction and analysis at Biosciences Eastern and Central Africa- International Livestock Research Institute (BeCA-ILRI). Only healthy and recently flushed leaves from the previous season were sampled and placed in DNA extraction kit and dried using Silica gel before shipping to BeCA-ILRI.

After leaf tissue sampling, the genotypes were further managed in the nursery for another 6 months to allow them to heal and later planted in a multi-locational trial (breeding seed orchard) located in Lira (NgettaZARDI) and Serere (National Semi Arid Resources Research Institute (NASARRI), using Random Complete Block Design (RCBD) in the month of October 2019. The trials were maintained as germplasm collection for future breeding programme in Uganda.

2.1.1. Shea Oil Extraction Procedure

Oil content was determined using Soxhlet extraction [21], the American Official Agricultural Chemists' method for determination of oil content in plant materials in the months of September and October 2020. Oil was extracted with continuous reflux of petroleum ether over crushed dried Shea nut powder in a Soxhlet extractor. The oil contents of each seed lot were extracted in triplicates and presented in percentage of its dry matter content.

2.1.2. DNA Extraction and SNP Discovery by DArTseq™ Technology

Total genomic DNA from silica dried leaf samples were extracted at BeCA-ILRI following the CetylTrimethylAmmonium Bromide (CTAB)/chloroform/isoamyl alcohol method [22]. DNA samples were processed in digestion/ligation reactions as described by Hale et al. [8]. The DNA was quality checked using standard processes involving 0.8% agarose gel electrophoresis, optical measurements for 260 and 280 nm using a NanoDrop 2000 spectrophotometer (ND-2000 V3.5, NanoDrop Technologies, Inc., Wilmington, DE, USA) and quantification using a Qubit™ 3.0 Fluorometer (Thermo Fisher Scientific, Grand Island, NY, USA). The libraries were prepared for 752 individuals using the PstI-SphI complexity reduction method [23] and partial-genome sequenced using proprietary DArTseq (1.0) methodology [19] on a HiSeq2500 Sequencer (Illumina Inc., San Diego, CA, USA) with 72 bases read length [24,25].

Sequences generated from each lane were processed using proprietary DArT analytical pipelines. DArT-Seq™ technology relies on a complexity reduction method using restriction enzymes that are sensitive to DNA methylated sites and repetitive DNA [24]. In the primary pipeline, the FASTQ files were first processed to filter poor-quality sequences, applying more selection criteria to the barcode region compared to the rest of the sequence. Approximately 2,500,000 (±7%) sequences per barcode/sample were used in marker calling. Finally, identical sequences were collapsed into "fastqcall files." These files were used in the secondary pipeline for DArT P/L's proprietary SNP and SilicoDArT (Presence/Absence Markers in genomic representations) (present = 1 vs. absent = 0) calling algorithms (DArTsoft14). The analytical pipeline processed the sequence data. The reads were then aligned to the shea_V1 reference genome publicly available from the ORCAE database (https://bioinformatics.psb.ugent.be/orcae) (accessed on 30 December 2021), using BWA-MEM/VarDict mapper for mapping of reads against the reference genome [8].

2.2. Data Analysis

2.2.1. Seed Trait Data Analysis

The seed trait data were analysed using *"agricolae"* package in R software v 4.0 [26]. Analysis of variance (ANOVA) was performed to determine the variations within and among the genotypes. The *"corr"* function in R software v.4.0 (R Core Team, 2022) was used to calculate correlation coefficients between the studied traits and presented in graphical form.

2.2.2. Genome-Wide Association Analysis and Gene Annotation Identification

A multi-locus random-SNP-effect mixed linear model (mrMLM) [26] was implemented in R statistical software using the mixed model equation for GWAS presented in Equation (1), in accordance to Yu et al. [27], using additive, general; dominant alternative and dominant reference gene action models for trait association study [28]. This current study selected mrMLM method to avoid bottlenecks in stringent correction using other control measures (false discovery rate (FDR) and Bonferroni correction) against false positive rate [29]. The mrMLM uses a less stringent significance threshold considering a critical probability value or log of odds (LOD) making it possible to identify any possible loci of importance.

$$Y = Xb + Zu + e \tag{1}$$

where:

Y = the vector of the phenotypic observations estimated for the traits studied;
X = the SNP markers (fixed effect) matrix;
Z = the random kinship (co-ancestry) matrix;
b = a vector representing the estimated SNP effects;
u = a vector representing random additive genetic effects, and
e = the vector for random residual errors.

The phenotypic variation explained by the model for a trait and a particular SNP was determined using stepwise regression implemented in the *"lme4"* R package. The SNP loci in significant association with traits were determined by adjusted *p*-value using Bonferroni correction [30]. Quantile–quantile (QQ) plots were generated by plotting the negative logarithms (−log10) of the *p*-values against their expected *p*-values to test the appropriateness of the GWAS model with the null hypothesis of no association and to determine how well the models accounted for the population structure.

To account for the putative genes linked to traits, a window range of 5 kb (upstream and downstream) was defined [31]; and genes were searched from the *V. paradoxa* Whole Genome v2.0 Assembly and Annotation v2.1 [32] in the ORCAE database (https://bioinformatics.psb.ugent.be/orcae, accessed on the 30 November 2022) [3], with a search for candidate genes associated with oil yield traits. The gene name, description, and AGPv4 coordinates with their protein, were then retrieved from the *Vitellaria paradoxa* reference genome database. The putative functional candidate genes linked to the associated SNPs were then annotated in line with any initially annotated genes from other species.

A Linkage disequilibrium (LD) heat map was generated for the entire genome, with heterozygous calls ignored and a default sliding window of 50 used in tassel software. LD decay rate was then evaluated on a chromosome-by-chromosome basis. A measure of LD (r^2) and pairwise distance between SNPs were generated in TASSEL and exported to R version 4.3, where scripts were written to generate LD decay plots for significant

LD pairs. Mean LD per chromosome was calculated after every 20 kb interval, and the average genome-wide decay rate estimated by averaging LD in each interval across all chromosomes. A line graph was used to clearly display an overlay of chromosome-specific and the mean genome-wide LD decay rates.

3. Results

3.1. Phenotypic Variation for the Shea Tree Traits

The traits mean values, standard deviations and the phenotypic data range of a collection of 374 open pollinated seeds from 180 shea trees from Uganda's parklands are presented in Table 1.

Table 1. Summary statistics for the studied traits.

Traits	Mean ± (SD [a])	Minimum	Maximum
Kernel dry matter oil content (% [b])	53.53 ± 2.28	39.05	69.77
Kernel length (cm [c])	3.19 ± 0.34	1.90	8.43
Kernel width (cm)	3.61 ± 0.43	2.23	4.97
Kernel weight (mg [d])	10.30 ± 0.30	2.00	18.8

[a] Standard Deviation; [b] Percentage; [c] Centimetre; [d] Milligram.

The mean seed oil content of 180 shea genotypes was 53.53% with a range of 39.05–69.77%. A relatively heavy kernels (18.81) and very low weight genotypes were also observed (Table 1).

Analysis of variance showed that genotype, environment, and their interaction (genotype-environment) were highly significant for kernel oil content (Table 2). Variation in kernel weight and its axial dimensions were significantly influenced by genotype and the environment. However, the interaction of genotype and environment had no significant effect on kernel weight and its axial dimension (Table 2).

Table 2. Summary analysis of variance for the studied traits.

Source of Variation	Df [a]	KOC [b]	KL [c]	KW [d]	KWt [e]
Replications	2	4.81	0.01307	0.0249	0.08108
Environment	4	1840.82 ***	0.694 ***	0.82403 ***	0.90574 ***
Genotypes	373	60.42 ***	1.45026 ***	2.54701 ***	0.9112 ***
Genotype x Environment	1492	35.9 **	0.01524	20.69	0.01666
Residuals	3738	8.61	0.0159	0.01553	0.02156

[a] Degrees of freedom; [b] Kernel dry matter oil content (%); [c] kernel length (cm); [d] kernel width (cm); [e] Kernel weight (mg) and levels of significance '***' 0.001 '**' 0.01 '*' 0.05.

Seed oil content showed a significant positive correlation with kernel width ($r = 0.1$, $p \leq 0.001$). However, it negatively correlated with kernel weight (−0.01) and kernel length (−0.09) (Figure 1). The result further revealed a moderate (0.44) correlations between kernel width and kernel weight, and kernel width and kernel oil content (0.1), whereas oil content is negatively correlated with kernel weight (−0.1) and kernel length (−0.9) (Figure 1).

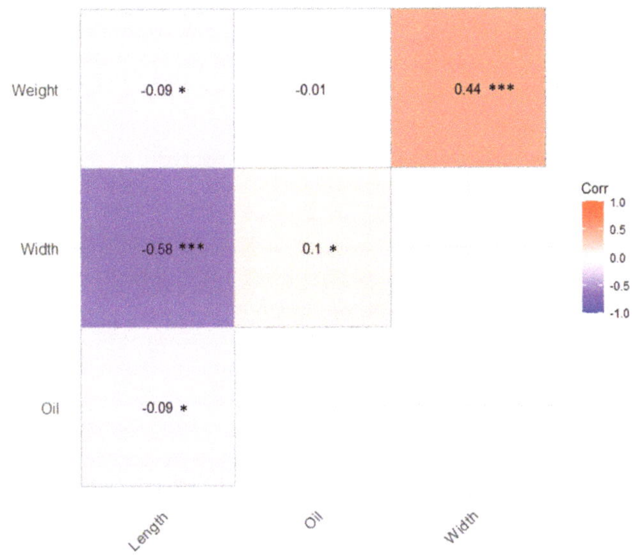

Figure 1. Correlation among four traits (Length = Kernel length, Width = Kernel width and Weight = Kernel Weight and Oil = Kernel oil content) of the 374 Shea tree lines. Colour in the boxes indicate proportion of correlations. * $p < 0.05$; ** $p < 0.01$; *** $p < 0.001$.

3.2. Marker Coverage and SNP Distribution

The SNP calling pipeline generated 30,733 highly polymorphic SNP markers, of which 27,063 (88.1%) remained unmapped on the 12 *Vitellaria paradoxa* chromosomes. Only 7530 SNP markers (27.8%) of the mapped SNP markers were retained after filtering with >20% of missing data, <0.05 minor allele frequency (MAF) and utilized as input for the GWAS analysis.

Chromosome two had the highest number of markers (960 SNPs; Chr size = 74.5 Mb, ~13 SNPs/Mb) followed by chromosomes one (805 SNPs; Chr size = 82 Mb; ~10 SNPs/Mb), chromosome ten (780 SNPs; Chr size = 50 Mb; 10 SNPs/Mb), five and eight (650 SNPs; Chr size = 56.5 Mb; ~11 SNPs/Mb, and 645 SNPs; Chr size = 58 Mb; ~12 SNPs/Mb respectively). Meanwhile, chromosomes four (425 Chr size = 37 Mb; ~ 12 SNPs/Mb) and chromosome three (430 SNPs; Chr size = 38.6 Mb; 11 SNPs/Mb) had the lowest number of markers (Figure 2 and Table 3). This indicates a non-random distribution of SNPs with varying SNP frequencies on the 12 chromosomes of shea tree genome in Uganda. Further population structure and SNP data (Table 3) information are available in Odoi et al. [33].

Minor allele frequency (MAF) among the 7530 SNP markers varied from 0.03 to 0.50. The study further revealed a high level of heterozygosity within individuals (0.26) and markers (0.32) indicating a high non-random association of alleles at different loci that offer opportunity for association studies and allele transfer through marker-assisted selection of the population. The filtered markers were similar in their Polymorphic Information Content (PIC), ranging from 0.258 (chromosome 4) to 0.269 (chromosome 12) with a mean PIC of 0.26 across the chromosomes (Table 3).

There was a general high gene diversity (0.32) across the chromosomes with chromosome 12 being the highest (0.33) and chromosomes 4, 1 and 6 being the lowest (0.31 respectively). Structure analysis revealed that shea tree populations in Uganda are genetically grouped into two clusters of Eastern group and West Nile/Northern Uganda group. The Eastern cluster contributed the highest (57%) proportion of individuals and West Nile/Northern Uganda cluster (43%).

Out of the 12 chromosomes in the shea genome (Figure 2), only two (Chromosome 1 and 8) revealed significant loci. The result of Linkage disequilibrium (LD) indicated that 187,487 loci pairs in a physical distance of 605,450 bp. Of the total loci, 3.62% (6795) of them were in significant ($p < 0.01$) LD. The results further revealed that 87 (1.28%) loci pairs had $r^2 = 1$ (were in complete LD).

Figure 2. The number and size of SNPs within 1 Mb window size of *V. paradoxa* Subsp. *nilotica* genome.

Table 3. Number of SNPs for each chromosome before and after filtration and the average polymorphism information content for *V. paradoxa* subsp. *nilotica*.

Chromosomes	All SNPs [a]	Filtered SNPs	Chr [b] Size (Mbs)	PIC [c]	Gene Div [d]
1	2893	805	82	0.262	0.32
2	3450	960	74.5	0.260	0.32
3	1545	430	38.6	0.261	0.32
4	1527	425	37	0.258	0.31
5	2336	650	56.5	0.261	0.32
6	2210	615	58	0.259	0.31
7	2088	581	57.3	0.262	0.32
8	2318	645	48	0.260	0.32
9	2124	591	56.5	0.262	0.32
10	2803	780	50	0.265	0.32
11	1791	498	47.1	0.265	0.32
12	1978	550	46.9	0.269	0.33
Total/Mean	27,063	7530	652.4	0.260	0.32

[a] Single Nucluotide Polymorphism; [b] Chromosome; [c] Polymorphic information content and [d] gene diversity.

3.3. Marker Association for the Studied Traits

The association analysis was performed on shea seed-related traits and 16 significant markers were identified on chromosomes 1, 2, 3, 5, 6, 7, 8, 9, 10, 11 and 12 (Table 4 and Figure 3). Quantile-Quantile plots produced by displaying $-\log10$ p-values against individual p-values revealed suitability of GWAS for the trait's connection in the shea tree genotypes. The association analysis was performed for percent oil content of each shea tree line in a location using the *V. paradoxa* reference genome (https://bioinformatics.psb.ugent.be/orcae) (accessed on 8 March 2022). There were differences between the observed and expected values of the target traits, indicating a link between the phenotypic and SNP markers as indicated in Quantile-Quantile plots.

The seven SNP markers linked to shea nut oil yield (S1_60237300, S3_14843482, S4_32032310, S5_6275145, S8_41696703, S9_32689981 and S11_43126044) were located on chromosomes 1, 3, 4, 5, 8, 9 and 11 (Table 4) and were associated with high nut percent oil content estimated on dry matter basis. These seven loci explained an overall phenotypic variance of 12.4%, however, makers S8_41696703 and S9_32689981 had negative effects on seed oil content, although they explained the most (13.31% and 11.52% respectively) of phenotypic variation.

This current study revealed six significant SNP markers linked with shea kernel length (S3_11153087, S5_15524578, S6_46530240, S8_11121701, S11_8320549 and S12_32853547) located on chromosomes 3, 5, 6, 8, 11 and 12 (Table 4; Figure 3). The proportion of phenotypic variance explained by significant QTNs ranged from 6.5% in marker S5_15524578 to 14.6% in S6_46530240. The total phenotypic variance expressed by the trait was 0.095.

The GWAS revealed 8 genomic regions that were significant associated with kernel width. The 8 significant SNP markers linked to shea kernel width (S1_32402910, S2_47786838, S2_64059706, S7_3025298, S9_43700743, S10_50604452, S12_32853547 and S12_7613999) were located on chromosomes 1, 2, 7, 9, 10 and 12 (Table 4). Marker S12_32853547 contributed most (13.14%) of the phenotypic variation compared to the rest (ranging from 4.5%to 9.75%) (Table 4). The total phenotypic variation explained by the trait was 0.17.

In two significant SNPs (S1_30720144 and S8_43605016) located on chromosomes 1 and 8. Marker S8_43605016 contributed most (15.79%) of the phenotypic variation compared to S1_30720144 (9.21%) (Table 4). The total phenotypic variance in this trait was 0.061 Table 4: Figure 3).

Table 4. List of significant markers in a panel of 374 *Vitellaria paradoxa* genotypes indicating the genomic regions associated with studied traits.

Trait	$P\sigma$ [a]	Marker	Chr [b]	Position (bp)	Alleles	QTN Effect	LOD Score	−log10 [c]	r^2 [d]	MAF [e]
Oil content	4.03	S1_60237300	1	60237300	AA	0.83	3.39	4.11	6.61	0.12
		S3_14843482	3	14843482	AA	−1.06	5.67	6.49	11.80	0.14
		S4_32032310	4	32032310	AA	0.74	3.07	3.77	6.76	0.19
		S5_6275145	5	6275145	AA	0.68	3.21	3.92	5.11	0.15
		S8_41696703	8	41696703	TT	−1.06	5.93	6.76	13.31	0.17
		S9_32689981	9	32689981	CC	−1.22	5.38	6.19	11.52	0.09
		S11_43126044	11	43126044	CC	0.81	4.28	5.05	8.18	0.31
kernel length	0.095	S3_11153087	3	11153087	TT	−0.13	3.44	4.16	8.19	0.12
		S5_15524578	5	15524578	AA	0.10	3.37	4.09	6.51	0.32
		S6_46530240	6	46530240	TT	−0.25	4.71	5.49	14.55	0.05
		S8_11121701	8	11121701	GG	−0.14	3.16	3.87	9.08	0.10
		S11_8320549	11	8320549	CC	−0.13	3.74	4.48	7.28	0.10
		S12_32853547	12	32853547	CC	−0.18	3.96	4.71	9.31	0.06
kernel width	0.169	S1_32402910	1	32402910	CC	−0.19	4.42	5.20	9.75	0.12
		S2_47786838	2	47786838	CC	0.16	4.99	5.79	9.01	0.26
		S2_64059706	2	64059706	AA	0.17	4.73	5.52	8.28	0.13
		S7_3025298	7	3025298	CC	0.15	3.22	3.92	5.29	0.10
		S9_43700743	9	43700743	AA	−0.18	3.30	4.01	7.77	0.11
		S10_50604452	10	50604452	GG	0.19	3.81	4.55	8.69	0.10
		S12_32853547	12	32853547	CC	0.29	7.02	7.89	13.14	0.06
		S12_7613999	12	7613999	TT	0.12	3.44	4.17	4.47	0.20
kernel weight	0.061	S1_30720144	1	30720144	CC	−0.08	3.06	3.76	9.20	0.22
		S8_43605016	8	43605016	CC	−0.11	3.29	4.00	15.70	0.18

[a] Phenotypic variance [b] Chromosome, [c] the negative logarithms (−log10) of the *p*-values [d] squared correlation coefficient [e] minimum allele frequency.

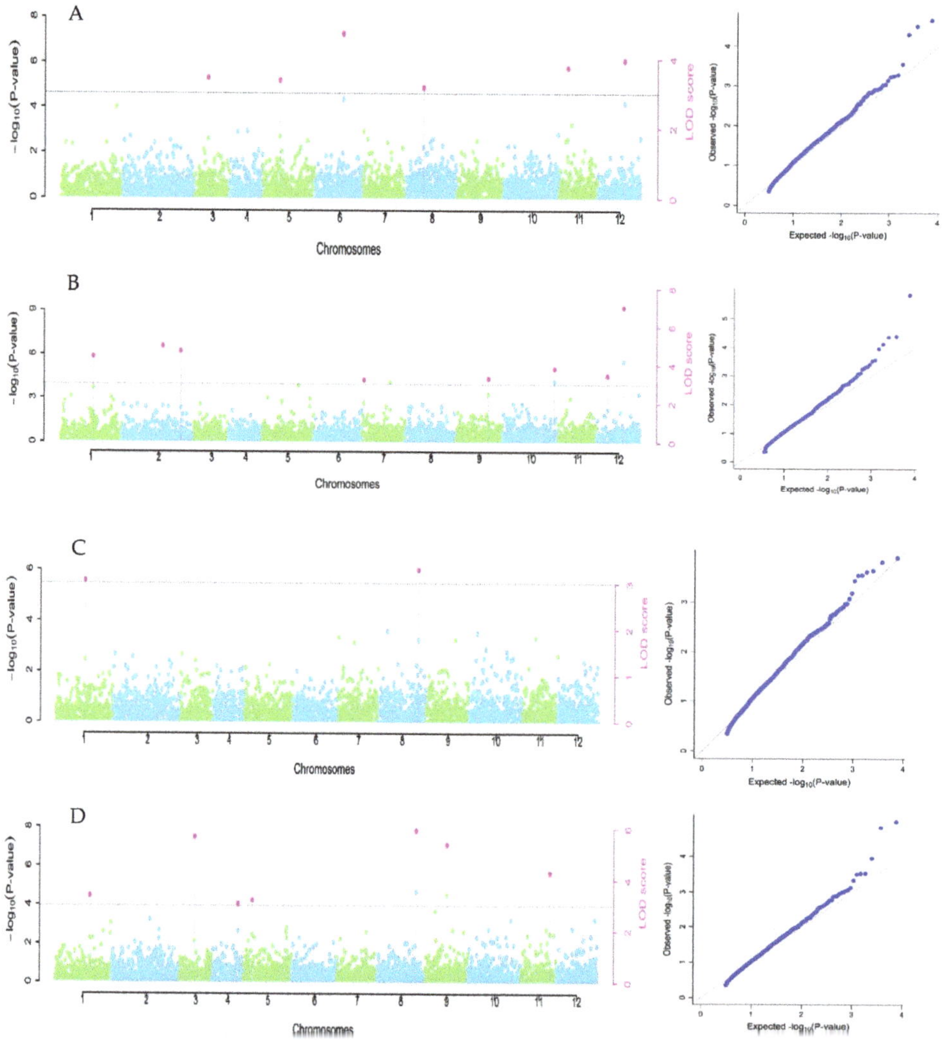

Figure 3. Genome-wide association of Kernel dry matter oil content in a panel of 374 *Vitellaria paradoxa* genotypes with 7530 SNP markers for kernel length (**A**); kernel width (**B**); kernel weight (**C**), and kernel dry matter oil content (**D**). The y-axis representing the p-value of the marker-trait association on a $-\log10$ scale and the x-axis relates to the 12 shea tree chromosomes. The dots above the horizontal 5% Bonferroni threshold light dotted line indicates SNPs associated with QTL that condition the kernel traits.

Variations in the seed traits explained by the individual SNP markers (r^2) varied from 4.47% in kernel width to 15.79% in kernel weight for the significant SNPs, indicating that they represent major QTLs associated with oil yield and kernel physical parameters Alleles 'A' of marker S1_60237300; 'T' of marker S11_43126044; 'A' of markers S4_32032310 and S5_6275145 in oil yield, had the highest positive QTN effect (0.8255, 0.8098, 0.737 and 0.683 respectively) revealing higher association with increasing oil yield. Although most of the seed related traits indicated negative QTN effects, the allele which had the highest (0.2942) positive QTN effect was allele 'C' in marker S12_32853547.

3.4. Potential Candidate Genes

A total of 23 candidate genes were identified by linking the significant SNP regions with the *V. paradoxa* genome (Table 5). The annotation result revealed six putative genes associated with seed length traits. Among these were: Protein metabolism and gluconeogenesis on chromosome 12 and Protein translocation on chromosome 11. The proteins are well known to play important role in mediating plant seed oil biosynthesis [34] and early seedling morphogenesis and development.

From the kernel width, eight putative genes were discovered, of which three (Zinc Finger Transcription Factor located in chromosome 3, protein binding located on chromosome 9 and Protein metabolism and gluconeogenesis located on chromosome 12) had linkage with shea seed oil biosynthesis pathways. Zinc Finger has been associated with playing a key role in plant seed oil biosynthesis and accumulation [35]. All the two identified genes (ATP hydrolase located on chromosome 1 and Protein Kinase on chromosome 8) in kernel weight trait are important in the biochemical pathways of plant seed oil synthesis (Table 5). The hydrolysis process is performed by the FATB acyl-ACP thioesterase or by 3-ketoacyl-ACP synthase II (KASII).

Table 5. Gene annotation for the significant SNPs for shea seed related traits.

Traits	Marker	Chr [a]	Pos [b]	Gene ID	GO. [c]	Function
Kernel length	S3_11153087	3	11153087	Vitpa03g07900	IPR006968	UVB-sensing and in early seedling morphogenesis and development
	S5_15524578	5	15524578	Vitpa05g09840	GO:0005515	ion transportation and signal transduction
	S6_46530240	6	46530240	Vitpa06g28930	PTHR23155	Disease resistance (R)
	S8_11121701	8	11121701	Vitpa08g10570	GO:0004017	Predicts residues in protein biosythesis
	S11_8320549	11	8320549	Vitpa11g07160	PTHR33052	Protein translocation
	S12_32853547	12	32853547	Vitpa12g19540	GO:0003824	Protein metabolism and gluconeogenesis
Kernel width	S1_32402910	1	32402910	Vitpa01g21080	GO:0005515	Consensus disorder prediction
	S2_47786838	2	47786838	Vitpa02g27300	GO:0043190	Glutathione synthetase ATP-binding
	S2_64059706	2	64059706	Vitpa02g39460		Zinc finger
	S7_3025298	7	3025298	Vitpa07g02460	GO:0005515	Calcium signaling
	S9_43700743	9	43700743	Vitpa09g19440	PTHR14859	Protein binding
	S10_50604452	10	50604452	Vitpa10g25960	GO:0003677	Chromosome cohesion
	S12_32853547	12	32853547	Vitpa12g19540	GO:0003824	Protein metabolism and gluconeogenesis
	S12_7613999	12	7613999	Vitpa12g07520	GO:0055114	Catalyze the oxidation of alcohols to aldehydes and ketones
Kernel weight	S1_30720144	1	30720144	Vitpa01g20620	GO:0003676	Hydrolyze ATP
	S8_43605016	8	43605016	Vitpa08g25310	GO:0004672	Predict protein residues as disordered

[a] Chromosome, [b] Marker chromosome position and [c] Gene ontology.

This study further identified seven gene/protein families associated with the percent dry matter oil content in shea nuts: Acyl-ACP Thioesterase Fat B (FATB); Acyl-CoA-binding protein (ACBP); Long Chain Acyl-CoA Synthetase (LACS); Fatty acid exporter (FAX2); (3-ketoacyl-ACP synthase II (KASII) and Fatty acid desaturases (FADs) on chromosomes 1, 3, 8, 9, and 11 (Table 6).

Acyl-CoA-binding protein (ACBP) was identified on chromosomes 3 at loci S3_14843482 and chromosome 5 at loci S5_6275145 that govern plant seed oil accumulation (Table 6). The genes are 1 Mbs from their respective SNPs. Candidate Gene (CG) selection for shea nut oil accumulation is presented (Supplementary Table S2). The genes were annotated with protein-coding genes, using GO.OBO v2.1. The functions of these genes in enhancing shea oil content are explained in Table 6.

Table 6. Gene annotation for the significant SNPs for oil content traits.

Traits	Marker	Chr [a]	Pos [b]	Gene ID [c]	GO. [d]	Function
Oil content	S1_60237300	1	62536299	Vitpa01g27780 (Acyl-ACP Thioesterase Fat B (FATB))	GO:0004553	Consensus disorder prediction
	S3_14843482	3	14843482	Vitpa03g10720 (Acyl-CoA-binding protein (ACBP))	GO:0005515	Protein binding
	S4_32032310	4	32032310	Vitpa04g14070 Long Chain Acyl-CoA Synthetase (LACS))	G3DSA	Oxidoreductase activity
	S5_6275145	5	6275145	Vitpa05g04280 (Acyl-CoA-binding protein (ACBP))	GO:0000160	Transcriptional regulation of oil biosynthesis in seed plants
	S8_41696703	8	41696703	Vitpa08g23790 (Fatty acid exporter (FAX2))	GO:0008168	methyltransferase activity
	S9_32689981	9	32689981	Vitpa09g14250 (3-ketoacyl-ACP synthase II (KASII))	GO:0004672	Early noduling
	S11_43126044	11	43126044	Vitpa11g24760 (Fatty acid desaturases (FADs))		abiotic stress reduction

[a] Chromosome, [b] Chromosome position, [c] Gene identification and [d] Gene Ontology.

3.5. Linkage Disequilibrium (LD)

The distance of the first part of the LD decay before correlation coefficient, r^2 values reach zero was 2,312,772 bp, comprising of 375,037 marker pairs (Table 7). The r^2, decayed within 1–2 Mbps to a value < 0.01.

Chromosome 2 had the highest (46,764 marker pairs) LD followed by chromosome 1 (39,461 marker pairs), while chromosome 4 had the lowest (21,698 marker pairs). The total number of significant marker pairs was 11,940, with chromosome 2 having the most (1330) marker pairs and chromosome 4 (626) having the list.

Table 7. Distribution of LD marker pairs according to chromosomes.

Chromosome	Chr1	Chr2	Chr3	Chr4	Chr5	Chr6	Chr7	Chr8	Chr9	Chr10	Chr11	Chr12
# marker pairs	39,461	46,764	22,990	21,698	34,482	30,834	28,187	32,786	30,232	35,784	24,736	27,091

"#" in Table 7 represents the word "number".

The association analysis of the 374 highly heterozygous shea trees and the 7530 quality SNPs resulted in two most significant SNPs. Variations in the seed traits explained by the individual SNP markers (r^2) varied from 4.47 to 15% for the significant SNPs. Allele 'A' of S1_60237300 marker had the highest allele effect (0.83) revealing higher association with increasing oil yield in shea tree, followed (0.81) by allele C in marker S11_43126044 and allele A (0.68) in S5_6275145 marker. Furthermore, for kernel width, allele 'C' in S12_32853547 also had a moderate effect (0.29). None the less, allele 'T' in marker S6_46530240 revealed the highest (−0.25) negative effect for the studied traits. The LD of significant SNP loci revealed six loci, three each on chromosomes 1 and 8, indicate that the markers had higher LD ($r^2 > 0.8$). The markers in the rest of the chromosomes had a considerably low LD ($r^2 < 0.5$).

4. Discussion

4.1. Phenotypic Data

Shortening juvenile maturity period for early fruiting and increasing oil yield per acre and quality aspects in shea oil are the major concerns in the shea industry. This study aimed at selecting shea parent materials for future breeding programme and bring about farmer solutions by establishing Multi-location Breeding Seed Orchards as a short-term remedy to quality source of shea tree planting materials.

There was variation in the seed trait characteristics in the shea accessions. The results for seed traits indicated a significant variation within the populations and a non-significant variation among the populations. Such non-significant variation observed among the populations is important in breeding for varieties that can easily be adapt across all the geographical range in Uganda. Furthermore, any newly bred variety shall be acceptable by all the communities within the shea parkland in Uganda. With the reliable heritability, selecting traits with marker association for high oil yield in shea tree will result in good genetic progress of the species. Earlier studies by Gwali et al. [36] and Okullo et al. [37] reported similar oil content (52.26%) in the species with this study (53.5%). The results of this study were slightly higher due to the participatory selection which suggests a potential genetic gain from selection given several genotypes with known higher yield (69.77%). Such results can be important in assessing the G $\times$ E interactions for some traits.

4.2. Candidate Gene Scan in the Oil Content Traits

The shea genome revealed associated SNP markers, important for identification of QTL regions controlling the variations of the quantitative traits [1]. Most important of this was the identification of the seven significant SNPs located close to genes that encode different proteins related to plant metabolic mechanisms and transport of biosynthetic products and materials.

GWAS can increase detectability of genomic association in plants [38]. In fact, GWAS has gained increasing popularity as a tool for analysing complex traits in plants [39]. It has been used to reveal the genes controlling polygenic traits including the genetic loci associated with the trait of interest in fruit trees [20]. Kumar et al. [40] used Mixed Linear Model statistical model for GWAS to study six commercial fruit traits in apple seedlings suggesting the potential of the tool in shortening the breeding cycle of tree species like shea.

The advances in omics technologies have enabled researchers to identify candidate genes that promote improvement of associated traits of commercial importance in plants. Earlier very few studies were conducted in determining these functional genes in shea tree [6,41,42]. However, recent sequencing of shea reference genome [8] and identification of genes in shea tree [1], has paved new avenues of genomic studies in the species. Studies for biochemical pathways of oil synthesis in plant seeds have been advanced [43] and several gene expression and enzyme activities in plant seed oil accumulation fronted [44]. Interestingly, 45 seed oil biosynthesis genes were reported in shea tree genome [8]. This study discovered 23 such genes that are potentially associated with shea nut oil biosynthesis pathways (Tables 5 and 6). Of all, acetyl-CoA carboxylase (ACC) is notably the major enzyme catalyst in shea oil biosynthesis [1]. Earlier studies revealed 9 gene copies of Ketoacyl-ACP synthase (KAS) in shea, 6 of these were also reported in *Theobroma cacao*, suggesting their contribution to the increased lipid content in shea than in cocoa. Other genes with higher number of copies in shea include: FAD2, FAD3, and LACS genes [34]. The biological effect of LACs genes discovered on chromosome 4 includes modification of fatty acids chain lengths along the plant oil biosynthesis pathways [45].

The validity of the interrelations among the traits of study was assessed using correlation matrix. It was observed that oil content in *V. paradoxa* was only moderately correlated with kernel width. As observed in the biochemical functions of the genes conditioning the seed related traits that condition seed development and seedling germination. In concordance with this study, Jasinski et al. [46] reported that plant seed oils in angiosperms act as an important reserve of carbon and energy soon after seedling germination until it starts photosynthesis. The presence of proteins suggests their role in promoting shea bioactive functions that condition high oil yield in the species. Previous studies in shea by Lovett and Haq [3] revealed similar proteins that play a major role in oil biosynthesis pathways in oil plant seeds. In fact, Wei et al. [1] predicted presence of more genes associated with oil metabolism in the shea tree genome. Another study Hale, et al. [8] predicted expansion of gene families involved in stearic acid biosynthesis in shea tree which agrees with this current study.

The significant candidate gene for oil content in this study, Acyl-CoA-binding protein (ACBP) was located on chromosome 3 and 5 associated to markers S3_14843482 and S5_6275145 with annotated transcriptional regulation of oil biosynthesis in seed plants. The enzyme plays a role during early fruit formation and play multiple functions such as: tissue growth, cellular trafficking, and physiological processes [47]. The enzymes are usually in the nucleus, are expressed predominantly in developing seeds during maturation. Similar findings were also reported in Arabidopsis thaliana seeds [35]. Moreover, the strong association with annotated function and Acyl-CoA-binding protein (ACBP) genes could be taken advantage of to breed for high oil yield shea tree varieties in Uganda. The biological effect of ACBP includes lipid metabolism, cellular signalling for stress management and disease resistance in plants [48]. This gene encodes metal ion binding enzyme, mostly carbonic anhydrase and alcohol dehydrogenase enzymes that contain zinc as part of their molecule. This zinc finger gene family has been reported to play a major role in oil biosynthesis pathways in the oil palm [48].

The third significant candidate gene was Ketoacyl-ACP synthase (KAS) gene. The gene plays a major role in lipid biosynthesis pathways in shea nuts, thereby increasing oil content in the species [8]. Similar findings on Chinese seed oil shrub, *Paeonia lactiflora* have been advanced [49]. KAS II for example is key in the biosynthesis pathways of fatty acids in plant seeds [50] and early nudling. The Fatty acid exporter (FAX2) genes play a major role in biosynthesis transportation and significantly increases oil content in shea tree. In another study, Janik et al. [51] reported the involvement of FAX in *Chlamydomonas reinhardtii* oil synthesis, similar to this current study. On the other hand, Acyl-ACP Thioesterase Fat B (FATB) was also discovered in other plants like *Koelreuteria paniculata* known to be involved in the synthesis of saturated fatty acids in the species [52], which is in line with this current study. Further still, FADS genes reported in this study, is responsible for the synthesis of unsaturated fatty acids and important for plant development and response to biotic and abiotic stresses [53]. The report therefore confirms the findings in this current study for the role played by the genes in significantly controlling high oil yield in *V. paradoxa* Subsp. *nilotica*.

4.3. Candidate Gene Scan within the Seed Related Traits

The seed related traits with significant SNPs under this study were having linkage with oil yield in shea nuts. The proteins responsible for oil biosynthesis identified in kernel length trait was associated to marker S8_11121701 in chromosome 8. In kernel width trait, S1_32402910 marker discovered on chromosomes 1 had proteins which are linked with processes involved in plant seed oil biosynthesis pathways [24]. For kernel weight trait, S1_30720144 and S8_43605016 markers in chromosomes 1 and 8 were associated with the proteins responsible for oil biosynthesis. In fact, Wei et al. [1] reported similar results with QTLs identified at different locations of shea tree genome. The proteins play a major role in ATP hydrolysis and prediction of protein residues as disordered, during plant seed oil biosynthesis processes. The first evidence was reported by Botha et al [54] linking the functions of the genes to seed development and early seedling growth in *Ricinus communis* oil seeds. The genes reportedly play a major role during seed drying by concentrating inorganic phosphate while de-concentrating the extracellular pyrophosphate which inhibits formation of minerals [55].

4.4. Linkage Disequilibrium (LD)

The LD reveals the evolutionary and demographic events of a population and in mapping genes that are associated with quantitative traits. The implication of this association is that the marker loci contain a causal variant in LD with the identified marker by GWAS. This is further revealed by the small blocks in heat map where the causal variant(s) can be sought. Therefore, it is important to increase our understanding of co-evolution of linked sets of genes. A wide range of LD ($r^2 > 0.2$) in the shea tree population used in this study, was also found in citrus [56]. Such a range of LD is expected in heterozygous outcrossing species like shea tree [56]. The mean r^2 (0.2) in the shea tree population indicated that the markers in the shea tree population is sufficient for genomic selection as LD is maintained by selection. This study describes the potential candidate genes associated with oil yield in shea tree. It further describes the locations of these significant genes in the chromosomes for any further verification. The significant association was discovered on chromosome 1 and 8 for seed related and oil yield traits, explaining 58% of the phenotypic variation.

Inbreeding creates LD owing to the recent common ancestry by increasing the co-variance between alleles at different loci. This, therefore, offers opportunities to design association studies and allele transfer using marker-assisted selection [57,58]. LD therefore presents an opportunity in this study in that if an upper positive selection of preferred traits in shea tree is conducted, it will accelerate the frequency of alleles conferring the preferred trait during breeding. This is because as the linked loci strongly remain in LD with that allele.

4.5. Marker Assisted Selection in Shea Tree

The oil content candidate genes identified in this present study will be cross validated in the established multi-locational trials in NgetaZARDI and NASARRI to determine the ideal molecular markers for enhanced shea tree oil content breeding programs in the country. This is possible by stacking the novel genes into the shea tree genotypes with high oil content using marker-assisted selection. A combination of novel QTLs can further enhance oil content in the shea tree. Furthermore, determination of the allelic status at the markers with significant alleles for oil content will enable the selection of those significant markers for shea oil yield improvement in Uganda. The variations observed in the traits within the location but not across confirms that the species is highly outcrossing [42] or segregating population. The Analysis of variance (ANOVA) in Table 3 indicates a significant variation within the population and this further re-affirms the level of variation in the species. The result of this study points to potential QTNs that explain the genetic variations in the population. In this study, the putative major QTN for oil content explains up to 58% of the phenotypic variance in the species.

Developing MAS options that use the identified molecular markers linked to traits of interest is of importance for speeding the selection process in shea tree with high oil content [59]. The use of significant SNP markers identified through GWAS analysis are important for performing MAS for shea tree breeding. In fact, the application of MAS in shea tree breeding is now made easy with the availability of genomic information on the species [8] coupled with sequencing transcriptome that now makes it possible to align them with the identified markers of interest [1,8]. The six identified markers (S1_30720144, S1_32402910, S1_60237300, S8_11121701, S8_41696703 and S8_43605016) in this study could be applied in MAS for enhanced oil content in *V. paradoxa* Subsp. *niltica*. The MAS can play a very important role in this kind of trait useful for early nursery selection of late expressing traits in the species, and therefore, by performing MAS at seedling stage (far earlier than the juvenile maturity) will greatly reduce the breeding circle.

In this current study, the application of MAS will enable the selection of S1_30720144, S1_32402910, S1_60237300, S8_11121701, S8_41696703 and S8_43605016 markers linked with high oil content genes in the shea nuts. Selection of genotypes with a combination of preferred traits accumulated in one accession would therefore augment the process of shea tree improvement. More value to the communities as an upstream selection would also require prioritizing the genotypes with significant SNPs but from sweet pulped ethnovariety to meet the community's food and nutrition requirement [60,61]. The availability of markers linked to the identified genes will even make it possible to take the advantage of MAS in identifying heterozygous genotypes and therefore apply positive MAS selection for the alleles resulting in a very informative phenotypic traits selected for. On the other hand, MAS could also be applied in negative selection in order to introgress the target trait.

5. Conclusions

The study of marker trait association presents an important step towards identifying the genomic regions associated with the traits of interest to further marker-assisted breeding in shea tree. The current study identified 23 putative markers associated with oil accumulation in shea nut. Candidate genes located on chromosomes 1 and 8 were the most important genes in oil biosynthesis and accumulation in *V. paradoxa*. It is important to note in this study that the position of the seed traits related candidate genes were in agreement with the locations of the oil yield hotspots on chromosomes 1 and 8. This is in support of the need for application of MAS in shea tree and presents the first ever breakthrough in identification of chromosomes 1 and 8 hotspots in the improvement and breeding of shea tree in Uganda for increased oil yield. This study therefore presents the first ever genomic information on associated genes responsible for *V. paradoxa* Subspecies *nilotica* nut oil biosynthesis. The results therefore establish the foundation for explaining the molecular mechanisms of oil biosynthesis for *V. paradoxa* Subspecies *nilotica*. The markers and their linked genes provide a significant resource for improving oil content in the species. The study therefore sets pace for genomic assisted breeding in *V. paradoxa* Subsp. *nilotica* and also broadens our understanding in the role of genomic approaches in advancing yield component traits. The findings of this study will contribute to the initiation of shea breeding for increased oil yield in Uganda. This information could also be used for future gene pyramiding, increasing genetic gain, trait introgression, marker-assisted selection, and selection of parental lines for multiplication and generation of putative genotypes for shea tree breeding programs in Uganda. The study further presents gaps for future validation of the hot spot regions identified on chromosomes 1 and 8.

Supplementary Materials: The following supporting information can be downloaded at: https://www.mdpi.com/article/10.3390/horticulturae9070811/s1, Table S1: Passport data for the selected shea tree families used in this study, with details of their location, tree identification number, geographical coordinated and the details of the farmer on whose farm the tree is located. Table S2: Candidate genes at QTL region searched within approximately ± 20 Kb region of significant SNP markers. The identified genes is believed to be playing an important part in oil yield variation in shea tree.

Author Contributions: J.B.O. and S.G. conceived and designed the study. J.B.O., E.A.A., S.G. and T.L.O. wrote the original manuscript. J.B.O. and E.A.A. analysed the data. P.H., R.E., M.T.B. and A.D. helped to edit the manuscript. All authors have read and agreed to the published version of the manuscript.

Funding: The authors would like to extend their sincere thanks to the following institutions/organizations for funding this study: The World Agroforestry (CIFOR-ICRAF) provided funding under the Genebank Platform, for field data and sample collection under Grant/Award Number: GCDT-1213; Makerere Regional Centre for Crop Improvement (MaRCCI) provided the student research grant for genetic analysis under World Bank fund through ACE II, the Integrated Genotyping Support and Service (IGSS) offered waiver to laboratory analysis expenses and the Intra-Africa Academic Mobility Project for Training Scientists in Crop Improvement for Food Security in Africa (SCIFSA) offered fund type; TG1 PhD Credit Seeking mobility as part of Intra-Africa Academic Mobility Scheme of the European Union that facilitated data analysis, completion of article write up and online publication of this article.

Data Availability Statement: The data presented in this study are available on request from the corresponding author. The data are not publicly available because it will still be used in another ongoing study as part of the whole Ph.D. study.

Acknowledgments: The authors would wish to thank the management and administration of the Cocoa Research Institute of Ghana (CRIG) for their moral support and hosting the principal author of this paper during it write up and submission. The University of Ghana, West African Centre for Crop Improvement (WACCI) is here by also greatly acknowledged for coordination of the principal authors' stay in Ghana to finalize this paper. Finally, great thanks go to the National Agricultural Research Organization (NARO), through the Director of Research (Hillary Agaba), National Forestry

Resources Research Institute (NaFORRI) for the immense support rendered that made data collection, analysis and write up of this paper be a success.

Conflicts of Interest: The authors declare that the research was conducted in the absence of any commercial or financial relationships that could be construed as a potential conflict of interest.

References

1. Wei, Y.; Ji, B.; Siewers, V.; Xu, D.; Halkier, B.A.; Nielsen, J. Identification of genes involved in shea butter biosynthesis from Vitellaria paradoxa fruits through transcriptomics and functional heterologous expression. *Appl. Microbiol. Biotechnol.* **2019**, *103*, 3727–3736. [CrossRef] [PubMed]
2. Naughton, C.C.; Lovett, P.N.; Mihelcic, J. Land suitability modeling of shea (*Vitellaria paradoxa*) distribution across sub-Saharan Africa. *Appl. Geogr.* **2015**, *58*, 217–227. [CrossRef]
3. Lovett, P.; Haq, N. Evidence for anthropic selection of the Sheanut tree (*Vitellaria paradoxa*). *Agrofor. Syst.* **2000**, *48*, 273–288. [CrossRef]
4. Choungo Nguekeng, P.B.; Hendre, P.; Tchoundjeu, Z.; Kalousová, M.; TchanouTchapda, A.V.; Kyereh, D.; Masters, E.; Lojka, B. The Current State of Knowledge of Shea Butter Tree (*Vitellaria paradoxa* C.F. Gaertner.) for Nutritional Value and Tree Improvement in West and CentralAfrica. *Forests* **2021**, *12*, 1740. [CrossRef]
5. Hemsley, J.H. Sapotaceae. In *Flora of Tropical East Africa*; Milne, E., Polhill, R.M., Eds.; Crown Agents for Overseas Governments and Administrations: London, UK, 1968; pp. 47–50.
6. Issaka, A.; Konstantin, V.K.; Reiner, F. Morphological and genetic diversity of shea tree (Vitellaria paradoxa) in the savannah regions of Ghana. *Genet. Resour. Crop Evol.* **2017**, *64*, 1253–1268.
7. Allal, F.; Piombo, G.; Kelly, B.A.; Okullo, J.B.L.; Thiam, M.; Diallo, O.B.; Nyarko, G.; Davrieux, F.; Lovett, P.N.; Bouvet, J.-M. Fatty acid and tocopherol patterns of variation within the natural range of the shea tree (*Vitellaria paradoxa*). *Agrofor. Syst.* **2013**, *87*, 1065–1082. [CrossRef]
8. Hale, I.; Ma, X.; Melo, A.T.O.; Padi, F.K.; Hendre, P.S.; Kingan, S.B.; Sullivan, S.T.; Chen, S.; Boffa, J.-M.; Muchugi, A.; et al. Genomic Resources to Guide Improvement of the Shea Tree. *Front. Plant Sci.* **2021**, *12*, 720670. [CrossRef]
9. Cardi, C.; Vaillant, A.; Sanou, H.; Bokary Kelly, A.; Bouvet, J.-M. Characterization of microsatellite markers in the shea tree (Vitellaria paradoxa C. F Gaertn) in Mali. *Mol. Resour.* **2005**, *5*, 524–526. [CrossRef]
10. Odoi, J.B.; Odong, T.L.; Okia, C.A.; Edema, R.; Muchugi, A.; Gwali, S. Variation in phenotypic traits of high oil yielding and early maturing shea trees (Vitellaria paradoxa) selected using local knowledge. *J. Agric. Nat. Resour. Sci.* **2020**, *7*, 34–42.
11. Global Market Insight. *Market Research Report 2021*; TechNavio | PRODUCT CODE: 1044351; Global Market Insight: Elmhurst, IL, USA, 2021.
12. Abdul-Mumeen, I.; Beauty, D.; Adam, A. Shea butter extraction technologies: Current status and future perspective. *Afr. J. Biochem. Res.* **2019**, *13*, 9–22.
13. Global Shea Alliance. *Shea Production and Market*; Global Shea Alliance: Accra, Ghana, 2021.
14. Boffa, J.-M. *Opportunities and Challenges in the Improvement of the Shea (Vitellaria paradoxa) Resource and Its Management*; Occasional Paper 24; World Agroforestry Centre: Nairobi, Kenya, 2015.
15. Aleza, K.; Villamor, G.B.; Nyarko, B.K.; Wala, K.; Akpagana, K. Shea (*Vitellaria paradoxa* Gaertn C. F.) fruit yield assessment and management by farm households in the Atacora district of Benin. *PLoS ONE* **2018**, *13*, e0190234. [CrossRef]
16. Yao, S.D.M.; Diarrassouba, N.; Diallo, R.; Koffi, E.-B.Z.; Dago, D.N.; Fofana, I.J. Effects of Sowing Depth and Seed Orientation on the Germination and Seedling Growth in Shea Tree (*Vitellaria paradoxa* C.F. Gaertn.) for Rootstock Production in Nursery. *Res. Plant Sci.* **2021**, *9*, 13–22. [CrossRef]
17. Chimsah, F.A. Shea Sapling Management and Grafting. In *The Way forward to Shea Domestication A Case Research from the University for Development Studies*; Tamale: Ghana, West Africa, 2012.
18. Pilipović, A.; Orlović, S.; Kovačević, B.; Galović, V.; Stojnić, S. Selection and Breeding of Fast-Growing Trees for Multiple Purposes in Serbia. In *Forests of Southeast Europe Under a Changing Climate*; Advances in Global Change Research; Šijačić-Nikolić, M., Milovanović, J., Nonić, M., Eds.; Springer: Cham, Switzerland, 2019; Volume 65. [CrossRef]
19. Kilian, A.; Wenzl, P.; Huttner, E.; Carling, J.; Xia, L.; Blois, H.; Caig, V.; Heller-Uszynska, K.; Jaccoud, D.; Hopper, C.; et al. Diversity Arrays Technology: A Generic Genome Profiling Technology on Open Platforms. In *Data Production and Analysis in Population Genomics*; Methods in Molecular Biology (Methods and Protocols); Pompanon, F., Bonin, A., Eds.; Humana Press: Totowa, NJ, USA, 2012; pp. 67–89.
20. Zahid, G.; Aka Kaçar, Y.; Dönmez, D.; Küden, A.; Giordani, T. Perspectives and recent progress of genome-wide association studies (GWAS) in fruits. *Mol. Biol. Rep.* **2022**, *49*, 5341–5352. [CrossRef] [PubMed]
21. AOAC. *Association of Official Analytical Chemist, Official Methods of Analysis*, 19th ed.; AOAC: Washington, DC, USA, 2012; 130p.
22. Doyle, J.J.; Doyle, J.L. A rapid DNA isolation procedure for small quantities of fresh leaf tissue. *Phytochem. Bull.* **1987**, *19*, 1–15.
23. Sansaloni, C.; Petroli, C.; Jaccoud, D.; Carling, J.; Detering, F.; Grattapaglia, D.; Kilian, A. Diversity Arrays Technology (DArT) and next-generation sequencing combined: Genome-wide, high throughput, highly informative genotyping for molecular breeding of Eucalyptus. *BMC Proc.* **2011**, *5*, P54. [CrossRef]

24. Elshire, R.J.; Glaubitz, J.C.; Sun, Q.; Poland, J.A.; Kawamoto, K.; Buckler, E.S.; Mitchell, S.E. A Robust, Simple Genotyping-by-Sequencing (GBS) Approach for High Diversity Species. *PLoS ONE* **2011**, *6*, e19379. [CrossRef] [PubMed]
25. Raman, H.; Raman, R.; Kilian, A.; Detering, F.; Carling, J.; Coombes, N.; Diffey, S.; Kadkol, G.; Edwards, D.; McCully, M.; et al. Genome-wide delineation of natural variation for pod shatter resistance in Brassica napus. *PLoS ONE* **2014**, *9*, e101673. [CrossRef] [PubMed]
26. R Core Team. *R: A Language and Environment for Statistical Computing*; R Foundation for Statistical Computing: Vienna, Austria, 2020; Available online: https://www.R-project.org/ (accessed on 22 March 2022).
27. Yu, J.; Pressoir, G.; Briggs, W.H.; Vroh Bi, I.; Yamasaki, M.; Doebley, J.F.; Buckler, E.S.A. Unified mixed-model method for association mapping that accounts for multiple levels of relatedness. *Nat. Genet.* **2006**, *38*, 203–208. [CrossRef]
28. Rosyara, U.R.; De Jong, W.S.; Douches, D.S.; Endelman, J.B. Software for genome-wide association studies in autopolyploids and its application to potato. *Plant Genome* **2016**, *9*, 1–10. [CrossRef]
29. Zhang, Y.W.; Lwaka Tamba, C.; Wen, Y.J.; Li, P.; Ren, W.L.; Ni, Y.L.; Gao, J.; Zhang, Y.M. mrMLM v4.0: An R platform for multi-locus genome-wide association studies. *Genom. Proteom. Bioinform.* **2020**, *18*, 481–487. [CrossRef]
30. Benjamini, Y.; Hochberg, Y. Controlling the false discovery Rate: A practical and powerful approach to multiple testing. *J. R. Stat. Soc. Ser. B (Methodol.)* **1995**, *57*, 289–300. [CrossRef]
31. Gatarira, C.; Agre, P.; Matsumoto, R.; Edemodu, A.; Adetimirin, V.; Bhattacharjee, R.; Asiedu, R.; Asfaw, A. Genome-Wide Association Analysis for Tuber Dry Matter and Oxidative Browning in Water Yam (*Dioscorea alata* L.). *Plants* **2020**, *9*, 969. [CrossRef] [PubMed]
32. Verde, I.; Jenkins, J.; Dondini, L.; Micali, S.; Pagliarani, G.; Vendramin, E.; Paris, R.; Aramini, V.; Gazza, L.; Rossini, L.; et al. The peach v2.0 release: High-resolution linkage mapping and deep resequencing improve chromosome-scale assembly and contiguity. *BMC Genom.* **2017**, *18*, 225. [CrossRef] [PubMed]
33. Odoi, J.B.; Adjei, E.A.; Hendre, P.; Nantongo, J.S.; Ozimati, A.A.; Badji, A.; Nakabonge, G.; Edema, R.; Gwali, S.; Odong, T.L. Genetic diversity and population structure among Ugandan shea tree (*Vitellariaparadoxa* subsp. nilotica) accessions based on DarTSeq markers. *Crop Sci.* **2023**, 1–13. [CrossRef]
34. Ma, W.; Kong, Q.; Mantyla, J.J.; Yang, Y.; Ohlrogge, J.B.; Benning, C. 14-3-3 protein mediates plant seed oil biosynthesis through interaction with AtWRI1. *Plant J.* **2016**, *88*, 228–235. [CrossRef] [PubMed]
35. Yang, Y.; Kong, Q.; Lim, A.R.Q.; Lu, S.; Zhao, H.; Guo, L.; Yuan, L.; Ma, W. Transcriptional regulation of oil biosynthesis in seed plants: Current understanding, applications, and perspectives. *Plant Commun.* **2022**, *3*, 100328. [CrossRef]
36. Gwali, S.; Nakabonge, G.; Okullo, J.B.L.; Eilu, G.; Forestier-Chironc, N.; Piombod, G.; Davrieux, F. Fat content and fatty acid profiles of shea tree (*Vitellaria paradoxa* subspecies *nilotica*) ethno-varieties in Uganda. *For. Trees Livelihoods* **2012**, *21*, 267–278. [CrossRef]
37. Okullo, J.B.L.; Omujai, F.; Agea, J.G.; Vuzi, P.C.; Namutebi, A.; Okello, J.B.; Nyanzi, S.A. Physico-chemical characteristics of Shea butter (Vitellaria Paradoxa C. F. Gaertu) oil from the shea districts of Uganda. *AJFAND Afr. J. Food Agric. Nutr. Dev.* **2010**, *10*, 2070–2084.
38. Mattia, M.R.; Du, D.; Yu, Q.; Kahn, T.; Roose, M.; Hiraoka, Y.; Wang, Y.; Munoz, P.; Gmitter, F.G., Jr. Genome-Wide Association Study of Healthful Flavonoids among Diverse Mandarin Accessions. *Plants* **2022**, *11*, 317. [CrossRef]
39. Hall, D.; Tegstrom, C.; Ingvarsson, P.K. Using association mapping to dissect the genetic basis of complex traits in plants. *Brief. Funct. Genom. Proteom.* **2010**, *9*, 157–165. [CrossRef]
40. Kumar, S.; Chagne, D.; Bink, M.C.A.M.; Volz, R.K.; Whitworth, C.; Carlisle, C. Genomic selection for fruit quality traits in apple (Malus 3 domestica Borkh.). *PLoS ONE* **2012**, *7*, e36674. [CrossRef]
41. Gwali, S.; Vaillant, A.; Nakabonge, G.; Okullo, J.B.L.; Eilu, G.; Muchugi, A.; Jean-Marc Bouvet, J.-M. Genetic diversity in shea tree (Vitellaria paradoxa subspecies nilotica) ethno-varieties in Uganda assessed with microsatellite markers. *For. Trees Livelihoods* **2014**, *24*, 163–175. [CrossRef]
42. Fontaine, C.; Lovett, P.N.; Sanou, H.; Maley, J.; Bouvet, J.-M. Genetic diversity of the shea tree (*Vitellaria paradoxa* C.F. Gaertn), detected by RAPD and chloroplast microsatellite markers. *Heredity* **2004**, *93*, 639–648. [CrossRef] [PubMed]
43. Bates, P.D.; Stymne, S.; Ohlrogge, J. Biochemical pathways in seed oil synthesis. *Curr. Opin. Plant Biol.* **2013**, *16*, 358–364. [CrossRef]
44. Xianghan, L.; Tianxiang, T.; Chao, S.; Libo, S.; Hui, Z.; Chuanli, Z.; Liping, L.; Liangbin, L. Several Key Enzymes in Oil Synthesis of the Brassica napus. *J. Chin. Cereals Oils Assoc.* **2017**, *12*, 100–104.
45. Zhao, H.; Kosma, D.K.; Lü, S. Functional Role of Long-Chain Acyl-CoA Synthetases in Plant Development and Stress Responses. *Front. Plant Sci.* **2021**, *12*, 640996. [CrossRef]
46. Jasinski, S.; Chardon, F.; Nesi, N.; Lécureuil, A.; Guerche, P. Improving seed oil and protein content in Brassicaceae: Some new genetic insights from Arabidopsis thaliana. *OCL* **2018**, *25*, D603. [CrossRef]
47. Osorio-Guarín, J.A.; Garzón-Martínez, G.A.; Delgadillo-Duran, P.; Bastidas, S.; Moreno, L.P.; Enciso-Rodríguez, F.E.; Cornejo, O.E.; Barrero, L.S. Genome-wide association study (GWAS) for morphological and yield-related traits in an oil palm hybrid (Elaeis oleifera x Elaeis guineensis) population. *BMC Plant Biol.* **2019**, *19*, 533. [CrossRef]
48. Raboanatahiry, N.; Wang, B.; Yu, L.; Li, M. Functional and Structural Diversity of Acyl-coA Binding Proteins in Oil Crops. *Front. Genet.* **2018**, *9*, 182. [CrossRef] [PubMed]

49. Meng, J.-S.; Tang, Y.-H.; Sun, J.; Zhao, D.-Q.; Zhang, K.-L.; Tao, J. Identification of genes associated with the biosynthesis of unsaturated fatty acid and oil accumulation in herbaceous peony 'Hangshao' (Paeonia lactiflora 'Hangshao') seeds based on transcriptome analysis. *BMC Genom.* **2021**, *22*, 94. [CrossRef] [PubMed]
50. Wu, G.-Z.; Xue, H.-W. Arabidopsis b-Ketoacyl-[Acyl Carrier Protein] Synthase I Is Crucial for Fatty Acid Synthesis and Plays a Role in Chloroplast Division and Embryo Development. *Plant Cell* **2010**, *22*, 3726–3744. [CrossRef] [PubMed]
51. Janick, P.; Huleux, M.; Spaniol, B.; Sommer, F.; Neunzig, J.; Schroda, M.; Li-Beisson, Y.; Philippar, K. Fatty acid export (FAX) proteins contribute to oil production in the green microalga Chlamydomonas reinhardtii. *Front. Mol. Biosci.* **2022**, *9*, 939834. [CrossRef]
52. Martins-Noguerol, R.; DeAndres-Gil, C.; Garces, R.; Salas, J.J.; Martínez-Force, E.; Moreno-Perez, A.J. Characterization of the acyl-ACP thioesterases from Koelreuteria paniculata reveals a new type of FatB thioesterase. *Heliyon* **2020**, *6*, e05237. [CrossRef] [PubMed]
53. Hajiahmadi, Z.; Abedi, A.; Wei, H.; Sun, W.; Ruan, H.; Zhuge, Q.; Movahedi, A. Identification, evolution, expression, and docking studies of fatty acid desaturase genes in wheat (*Triticum aestivum* L.). *BMC Genom.* **2020**, *21*, 778. [CrossRef] [PubMed]
54. Botha, F.; Dennis, D. Phosphoglyceromutase activity and concentration in the endosperm of developing and germinating Ricinus communis seeds. *Biol. Chem.* **1987**. [CrossRef]
55. Golub, E.E.; Boesze-Battaglia, K. The role of alkaline phosphatase in mineralization. *Curr. Opin. Orthop.* **2007**, *18*, 444–448. [CrossRef]
56. Minamikawa, M.F.; Nonaka, K.; Kaminuma, E.; Kajiya-Kanegae, H.; Onogi, A.; Goto, S.; Yoshioka, T.; Imai, A.; Hamada, H.; Hayashi, T.; et al. Genome-wide association study and genomic prediction in citrus: Potential of genomics-assisted breeding for fruit quality traits. *Sci. Rep.* **2017**, *7*, 4721. [CrossRef]
57. Kim, S.; Plagnol, V.; Hu, T.T.; Toomajian, C.; Clark, R.M.; Ossowski, S.; Ecker, J.R.; Weigel, D.; Nordborg, M. Recombination and linkage disequilibrium in Arabidopsis thaliana. *Nat. Genet.* **2007**, *39*, 1151–1155. [CrossRef]
58. Thomson, M.J.; Ismail, A.M.; McCouch, S.R.; Mackill, D.J. *Abiotic Stress Adaptation in Plants;* Pareek, A., Sopory, S.K., Bohnert, H.J., Eds.; Springer: Dordrecht, The Netherlands, 2009; pp. 451–469.
59. Tartarini, S.; Sansavini, S. Advances in the use of molecular markers in Pome fruit breeding. In Proceedings of the XXVIth International Horticultural Conference and Exhibition, Toronto, ON, Canada, 11–17 August 2002; p. 622.
60. Odoi, J.B.; Muchugi, A.; Okia, C.A.; Gwali, S.; Odong, T.L. Local knowledge, identification and selection of shea tree (Vitellaria paradoxa) ethnovarieties for pre-breeding in Uganda. *J. Agric. Nat. Resour. Sci.* **2020**, *7*, 22–33.
61. Agúndez, D.; Nouhoheflin, T.; Coulibaly, O.; Soliño, M.; Alía, R. Local Preferences for Shea Nut and Butter Production in Northern Benin: Preliminary Results. *Forests* **2020**, *11*, 13. [CrossRef]

Disclaimer/Publisher's Note: The statements, opinions and data contained in all publications are solely those of the individual author(s) and contributor(s) and not of MDPI and/or the editor(s). MDPI and/or the editor(s) disclaim responsibility for any injury to people or property resulting from any ideas, methods, instructions or products referred to in the content.

horticulturae

MDPI

Article

Construction of a High-Density Genetic Linkage Map Based on Bin Markers and Mapping of QTLs Associated with Fruit Size in Jujube (*Ziziphus jujuba* Mill.)

Tianfa Guo [1,2,†], Qianqian Qiu [1,2,†], Fenfen Yan [1,2], Zhongtang Wang [3], Jingkai Bao [1,2], Zhi Yang [1,2], Yilei Xia [1,2], Jiurui Wang [4], Cuiyun Wu [1,2,*] and Mengjun Liu [1,2,5,*]

1 The National and Local Joint Engineering Laboratory of High Efficiency and Superior-Quality Cultivation and Fruit Deep Processing Technology of Characteristic Fruit Trees in Southern Xinjiang, Aral 843300, China; gtfzky@taru.edu.cn (T.G.); zkyqqq1025@126.com (Q.Q.); yanfening@163.com (F.Y.); baomouren1997@163.com (J.B.); yangzhi950103@outlook.com (Z.Y.); xyl128@126.com (Y.X.)
2 College of Horticulture and Forestry, Tarim University, Aral 843300, China
3 Shandong Institute of Pomology, Tai'an 271000, China; sdgss213@163.com
4 College of Forestry, Hebei Agricultural University, Baoding 071000, China; wjrjujube@126.com
5 College of Horticulture, Hebei Agricultural University, Baoding 071000, China
* Correspondence: wcyby@taru.edu.cn (C.W.); lmj1234567@aliyun.com (M.L.); Tel.: +86-136-575-68050 (C.W.); +86-139-322-62298 (M.L.)
† These authors contributed equally to this work.

Abstract: Jujube (*Ziziphus jujuba* Mill.) is a fruit tree that is gaining increasing importance in drought-affected regions worldwide. The fruit size is an important quantitative agronomic trait that affects not only the fruit yield and attractiveness but also consumer preference. Genetic enhancement of fruit appearance is a fundamental goal of jujube breeding programs. The genetic control of jujube fruit size traits is highly quantitative, and development of high-density genetic maps can facilitate fine mapping of quantitative trait loci (QTLs) and gene identification. However, studies regarding the construction of high-density molecular linkage maps and identification of quantitative trait loci (QTLs) targeting fruit size in jujube are limited. In this study, we performed whole-genome resequencing of the jujube cultivars "JMS2" and "Xing16" and their 165 F_1 progenies to identify genome-wide single-nucleotide polymorphism (SNP) markers and constructed a high-density bin map of jujube that can be used to assist in the selection of multiple traits in jujube breeding. This analysis yielded a total of 116,312 SNPs and a genetic bin map of 2398 bin markers spanning 1074.33 cM with an average adjacent interval of 0.45 cM. A quantitative genetic analysis identified 15 QTLs related to fruit size and the observed phenotypic variation associated with a single QTL ranged from 9.5 to 13.3%. Through the screening of overlapping and stable QTL regions, we identified 113 candidate genes related to fruit size. These genes were ascertained to be involved in cell division, cell wall metabolism, synthesis of phytohormones (ABA, IAA, and auxin), and encoding of enzymes and transcription factors. These candidate genomic regions will facilitate marker-assisted breeding of fruits with different sizes and shapes and lay a foundation for future breeding and manipulation of fruit size and shape in jujube.

Keywords: jujube; whole-genome resequencing; bin map; QTL; fruit size

1. Introduction

Chinese jujube (*Ziziphus jujuba* Mill., 2n = 2x = 24) belongs to the genus *Ziziphus* and is one of the most economically important members of the Rhamnaceae family [1–4]. Native to China, it is one of the oldest cultivated fruit trees in the world with over 7000 years of cultivation history and is now a major dry fruit crop [5]. At present, more than 90% of jujube production is concentrated in six provinces: Xinjiang, Hebei, Shandong, Shanxi, Shaanxi, and Henan. It is the foremost dry fruit in terms of production and the main income source of ~20 million farmers in China [2,3]. It is well adapted to various biotic and abiotic stresses, especially drought and salinity, and is considered an ideal cash crop for arid regions, such as Xinjiang. Although jujube cultivation is several thousand years old [5,6], jujube fruit quality traits, such as the size and flesh flavor, still require significant improvement to satisfy consumer preference. Understanding the molecular mechanism of genes controlling fruit quality is the key to developing jujube cultivars with improved fruit quality.

Highly saturated genetic linkage maps are essential for the fine localization of genes, marker-assisted selection (MAS), and structural and functional genomics. Marker-assisted quantitative trait locus (QTL) studies have used genetic linkage maps to dissect the genetic basis of complex traits in several horticulture plants [7–9]. The first genetic map of jujube, developed from 72 progeny of the cultivars "Dongzao" and "linyilizao", was constructed using random amplified polymorphic DNA (RAPD) makers [10]. In this case, the use of RAPD markers led to a separate map for each of the parents, which had unequal number of linkage groups. Due to the lack of genomic information, the number of available markers in the map was small and the repeatability was poor. The inability to distinguish between homo- and heterozygosity meant this method was unable to capture complete genetic information in jujube. Later studies used codominant AFLP and SSR markers for genetic map construction [11–13], but the resulting map density was too limited for fine mapping of traits of interest in jujube.

With the availability of draft jujube genome assemblies and genome resequencing data based on next-generation sequencing platforms [14], a genetic linkage map of jujube with the most comprehensive genome coverage, the largest number of markers, and the largest marker density has been constructed [15,16]. The markers can improve genetic resolution to fine map qualitative traits and facilitate map-based gene cloning in jujube [17]. However, the tools and approaches required to develop genetic maps from millions of markers from genome resequencing projects are not available, and some data reduction approaches are useful to obtain the most informative markers for further analysis [18].

Bin markers are continuous SNPs occurring at nonrecombining intervals in the genome. They have the advantages of being computationally less intensive and highly accurate, having better density, and being more precisely mapped and cost-effective [19]. In combination with whole-genome sequencing approaches, the bin mapping strategy helps to construct highly dense genome-wide linkage maps by capturing rare recombination events in segregating populations. For example, Peng et al. used a multiomics approach that integrated whole-genome resequencing-based quantitative trait locus (QTL) mapping with an F_1 population, population genomics analysis using germplasm accessions, and transcriptome analysis to identify genomic regions that are potentially associated with fruit weight in loquat [20]. Therefore, the bin mapping approach is highly suited to reducing the sequencing datasets by keeping the most informative markers for high-density genetic map construction and QTL identification [18].

Fruit size is an integral part of fruit quality and directly influences the commodity value and economic return of fruit crops. Linkage-map-based identification of molecular markers or genes of interest and molecular-marker-assisted breeding hold enormous promise. Thus, their application is urgently needed to improve the existing cultivars of jujube. Despite the importance of fruit size, its underlying molecular mechanisms remain understudied in jujube. Studies have shown that fruit size traits are controlled by polygenes with weak inheritability [21] and are directly or indirectly regulated by one or more hormones [22,23].

In this study, we report a high-density bin-marker-based genetic map on fruit size in Chinese jujube. The stably inherited major QTLs were screened, and the candidate genes determining fruit size were mined. The results obtained lay a foundation for employing MAS in breeding programs, fine mapping, and cloning of crucial genes in jujube.

2. Materials and Methods

2.1. Test Materials and Phenotypic Determination

Because it is difficult to construct F_2 populations in fruit trees, F_1 generation is generally used as the material for constructing genetic maps. For example, in apples [24], pears [18,25], grapes [26], and apricots [27], F_1 hybrids have been used as materials for mapping. In this study, an F_1 segregating population consisting of 165 progenies obtained from a "JMS2" × "Xing16" cross conducted by Prof. Liu Mengjun and team at the Hebei Agricultural University, Hebei, China, was used. They were planted with a spacing of 1 × 3 m in a jujube orchard (80°28′ E, 40°59′ N) located in Aral City, China, in 2018. In 2019, 2020, and 2021, 30 representative fruits with uniform size, no pests and diseases, and normal development at the fully red mature stage were picked from each progeny and each of their two parents as representative samples. SFW, FLD, and FTD were measured according to the method described in the Germplasm Resources of Chinese Jujube [6].

2.2. Extraction of Genomic DNA, Library Construction, and Sequencing

In mid-May 2020, the healthy young leaves of each progeny and the two parents were harvested, cleaned, placed into numbered cryovials, flash-frozen in liquid N, and stored in a −80 °C ultralow temperature laboratory freezer. Total genomic DNA was extracted by the CTAB method and randomly digested into 150 bp long fragments [28]. The library construction involved end repair, the addition of polyA to the 3′ end, ligation of sequencing adapters, purification, and PCR-based amplification. After stringent quality checks, the libraries were paired end to end and sequenced using an HiSeqTM2500 platform (Illumina, San Diego, CA, USA).

2.3. SNP Marker Screening and Genotyping

The raw reads were filtered to obtain clean reads by eliminating the adapters, reads with >10% of bases, and reads of low quality [29]. The cleaned reads were aligned to the jujube reference genome of "Dongzao" [14] using the Burrows–Wheeler Aligner (BWA) software [30]. The read pairing information and flags on the BWA (sequence alignment/map format) records were first cleaned up using the Duplicates tool (Picard: http://sourceforgr.net/projects/picard/ (accessed on 18 September 2020)) to shield the effect of PCR duplication. InDel realignment was performed using GATK [31], that is, the sites near the insertion–deletion alignment results were partially realigned to correct the alignment error caused by the insertion–deletion. Base recalibration was performed using GATK to correct the base mass value. GATK was used for variant calling, including SNP and InDel [32]. Based on the minimum recombination fragment identified in each progeny, the SNP segments that did not undergo recombination in each progeny were combined, thus forming a bin marker. All of the sequences of the bin markers that were used to construct the linkage map were aligned to the physical sequences of the reference genome. In order to ensure the quality of the bin mapping, the genotype homozygosity; parental marker depths of 1×, 2×, and 3×; and markers located on nonchromosomal markers were not considered. High-depth sequencing parents were used to fill in relatively correct genotypes and correct the genotypes of low-depth offspring to ensure the correctness of offspring typing.

2.4. Construction of Genetic Linkage Map

In order to ensure the quality of the map, the markers were filtered and screened using the following methods: (1) remove markers with homozygous parents; (2) ensure parental marker depth is not less than 4×; (3) remove nonchromosomal markers. According to the parental genotyping of the offspring, high-depth parental sequencing ensures the correctness of offspring typing. The linkage group was divided by the chromosome of the marker, and the genetic linkage test was performed on the two markers. The linkage phase was determined according to the recombination rate of markers. Wrong genotypes were corrected using genotypes with relatively definite linkage. After marker filling and correction, the bin was divided according to the recombination of offspring. The samples were arranged neatly according to the physical position of the chromosome. When there was a typing change in any sample, it was considered that there was a recombination breakpoint. The SNP between the recombination breakpoints was classified as bin, and there was no recombination time in bin. Finally, the genetic map was constructed using the bin as a mapping marker. The bin was divided into 15 linkage groups based on known information. The linear arrangement of markers in the linkage group was analyzed by HighMap (http://highmap.biomarker.com.cn (accessed on 24 February 2021)) [33] software, and the genetic distance between adjacent markers was estimated.

2.5. Gene Mapping and Nomenclature of QTLs

The CIM method in R/QTL [34] was used to locate the QTLs related to SFW, FLD, and FTD in the F_1 population, and those with an LOD $\geq$3 were considered to be effective in determining fruit size [35,36]. The QTL determining a specific phenotypic trait was named using the following pattern: abbreviation of the English name of the trait + year + the number of the LG it is mapped to + the code number of QTL, e.g., FW19.1.1 indicates that the trait of SFW identified in 2019 was located in LG1 and belonged to QTL1. If the same trait overlapped in the same location repeatedly over a period of two or more years and demonstrated an LOD $\geq$ 3.0 and a PVE $\geq$ 10%, the QTL identified at that location was considered to exist stably. Thus, the genes in these stable QTLs had a higher probability of regulating the fruit size and could be screened to serve as candidate genes for this trait.

2.6. Screening and Annotation of Candidate Genes

The markers within the acceptable confidence interval associated with the stable QTLs described in the previous section were compared with the genome of "Dongzao" as a reference [GCF_000826755.1_ZizJuj_1.1] (https://www.ncbi.nlm.nih.gov/genome/15586?genome_assembly_id=219393 (accessed on 1 February 2021)). Functional annotation and alignment were performed using the relevant sequences obtained from the COG, GO, KEGG, Swissprot, and Nr5 databases. Based on the screening results of the functional annotation studies, the candidate genes not related to the trait of fruit size were excluded but those related to fruit size were identified as putative genes.

2.7. Data Processing and Analysis

The software OriginPro 8.5 (OriginLab, Northampton, MA, USA) was used to draw the frequency distribution histogram of the SFW, FLD, and FTD parameters. SPSS 17.0 (IBM, San Jose, CA, USA) was used to determine if the parameters of fruit size conformed to a pattern of normal distribution. The genetic transmission ability (*Ta*) of the parents was calculated using the following formula: *Ta* = the average value of the trait in the hybrid offspring/the average value of the trait in both the parents × 100%.

3. Results

3.1. Identification of SNP Markers Based on Whole-Genome Resequencing Data

A total of 386.5 GB of filtered data were obtained through the whole-genome resequencing (WGRS) of the female parent "JMS2," the male parent "Xing16", and their 165 progenies, with 13.93, 9.92, and 362.65 GB of individual data, respectively, with an average Q30 and GC content of 93.21 and 33.80% (Table 1). The sequencing reads of all three groups demonstrated an alignment rate of >90% and average coverage depths of 28×, 19×, and 4.14×, respectively, compared to the genome of the jujube cultivar "Dongzao" used as a reference [14]. The genome coverage was >80% (covering at least 1×) for the two parents and 76.33% (covering at least 1×) for the progeny. The results obtained indicated that the sequenced DNA samples had a low error rate.

Table 1. Resequencing data statistics of "JMS2" × "Xing16" F_1 segregating population.

Sample	Total Clean Bases (Gb)	Q30 Proportion (%)	GC Proportion (%)	Average Sequencing Depth (×)	Mapped (%)	Coverage Ratio (%)
Female parent	13.93	93.19	33.71	28×	97.91	87.31
Male parent	9.92	93.61	33.74	19×	97.84	86.24
Offspring	362.65	92.82	33.96	4.14×	97.04	76.33

A total of 1,569,033 SNPs were detected through a comparative study between the genomes of the parents with a transition/transversion ratio (Ti/Tv) of 1.75 (Table 2). The number of heterozygous and homozygous SNPs in "JMS2" were 615,822 and 953,211, respectively, while those in "Xing16" were 713,815 and 855,218, respectively. A total of 148,738 SNPs with a depth not less than 4× were found suitable for coupling (CP). The parents and their progeny were rigorously screened and filtered for the four genotype-specific markers "nn × np" (n = 41,094), "lm × ll" (n = 51,657), "hk × hk" (n = 23,405), and "ef × eg" (n = 156), which accounted for 78.2% of the total number of markers identified. Finally, 116,312 SNPs were retained for constructing bin markers using filtered data by referring to a previously described method [37].

Table 2. Statistics of SNPs obtained from "JMS2" and "Xing16" detection.

Parents	SNP Number	Transition Number	Transverison Number	Ti/Tv Ratio	Heterozygous SNP Number	Homozygous SNP Number
Female parent	1,569,033	998,634	570,399	1.75	615,822	953,211
Male parent	1,569,033	998,634	570,399	1.75	713,815	855,218

3.2. Construction of a High-Density Genetic Map Using the F_1 Segregating Population Obtained from "JMS2" × "Xing16"

Based on the genomic sequence of the jujube cultivar "Dongzao" as a reference (https://www.ncbi.nlm.nih.gov/genome/15586?genome_assembly_id=219393 (accessed on 24 February 2021)), the 116,312 SNPs were combined to form 2398 bin markers (each containing an average of 49 SNP markers) using consecutive SNPs derived from all offspring of the same parent. These bin markers were then used for the molecular mapping of selected genes. A genetic map was constructed with a total map distance of 1074.33 cM containing 12 LGs and an average distance of 0.45 cM between adjacent markers. The 12 LGs ranged in length from 67.43 (LG9) to 124.86 (LG6) cM, the number of bin markers ranged from 135 (LG9) to 263 (LG1) cM, and the average intertag distance ranged from 0.38 (LG1) to 0.52 (LG3) cM. The largest intertag distance was 7.76 cM (LG1), followed by LG2, 3, and 9 (7.06 cM in all). The probability of gaps < 5 cM in length in the 12 LGs ranged from 99.25 to 100%, and the average length of 99.61% of the gaps between consecutive markers was <5 cM. LG6 and 9 were the longest and shortest LGs, respectively (Figure 1), and the largest gap of 82–90 cM was in LG1 (Table 3).

Table 3. Distribution of bin markers on the high-density genetic map constructed.

Linkage Group	Number of Bin Markers	Number of SNP Markers	Genetic Length (cM)	Average Distance (cM)	Max Gap (cM)	Gap < 5 cM (%)
LG1	263	16,203	99.45	0.38	7.76	99.62
LG2	211	10,638	96.26	0.46	7.06	99.52
LG3	188	9802	98.36	0.52	7.06	99.47
LG4	228	10,716	92.54	0.41	4.05	100.00
LG5	182	11,390	77.14	0.42	2.46	100.00

Table 3. Cont.

Linkage Group	Number of Bin Markers	Number of SNP Markers	Genetic Length (cM)	Average Distance (cM)	Max Gap (cM)	Gap < 5 cM (%)
LG6	249	9032	124.86	0.50	5.37	99.60
LG7	165	9878	67.46	0.41	1.83	100.00
LG8	221	8205	96.89	0.44	5.70	99.55
LG9	135	9366	67.43	0.50	7.06	99.25
LG10	206	7228	99.12	0.48	6.38	99.51
LG11	151	6948	71.71	0.47	5.04	99.33
LG12	199	6906	83.11	0.42	5.04	99.49
Totals	2398	116,312	1074.33	-	-	-
Overall average	-	-	-	0.45	-	99.61
Max	-	-	-	-	7.76	-

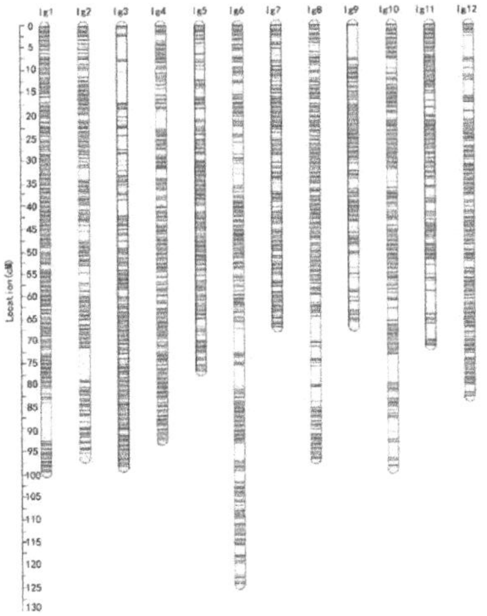

Figure 1. The high-density genetic linkage map. Distribution of bin markers on 12 LGs. A black bar indicates a bin marker. The LG number is shown on the x-axis, and the genetic distance in cM is shown on the y-axis.

3.3. Phenotypic Analysis of Traits Relating to Fruit Size in the F_1 Segregating Population

The values of the three indicators of fruit size in the two parents "JMS2" and "Xing16" and the F_1 segregating population derived from them were determined over a period of three years: 2019, 2020, and 2021 (Table 4). The average values of fruit size in "JMS2" were significantly higher than those in "Xing16" in all three years. The transmission of parental genetic information showed the following pattern with regard to importance: FTD > FLD > SFW. The absolute values of skewness and kurtosis in the F_1 population were <2, which conformed to a pattern of continuous normal distribution. The results obtained were consistent with the number distribution plot, showing that the three typical quantitative trait characteristics selected were suitable for use in QTL mapping (Figure 2). In addition, three years of phenotypic data showed a significant positive correlation between SFW, FLD, and FTD, indicating that the single fruit weight and the vertical and horizontal diameter can influence each other (Table 5).

Table 4. Statistics of fruit size traits in two parents and their F_1 population over three years.

Traits	Year	Parents		F_1 population				
		JMS2	Xing16	Range	Mean ± SD	Ta %	Skewness	Kurtosis
Single fruit weight (g)	2019	11.00	2.76	1.02–8.45	4.30 ± 1.38	62.52	0.200	0.287
	2020	10.50	2.81	1.46–13.41	5.47 ± 1.99	82.22	0.989	1.817
	2021	7.82	2.52	1.67–9.01	4.45 ± 1.38	86.09	0.574	0.557
Fruit longitudinal diameter (mm)	2019	34.47	17.76	15.04–27.19	20.87 ± 2.64	79.92	0.020	−0.509
	2020	35.04	18.49	14.32–32.49	22.33 ± 3.18	83.44	0.200	0.108
	2021	32.58	17.45	14.88–31.34	21.40 ± 2.71	85.54	0.566	1.205
Fruit transverse diameter (mm)	2019	25.58	16.16	12.27–23.57	19.22 ± 2.44	92.10	−0.512	0.111
	2020	24.88	16.56	13.60–28.20	20.83 ± 2.74	100.53	0.146	0.067
	2021	21.70	16.44	13.71–24.49	19.27 ± 2.24	101.04	0.566	1.205

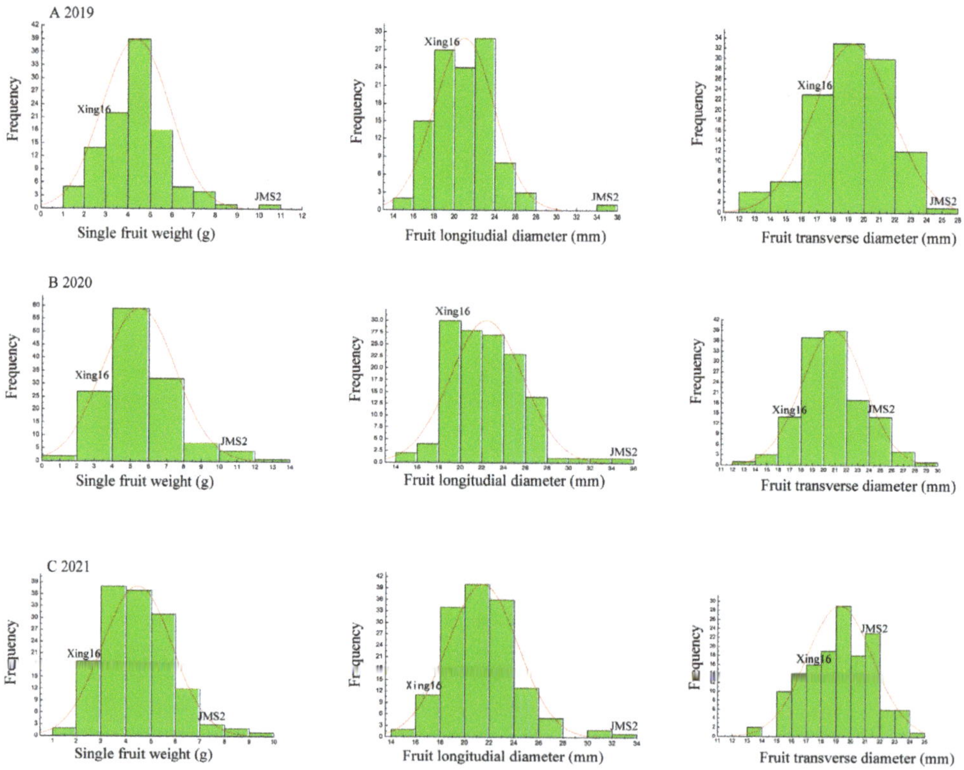

Figure 2. Frequency distribution of fruit size traits in the parents and the F_1 population over a three-year duration. Frequency distributions of SFW, FLD, and FTD in progenies of the "JMS2" and "Xing16" hybrid cross. The phenotypic data were collected in (**A**) 2019, (**B**) 2020, and (**C**) 2021. The vertical coordinates correspond to the columns in which "JMS2" and "Xing16" are located in the figure and indicate the range of the phenotypic values of SFW, FLD, and FTD in the parents. The horizontal coordinates indicate the number of progenies occurring within the range of phenotypic values.

Table 5. Correlation analysis of traits relating to fruit size in the F_1 segregating population.

Year	FW vs. FLD	FW vs. FTD	FLD vs. FTD
2019	0.8360 **	0.9280 **	0.7512 **
2020	0.8771 **	0.9424 **	0.7972 **
2021	0.8291 **	0.9370 **	0.6832 **

Double asterisks (**) indicated extremely significant correlation at $p < 0.01$ level; SFW stands for Single fruit weight; FLD stands for Fruit longitudina diameter; FTD stands for Fruit transverse diameter.

3.4. QTL Analysis of Traits Relating to Fruit Size in the F_1 Segregating Population Derived from "JMS2" × "Xing16"

Based on the high-density bin marker map described in the previous section and the phenotypic data for SFW, FLD, and FTD obtained over three years, a total of 19 QTLs associated with these three fruit size traits were mapped to seven chromosomes/LGs with a phenotypic interpretation range of 9.5–13.3% (Table 6). The number of QTLs detected on LG2, 4, 6,7, 8, 11, and 12 was 7, 1, 3, 1, 2, 3, and 2, respectively. In 2019 and 2020, QTL FTD2.1 was repeatedly detected on LG2. Its yearly phenotypic interpretation rates were 13.3 and 10.5% and its annual LOD thresholds were 3.08 and 3.35, respectively, due to which it was recognized as the most effective QTL in determining fruit size. All the other QTLs could only be detected in 2019 or 2020. However, the LOD thresholds of FW21.2.2, FLD21.2.1, and FTD21.8.1 were all >3.5 and their contribution rates were >10%, due to which they too were regarded to be equally effective QTLs.

Table 6. QTL mapping of traits relating to fruit size in the F_1 segregation population.

Trait	QTLs	Intervals on Maps (cM)	LOD	Peak Marker Position (cM)	PVE (%)	Containing Markers
Single fruit weight	FW20.11.1	20.475–21.382	3.15	20.475	10.7	2
	FW21.2.1	91.738–92.944	3.44	92.644	10.8	3
	FW21.2.2	94.45–96.259	3.63	94.45	11.3	4
	FW21.12.1	28.383	3.02	28.383	9.5	1
Fruit transverse diameter	FTD19.2.1	91.437–91.738	3.08	91.437	13.3	2
	FTD20.4.1	18.968–23.921	3.27	23.02	11.1	5
	FTD20.6.1	111.226	3.02	111.226	10.3	1
	FTD20.6.2	113.029–113.93	3.15	113.33	10.7	4
	FTD20.6.3	122.75	3.2	122.75	10.9	1
	FTD20.11.1	21.382	3.04	21.382	10.3	1
	FTD20.11.2	47.29–47.891	3.04	47.891	10.4	3
	FTD21.2.1	91.738–92.944	3.35	92.644	10.5	3
	FTD21.2.2	94.45	3.07	94.45	9.7	1
	FTD21.2.3	95.352–96.259	3.15	95.352	9.9	2
	FTD21.8.1	64.38–86.663	4.51	71.891	13.9	18
	FTD21.8.2	92.085	3.00	92.085	9.5	1
	FTD21.12.1	27.783–28.683	3.47	28.383	10.9	4
Fruit longitudinal diameter	FLD21.2.1	91.137–96.259	3.91	94.45	12.1	11
	FLD21.7.1	38.567–39.468	3.10	38.867	9.7	4

The QTLs were, however, not evenly distributed among all the LGs/chromosomes. Certain LGs contained QTLs influencing multiple indicators of fruit size, with some even clustered in the same regions (Figure 3). There was an overlap between FTD21.1, FW21.1, and FLD21.1 on LG2 with a length of 91.738–92.944 cM, which was linked to all three traits SFW, FLD, and FTD simultaneously. Another overlap zone between FW21.2, FTD21.2, and FLD21.1 was found on LG2 with a length of 94.45–96.259 cM. This region is associated with fruit weight and size. One overlap zone between FW21.1 and FTD21.1 with a length of 28.383 cM on LG12 and another one between FW20.1 and FTD20.1 with a length of 21.382 cM on LG11 were identified with SFW and FTD, respectively. These genetic regions identified with QTLs demonstrating stable effects are worthy of attention in follow-up studies (Table 7).

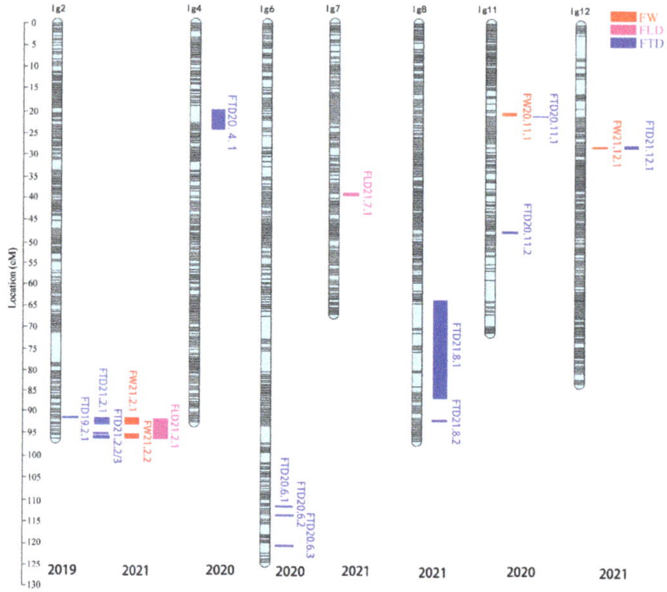

Figure 3. The QTLs distribution map of fruit size-related traits. The CIM method in R/qtl was used to locate the QTLs regarding SFW, FLD, and FTD in the F_1 population, and loci with LOD ≥3 were considered to be effective. The QTL determining a specific phenotypic trait was named using the following pattern: abbreviation of the English name of the trait + year + the number of the LG it is mapped to + the code number of the QTL. The letters LG on the top of the linkage maps represent "linkage group", and the number following LG indicates the number of the linkage group. The QTLs for SFW, FLD, and FTD are marked with red, pink, and blue colored bars, respectively.

Table 7. QTL cluster information of traits relating to fruit size.

Traits	Corresponding QTL	Coincidence Interval (cM)	Marker Number	LG
Fruit transverse diameter, Single fruit weight, Fruit longitudinal diameter	FTD21.1, FW21.1, FLD21.1	91.738–92.944	3	2
Single fruit weight, Fruit transverse diameter, Fruit longitudinal diameter	FW21.2, FTD21.2, FLD21.1	94.450–96.259	4	2
Fruit transverse diameter, Single fruit weight	FTD21.1, FW21.1	28.383	1	12
Fruit transverse diameter, Single fruit weight	FTD20.1, FW20.1	21.382	1	11

3.5. Prediction of the Putative Functions of Candidate Genes

The candidate genes influencing fruit size were identified based on the markers mapped within the 15 QTLs and their physical location on the "Dongzao" genome. The genes were mined from five databases, namely, Clusters of Orthologous Groups (COG, [38]), Gene Ontology (GO, http://geneontology.org/ (accessed on 22 July 2021 and 7 July 2022)), Kyoto Encyclopedia of Genes and Genomes (KEGG, Kanehisa Laboratories, Kyoto University), Swissprot (http://www.uniprot.org/ (accessed on 22 July 2021 and 7 July 2022)), and

Nr (NCBI), to determine the tag information, markers, and genes in the associated regions. The candidate QTLs that were identified in any two years of the study period concerning at least two of the three indicators concerning fruit size or with LOD threshold $\geq$3.5 and PVE $\geq$10% in any one year were selected for gene mining. A total of 727 genes were mined (Table 8).

Table 8. Genes in the associated regions by comparing the databases.

LG	Coincidence Interval (cM)	QTL Lloci	Gene Number	COG Anno	GO Anno	KEGG Anno	Swissprot Anno	Nr Anno
2	91.738	FTD19.1	9	3	4	5	9	9
		FTD21.1	19	0	10	10	14	19
2	91.738–92.944	FTD21.1	19	0	10	10	14	19
		FW21.1	19	0	10	10	14	19
		FLD21.1	99	33	60	40	71	99
2	94.450–96.259	FW21.2	61	21	36	21	47	61
		FTD21.2	41	12	24	16	32	41
		FLD21.1	99	33	60	40	71	99
12	28.383	FTD21.1	30	12	25	14	28	30
		FW21.1	3	2	3	2	2	3
11	21.382	FTD20.1	0	0	0	0	0	0
		FW20.1	1	0	1	0	1	1
2	94.45–96.259	FW21.2	61	21	36	21	47	61
2	91.137–96.259	FLD21.1	99	33	60	40	71	99
8	64.38–86.663	FTD21.1	167	55	111	68	125	167
Total	-	-	727	225	450	297	546	727

According to the map positions of the candidate QTLs, 15 candidate genomic regions affecting traits relating to fruit size were identified, the relevant candidate genes were identified, and their functions were predicted in GO and KEGG (Table 9). Consequently, a total of 113 candidate genes possibly associated with the regulation of fruit size were identified. These genes were found to be involved in the processes of cell division, cell cycle, cell wall metabolism, and synthesis of phytohormones (ABA, IAA, and auxin) as well as encoding of enzymes, transcription factors (TFs), and zinc finger proteins (ZFPs). The genes *LOC107410242* and *LOC107409642* were related to the morphogenesis of anatomical structure and tissue development as well as regulation of gene expression and cellular and developmental processes. The genes *LOC107409953*, *LOC112490770*, *LOC107409998*, *LOC107409919*, *LOC107423923*, and *LOC107423861* were related to encoding LRR receptor-like serine/threonine protein kinases. *LOC107410143* and *LOC107423897* were associated with LRR receptor-like proteins and kinases, respectively. *LOC107410070* and *LOC107423821* were related to the cell cycle. *LOC107423912*, *LOC107423913*, *LOC107431775*, *LOC107423905*, and *LOC107423928* were related to the synthesis of phytohormones such as IAA, auxins, and cytokinins. *LOC107409704*, *LOC107423900*, *LOC107431772*, *LOC107423814*, *LOC107423857*, *LOC107423922*, *LOC107423848*, *LOC107423846*, *LOC107423858*, *LOC107423925*, *LOC107423856*, *LOC107423903*, and *LOC107423881* were related to the cell wall formation or metabolic enzyme activity. *LOC107423820* and *LOC107423815* were associated with pectin esterase. *LOC107423811* was related to the TF BHLH. *LOC107409987* was associated with a NAC domain-containing protein. *LOC107409426* was related to ZFPs. *LOC107423816*, *LOC107423801*, *LOC107423813*, and *LOC107431773* were related to E3 ubiquitin protein ligase. These genes may be involved in the regulation of fruit size in jujube.

Table 9. The genes putatively associated with jujube fruit development.

Range (cM)	QTL Name	Candidate Genes	Candidate Gene ID	Gene Annotation
91.738	FTD19.2.1	-	-	-
	FTD21.2.1	*LOC107410242*	rna-XM_025070976.1	anatomical structure morphogenesis;tissue development; regulation of gene expression; cellular process; developmental process; regulation of cellular process
		LOC107409642	rna-XM_016017070.2	regulation of cellular process
91.738–92.944	FW21.2.1	*LOC107410242*	rna-XM_025070976.1	anatomical structure morphogenesis; tissue development; regulation of gene expression; cellular process; developmental process; regulation of cellular process
		LOC107409642	rna-XM_016017070.2	regulation of cellular process
	FLD21.2.1	*LOC107409704*	rna-XM_016017148.2	Cell wall
		LOC107409953	rna-XM_016017375.2	LRR receptor-like serine/threonine-protein At3g47570
		LOC112490770	rna-XM_025071274.1	LRR receptor-like serine/threonine-protein EFR
		LOC107410143	rna-XM_016017551.2	Leucine-rich repeat-containing protein
		LOC107410070	rna-XM_016017481.1	regulation of mitotic cell cycle
		LOC107409704	rna-XM_016017134.2	Cell wall
		LOC107409998	rna-XM_016017417.2	LRR receptor-like serine/threonine-protein EFR
		LOC107410070	rna-XM_025070726.1	regulation of mitotic cell cycle
		LOC107410070	rna-XM_016017502.2	regulation of mitotic cell cycle
		LOC107410242	rna-XM_025070976.1	anatomical structure morphogenesis; tissue development; regulation of gene expression; cellular process; developmental process; regulation of cellular process
		LOC107409704	rna-XM_016017140.2	Cell wall
		LOC107409426	rna-XM_016016863.2	Zinc finger protein
		LOC107410070	rna-XM_016017488.1	regulation of mitotic cell cycle
		LOC107409919	rna-XM_016017343.2	LRR receptor-like serine/threonine-protein At3g47570
		LOC107409987	rna-XM_016017405.2	NAC domain-containing protein
		LOC107410070	rna-XM_025070727.1	regulation of mitotic cell cycle
		LOC107410143	rna-XM_016017558.2	Leucine-rich repeat-containing protein
		LOC107409642	rna-XM_016017070.2	regulation of cellular process
94.450–96.259	FW21.2.2	*LOC107409953*	rna-XM_016017375.2	LRR receptor-like serine/threonine-protein At3g47570
		LOC107409987	rna-XM_016017405.2	NAC domain-containing protein
		LOC112490770	rna-XM_025071274.1	LRR receptor-like serine/threonine-protein EFR
		LOC107409919	rna-XM_016017343.2	LRR receptor-like serine/threonine-protein At3g47570
		LOC107409426	rna-XM_016016863.2	Zinc finger protein
		LOC107410143	rna-XM_016017551.2	Leucine-rich repeat-containing protein
		LOC107409998	rna-XM_016017417.2	LRR receptor-like serine/threonine-protein EFR
		LOC107410143	rna-XM_016017558.2	Leucine-rich repeat-containing protein
	FTD21.2.2	*LOC107409426*	rna-XM_016016863.2	Zinc finger protein
		LOC107410143	rna-XM_016017551.2	Leucine-rich repeat-containing protein
		LOC107409953	rna-XM_016017375.2	LRR receptor-like serine/threonine-protein At3g47570
		LOC107410143	rna-XM_016017558.2	Leucine-rich repeat-containing protein
	FLD21.2.1	*LOC107409704*	rna-XM_016017148.2	Cell wall
		LOC107409953	rna-XM_016017375.2	LRR receptor-like serine/threonine-protein At3g47570
		LOC112490770	rna-XM_025071274.1	LRR receptor-like serine/threonine-protein EFR
		LOC107410143	rna-XM_016017551.2	Leucine-rich repeat-containing protein
		LOC107410070	rna-XM_016017481.1	regulation of mitotic cell cycle
		LOC107409704	rna-XM_016017134.2	Cell wall

Table 9. *Cont.*

Range (cM)	QTL Name	Candidate Genes	Candidate Gene ID	Gene Annotation
		LOC107409998	rna-XM_016017417.2	LRR receptor-like serine/threonine-protein EFR
		LOC107410070	rna-XM_025070726.1	regulation of mitotic cell cycle
		LOC107410070	rna-XM_016017502.2	regulation of mitotic cell cycle
		LOC107410242	rna-XM_025070976.1	anatomical structure morphogenesis; tissue development; regulation of gene expression; cellular process; developmental process; regulation of cellular process
		LOC107409704	rna-XM_016017140.2	Cell wall
		LOC107409426	rna-XM_016016863.2	Zinc finger protein
		LOC107410070	rna-XM_016017488.1	regulation of mitotic cell cycle
		LOC107409919	rna-XM_016017343.2	LRR receptor-like serine/threonine-protein At3g47570
		LOC107409987	rna-XM_016017405.2	NAC domain-containing protein
		LOC107410070	rna-XM_025070727.1	regulation of mitotic cell cycle
		LOC107410143	rna-XM_016017558.2	Leucine-rich repeat-containing protein
		LOC107409642	rna-XM_016017070.2	regulation of cellular process
28.383	FTD21.12.1	*LOC107431773*	rna-XM_016042763.2	E3 ubiquitin-protein ligase
		LOC107431775	rna-XM_025079541.1	Cindole-3-acetic acid amido synthetase activity
		LOC107431773	rna-XM_016042765.2	E3 ubiquitin-protein ligase
		LOC107431773	rna-XM_025066560.1	E3 ubiquitin-protein ligase
		LOC107431775	rna-XM_016042766.2	indole-3-acetic acid amido synthetase activity
		LOC107431773	rna-XM_025066561.1	E3 ubiquitin-protein ligase
		LOC107431772	rna-XM_016042762.2	plant-type secondary cell wall biogenesis
		LOC107431773	rna-XM_016042764.2	E3 ubiquitin-protein ligase
	FW21.12.1	-	-	-
94.450–96.259	FW21.2.2	*LOC107409953*	rna-XM_016017375.2	LRR receptor-like serine/threonine-protein At3g47570
		LOC112490770	rna-XM_025071274.1	LRR receptor-like serine/threonine-protein EFR
		LOC107409919	rna-XM_016017343.2	LRR receptor-like serine/threonine-protein At3g47570
		LOC107410143	rna-XM_016017551.2	Leucine-rich repeat-containing protein
		LOC107409998	rna-XM_016017417.2	LRR receptor-like serine/threonine-protein EFR
		LOC107410143	rna-XM_016017558.2	Leucine-rich repeat-containing protein
91.137–96.259	FLD21.2.1	*LOC107409704*	rna-XM_016017148.2	Cell wall
		LOC107409953	rna-XM_016017375.2	LRR receptor-like serine/threonine-protein At3g47570
		LOC112490770	rna-XM_025071274.1	LRR receptor-like serine/threonine-protein EFR
		LOC107410143	rna-XM_016017551.2	Leucine-rich repeat-containing protein
		LOC107410070	rna-XM_016017481.1	regulation of mitotic cell cycle
		LOC107409704	rna-XM_016017134.2	Cell wall
		LOC107409998	rna-XM_016017417.2	LRR receptor-like serine/threonine-protein EFR
		LOC107410070	rna-XM_025070726.1	regulation of mitotic cell cycle
		LOC107410070	rna-XM_016017502.2	regulation of mitotic cell cycle
		LOC107410242	rna-XM_025070976.1	anatomical structure morphogenesis; tissue development; regulation of gene expression; cellular process; developmental process; regulation of cellular process
		LOC107409704	rna-XM_016017140.2	Cell wall
		LOC107409426	rna-XM_016016863.2	Zinc finger protein
		LOC107410070	rna-XM_016017488.1	regulation of mitotic cell cycle
		LOC107409919	rna-XM_016017343.2	LRR receptor-like serine/threonine-protein At3g47570
		LOC107410070	rna-XM_025070727.1	regulation of mitotic cell cycle
		LOC107410143	rna-XM_016017558.2	Leucine-rich repeat-containing protein

Table 9. *Cont.*

Range (cM)	QTL Name	Candidate Genes	Candidate Gene ID	Gene Annotation
64.380–86.663	FTD21.8.1	*LOC107423900*	rna-XM_016033552.2	cell wall; auxin-activated signaling pathway
		LOC107423928	rna-XM_025076890.1	cytokinin metabolic process
		LOC107423913	rna-XM_016033565.2	indoleacetic acid biosynthetic process; response to auxin
		LOC107423813	rna-XM_016033450.2	E3 ubiquitin-protein ligase
		LOC107423814	rna-XM_016033451.1	cell wall; cell wall modification
		LOC107423928	rna-XM_025076888.1	cytokinin metabolic process
		LOC107423861	rna-XM_016033508.2	protein serine/threonine kinase activity
		LOC107423912	rna-XM_016033564.2	indoleacetic acid biosynthetic process; response to auxin
		LOC107423811	rna-XM_016033446.2	Transcription factor bHLH
		LOC107423821	rna-XM_025076730.1	mitotic cell cycle; regulation of cell cycle
		LOC107423857	rna-XM_016033503.2	cell wall
		LOC107423801	rna-XM_016033440.1	E3 ubiquitin-protein ligase
		LOC107423922	rna-XM_016033576.2	plant-type cell wall biogenesis; regulation of meristem growth
		LOC107423928	rna-XM_016033582.2	cytokinin metabolic process
		LOC107423848	rna-XM_016033491.2	cell wall; Pectin methylesterase
		LOC107423928	rna-XM_025076887.1	cytokinin metabolic process
		LOC107423928	rna-XM_025076889.1	cytokinin metabolic process
		LOC107423846	rna-XM_016033490.2	cell wall; pectinesterase activity; cell wall modification; pectin catabolic process
		LOC107423858	rna-XM_025076143.1	cell wall
		LOC107423925	rna-XM_016033579.2	cell wall
		LOC107423923	rna-XM_016033577.2	protein serine/threonine kinase activity
		LOC107423815	rna-XM_025076224.1	putative pectinesterase/pectinesterase inhibitor 45-like
		LOC107423897	rna-XM_016033547.2	Leucine-rich repeat receptor-like protein kinase
		LOC107423905	rna-XM_016033556.2	response to auxin
		LOC107423856	rna-XM_025076290.1	cell wall
		LOC107423846	rna-XM_016033489.2	cell wall; Pectinesterase PPE8B
		LOC107423903	rna-XM_016033555.1	cell wall
		LOC107423821	rna-XM_016033457.2	mitotic cell cycle; meiotic cell cycle; regulation of cell cycle
		LOC107423816	rna-XM_016033453.2	E3 ubiquitin-protein ligase
		LOC107423881	rna-XM_016033528.2	Cell wall
		LOC107423820	rna-XM_016033456.2	Pectinesterase 54
21.382	FTD20.11.1	-	-	-
	FW20.11.1	-	-	-

4. Discussion

4.1. Advantages of Constructing a High-Density Genetic Linkage Map Based on Bin Markers

In this study, the genomes of "JMS2" and "Xing16" and their F_1 progeny (165 in number) were resequenced. The average coverage depth of the parental genomes was >20× with an average genome coverage > 90%, while the average coverage depth of the offspring was 4.14× with an average genome coverage > 76.33%. A total of 2398 recombination bin markers comprising 116,312 SNP markers were mapped onto 12 LGs. The total length of the linkage map was 1074.33 cM with an average bin intermarker distance of 0.45 cM. The map presented in this study identified manifold SNP markers (116,312) in comparison to those in the six genetic linkage maps already available (2540–8158) with the highest sequencing depth. In addition, the genetic map presented used a bin-marker-based mapping to classify consecutive SNP markers that did not undergo recombination as bins, thus avoiding the probability of errors occurring in the SNP calculation due to the detection of numerous loci and collection of a large amount of data. Bin markers have been used to develop high-density genetic maps in many crops, such as radish [39], brassica napus [40], and melon [41], but not so much on fruit trees, such as hawthorn [42], pear [18], and grapes [26,43]. Our study found the highest number of SNP markers that had shorter intermarker gaps, were of high quality, and demonstrated enhanced precision in jujube. This will be more conducive to determining the location of trait candidate gene segments.

4.2. Mapping of QTLs Associated with Traits Relating to Fruit Size in Jujube

The identification of QTLs affecting fruit size traits in jujube has been less studied when compared to other fruit tree species. In this study, 15 QTLs associated with fruit size traits were mapped, and candidate QTLs that were identified through molecular analyses in at least two years of a three-year study, related to at least two of the three indicators of fruit size (SFW, FLD, and FTD), or had an LOD threshold $\geq$ 3.5 and a PVE $\geq$10% in any one year were selected for gene mining. A total of 113 genes were identified to be located on LG2, 8, and 12. Previous studies have also mapped QTLs related to jujube fruit size, but the results were inconsistent [44,45]. This may be due to differences in groups, environment, and other aspects. Although the location of the QTL varies greatly, we can analyze the function of these 113 genes in other species and may find some candidate genes related to fruit development, laying the foundation for jujube breeding (Table 9).

4.3. In Silico Prediction of the Putative Functions of the Candidate Genes Determining Fruit Size

Rapid cell division, cell elongation, and duration of the cell cycle determine the final size, shape, and weight of the fruit. The family of serine/threonine protein kinases and cell cycle proteins together with phosphorylated compounds act at two checkpoints to initiate DNA replication and mitosis to regulate the cell cycle [46]. During the growth and development of pear fruits, pectin esterase is associated with the relaxation and extension of the cell wall pulp, which may lead to an increase in the number of cells. Similarly, cellulase facilitates cell division when required while regulating their growth and development [47]. Cell wall biogenesis also plays an important role in cell expansion and unidirectional elongation [48]. A homolog of an *Arabidopsis* ubiquitin-specific protease, *ZjDA3,* was found to be a negative regulator of fruit size in jujube [49]. In addition, the MYB TFs are also known to demonstrate regulatory effects on fruit development. The *R2R3-MYB* TF was found to alter the size and shape of cells in the fruits and leaves of tomatoes [50].

Plant hormones directly regulate the fruit size and shape growth and development by altering the expression of early response genes in horticultural plants. The high auxin content depressed the expression of *MdAux/IAA2*, and the downregulated expression of *MdAux/IAA2* led to the formation of a large fruit size apple [22]. An important role for auxin in the regulation of fruit development, especially at the fruit enlargement stage, and three single nucleotide polymorphism (SNP) markers were also closely associated with fruit weight in loquat [20]. Studies on the direct regulation of fruit size and shape by hormones are more common in tomato and cucumber. ABA participates in the CsTRM5-mediated cell expansion during fruit elongation [51], and the auxin-responsive protein CsARP1 promotes cell expansion and fruit elongation in cucumber [23]. In a study of strawberry fruit shape, different expression genes were mainly enriched in DNA replication, cell cycle, plant hormone synthesis, and signal transduction, including auxin-related genes in elongated strawberry fruit [52]. Six candidate genes for fruit size were found to be involved in the regulation of the cell cycle and hormone biosynthesis pathways, including *LOC107404981* and *LOC107406728*, which may be involved in the molecular regulation of fruit size in jujube [53]. These studies have shown that fruit size and shape are inseparable from plant hormones and can provide certain reference value for further research. In this study, we identified 113 candidate genes regulating fruit size in jujube. These genes were determined to be involved in the regulation of cell division, cell cycle, and cell wall metabolism; biosynthesis of phytohormones (ABA, IAA, and auxin); and encoding of enzymes, transcription factors, and ZFPs.

In conclusion, we carried out high-density genetic bin mapping for the identification of reliable QTLs and candidate genes in a single F_1 hybrid population in jujube. These results provide a foundation for further research on fruit size in jujube and also provide a theoretical basis for molecular breeding of new jujube varieties.

Author Contributions: C.W. and M.L. conceived and designed the experiments. F.Y. provided support with experimental materials. Q.Q., T.G., Z.W. and J.W. analyzed the data. T.G., J.B., Z.Y. and Y.X. were involved in the experiment and provided technical and theoretical support for this work. Q.Q. and T.G. wrote the paper. C.W. and M.L. revised the intellectual content of this paper. All authors have read and agreed to the published version of the manuscript.

Funding: This research was supported by the Major Scientific and Technological Projects of Xinjiang Production and Construction Corps (2017DB006), Innovation and Entrepreneurship Platform and Base Construction Project of Xinjiang Production and Construction Corps (2019CB001), and earmarked fund of Xinjiang Jujube Industrial Technology System (XJCYTX-01). We wish to express our thanks for their financial support.

Data Availability Statement: The data presented in the study are deposited in the NCBI repository, accession number is PRJNA996341 (https://www.ncbi.nlm.nih.gov/sra/PRJNA996341).

Acknowledgments: The fruit of jujube cultivars "JMS2" × "Xing16" and their F_1 segregating population consisting of 165 progenies were obtained from Hebei Agricultural University, Hebei, China.

Conflicts of Interest: The authors declare no competing interest.

References

1. Liu, M.; Wang, J.; Wang, L.; Liu, P.; Zhao, J.; Zhao, Z.; Yao, S.; Stănică, F.; Liu, Z.; Wang, L. The historical and current research progress on jujube—A superfruit for the future. *Hortic. Res.* **2020**, *7*, 119. [CrossRef] [PubMed]
2. Liu, M.; Wang, J. Fruit scientific research in New China in the past 70 years: Chinese jujube. *J. Fruit Sci.* **2019**, *36*, 1369–1381.
3. Meng-jun, L.; Jiu-rui, W.; Ping, L.; Jin, Z.; Zhi-hui, Z.; Li, D.; Xian-Song, L.; Zhi-guo, L. Historical achievements and frontier advances in the production and research of Chinese jujube (*Ziziphus jujuba*) in China. *Acta Hortic. Sin.* **2015**, *42*, 1683.
4. Richardson, J.E.; Fay, M.F.; Cronk, Q.C.; Bowman, D.; Chase, M.W. A phylogenetic analysis of Rhamnaceae using rbcL and trnL-F plastid DNA sequences. *Am. J. Bot.* **2000**, *87*, 1309–1324. [CrossRef]
5. Qu, Z.-Z.; Wang, Y.-H. *Chinese Fruit Trees Record-Chinese Jujube*; The Forestry Publishing House of China: Beijing, China, 1993.
6. Liu, M.J.; Wang, M. *Germplasm Resources of Chinese Jujube*; China Forestry Pubilishing House: Beijing, China, 2009.
7. Yang, H.; Zou, Y.; Li, X.; Zhang, M.; Zhu, Z.; Xu, R.; Xu, J.; Deng, X.; Cheng, Y. QTL analysis reveals the effect of CER1-1 and CER1-3 to reduce fruit water loss by increasing cuticular wax alkanes in citrus fruit. *Postharvest Biol. Technol.* **2022**, *185*, 111771. [CrossRef]
8. Calle, A.; Wünsch, A. Multiple-population QTL mapping of maturity and fruit-quality traits reveals LG4 region as a breeding target in sweet cherry (*Prunus avium* L.). *Hortic. Res.* **2020**, *7*, 127. [CrossRef]
9. Wu, Y.; Popovsky-Sarid, S.; Tikunov, Y.; Borovsky, Y.; Baruch, K.; Visser, R.G.; Paran, I.; Bovy, A. *CaMYB12-like* underlies a major QTL for flavonoid content in pepper (*Capsicum annuum*) fruit. *New Phytol.* **2023**, *237*, 2255–2267. [CrossRef] [PubMed]
10. Lu, L.Y. *Study on Identifieation of Hybrids and Hereditary Variation of Natural Pollinated Chinese Jujube Seedlings*; Agricultural University of Hebei: Baoding, China, 2003.
11. Xu, L.S. QTL Mapping of Fruit Traits and Superior Genotypes Selecting in Chinese Jujube (*Ziziphus jujuba* Mill.). Ph.D. Thesis, Agricultural University of Hebei, Baoding, China, 2012.
12. Qi, J.; Dong, Z.; Mao, Y.M.; Shen, L.Y.; Zhang, Y.X.; Liu, J.; Wang, X.L. Construction of a Dense Genetic Linkage Map and QTL Analysis of Trunk Diameter in Chinese Jujube. *Sci. Silvar Sin.* **2009**, *45*, 44–49.
13. Shen, L.Y. Construction of Genetic Linkage Map and Mapping QTLs for Some Traits in Chinese Jujube (*Ziziphus jujuba* Mill.). Ph.D. Thesis, Agricultural University of Hebei, Baoding, China, 2005.
14. Liu, M.-J.; Zhao, J.; Cai, Q.-L.; Liu, G.-C.; Wang, J.-R.; Zhao, Z.-H.; Liu, P.; Dai, L.; Yan, G.; Wang, W.-J. The complex jujube genome provides insights into fruit tree biology. *Nat. Commun.* **2014**, *5*, 5315. [CrossRef] [PubMed]
15. Yan, F.; Luo, Y.; Bao, J.; Pan, Y.; Wang, J.; Wu, C.; Liu, M. Construction of a highly saturated genetic map and identification of quantitative trait loci for leaf traits in jujube. *Front. Plant Sci.* **2022**, *13*, 1001850. [CrossRef]
16. Zhang, Z.; Wei, T.; Zhong, Y.; Li, X.; Huang, J. Construction of a high-density genetic map of *Ziziphus jujuba* Mill. using genotyping by sequencing technology. *Tree Genet. Genomes* **2016**, *12*, 76. [CrossRef]
17. Hou, L.; Chen, W.; Zhang, Z.; Pang, X.; Li, Y. Genome-wide association studies of fruit quality traits in jujube germplasm collections using genotyping-by-sequencing. *Plant Genome* **2020**, *13*, e20036. [CrossRef] [PubMed]

18. Qin, M.-F.; Li, L.-T.; Singh, J.; Sun, M.-Y.; Bai, B.; Li, S.-W.; Ni, J.-P.; Zhang, J.-Y.; Zhang, X.; Wei, W.-L. Construction of a high-density bin-map and identification of fruit quality-related quantitative trait loci and functional genes in pear. *Hortic. Res.* **2022**, *9*, uhac141. [CrossRef] [PubMed]
19. Vision, T.J.; Brown, D.G.; Shmoys, D.B.; Durrett, R.T.; Tanksley, S.D. Selective mapping: A strategy for optimizing the construction of high-density linkage maps. *Genetics* **2000**, *155*, 407–420. [CrossRef]
20. Peng, Z.; Zhao, C.; Li, S.; Guo, Y.; Xu, H.; Hu, G.; Liu, Z.; Chen, X.; Chen, J.; Lin, S.; et al. Integration of genomics, transcriptomics and metabolomics identifies candidate loci underlying fruit weight in loquat. *Hortic. Res.* **2022**, *9*, uhac037. [CrossRef] [PubMed]
21. Wang, Y.; Aizezi, S.; Li, Y.L.; Sun, F.; Wu, G.H. Inheritance trend of the fruit traits of 'huozhouheiyu' grape. *Acta Bot. Boreal.-Occident. Sin.* **2015**, *35*, 275–281.
22. Bu, H.; Sun, X.; Yue, P.; Qiao, J.; Sun, J.; Wang, A.; Yuan, H.; Yu, W. The MdAux/IAA2 Transcription Repressor Regulates Cell and Fruit Size in Apple Fruit. *Int. J. Mol. Sci.* **2022**, *23*, 9454. [CrossRef]
23. Che, G.; Song, W.; Zhang, X. Gene network associates with CsCRC regulating fruit elongation in cucumber. *Veg. Res.* **2023**, *3*, 7. [CrossRef]
24. Ma, B.; Zhao, S.; Wu, B.; Wang, D.; Peng, Q.; Owiti, A.; Fang, T.; Liao, L.; Ogutu, C.; Korban, S.S.; et al. Construction of a high density linkage map and its application in the identification of QTLs for soluble sugar and organic acid components in apple. *Tree Genet. Genomes* **2015**, *12*, 1. [CrossRef]
25. Wang, L.; Li, X.; Wang, L.; Xue, H.; Wu, J.; Yin, H.; Zhang, S. Construction of a high-density genetic linkage map in pear (*Pyrus communis* × *Pyrus pyrifolia nakai*) using SSRs and SNPs developed by SLAF-seq. *Sci. Hortic.* **2017**, *218*, 198–204. [CrossRef]
26. Jiang, J.; Fan, X.; Zhang, Y.; Tang, X.; Li, X.; Liu, C.; Zhang, Z. Construction of a High-Density Genetic Map and Mapping of Firmness in Grapes (*Vitis vinifera* L.) Based on Whole-Genome Resequencing. *Int. J. Mol. Sci.* **2020**, *21*, 797. [CrossRef] [PubMed]
27. Zhang, Q.; Liu, J.; Liu, W.; Liu, N.; Zhang, Y.; Xu, M.; Liu, S.; Ma, X.; Zhang, Y. Construction of a High-Density Genetic Map and Identification of Quantitative Trait Loci Linked to Fruit Quality Traits in Apricots Using Specific-Locus Amplified Fragment Sequencing. *Front Plant Sci* **2022**, *13*, 798700. [CrossRef] [PubMed]
28. Paterson, A.H.; Brubaker, C.L.; Wendel, J.F. A rapid method for extraction of cotton (*Gossypium* spp.) genomic DNA suitable for RFLP or PCR analysis. *Plant Mol. Biol. Report.* **1993**, *11*, 122–127. [CrossRef]
29. Bolger, A.M.; Lohse, M.; Usadel, B. Trimmomatic: A flexible trimmer for Illumina sequence data. *Bioinformatics* **2014**, *30*, 2114–2120. [CrossRef]
30. Li, H.; Durbin, R. Fast and accurate short read alignment with Burrows–Wheeler transform. *Bioinformatics* **2009**, *25*, 1754–1760. [CrossRef]
31. McKenna, A.; Hanna, M.; Banks, E.; Sivachenko, A.; Cibulskis, K.; Kernytsky, A.; Garimella, K.; Altshuler, D.; Gabriel, S.; Daly, M. The Genome Analysis Toolkit: A MapReduce framework for analyzing next-generation DNA sequencing data. *Genome Res.* **2010**, *20*, 1297–1303. [CrossRef]
32. Cingolani, P.; Platts, A.; Wang, L.L.; Coon, M.; Nguyen, T.; Wang, L.; Land, S.J.; Lu, X.; Ruden, D.M. A program for annotating and predicting the effects of single nucleotide polymorphisms, SnpEff: SNPs in the genome of Drosophila melanogaster strain w^{1118}; iso-2; iso-3. *fly* **2012**, *6*, 80–92. [CrossRef]
33. Liu, D.; Ma, C.; Hong, W.; Huang, L.; Liu, M.; Liu, H.; Zeng, H.; Deng, D.; Xin, H.; Song, J. Construction and analysis of high-density linkage map using high-throughput sequencing data. *PLoS ONE* **2014**, *9*, e98855. [CrossRef]
34. Yandell, B.S.; Mehta, T.; Banerjee, S.; Shriner, D.; Venkataraman, R.; Moon, J.Y.; Neely, W.W.; Wu, H.; Von Smith, R.; Yi, N. R/qtlbim: QTL with Bayesian interval mapping in experimental crosses. *Bioinformatics* **2007**, *23*, 641–643. [CrossRef]
35. Zhao, J.; Xu, Y.; Li, H.; Yin, Y.; An, W.; Li, Y.; Wang, Y.; Fan, Y.; Wan, R.; Guo, X. A SNP-based high-density genetic map of leaf and fruit related quantitative trait loci in wolfberry (*Lycium* Linn.). *Front. Plant Sci.* **2019**, *10*, 977. [CrossRef]
36. Pereira, L.; Ruggieri, V.; Pérez, S.; Alexiou, K.G.; Fernández, M.; Jahrmann, T.; Pujol, M.; Garcia-Mas, J. QTL mapping of melon fruit quality traits using a high-density GBS-based genetic map. *BMC Plant Biol.* **2018**, *18*, 324. [CrossRef] [PubMed]
37. Huang, X.; Feng, Q.; Qian, Q.; Zhao, Q.; Wang, L.; Wang, A.; Guan, J.; Fan, D.; Weng, Q.; Huang, T. High-throughput genotyping by whole-genome resequencing. *Genome Res.* **2009**, *19*, 1068–1076. [CrossRef] [PubMed]
38. Tatusov, R.L.; Koonin, E.V.; Lipman, D.J. A Genomic Perspective on Protein Families. *Science* **1997**, *278*, 631–637. [CrossRef] [PubMed]
39. Luo, X.; Xu, L.; Wang, Y.; Dong, J.; Chen, Y.; Tang, M.; Fan, L.; Zhu, Y.; Liu, L. An ultra-high-density genetic map provides insights into genome synteny, recombination landscape and taproot skin colour in radish (*Raphanus sativus* L.). *Plant Biotechnol. J.* **2020**, *18*, 274–286. [CrossRef] [PubMed]
40. Dong, Z.; Alam, M.K.; Xie, M.; Yang, L.; Liu, J.; Helal, M.; Huang, J.; Cheng, X.; Liu, Y.; Tong, C. Mapping of a major QTL controlling plant height using a high-density genetic map and QTL-seq methods based on whole-genome resequencing in Brassica napus. *G3* **2021**, *11*, jkab118. [CrossRef] [PubMed]
41. Lian, Q.; Fu, Q.; Xu, Y.; Hu, Z.; Zheng, J.; Zhang, A.; He, Y.; Wang, C.; Xu, C.; Chen, B. QTLs and candidate genes analyses for fruit size under domestication and differentiation in melon (*Cucumis melo* L.) based on high resolution maps. *BMC Plant Biol.* **2021**, *21*, 126. [CrossRef]
42. Zhao, Y.; Zhao, Y.; Guo, Y.; Su, K.; Shi, X.; Liu, D.; Zhang, J. High-density genetic linkage-map construction of hawthorn and QTL mapping for important fruit traits. *PLoS ONE* **2020**, *15*, e0229020. [CrossRef]

43. Shi, G.; Sun, D.; Wang, Z.; Liu, X.; Guo, J.; Zhang, S.; Zhao, Y.; Ai, J. Construction of a resequencing-based high-density genetic map for grape using an interspecific population (*Vitis amurensis* × *Vitis vinifera*). *Hortic. Environ. Biotechnol.* **2022**, *63*, 489–497. [CrossRef]
44. Bao, J.K. Construction of High-Density Genetic Map and QTL Mapping of Fruit Size and Sugar-Acid Traits on 'JMS2′ × 'Jiaocheng 5′ Jujuba Hybrid Progeny. Master's Thesis, Tarim University, Alar, China, 2022.
45. Wang, Z. High-density Genetic Map Construction and QTL Mapping of Agronomic Traits of *Ziziphus jujuba* Mill. Ph.D. Thesis, College of Forestry Northwest A&F University, Yangling, China, 2020.
46. Zhang, H.; Tan, J.; Zhang, M.; Huang, S.; Chen, X. Comparative transcriptomic analysis of two bottle gourd accessions differing in fruit size. *Genes* **2020**, *11*, 359. [CrossRef]
47. Liu, X. Studies on Mechanism of Fruit Growth and Development of Different Ripening-Season of Pears. Ph.D. Thesis, Sichuan Agriculture University, Ya'an, China, 2008.
48. Bashline, L.; Lei, L.; Li, S.; Gu, Y. Cell wall, cytoskeleton, and cell expansion in higher plants. *Mol. Plant* **2014**, *7*, 586–600. [CrossRef]
49. Guo, M.; Zhang, Z.; Cheng, Y.; Li, S.; Shao, P.; Yu, Q.; Wang, J.; Xu, G.; Zhang, X.; Liu, J. Comparative population genomics dissects the genetic basis of seven domestication traits in jujube. *Hortic. Res.* **2020**, *7*, 89. [CrossRef] [PubMed]
50. Mahjoub, A.; Hernould, M.; Joubès, J.; Decendit, A.; Mars, M.; Barrieu, F.; Hamdi, S.; Delrot, S. Overexpression of a grapevine R2R3-MYB factor in tomato affects vegetative development, flower morphology and flavonoid and terpenoid metabolism. *Plant Physiol. Biochem.* **2009**, *47*, 551–561. [CrossRef] [PubMed]
51. Xie, Y.; Liu, X.; Sun, C.; Song, X.; Li, X.; Cui, H.; Guo, J.; Liu, L.; Ying, A.; Zhang, Z.; et al. CsTRM5 regulates fruit shape via mediating cell division direction and cell expansion in cucumber. *Hortic. Res.* **2023**, *10*, uhad007. [CrossRef] [PubMed]
52. Li, Z.; Song, X.; Li, H.; Zhang, Z.; Zhang, J. Transcriptome and hormones analysis provide insights into elongated fruit somaclonal mutant in 'Akihime' strawberry. *Sci. Hortic.* **2023**, *309*, 111608. [CrossRef]
53. Yang, L.; Zhou, H.i.; Bo, W.h.; Li, Y.y.; Pang, X.m. Identification of genes related with jujube fruit size based on selective sweep analysis. *J. Beijing For. Univ.* **2019**, *41*, 30–36.

Disclaimer/Publisher's Note: The statements, opinions and data contained in all publications are solely those of the individual author(s) and contributor(s) and not of MDPI and/or the editor(s). MDPI and/or the editor(s) disclaim responsibility for any injury to people or property resulting from any ideas, methods, instructions or products referred to in the content.

horticulturae

Article

Callus Induction and Adventitious Root Regeneration of Cotyledon Explants in Peach Trees

Lingling Gao [1,2,3], Jingjing Liu [1,3], Liao Liao [1,2], Anqi Gao [1,3], Beatrice Nyambura Njuguna [1,3], Caiping Zhao [4], Beibei Zheng [1,2,*] and Yuepeng Han [1,2,*]

1 CAS Key Laboratory of Plant Germplasm Enhancement and Specialty Agriculture, Wuhan Botanical Garden, The Innovative Academy of Seed Design of Chinese Academy of Sciences, Wuhan 430074, China
2 Hubei Hongshan Laboratory, Wuhan 430070, China
3 University of Chinese Academy of Sciences, 19A Yuquanlu, Beijing 100049, China
4 College of Horticulture, Northwest Agriculture and Forestry University, Yangling 712100, China; zhcc@nwsuaf.edu.cn
* Correspondence: zhengbeibei@wbgcas.cn (B.Z.); yphan@wbgcas.cn (Y.H.)

Abstract: Callus induction is a key step in establishing plant regeneration and genetic transformation. In this study, we present a comprehensive large-scale investigation of the callus induction rate (CIR) in peach trees, which revealed significant variability within the peach germplasm. Notably, the late-maturing cultivars exhibited significantly higher levels of CIR. Moreover, cultivars characterized as having high CIR exhibited potential for the development of adventitious roots (ARs) during callus induction, and a positive correlation was observed between CIR and the ability to regenerate ARs. However, long-term subculture callus lost root regeneration capacity due to changes in cellular morphology and starch and flavonoid content. Additionally, *PpLBD1* was identified as a good candidate gene involved in the regulation of callus adventitious rooting in peach trees. Our results provide an insight into the mechanisms underlying callus induction and adventitious root development and will be helpful for developing regeneration systems in peach trees.

Keywords: *Prunus persica*; cotyledon; adventitious root regeneration; long-term subculture

1. Introduction

Peaches (*Prunus persica* L.) are widely cultivated in temperate regions as an important economic fruit crop. The trees' juvenility period is relatively short and the species has a small haploid genome size, approximately 230 Mb [1,2]. However, it is well known that the genetic transformation of the species is very difficult [3]. To date, a stable genetic transformation system is not available for peaches, which significantly hinders its application as a genetic research model for Rosaceae fruit trees [4].

Callus induction and subsequent plantlet regeneration are the essential prerequisite for genetic transformation [5]. Callus induction depends on cell fate transition that turns the somatic state into pluripotency [6]. Callus can be derived from many kinds of explants such as leaves, mature embryos, stems, and immature embryos, but different types of calli have a significant difference in embryogenic potential [7]. In peach trees, callus induced from immature embryos is able to develop shoots, but mature embryos show callus formation without shoot development [8]. However, in apple trees, embryogenic callus can be efficiently induced from leaf explants from the cultivar Royal Gala, which facilitates the establishment of a transformation system [9]. As callus induction materials, cotyledon explants have been proved to have high frequency regeneration due to their higher content of meristematic tissue, lower level of contamination, and improved browning [10–13]. Overall, leaf and cotyledon explants are the main sources of callus induction and subsequent plantlet regeneration [14,15]

In peach trees, callus induction from leaf and cotyledon explants has been widely reported, but there are few reports of successful plantlet regeneration from callus culture [16–18]. Recently, Xu et al. [19] developed a fast and efficient root transgenic system through *Agrobacterium tumefaciens*-mediated transformation in peach trees. However, transgenic roots induced by *Agrobacterium* are difficult to develop lateral roots (LRs) as shown in our study [19]. Thus, root differentiation seems to be a barrier for plant regeneration in peach trees. Moreover, adventitious root emergence is a crucial step in stem cutting propagation of fruit tree rootstocks. However, rooting difficulty of stem cuttings has been reported to be a serious problem in asexual propagation of peach rootstocks [20]. There are few reports to date on the adventitious root development and growth mechanism of peach trees [16,21].

Adventitious roots are essential for plant growth due to their functions of expanding plant absorption area and supporting plants [22]. The mechanism of adventitious root development in model plants such as *Arabidopsis thaliana* (L.) Heynh. and *Oryza sativa* L. have been extensively studied. For instance, in *A. thaliana*, members of the lateral organ boundary domain *(LBD)* gene family play an important role in plant organ development [23]. Auxin response factors (ARFs) such as *AtARF7* and *AtARF19* are involved in adventitious root occurrence by triggering the transcription of *AtLBD16* and *AtLBD18* [24]. However, auxin/indole-3-acetic acid protein (AUX/IAA) interacts with ARFs to inhibit ARF-induced AR formation [25]. In rice, the LBD gene crown rootless1 (*CRL1*) is required for adventitious root origination [26,27]. In addition to LBD and ARF transcription factors (TFs), the basic helix–loop–helix (bHLH) TF family was also reported to regulate root formation [28,29].

Although numerous peach somatic tissues, for instance, leaves, stems and calyxes, have been successfully applied to induce callus [30–32], the genetic mechanisms underlying callus induction have not yet been illustrated. In this study, we evaluated diversity in the callus induction rate among peach cultivars. Interestingly, adventitious roots were often developed during callus culture, and there was a positive correlation between callus induction rate and root regeneration capacity. Moreover, we identified candidate genes for callus-based adventitious root formation. Our results not only provide insight into the mechanism of adventitious root formation, but will also contribute to the establishment of a genetic transformation model in peach trees.

2. Material and Methods

2.1. Plant Material and Growth Conditions

The peach cultivar ZYT used in this study was grown at Hubei Academy of Agricultural Sciences, Wuhan, Hubei Province, China, while the other 99 peach cultivars, including 2 wild relatives, 8 landrace accessions, and 89 modern cultivars (Table S1), were grown at Northwest A&F University, Yangling, Shanxi Province. The surface sterilization method used was optimized based on a previous study [19]. Firstly, the pits of ripe fruits were collected by peeling away the flesh with a knife and subsequently immersed in a solution of sodium hypochlorite and water (v/v = 1:3) for 0.5 h. Later, the pits were thoroughly rinsed three times with distilled water, left to air-dry at room temperature, and then stored in the refrigerator at 4 °C. Two months later, the endocarp was removed and the seeds were sterilized with 70% ethanol for 2 min. The seeds were transferred to a sterile triangular flask and submerged in a solution of sodium hypochlorite and water (v/v = 1:3) for a duration of 0.5 h, all while placed on an ultraclean platform. After 10 rinses with sterile water, the seeds were incubated in sterile water at 25 °C for 15 h.

The coat of the seeds was removed with a scalpel, and the embryos and hypocotyls were discarded. The cotyledons were cut into rectangular pieces of 1 cm × 0.5 cm. The pieces were then placed in callus induction medium (CIM) as previously reported [30]. The pH of the medium was adjusted to 5.8 and sterilized using an autoclave at 121 °C for 20 min. Three biological replicates of each cultivar were created, and each replicate contained at least two cotyledons.

2.2. Estimation of Callus Induction Rate and Rooting Rate in Peach trees

After 3 weeks of culture in CIM, the cotyledon was photographed using a Nikon SMZ25 microscope. The micrographs were used to estimate the size of the cotyledon and its callus with ImageJ v1.8.0. CIR was calculated using the following formula: CIR $=\frac{\text{callus area} \times 100\%}{\text{original cotyledon area}}$. After being cultured in CIM for 6 weeks, the number of callus-induced roots was recorded. Five-year-old subcultured ZYT callus of was used as a control.

2.3. Histological and Ultrastructural Analyses

Callus induced from cotyledons at different stages was examined via scanning electron microscope (SEM). After fixation with 2.5% glutaraldehyde, the callus was freeze-dried. A Leica EMACE ion-sputtering coater (Leica, Wetzlar, Germany) was used to coat a layer of gold powder onto the surface of the callus. Then, a desktop scanning electron microscope (Hitachi, TM3030, Tokyo, Japan) was used to scan the surface structure of the callus.

For analysis via transmission electron microscopy (TEM), the callus was fixed with 2.5% glutaraldehyde and then submerged in ethanol with differing gradient concentrations for 3 h. The fixed callus was cut into slices and stained, first with aqueous uranyl acetate and then with lead citrate. The stained section was scanned using transmission electron microscope (JEOL, Inc., Boston, MA, USA).

2.4. Measurement of Plant Metabolites

Fresh callus (100 mg) was ground into powder in liquid nitrogen, then the target product was extracted using a mixture of methanol:water (v/v = 80:20) at 4 °C. Flavonoid content and total phenol content were measured using the Plant Flavonoids Assay Kit (BC1335, Solarbio, Beijing, China) and the Total Phenol Assay Kit (BC1345, Solarbio, Beijing, China), respectively. The measurement of soluble sugars was based on our previous report [33].

2.5. RNA Extraction, Quantitative Real-Time PCR and Transcriptome Analysis

Callus samples of approximately 100 mg were collected and ground in liquid nitrogen for RNA extraction. RNA Rapid Extraction Kit (Megan, Guangzhou, China) was used for total RNA isolation. A Primer Script™ RT reagent Kit with gDNA Eraser (TransGen, Beijing, China) was used for first-strand cDNA synthesis. Quantitative real-time RT-PCR (qRT-PCR) was performed using SYBR premix Ex Taq II Kit (YEASEN, Shanghai, China) according to the manufacturer's instructions to yield a final volume of 20 µL. qRT-PCR was checked via StepOne Plus (ABI), and the internal reference gene was *PpTEF2* [34]. The amplification procedure was 95 °C for 30 s, 95 °C 5 s, and 60 °C 5 s, over 40 cycles. Relative gene expression level was measured according to the cycle threshold (Ct) $2^{-\Delta\Delta Ct}$ method [35]. Three replications were conducted for each treatment. Table S7 lists the primers used in the experiment.

2.6. Transcriptome Analysis and Identification of Differentially Expressed Genes (DEGs)

RNA samples for library construction were prepared following the abovementioned method, with each sample consisting of three biological replicates. RNA sequencing was performed using the Illumina HiSeq2500 platform. Clean reads were obtained by removing adapter sequences and low-quality reads. The HISAT2 v2.1.0 software program was utilized to align the clean reads to the reference genome [2]. The DESeq2 v1.22.2 software program was used to analyze the differential expression of each group [36,37]. Fragments per kilobase of transcript per million fragments mapped (FPKM) was utilized as an index to quantify the level of gene expression. Genes with $|\log2\text{fold change}| \geq 1$ and a false discovery rate (FDR) < 0.05 were considered DEGs, and FDR was estimated according to a previous report [38].

3. Results

3.1. Difference in Callus Induction Rate and Callus-Based Root Regeneration among Peach Cultivars

To quantitatively assess the impact of genotype on callus induction, the size of callus induced from the designated cotyledon callus area together with the original cotyledon area were estimated (Figure 1A). The ratio of callus area to original cotyledon area was used to estimate CIR. A total of 100 peach cultivars were analyzed, and CIR showed a great variation among cultivars, ranging from 0–135% (Table S1). Based on their rates of callus induction, peach cultivars were divided into six classes: I (≥100%), II (70–99%), III (40–69%), IV (20–39%), V (1–19%) and VI (0%) (Figure 1B). Approximately 27% of cultivars showed no callus induction capacity, while one cultivar, BJLY, had the highest rate of callus induction. Both class II and class III cultivars exhibited relatively vigorous callus growth, accounting for 13% and 25%, respectively, of all accessions tested. By contrast, class IV and class V cultivars both had slow callus growth, and accounted for 16% and 17% of all tested accessions, respectively (Figure 1C).

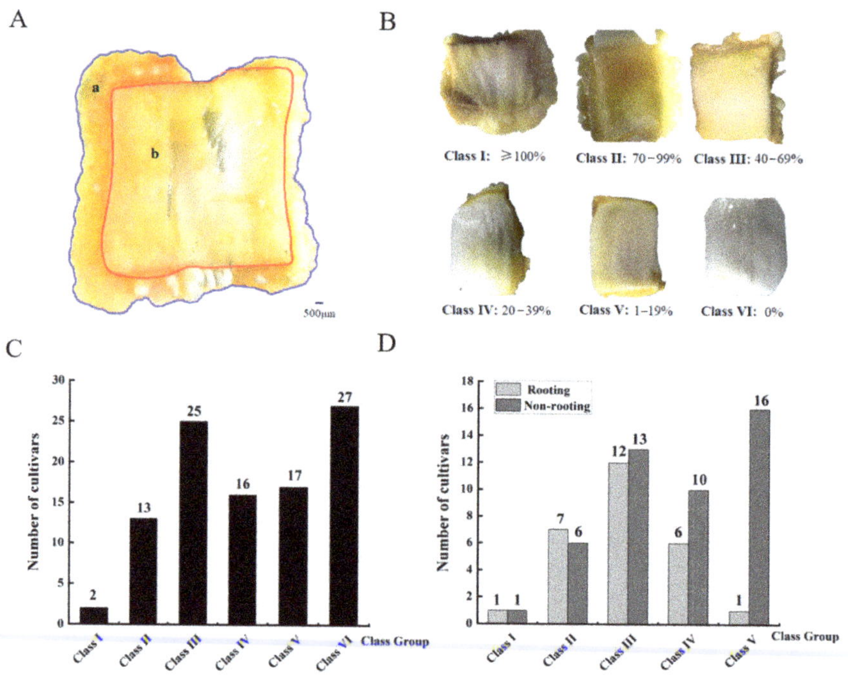

Figure 1. Relationship between callus induction rate and rooting ability. (**A**) Evaluation of rooting ability of callus from one hundred peach cultivars. Callus induction rate (CIR) was calculated based on the ratio of callus area (a) to original cotyledon area (b). (**B**) Classification of callus induction rate. (**C**) The numbers of cultivars in different CIR classes. (**D**) The numbers of cultivars exhibiting root regeneration ability in different CIR classes.

Notably, one or more ARs were often developed from the cotyledon-derived callus. An example of ARs developed from the callus of four cultivars is shown in Figure 2A. Approximately 50%, 54%, 48%, 38%, 6% of class I, class II, class III, class IV and class V cultivars, respectively, were capable of developing adventitious roots from cotyledon-derived callus, indicating an overall positive correlation between CIR and root regeneration capacity (Figure 1D). In addition, approximately 60% of the cotyledons from the ZYT cultivar developed adventitious roots during the process of callus induction. Interestingly, adventitious roots from the cotyledon-derived ZYT callus had a more vigorous growth compared with those from the ZYT seedlings (Figure 2A,B), suggesting that ZYT has a strong propensity for adventitious rooting from callus.

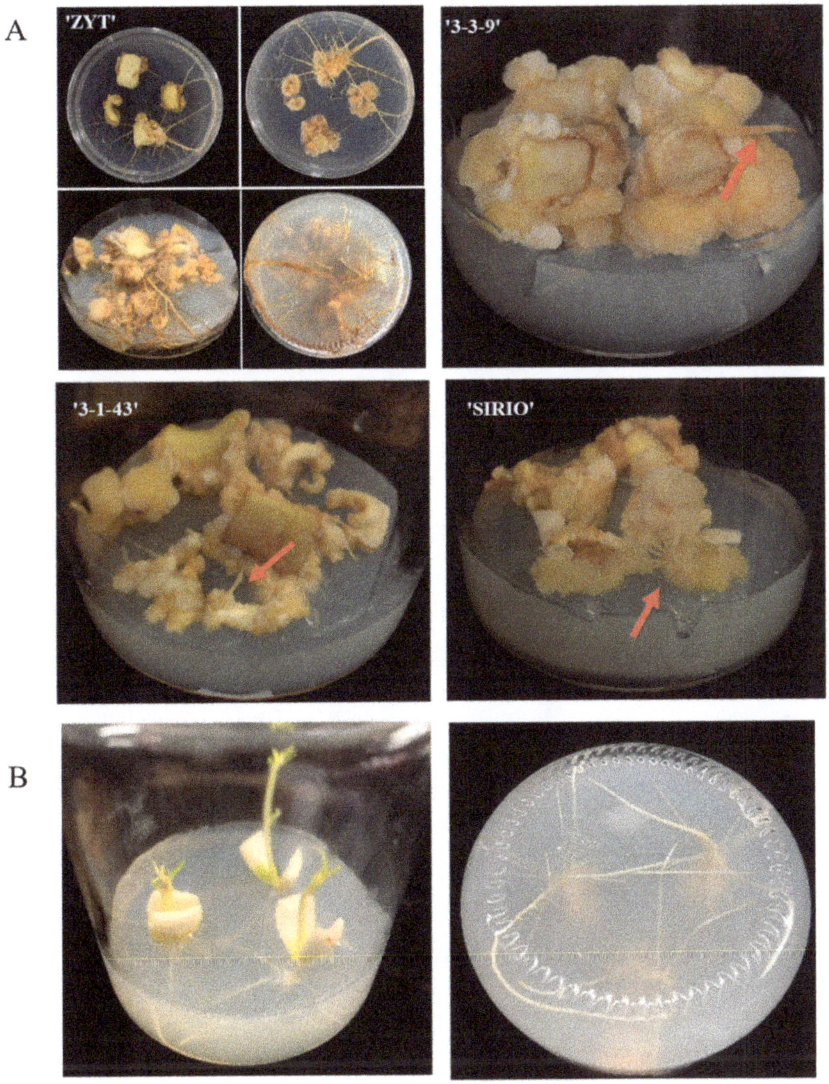

Figure 2. Adventitious root formation from cotyledon-derived callus from different peach cultivars. (**A**) Adventitious roots from cotyledon-derived callus from different peach cultivars cultured in CIM for 6 weeks. Note: the arrow shows that adventitious roots. (**B**) ZYT seedlings after cultivated on CIM for 6 weeks. The plastic petri dish is 90 mm in diameter, while the cultivation vial is 75 mm in diameter and 108 mm in height.

3.2. Identification of Candidate Genes Related to Adventitious Root Formation during Callus Culture

The CIR analysis revealed that cultivars ZH and QQ, like cultivar ZYT, were prone to developing adventitious roots from cotyledon-derived callus. Since long-term subcultured callus is characterized by a lack of regeneration ability [39], the five-year-old subcultured ZYT callus designated CC5Y-ZY was unable to regenerate adventitious roots. To investigate the mechanism underlying callus-based root regeneration in peach trees, we compared the difference in transcriptome profile between the cotyledon-derived calli of ZYT, ZH, and QQ after 3 and 6 weeks of culture and CC5Y-ZY. For ease of description, cotyledon-derived calli with potential rooting capacity from ZYT, ZH and QQ after 3 weeks of culture in CIM were designated CC3W-ZY, CC3W-ZH, and CC3W-QQ, respectively, while the root-bearing calli from ZYT, ZH, and QQ after 6 weeks of culture on CIM were termed CC6WR-ZY, CC6WR-ZH, and CC6WR-QQ, respectively. The transcriptomes of CC3W-ZY/ZH/QQ, CC6WR-ZY/ZH/QQ, and CC5Y-ZY were sequenced, with each containing three biological replicates. After the adapter sequence and low-quality reads were removed, 143.33 Gb clean data were acquired from 21 RNA-seq libraries (Table S2). On average, approximately 93.8% of the clean reads were uniquely mapped to the reference genome Lovell v.2.0 [2] (Table S3).

A comparison of the transcriptomes from CC3W-ZY/ZH/QQ to those from CC5Y revealed 1773 and 1512 commonly up-regulated and down-regulated differentially expressed genes (DEGs), respectively (Figure 3A). Likewise, comparison of the transcriptomes from CC6WR-ZY/ZH/QQ with those from CC5Y revealed 2766 and 1605 commonly up-regulated and down-regulated DEGs, respectively (Figure 3A). Overall, 1621 commonly up-regulated DEGs and 1120 commonly down-regulated DEGs were expressed differentially in the pluripotent calli of the three cultivars tested and the long-term subcultured callus in the absence of totipotency (Table S4). Thus, these 2741 DEGs were expected to contain genes related to the formation of callus and adventitious rooting. Annotation of the 2741 DEGs was performed using the online program plantTFDB available at: http://planttfdb.gao-lab.org/ (accessed on 7 April 2022). As a result, 184 and 62 transcription factors (TFs) were identified in the commonly up-regulated and down-regulated DEGs, respectively (Table S5).

Of the 184 up-regulated TFs, 14 are homologous to previously reported positive regulators of root development in *Arabidopsis*, including *PpMYB6* (Prupe.7G216000), *PpRAX2* (Prupe.7G128800), *PpMYB114* (Prupe.3G042100), *PpANL2* (Prupe.2G291400), *PpARR2* (Prupe.6G254900), *PpLRP1* (Prupe.2G036500), *PpSRS3* (Prupe.8G088300), *PpbHLH123 (Prupe.4G196600), PpbHLH68 (Prupe.7G225800), PpDof1.4* (Prupe.4G042500), *PpDof2.4* (Prupe.6G079500), *PpERF109* (Prupe.8G125100), *PpLBD1* (Prupe.2G191100), and *PpWRKY75* (Prupe.1G223200) (Figure 3B; Table S6). The three *MYB* genes, *PpMYB6*, *PpRAX2*, and *PpMYB114*, are homologous to *AtMYB68*, involved in root development [40]; *AtMYB36*, controlling LR primordium (LRP) development [41]; and *AtMYB66*, related to the formation of root hair cell [42], respectively. Two *bHLH* genes, *PpbHLH123* and *PpbHLH68*, are homologs of *AtPFA3* and *AtPFA6*, which both determine lateral root initiation [43]. Two *Dof* TFs, *PpDof1.4* and *PpDof2.4*, are homologs of *AtURP3* and *AtDof2.4*, which are associated with the promotion of radial growth in the root apical meristem [44]. Two RING-like zinc finger TFs, *PpLRP1* and *PpSRS3*, are homologs of *AtLRP1* and *AtSTY1*, which control LRP development [45]. *PpANL2* is a homolog of the *Arabidopsis* HD-ZIP homeodomain protein gene *AtANL2*, which is concerned with root growth [46]. *PpARR2* is a homolog of the G2-like family gene *AtARR2*, which is related to the domination of root meristem growth [47]. An ethylene response factor (ERF) subfamily, B-3 of ERF/AP2 TF *PpERF109*, is homologous to *AtERF109*, which regulates lateral root formation by mediating cross-talk between jasmonic acid and auxin signaling [48], while *PpLBD1* is homologous to *AtLBD1*, which controls lateral organ fate [49]. The remaining gene, *PpWRKY75*, is a homolog of *AtWRKY75*, which is related to lateral root (LR) development [50].

Of the 62 down-regulated TFs, 6 are homologous to previously reported negative regulators of root development in *Arabidopsis*, including *PpPTI6* (Prupe.2G183200), *PpAIL6* (Prupe.6G203000), *PpARF17* (Prupe.1G507000), *PpREM1* (Prupe.8G122700), PpHAT4 (Prupe.5G064300), and *PpLBD4* (Prupe.4G138200) (Figure 3B; Table S6). *PpPTI6* is homologous to *Arabidopsis* CRF TFs, which modulate root and shoot growth [51] *PpAIL6* is a homolog of *AtPLT3*, the key regulator of the de novo organ model during *Arabidopsis* LR formation [52]. *PpARF17* is homologous to the auxin-inducible repressor *AtARF17*, which negatively regulates AR formation [53]. *PpREM1* is homologous to *AtREM1*, which is involved in root gravitropic growth [54], while *PpHAT4* is homologous to *AtHAT4*, which inhibits root growth and branching [55]. *PpLBD4* is homologous to *AtLBD4*, a master regulator of root growth [56]. Taken together, above results suggested that 20 TFs have a potential function in the induction of callus-based adventitious roots in peach trees.

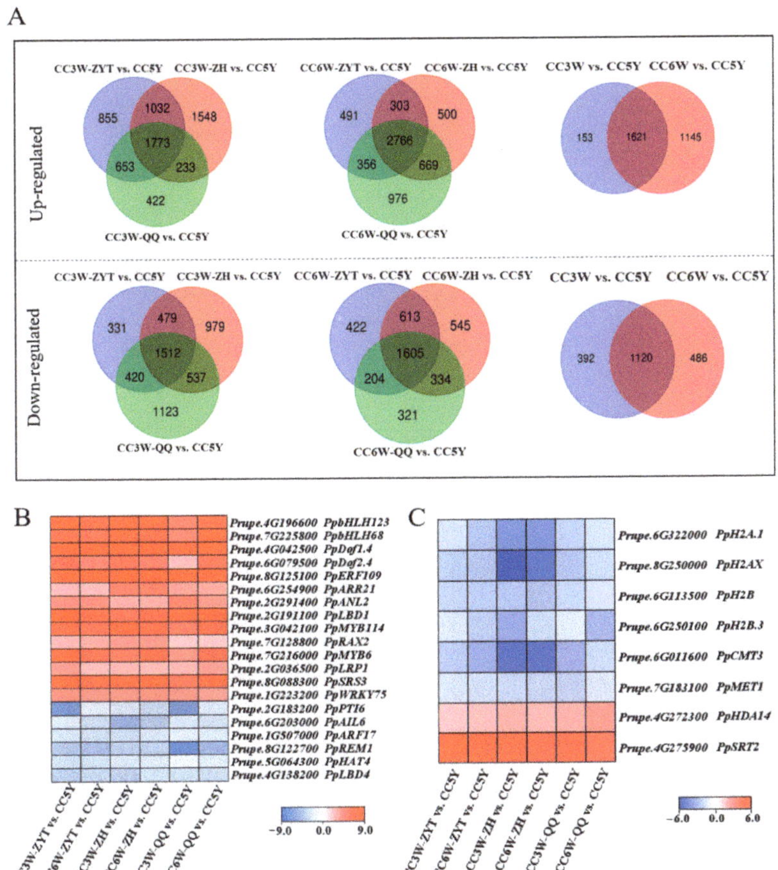

Figure 3. Identification of candidate genes related to the formation of callus adventitious rooting. (**A**) Venn diagram of DEGs between newly formed calli and long-term sub-cultured callus; (**B**) heat map display of transcription factors that were differentially expressed between newly formed calli and long-term sub-cultured callus; and (**C**) heat map display of DNA methylation and histone modification-related genes that were expressed differentially in newly formed calli and long-term sub-cultured callus.

Moreover, the abovementioned commonly up-regulated DEGs contained two histone deacetylase genes, *PpHDA14* and *PpSRT2* (Figure 3C) [57]. Similarly, the commonly down-regulated DEGs consisted of two DNA methyltransferase genes, *PpCMT3* and *PpMET1* [58], as well as four histone/histone variant genes, *PpH2A.1*, *PpH2AX*, *PpH2B*, and *PpH2B.3* [59] (Figure 3C). These results suggested a potential role in DNA methylation and histone modification in AR formation during peach callus culture.

3.3. Expression of Candidate Genes Associated with Callus Adventitious Rooting in the Cotyledon-Derived Callus

To confirm the relationship between callus adventitious rooting and the 20 abovementioned TFs, we investigated their expression in the cotyledon-derived callus of 8 cultivars, including 2, 2, and 4 of class I, class II, and class V cultivars, respectively. Of the 20 TFs, 6 showed high levels of expression in class I and class II cultivars, but were weakly expressed in class V cultivars (Figure 4), indicating a consistency between their expression and the callus rooting capacity. However, the remaining 14 TFs had no consistency between expression and callus rooting capacity (Figure S1). These results indicate that the six TFs *PpDof1.4*, *PpDof2.4*, *PpLBD1*, *PpSRS3*, *PpRAX2*, and *PpbHLH68* are good candidates for regulators correlated with the formation of ARs during callus culture in peach trees.

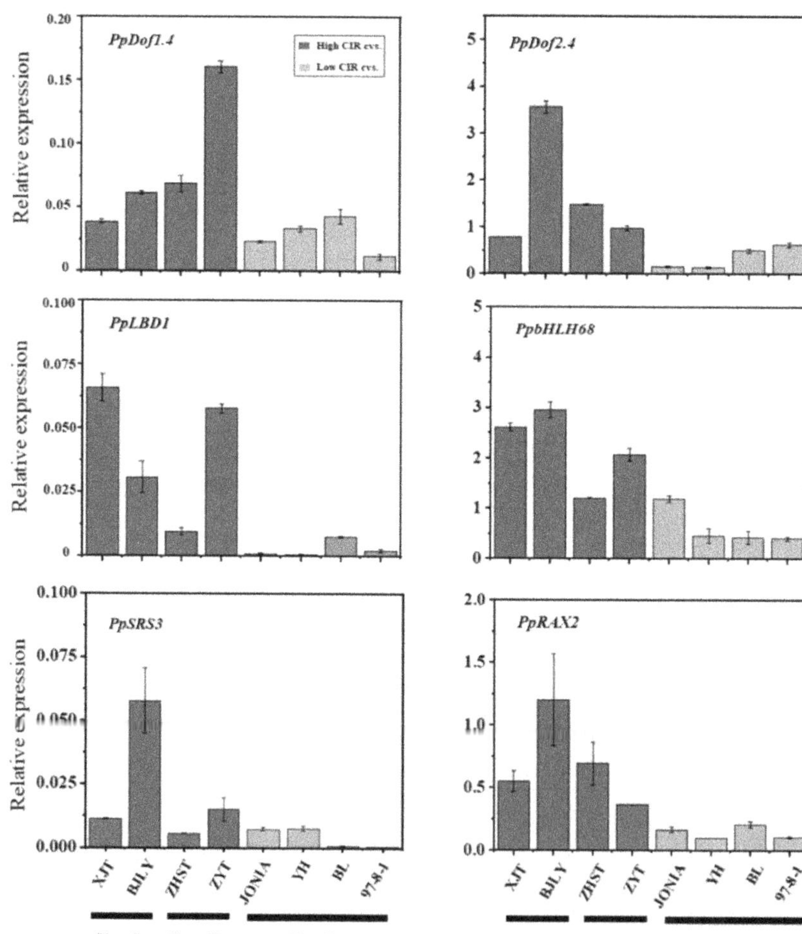

Figure 4. Expression levels of candidate genes associated with callus adventitious rooting in the cotyledon-derived callus. The genes expression as determined by Qrt-PCR.

3.4. Cellular Morphological Changes in the Cotyledon-Derived Callus during the Process of Root Differentiation

The abovementioned common DEGs consisted of genes relevant to sugar metabolism and flavonoid biosynthesis (Figure S2). Thus, we investigated metabolic changes in the cotyledon-derived callus during the process of root differentiation using high-performance liquid chromatography (HPLC) analysis. As a result, the soluble sugar contents of CC3W-ZY, CC6WR-ZY, and CC5Y-ZY all displayed the following trend: fructose > glucose > sucrose (Figure 5A). The CC5Y-ZY callus had the highest levels of fructose, glucose, and sucrose, corresponding to 7.91, 8.46 and 14.58 mg/g, respectively. By contrast, the contents of total phenols (TP) and flavonoids (FV) in CC5Y-ZY were extremely low, with 0.159 and 0.65 mg/g, respectively (Figure 5B). CC3W-ZY had the highest levels of TP and FV, while moderate levels of both TP and FV were observed in CC6WR-ZY.

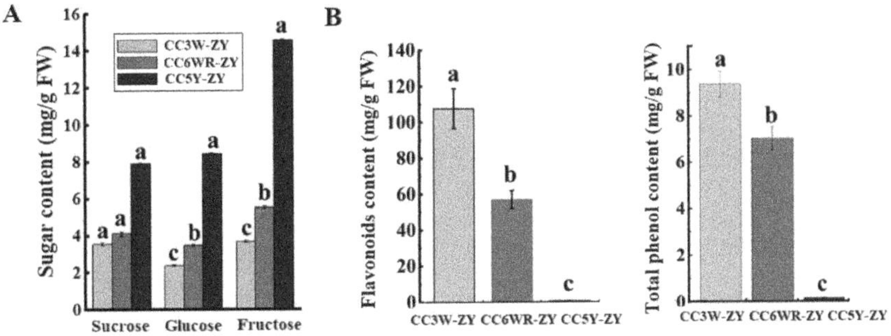

Figure 5. Metabolite content. (**A**) Soluble sugar content. (**B**) Total flavonoids and phenol content. The error bars show ± SE of at least three biological replicates, and significant difference at $p < 0.05$ is indicated by different lowercase letters based on LSD test.

Since the degradation of amyloplasts not only has an impact on sugar content, but is also associated with regeneration capacity [60], we investigated the cellular morphology of the cotyledon-derived callus at different stages. CC3W-ZY and CC6WR-ZY were yellow- and milky white-colored, respectively, while CC5Y-ZY had a glittering translucent appearance and was fragile (Figure 6A). Transmission electron microscopy results showed that CC3W-ZY cells had the smallest size and the largest number of amyloplasts, with abundant substances in the intercellular space (Figure 6B). CC6WR-ZY cells had a moderate size, fewer amyloplasts, and abundant phenolic compounds, with a moderate number of substances in the intercellular space. CC5Y-ZY cells were seriously vacuolated and contained many electron-dense inclusions, with no substances in the intercellular space. Notably, CC5Y-ZY cells had the largest number of mitochondria and an appearance of rough endoplasmic reticulum, suggesting an activation of cell metabolism and division. However, CC5Y-ZY cells lacked amyloplasts and obvious nuclei and nucleoli. These results suggested that the change in soluble sugars in the cotyledon-derived callus during the culture process is likely associated with the degradation of amyloplasts.

In addition, SEM analysis showed that CC3W-ZY was composed of plump round cells, while CC6WR-ZY cells had a columnar shape and loose structure (Figure 6C). CC5Y-ZY cells were irregular, elongated, and spiral-shaped. Both CC3W-ZY and CC6WR-ZY cells were tightly linked to each other with small intercellular spaces, but the reverse trend was observed in CC5Y-ZY cells. Altogether, these results suggest that the callus had undergone a significant change in cellular morphology during the culture process, and irregular morphology may be partially responsible for loss of regeneration capacity.

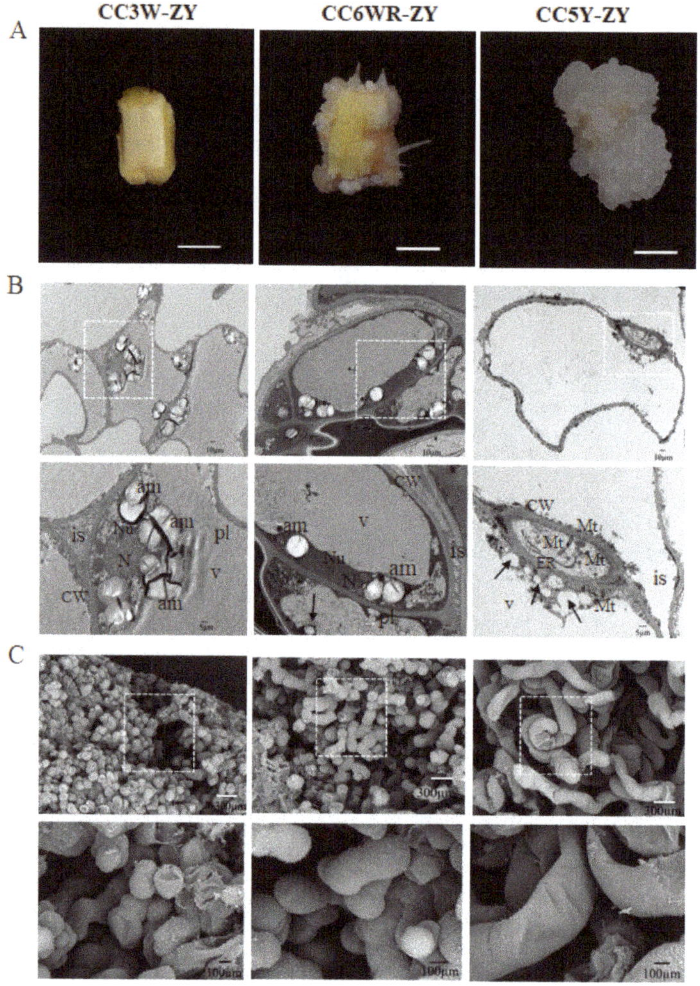

Figure 6. Morphology of CC3W-ZY, CC6WR-ZY, and CC5Y-ZY. (**A**) Appearance of ZYT cotyledon-derived callus culture in CIM at 3 weeks (CC3W-ZY), 6 weeks (CC6WR-ZY), and 5 years (CC5Y-ZY), respectively. (Bar = 1cm). (**B**) TEM images of three types of callus; magnification of the rectangle from subfigure (**B**) shows that cell ultra-structures are different. (**C**) SEM images of three callus types. Magnification of the rectangle from subfigure (**C**) shows that cell surfaces are different. Abbreviations: cell wall (CW), endoplasmic reticulum (ER), nucleus (N), nucleolus (Nu), mitochondria (Mt), phenolics (pl), vacuoles (v), amyloplasts (am), electron-dense inclusions (black arrows), intercellular space (is).

4. Discussion

4.1. Induction of Callus from Cotyledon and Its Root Regeneration in Peach trees

Callus can be rapidly propagated without the limitations of time and space and is thus considered one of the most important materials in verifying gene function [61]. Since callus induction is regarded as the critical stage for plantlet regeneration, callus induction rate is regarded as a critical index of regeneration ability [62]. As mentioned, callus has been successfully induced from leaf and stem explants in peach trees [30–32]; however, little is known about the influence of genotype on callus induction. In this project, we report the first large-scale investigation of CIR in peach trees. Our results showed that callus induction rate is genotype-dependent in peach trees, which is consistent with previous reports in cotton [63], sugarcane [64], sunflowers [65], and roses [66]. Notably, late-maturing cultivars had a higher callus induction rate (>70%), while the majority of early maturing cultivars showed no ability to induce callus from cotyledon explants. Callus tested in this study was induced via cotyledon, which is a vital part of the seed. The physiological state of seeds affects the growth of cotyledons and ultimately the formation of callus. The seeds of early maturing cultivars were shrunken, while the seeds of late-maturing cultivars were plump. The cotyledons of the shrunken seeds were thin, semi-transparent, and irregular, which are the manifestations of incomplete development, while the plump seeds had fleshy and milky white-colored cotyledons with an oval shape. Thus, the low callus induction rate of early maturing cultivars is likely attributable to incomplete seed development. These results are consistent with a previous report that the physiological state of explants has an important function in callus induction [67].

Moreover, our results showed that genotypes with higher CIR had a tendency to develop adventitious roots during callus induction. Overall, callus induction rate was positively correlated with adventitious root regeneration ability among the tested peach cultivars. In *Arabidopsis*, the induction of gene correlation with the formation of lateral root primordium displays an important effect in callus induction [68,69]. Therefore, it seems that genes such as *PpLRP1* and *PpLBD16/29*, both involved in the formation of lateral root primordium, may have critical roles in callus induction in peach trees. It is worth noting that two peach cultivars, ZYT and XJT, had high callus induction rates and root regeneration ability. These two cultivars may be ideal genotypes for the construction of the peach regeneration system in the future.

4.2. *PpLBD1 Is Likely Involved in Callus Adventitious Rooting in Peach Trees*

Callus cells have high plasticity when differentiating into adventitious shoot and root via direct somatic embryogenesis (SE) [6]. Previous studies have indicated that callus is induced via the LR development route [68,70]. The newly formed callus cells resemble root primordium-like cells that have potential adventitious root regeneration ability. Here, the five-year-old subcultured callus lost the potential to regenerate roots. Transcriptome comparison of the five-year-old subcultured callus with the newly formed callus revealed six candidate genes involved in callus adventitious rooting. These include an LBD family TF *PpLBD1*, which exhibited a high expression in the newly formed callus but was undetected in the five-year-old subcultured callus. Validation in different cultivars showed that *PpLBD1* might positively regulate the rooting of peach callus. Conversely, in this study, *PpLBD16* and *PpLBD29* showed a high expression in five-year-old subcultured callus, but a weak expression in the newly formed callus. The *LBD1* gene has been proven to control the development and architecture of lateral roots in *Medicago truncatula* and to regulate of secondary growth in *Populus* [71,72]. In *Arabidopsis*, LBD16/29 can form a complex with AtbZIP59 to activate transcription of the *FAD-BD* gene, thereby promoting the formation of callus [73]. Thus, *PpLBD1* may be a good candidate gene bound up with the regulation of callus adventitious rooting in peach trees, while *PpLBD16* and *PpLBD29* may be at least partially in charge of the strong callus proliferation ability of the five-year-old subcultured callus.

4.3. *The Effect of Metabolites on Callus Adventitious Rooting in Peach*

Starch is the most universal carbon form for energy substrate, and starch granules in callus serve as an energy source for the regeneration procedure [74–76]. In this study, starch granules were abundant in the newly formed callus with root regeneration ability, while the five-year-old subcultured callus contained few starch granules. This suggests that the deposition of starch granules is crucial for root regeneration from callus. Granule-bound starch synthase 1 (GBSS1) catalyzes one of the enzymatic steps of starch synthesis. *PpGBSS1* (Prupe.5G132800) exhibited a highly expression in the newly formed callus, but was weakly expressed in the five-year-old subcultured callus. Thus, activation of *GBSS1* is likely required for root regeneration from callus.

In addition to starch granules, flavonoid content showed a significant difference between the newly formed callus and the five-year-old subcultured callus. The regulatory roles of flavonoids in adventitious rooting of stem cuttings are different between plant species. For example, in cherry and chestnut, flavonoid accumulation inhibits AR regeneration in stem cuttings [77,78]. However, flavonoid biosynthesis has a positive effect on AR formation of stem cuttings in olive [79]. In *Chamaelaucium uncinatum*, flavonoids negatively and positively regulate the formation of ARs from hardwood and softwood cuttings, respectively [80]. In this study, the flavonoid content in the five-year-old subcultured callus was significantly lower than in the newly formed callus, suggesting that flavonoid accumulation might have a positive regulatory role in root regeneration from callus. Among the flavonoid biosynthesis genes, *PpPAL* (Prupe.6G235400) encoding phenylalanine ammonia lyase (PAL) displayed a diversity in expression between the newly formed callus and the five-year-old subcultured callus. This finding is in agreement with the previous report that PAL is the rate-limiting enzyme of flavonoid biosynthesis [81]. Hence, changes in *PpPAL* expression may cause differences in flavonoid levels and affect the formation of ARs in peach callus.

4.4. Epigenetic Regulation in Callus Adventitious Rooting in Peach

Cell morphology changes significantly during the process of subculture, caused by the loss of callus regeneration ability [82]. In *Theobroma cacao*, global DNA methylation levels show an increase in aged somatic embryos after long-term in vitro culture, but this increase can be inhibited via Aza treatment [83]. Thus, epigenetic regulation seems to show a crucial role in determining callus cell morphology. In this study, the cells of five-year-old callus exhibited irregular, elongated, and spiral-shaped morphology. Moreover, the expression of genes involved in DNA methylation (*PpCMT3* and *PpMET1*) and histone methylation (*PpH2A.1*, *PpH2AX*, *PpH2B*, and *PpH2B.3*) was found to be up-regulated in the five-year-old callus compared to the newly formed callus. This suggests that long-term subculture callus in peach trees has a significant change in epigenetic modifications, which coincides with a previous report that numerous of epigenetic modifications occur during the process of callus-based regeneration in plant [84]. Change in epigenetic landscape is known to be accompanied by transition of cell fate [85,86]. Here, our results indicate that aged callus cells with long-term in vitro culture lost root regeneration ability. Therefore, epigenetic regulation is likely associated with callus adventitious rooting capacity in peach trees.

5. Conclusions

In this study, a significant difference in CIR was found among peach germplasms. Notably, the cultivars ZYT and XJT can be recommended as ideal candidates for constructing regeneration system in peach. Moreover, our comparative transcriptome analysis revealed *PpLBD1* as a candidate gene implicated in the AR development. Overall, the results of this study not only provide valuable insights into the underlying mechanisms of regeneration, but are also useful for the development of regeneration systems in peach trees.

Supplementary Materials: The following supporting information can be downloaded at: https://www.mdpi.com/article/10.3390/horticulturae9080850/s1, Figure S1: Expression levels of candidate genes associated with callus adventitious rooting in the cotyledon-derived callus. The genes expression as determined by Qrt-PCR; Figure S2: Gene ontology (GO) enrichment; Table S1: Peach germplasm used in this study; Table S2: Transcriptome sequencing data in *Prunus persica* L.; Table S3: Transcriptome sequencing data in *Prunus persica* L.; Table S4 Total 2741 DEGs were divided into up- and down-regulated according to log2FC; Table S5 Transcription factors distribution in up- and down-regulated; Table S6 Candidate genes that may be related to adventitious roots formation of cotyledon-derived callus in peach; Table S7 Real-time quantitative PCR primers used in this study.

Author Contributions: L.G.: Conceptualization, Writing—original draft; Methodology and Validation. J.L.: Data analysis and project administration. L.L.: Data curation, Software and Visualization. A.G.: Validation and collect samples. B.N.N.: Validation and writing. C.Z.: Investigation, Provision and collection resources. B.Z.: Writing—review & editing, designed the experiments. Y.H.: Writing—review & editing and supervision. All authors have read and agreed to the published version of the manuscript.

Funding: This research was supported by funding received from National Natural Science Foundation of China (32272690 and 32272687), the National Key R&D Program of China (2019YFD1000800), and the China Agriculture Research System (grant no. CARS-30).

Institutional Review Board Statement: Not applicable.

Informed Consent Statement: Not applicable.

Data Availability Statement: Download the supplementary figure and table files used in this project online.

Conflicts of Interest: The authors declare no conflict of interest.

References

1. Abbott, A.G.; Lecouls, A.C.; Wang, Y.; Georgi, L.; Scorza, R.; Reigherd, G. Peach: The model genome for rosaceae genomics. *Acta Hortic.* **2002**, *592*, 199–209. [CrossRef]
2. Verde, I.; Jenkins, J.; Dondini, L.; Micali, S.; Pagliarani, G.; Vendramin, E.; Paris, R.; Aramini, V.; Gazza, L.; Rossini, L.; et al. The Peach v2. 0 release: High-resolution linkage mapping and deep resequencing improve chromosome-scale assembly and contiguity. *BMC Genom.* **2017**, *18*, 225. [CrossRef] [PubMed]
3. Ricci, A.; Sabbadini, S.; Prieto, H.; Padilla, I.M.; Dardick, C.; Li, Z.J.; Scorza, R.; Limera, C.; Mezzetti, B.; Perez-Jimenez, M.; et al. Genetic transformation in peach (*Prunus persica* L.): Challenges and ways forward. *Plants* **2020**, *9*, 971. [CrossRef] [PubMed]
4. San, B.; Li, Z.G.; Hu, Q.; Reighard, G.L.; Luo, H. Adventitious shoot regeneration from in vitro cultured leaf explants of peach rootstock Guardian is significantly enhanced by silver thiosulfate. *Plant Cell Tissue Organ Cult.* **2015**, *120*, 757–765. [CrossRef]
5. Ikeuchi, M.; Favero, D.S.; Sakamoto, Y.; Iwase, A.; Coleman, D.; Rymen, B.; Sugimoto, K. Molecular mechanisms of plant regeneration. *Annu. Rev. Plant Biol.* **2019**, *70*, 377–406. [CrossRef]
6. Ikeuchi, M.; Ogawa, Y.; Iwase, A.; Sugimoto, K. Plant regeneration: Cellular origins and molecular mechanisms. *Development* **2016**, *143*, 1442–1451. [CrossRef]
7. Wu, G.Y.; Wei, X.L.; Wang, X.; Wei, Y. Induction of somatic embryogenesis in different explants from *Ormosia henryi* Prain. *Plant Cell Tissue Organ Cult.* **2020**, *142*, 229–240. [CrossRef]
8. Hammerschlag, F.A.; Bauchan, G.R.; Scorza, R. Regeneration of peach plants from callus derived from immature embryos. *Theor. Appl. Genet.* **1985**, *70*, 248–251. [CrossRef]
9. Yao, J.L.; Cohen, D.; Atkinson, R.; Richardson, K.; Morris, B. Regeneration of transgenic plants from the commercial apple cultivar Royal Gala. *Plant Cell Rep.* **1995**, *14*, 407–412. [CrossRef]
10. Huang, H.; Wei, Y.; Zhai, Y.J.; Ouyang, K.X.; Chen, X.Y.; Bai, L.H. High frequency regeneration of plants via callus-mediated organogenesis from cotyledon and hypocotyl cultures in a multipurpose tropical tree (*Neolamarkia Cadamba*). *Sci. Rep.* **2020**, *10*, 4558. [CrossRef]
11. Zarinjoei, F.; Rahmani, M.S.; Shabanian, N. In vitro plant regeneration from cotyledon-derived callus cultures of leguminous tree *Gleditsia caspica* Desf. *New For.* **2014**, *45*, 829–841. [CrossRef]
12. Anandan, R.; Prakash, M.; Deenadhayalan, T.; Nivetha, R.; Kumar, N.S. Efficient invitro plant regeneration from cotyledon-derived callus cultures of sesame (*Sesamum indicum* L.) and genetic analysis of True-to-Type regenerants using RAPD and SSR markers. *S. Afr. J. Bot.* **2018**, *119*, 244–251. [CrossRef]
13. Parmar, N.; Kanwar, K.; Thakur, A.K. In Vitro Organogenesis from cotyledon derived callus cultures of *Punica granatum* L. cv. Kandhari Kabuli. *Natl. Acad. Sci. Lett.* **2012**, *35*, 215–220. [CrossRef]
14. Slusarkiewicz-Jarzina, A.; Ponitka, A.; Kaczmarek, Z. Influence of cultivar, explant source and plant growth regulator on callus induction and plant regeneration of *Cannabis sativa* L. *Acta Biol. Crac. Ser. Bot.* **2005**, *47*, 145–151.
15. Wang, C.; Ma, H.; Zhu, W.; Zhang, J.; Zhao, X.; Li, X. Seedling-derived leaf and root tip as alternative explants for callus induction and plant regeneration in maize. *Physiol Plant.* **2021**, *172*, 1570–1581. [CrossRef] [PubMed]
16. Gentile, A.; Monticelli, S.; Damiano, C. Adventitious shoot regeneration in peach [*Prunus persica* (L.) Batsch]. *Plant Cell Rep.* **2002**, *20*, 1011–1016. [CrossRef]
17. Smigocki, A.C.; Hammerschlag, F.A. Regeneration of plants from peach embryo cells infected with a shooty mutant strain of *Agrobacterium*. *J. Am. Soc. Hortic. Sci.* **1991**, *116*, 1092–1097. [CrossRef]
18. Perez-Clemente, R.M.; Perez-Sanjuan, A.; Garcia-Ferriz, L.; Beltran, J.P.; Canas, L.A. Transgenic peach plants (*Prunus persica* L.) produced by genetic transformation of embryo sections using the green fluorescent protein (GFP) as an in vivo marker. *Mol. Breed.* **2004**, *14*, 419–427. [CrossRef]

19. Xu, S.L.; Lai, E.H.; Zhao, L.; Cai, Y.M.; Ogutu, C.O.; Cherono, S.; Han, Y.P.; Zheng, B.B. Development of a fast and efficient root transgenic system for functional genomics and genetic engineering in peach. *Sci. Rep.* **2020**, *10*, 2836. [CrossRef]
20. Reighard, G.L.; Cain, D.W.; Newall, W.C. Rooting and survival potential of hardwood cuttings of 406 species, cultivars, and hybrids of *Prunus*. *Hortscience* **1990**, *25*, 517–518. [CrossRef]
21. Justamante, M.S.; Mhimdi, M.; Molina-Perez, M.; Albacete, A.; Moreno, M.A.; Mataix, I.; Perez-Perez, J.M. Effects of auxin (Indole-3-butyric Acid) on adventitious root formation in peach-based *Prunus* rootstocks. *Plants* **2022**, *11*, 913. [CrossRef] [PubMed]
22. Bellini, C.; Pacurar, D.I.; Perrone, I. Adventitious roots and lateral roots: Similarities and differences. *Annu. Rev. Plant Biol.* **2014**, *65*, 639. [CrossRef] [PubMed]
23. Shuai, B.; Reynaga-Pena, C.G.; Springer, P.S. The lateral organ boundaries gene defines a novel, plant-specific gene family. *Plant Physiol.* **2002**, *129*, 747–761. [CrossRef] [PubMed]
24. Lee, H.W.; Cho, C.; Pandey, S.K.; Park, Y.; Kim, M.J.; Kim, J. *LBD16* and *LBD18* acting downstream of *ARF7* and *ARF19* are involved in adventitious root formation in *Arabidopsis*. *BMC Plant Biol.* **2019**, *19*, 46. [CrossRef]
25. Yamauchi, T.; Tanaka, A.; Inahashi, H.; Nishizawa, N.K.; Tsutsumi, N.; Inukai, Y.; Nakazono, M.N. Fine control of aerenchyma and lateral root development through AUX/IAA-and ARF-dependent auxin signaling. *Proc. Natl. Acad. Sci. USA* **2019**, *116*, 20770–20775. [CrossRef]
26. Inukai, Y.; Sakamoto, T.; Ueguchi-Tanaka, M.; Shibata, Y.; Gomi, K.; Umemura, I.; Hasegawa, Y.; Ashikari, M.; Kitano, H.; Matsuoka, M. *Crown rootless1*, which is essential for crown root formation in rice, is a target of an AUXIN RESPONSE FACTOR in auxin signaling. *Plant Cell.* **2005**, *17*, 1387–1396. [CrossRef]
27. Liu, H.J.; Wang, S.F.; Yu, X.B.; Yu, J.; He, X.W.; Zhang, S.L.; Shou, H.X.; Wu, P. ARL1, a LOB-domain protein required for adventitious root formation in rice. *Plant J.* **2005**, *43*, 47–56. [CrossRef]
28. Bernhardt, C.; Lee, M.M.; Gonzalez, A.; Zhang, F.; Lloyd, A.; Schiefelbein, J. The bHLH genes *GLABRA3 (GL3)* and *ENHANCER OF GLABRA3 (EGL3)* specify epidermal cell fate in the *Arabidopsis* root. *Development* **2003**, *130*, 6431–6439. [CrossRef]
29. Ramsay, N.A.; Glover, B.J. MYB-bHLH-WD40 protein complex and the evolution of cellular diversity. *Trends Plant Sci.* **2005**, *10*, 63–70. [CrossRef]
30. Zheng, B.B.; Liu, J.J.; Gao, A.Q.; Chen, X.M.; Gao, L.L.; Liao, L.; Luo, B.W.; Ogutu, C.O.; Han, Y.P. Epigenetic reprogramming of H3K27me3 and DNA methylation during leaf-to-callus transition in peach. *Hortic. Res.* **2022**, *9*, uhac132. [CrossRef]
31. Zhou, H.C.; Li, M.; Zhao, X.; Fan, X.; Guo, A.G. Plant regeneration from in vitro leaves of the peach rootstock 'Nemaguard' (*Prunus persica* × *P. davidiana*). *Plant Cell Tissue Organ Cult.* **2010**, *101*, 79–87. [CrossRef]
32. Perez-Jimenez, M.; Lopez-Soto, M.B.; Cos-Terrer, J. In vitro callus induction from adult tissues of peach (*Prunus persica* L. Batsch). *Vitr. Cell. Dev. Biol. -Plant* **2013**, *49*, 79–84. [CrossRef]
33. Zhen, Q.L.; Fang, T.; Peng, Q.; Liao, L.; Zhao, L.; Owiti, A.; Han, Y.P. Developing gene-tagged molecular markers for evaluation of genetic association of apple SWEET genes with fruit sugar accumulation. *Hortic. Res.* **2018**, *5*, 14. [CrossRef] [PubMed]
34. Tong, Z.; Gao, Z.; Wang, F.; Zhou, J.; Zhang, Z. Selection of reliable reference genes for gene expression studies in peach using real-time PCR. *BMC Mol. Biol.* **2009**, *10*, 71. [CrossRef]
35. Livak, K.J.; Schmittgen, T.D. Analysis of relative gene expression data using real-time quantitative PCR and the $2^{-\Delta\Delta CT}$ method. *Methods* **2001**, *25*, 402–408. [CrossRef]
36. Love, M.I.; Huber, W.; Anders, S. Moderated estimation of fold change and dispersion for RNA-seq data with DESeq2. *Genome Biol.* **2014**, *15*, 1–21. [CrossRef]
37. Varet, H.; Brillet-Guéguen, L.; Coppée, J.Y.; Dillies, M.A. SARTools: A DESeq2-and EdgeR-based R pipeline for comprehensive differential analysis of RNA-Seq data. *PLoS ONE* **2016**, *11*, e0157022. [CrossRef]
38. Paul, P. Using ANOVA for gene selection from microarray studies of the nervous system. *Methods* **2003**, *31*, 282–289. [CrossRef]
39. Liu, L.; Fan, X.; Zhang, J.W.; Yan, M.L.; Bao, M.Z. Long-term cultured callus and the effect factor of high-frequency plantlet regeneration and somatic embryogenesis maintenance in *Zoysia japonica*. *In Vitro Cell. Dev. Biol.-Plant* **2009**, *45*, 673–680. [CrossRef]
40. Feng, C.P.; Andreasson, E.; Maslak, A.; Mock, H.P.; Mattsson, O.; Mundy, J. *Arabidopsis* MYB68 in development and responses to environmental cues. *Plant Sci.* **2004**, *167*, 1099–1107. [CrossRef]
41. Liberman, L.M.; Sparks, E.E.; Moreno-Risueno, M.A.; Petricka, J.J.; Benfey, P.N. MYB36 regulates the transition from proliferation to differentiation in the *Arabidopsis* root. *Proc. Natl. Acad. Sci. USA* **2015**, *112*, 12099–12104. [CrossRef]
42. Lee, M.M.; Schiefelbein, J. WEREWOLF, a MYB-related protein in *Arabidopsis*, is a position-dependent regulator of epidermal cell patterning. *Cell* **1999**, *99*, 473–483. [CrossRef] [PubMed]
43. Zhang, Y.; Mitsuda, N.; Yoshizumi, T.; Horii, Y.; Oshima, Y.; Ohme-Takagi, M.; Matsui, M.; Kakimoto, T. Two types of bHLH transcription factor determine the competence of the pericycle for lateral root initiation. *Nat. Plants* **2021**, *7*, 633–643. [CrossRef] [PubMed]
44. Miyashima, S.; Roszak, P.; Sevilem, I.; Toyokura, K.; Blob, B.; Heo, J.O.; Mellor, N.; Help-Rinta-Rahko, H.; Otero, S.; Smet, W.; et al. Mobile PEAR transcription factors integrate positional cues to prime cambial growth. *Nature* **2019**, *565*, 490–494. [CrossRef]
45. Smith, D.L.; Fedoroff, N.V. LRP1, a gene expressed in lateral and adventitious root primordia of *Arabidopsis*. *Plant Cell.* **1995**, *7*, 735–745. [CrossRef]
46. Kubo, H.; Peeters, A.J.M.; Aarts, M.G.M.; Pereira, A.; Koornneef, M. *ANTHOCYANINLESS2*, a homeobox gene affecting anthocyanin distribution and root development in *Arabidopsis*. *Plant Cell* **1999**, *11*, 1217–1226. [CrossRef]

47. Takahashi, N.; Kajihara, T.; Okamura, C.; Kim, Y.; Katagiri, Y.; Okushima, Y.; Matsunaga, S.; Hwang, I.; Umeda, M. Cytokinins control endocycle onset by promoting the expression of an APC/C activator in *Arabidopsis* roots. *Curr. Biol.* **2013**, *23*, 1812–1817. [CrossRef]
48. Cai, X.T.; Xu, P.; Zhao, P.X.; Liu, R.; Yu, L.H.; Xiang, C.B. *Arabidopsis* ERF109 mediates cross-talk between jasmonic acid and auxin biosynthesis during lateral root formation. *Nat. Commun.* **2014**, *5*, 5833. [CrossRef]
49. Ye, L.L.; Wang, X.; Lyu, M.; Siligato, R.; Eswaran, G.; Vainio, L.; Blomster, T.; Zhang, J.; Mahonen, A.P. Cytokinins initiate secondary growth in the *Arabidopsis* root through a set of LBD genes. *Curr. Biol.* **2021**, *31*, 3365–3373.e7. [CrossRef] [PubMed]
50. Devaiah, B.N.; Karthikeyan, A.S.; Raghothama, K.G. WRKY75 transcription factor is a modulator of phosphate acquisition and root development in *Arabidopsis*. *Plant Physiol.* **2007**, *143*, 1789–1801. [CrossRef]
51. Raines, T.; Shanks, C.; Cheng, C.Y.; McPherson, D.; Argueso, C.T.; Kim, H.J.; Franco-Zorrilla, J.M.; Lopez-Vidriero, I.; Solano, R.; Vankova, R.; et al. The cytokinin response factors modulate root and shoot growth and promote leaf senescence in *Arabidopsis*. *Plant J.* **2016**, *85*, 134–147. [CrossRef]
52. Du, Y.; Scheres, B. PLETHORA transcription factors orchestrate de novo organ patterning during *Arabidopsis* lateral root outgrowth. *Proc. Natl. Acad. Sci. USA* **2017**, *114*, 11709–11714. [CrossRef] [PubMed]
53. Sorin, C.; Bussell, J.D.; Camus, I.; Ljung, K.; Kowalczyk, M.; Geiss, G.; McKhann, H.; Garcion, C.; Vaucheret, H.; Sandberg, G.; et al. Auxin and light control of adventitious rooting in *Arabidopsis* require ARGONAUTE1. *Plant Cell* **2005**, *17*, 1343–1359. [CrossRef] [PubMed]
54. Ke, M.Y.; Ma, Z.M.; Wang, D.Y.; Sun, Y.B.; Wen, C.J.; Huang, D.Q.; Chen, Z.C.; Yang, L.; Tan, S.T.; Li, R.X.; et al. Salicylic acid regulates *PIN2* auxin transporter hyperclustering and root gravitropic growth via *Remorin*-depende-nt lipid nanodomain organisation in *Arabidopsis thaliana*. *New Phytol.* **2021**, *229*, 963–978. [CrossRef]
55. Kollmer, I.; Werner, T.; Schmulling, T. Ectopic expression of different cytokinin-regulated transcription factor genes of *Arabidopsis thaliana* alters plant growth and development. *J. Plant Physiol.* **2011**, *168*, 1320–1327. [CrossRef]
56. Smit, M.E.; McGregor, S.R.; Sun, H.; Gough, C.; Bagman, A.M.; Soyars, C.L.; Kroon, J.H.; Gaudinier, A.; Williams, C.J.; Yang, X.Y.; et al. A PXY-mediated transcriptional network integrates signaling mechanisms to control vascular development in *Arabidopsis*. *Plant Cell* **2020**, *32*, 319–335. [CrossRef]
57. Wu, R.; Citovsky, V. Adaptor proteins GIR1 and GIR2. II. Interaction with the co-repressor TOPLESS and promotion of histone deacetylation of target chromatin. *Biochem. Biophys. Res. Commun.* **2017**, *488*, 609–613. [CrossRef] [PubMed]
58. Johnson, L.M.; Bostick, M.; Zhang, X.; Kraft, E.; Henderson, I.; Callis, J.; Jacobsen, S.E. The SRA methyl-cytosine-binding domain links DNA and histone methylation. *Curr. Biol.* **2007**, *17*, 379–384. [CrossRef]
59. Panda, K.; McCue, A.D.; Slotkin, R.K. Arabidopsis RNA polymerase IV generates 21–22 nucleotide small RNAs that can participate in RNA-directed DNA methylation and may regulate genes. *Philos. Trans. R. Soc. B* **2020**, *375*, 20190417. [CrossRef]
60. Laparra, H.; Bronner, R.; Hahne, G. Amyloplasts as a possible indicator of morphogenic potential in sunflower protoplasts. *Plant Sci.* **1997**, *122*, 183–192. [CrossRef]
61. Efferth, T. Biotechnology applications of plant callus cultures. *Engineering* **2019**, *5*, 50–59. [CrossRef]
62. Al Ghasheem, N.; Stanica, F.; Peticila, A.G. Overview of studies on in vitro propagation of peach. *Acta Hortic.* **2021**, *1304*, 147–154. [CrossRef]
63. Michel, Z.; Hilaire, K.T.; Mongomake, K.; Georges, A.N.; Justin, K.Y. Effect of genotype, explants, growth regulators and sugars on callus induction in cotton (*Gossypium hirsutum* L.). *Aust. J. Crop Sci.* **2008**, *2*, 1–9.
64. Gandonou, C.H.; Errabii, T.; Abrini, J.; Idaomar, M.; Chibi, F.; Senhaji, S. Effect of genotype on callus induction and plant regeneration from leaf explants of sugarcane (*Saccharum* sp.). *Afr. J. Biotechnol.* **2005**, *4*, 1250–1255.
65. Ozyigit, I.I.; Gozukirmizi, N.; Semiz, B.D. Genotype dependent callus induction and shoot regeneration in sunflower (*Helianthus annuus* L.). *Afr. J. Biotechnol.* **2007**, *6*, 1498–1502.
66. Nguyen, T.H.N.; Tanzer, S.; Rudeck, J.; Winkelmann, T.; Debener, T. Genetic analysis of adventitious root formation in vivo and in vitro in a diversity panel of roses. *Sci. Hortic.* **2020**, *266*, 109277. [CrossRef]
67. Mohebodini, M.; Mokhtar, J.J.; Mahboudi, F.; Alizadeh, H. Effects of genotype, explant age and growth regulators on callus induction and direct shoot regeneration of Let-tuce (*Lactuca sativa* L.). *Aust. J. Crop Sci.* **2011**, *5*, 92–95.
68. Sugimoto, K.; Jiao, Y.; Meyerowitz, E.M. *Arabidopsis* regeneration from multiple tissues occurs via a root development pathway. *Dev. Cell* **2010**, *18*, 463–471. [CrossRef]
69. Ikeuchi, M.; Sugimoto, K.; Iwase, A. Plant callus: Mechanisms of induction and repression. *Plant Cell* **2013**, *25*, 3159–3173. [CrossRef]
70. Fan, M.Z.; Xu, C.Y.; Xu, K.; Hu, Y.X. LATERAL ORGAN BOUNDARIES DOMAIN transcription factors direct callus formation in *Arabidopsis* regeneration. *Cell Res.* **2012**, *22*, 1169–1180. [CrossRef]
71. Ariel, F.D.; Diet, A.; Crespi, M.; Chan, R.L. The LOB-like transcription factor MtLBD1 controls *Medicago truncatula* root architecture under salt stress. *Plant Signal. Behav.* **2010**, *5*, 1666–1668. [CrossRef]
72. Yordanov, Y.S.; Regan, S.; Busov, V. Members of the LATERAL ORGAN BOUNDARIES DOMAIN transcription factor family are involved in the regulation of secondary growth in *Populus*. *Plant Cell* **2010**, *22*, 3662–3677. [CrossRef]
73. Xu, C.Y.; Cao, H.F.; Zhang, Q.Q.; Wang, H.Z.; Xin, W.; Xu, E.J.; Zahng, S.Q.; Yu, D.X.; Hu, Y.X. Control of auxin-induced callus formation by bZIP59-LBD complex in *Arabidopsis* regeneration. *Nat. Plants* **2018**, *4*, 108–115. [CrossRef]

74. Carciofi, M.; Blennow, A.; Nielsen, M.M.; Holm, P.B.; Hebelstrup, K.H. Barley callus: A model system for bioengineering of starch in cereals. *Plant Methods* **2012**, *8*, 36. [CrossRef] [PubMed]
75. Czernicka, M.; Chlosta, I.; Keska, K.; Kozieradzka-Kiszkurno, M.; Abdullah, M.; Popielarska-Konieczna, M. Protuberances are organized distinct regions of long-term callus: Histological and transcriptomic analyses in kiwifruit. *Plant Cell Rep.* **2021**, *40*, 637–665. [CrossRef] [PubMed]
76. Feng, M.Q.; Lu, M.D.; Long, J.M.; Yin, Z.P.; Jiang, N.; Wang, P.B.; Liu, Y.; Guo, W.W.; Wu, X.M. miR156 regulates somatic embryogenesis by modulating starch accumulation in citrus. *J. Exp. Bot.* **2022**, *73*, 6170–6185. [CrossRef] [PubMed]
77. Osterc, G.; Trobec, M.; Usenik, V.; Solar, A.; Stampar, F. Changes in polyphenols in leafy cuttings during the root initiation phase regarding various cutting types at *Castanea*. *Phyton-Ann. Rei Bot.* **2004**, *44*, 109–119.
78. Trobec, M.; Stampar, F.; Veberic, R.; Osterc, G. Fluctuations of different endogenous phenolic compounds and cinnamic acid in the first days of the rooting process of cherry rootstock 'GiSelA 5'leafy cuttings. *J. Plant Physiol.* **2005**, *162*, 589–597. [CrossRef] [PubMed]
79. Denaxa, N.K.; Roussos, P.A.; Vemmos, S.N. Assigning a role to the endogenous phenolic compounds on adventitious root formation of olive stem cuttings. *J. Plant Growth Regul.* **2020**, *39*, 411–421. [CrossRef]
80. Curir, P.; Sulis, S.; Mariani, F.; Vansumere, C.F.; Marchesini, A.; Dolci, M. Influence of endogenous phenols on rootability of *chamaelaucium-uncinatum* schauer stem cuttings. *Sci. Hortic.* **1993**, *55*, 303–314. [CrossRef]
81. Pfeiffer, P.; Hegedus, A. Review of the molecular genetics of flavonoid biosynthesis in fruits. *Acta Aliment.* **2011**, *40*, 150–163. [CrossRef]
82. Bradai, F.; Pliego-Alfaro, F.; Sanchez-Romero, C. Long-term somatic embryogenesis in olive (*Olea europaea* L.): Influence on regeneration capability and quality of regenerated plants. *Sci. Hortic.* **2016**, *199*, 23–31. [CrossRef]
83. Pila Quinga, L.A.; Pacheco de Freitas Fraga, H.; do Nascimento Vieira, L.; Guerra, M.P. Epigenetics of long-term somatic embryogenesis in *Theobroma cacao* L.: DNA methylation and recovery of embryogenic potential. *Plant Cell Tissue Organ Cult.* **2017**, *131*, 295–305. [CrossRef]
84. Peng, X.; Zhang, T.T.; Zhang, J. Effect of subculture times on genetic fidelity, endogenous hormone level and pharmaceutical potential of *Tetrastigma hemsleyanum* callus. *Plant Cell Tissue Organ Cult.* **2015**, *122*, 67–77. [CrossRef]
85. Konar, S.; Adhikari, S.; Karmakar, J.; Ray, A.; Bandyopadhyay, T.K. Evaluation of subculture ages on organogenic response from root callus and SPAR based genetic fidelit-y assessment in the regenerants of *Hibiscus sabdariffa* L. *Ind. Crops Prod.* **2019**, *135*, 321–329. [CrossRef]
86. Pi, L.; Aichinger, E.; Graaff, E.; Llavata-Peris, C.I.; Weijers, D.; Hennig, L.; Groot, E.; Laux, T. Organizer-derived WOX5 signal maintains root columella stem cells through chromatin-mediated repression of *CDF4* expression. *Dev. Cell* **2015**, *33*, 576–588. [CrossRef]

Disclaimer/Publisher's Note: The statements, opinions and data contained in all publications are solely those of the individual author(s) and contributor(s) and not of MDPI and/or the editor(s). MDPI and/or the editor(s) disclaim responsibility for any injury to people or property resulting from any ideas, methods, instructions or products referred to in the content.

horticulturae

Article

Somatic Embryogenesis and Plant Regeneration from Stem Explants of Pomegranate

Jingting Wang [1,†], Xinhui Xia [2,†], Gaihua Qin [3,4], Jingwen Tang [1], Jun Wang [1], Wenhao Zhu [1], Ming Qian [1], Jiyu Li [3], Guangrong Cui [1], Yuchen Yang [2,*] and Jingjing Qian [1,*]

1 College of Agriculture, Anhui Science and Technology University, Fengyang 233100, China; wangjingting0815@163.com (J.W.); tangjw0809@163.com (J.T.); wangjun2021106@163.com (J.W.); zhuwenhao2010@163.com (W.Z.); qianm@ahstu.edu.cn (M.Q.); cuigr64@sina.com (G.C.)
2 State Key Laboratory of Biocontrol, School of Ecology, Sun Yat-sen University, Shenzhen 518107, China; xiaxh23@mail2.sysu.edu.cn
3 Key Laboratory of Horticultural Crop Germplasm Innovation and Utilization (Co-Construction by Ministry and Province), Institute of Horticulture, Anhui Academy of Agricultural Sciences, Hefei 230031, China; qghahstu@163.com (G.Q.); lijiyugx@163.com (J.L.)
4 Key Laboratory of Horticultural Crop Genetic Improvement and Eco-Physiology of Anhui Province, Institute of Horticultural Research, Anhui Academy of Agricultural Sciences, Hefei 230001, China
* Correspondence: yangych68@mail.sysu.edu.cn (Y.Y.); qianjingjing19@126.com (J.Q.)
† These authors contributed equally to this work.

Abstract: Plant regeneration through somatic embryogenesis provides a solution for maintaining and genetically improving crop or fruit varieties with desirable agronomic traits. For the fruit tree pomegranate (*Punica granatum* L.), despite some successful applications, the existing somatic embryogenesis protocols are limited by low availability of explants and susceptibility to browning. To address these problems, in this study, we developed an effective system for induction of high-vigor pomegranate somatic embryos derived from stem explants. The usage of stem explants breaks through the difficulty in obtaining material, thus making our system suitable for widespread commercial production. To enhance the performance of our system, we identified the optimal explants, subculture cycles and combination of basal media and plant growth regulators for each step. The results showed that inoculating stem explants onto a Murashige and Skoog (MS) medium supplemented with 1.0 mg/L 6-benzylaminopurine (6-BA) and 1.0 mg/L 1-naphthaleneacetic acid (NAA) achieved the best induction rate and growth status of pomegranate calli (induction rate = ~72%), and MS medium containing 0.5 mg/L 6-BA and 1.0 mg/L NAA was the optimal condition for the induction of embryogenic calli and somatic embryos (induction rate = ~74% and 79%, respectively). The optimal subculture period for embryogenic calli was found to be 30–35 days. Strong roots were then induced in the developed somatic embryo seedlings, which survived and grew well after transplantation to the natural environment, indicating the good vitality of the induced pomegranate somatic embryos. Together, our system provides a solution to mass somatic embryo induction and plant regeneration of pomegranate and lays a foundation for future genetic transformation and bioengineering improvement of pomegranate with favorable agronomic traits.

Keywords: in vitro culture; fruit tree; callus induction; plant growth regulators; optimal culture condition

1. Introduction

Pomegranate (*Punica granatum* L.) is one of the most popular fruits worldwide; it has integrated economic, ecological, and social benefits, as well as ornamental and health applications [1,2]. In China, the pomegranate is widely cultivated throughout Sichuan and Yunnan in the southwest, Shanxi in the north, and Shandong and Anhui in the east [3]. However, due to limited technologies, there are many challenges in the cultivation and management of pomegranate trees, such as low vigor and vitality, that greatly restrict the production of pomegranate fruits, especially for rare varieties. Importantly, some pomegranate varieties have been found to present special characteristics, such as dwarfed stature, white berries, or high vitamin content, which are beneficial to their yield and quality [4]. Thus, screening and breeding varieties with favorable agronomic or economic traits have attracted great interests from the pomegranate industry.

Plant regeneration through somatic embryogenesis provides a solution to improve and breed desirable varieties [5]. Without fertilization, somatic embryos are generated by dedifferentiated plant somatic cells and present a capacity to eventually develop into complete plants [6]. Compared to zygotic embryos, somatic embryos have an edge in high heritability from the mother plant that allows it to avoid genetic segregation during sexual reproduction, and maintain desirable agronomic traits [7–9]. Furthermore, somatic embryogenesis can be used for genetic transformation, for example, via Clustered Regularly Interspaced Short Palindromic Repeats (CRISPR)-mediated genome editing, which facilitates the genetic improvement of favorable phenotypes [10,11]. Hence, somatic embryogenesis provides a valuable model for embryo cell biology and molecular biology research [12]. Somatic embryogenesis can be carried out via direct or indirect strategies. Under certain conditions, somatic embryos can be directly induced from explant embryogenesis. For the indirect strategy, a new somatic embryo can be induced from a primary somatic embryo, either with or without a callus stage, or from an embryogenic callus [13,14]. Compared to primary somatic embryos or embryogenic calli, secondary somatic embryoids exhibit greater vitality and longer-lasting regeneration ability, which are better for transformative genetic research and germplasm conservation [15]. In addition, indirect somatic embryogenesis induction is an effective way to obtain tissue-cultured plants for the species that are recalcitrant to somatic embryo formation. However, the induction of somatic embryogenesis is quite complex; thus, there are many factors, such as the genetic background and DNA methylation level of the mother plant, the type of explant, and the level of plant growth regulators during induction and culture, that affect the success rate of somatic embryogenesis [16,17]. Plant growth regulators play important roles in regulating cell growth, division, and differentiation during in vitro plant tissue culture. For example, auxin is usually used for eliciting plant rooting and vegetative propagation from stem and leaf cuttings, and cytokinin is necessary for promoting somatic embryogenesis [18–21]. However, since plant tissues cannot produce auxin and cytokinin under in vitro culture conditions, exogenous supplementations of auxin and cytokinin are critically required for ensuring plant embryogenic callus induction and somatic embryogenesis [22–24]. In pomegranate, despite some successful applications of somatic embryogenesis, its recalcitrant nature poses challenges to the somatic embryo induction for the major reason that the existing protocols mainly employ cotyledonary or leaf explants; however, the low availability of cotyledonary tissues limits the widespread commercial application, while the leaf-derived materials are prone to browning, leading to tissue culture failure [25]. Thus, a stable and high-yield somatic embryogenesis system is urgently required for pomegranate.

To address the abovementioned problem, in this study, we aim to establish an effective somatic embryogenesis system for pomegranate using easily acquired explants. Additionally, we further explored the optimal culture conditions for the induction and development of somatic embryos to improve the efficiency of somatic embryogenesis in pomegranate. Our system offers a solution to bridge the difficulties of mass somatic embryo induction and plant regeneration in pomegranate and also provides a foundation for the future genetic transformation and bioengineering improvement of pomegranate with desirable traits.

2. Materials and Methods

2.1. Cultivation of Tissue Culture Plantlets

The pomegranate plantlets for tissue culture were produced following the methods described in Qian et al. (2022) [26]. Briefly, in the spring of 2022 in Huaiyuan, Anhui Province, 2–3 cm shoot tips were cut off from the tender stems of thriving and disease-free plants of the pomegranate variety 'White Flower Jade Seeds'. After cultivation in water for 2–3 days, the growth points of the excised shoot tips were collected and rinsed in a beaker with clean water for 15 min. Then, they were disinfected with 75% alcohol for 30 s, washed with sterile water 3 times, disinfected with 0.1% $HgCl_2$ for 8 min, and washed with sterile water 4 times. Tissue sections of ~0.3–0.5 mm were collected from the sterilized materials and inoculated onto a woody plant medium (WPM) supplemented with 0.8 mg/L indole-3-butyric acid (IBA, chemical formula: $C_{12}H_{13}NO_2$), 30 g/L sucrose, and 6 g/L agar. The pH of the medium was adjusted to 6.5 with 1 M NaOH and sterilized at 120 °C and 105 kPa for 20 min.

2.2. Screening Optimum Combination of Media and Plant Growth Regulators for Callus Induction

The pomegranate tissue culture plantlets were then subcultured 3 times, and plantlets with similar growth conditions were selected and cut into ~0.5–0.8 cm length stem segments without bud points or leaves on the ultraclean workbench. The explants were inoculated horizontally on the induction medium. Under in vitro conditions, the main factors affecting callus formation include the types of explants, medium, and concentrations of exogenous plant growth regulators [27]. In the current study, we tested the effects of different combinations of media (WPM or MS medium) and plant growth regulators (1.0 mg/L 6-BA ($C_{12}H_{11}N_5$), 1.0 mg/L NAA ($C_{12}H_{10}O_2$), and 0.5 mg/L IBA) on callus induction to determine the appropriate culture conditions for inducing pomegranate callus [25]. For each combination, 30 biological replicates were performed. The culture was maintained at a constant temperature of 22 ± 2 °C and illumination intensity of 1600 xl by LED lights at a 16/8 h light/dark cycle. After 30 days, the callus induction status was recorded, and the callus induction rate was calculated as follows: callus induction rate = number of calli induced from explants/number of explants inoculated × 100%. Factorial Analysis of Variance (ANOVA) was performed to assess the statistical significance level among 6 different combinations of basal medium and plant growth regulators (see Table 1 for details) in SPSS Statistics v. 17.0 (SPSS Inc., Chicago, IL, USA). Duncan's Multiple Range test (DMRT) was then conducted to measure the specific difference between each pair of culture conditions.

Table 1. Induction rate and growth status of calli induced by different combinations of media and plant growth regulators.

Medium	Basal Medium	Plant Growth Regulators (mg/L)			Callus Induction Percentage (%)	Callus Growth Status
		6-BA	NAA	IBA		
A1	WPM	1.0	1.0	0.0	67.78 ± 6.94 [b]	Milky white, not dense
A2	WPM	1.0	0.0	0.5	81.11 ± 5.09 [a]	Light brown, dense
A3	WPM	0.0	1.0	0.5	65.56 ± 3.85 [b]	Grayish white, not dense
A4	MS	1.0	1.0	0.0	72.22 ± 5.09 [ab]	Light yellow, slightly dense
A5	MS	1.0	0.0	0.5	66.67 ± 3.33 [b]	White, dense
A6	MS	0.0	1.0	0.5	68.89 ± 6.94 [b]	Milky white, not dense

Notes: Different lowercase letters upward indicate statistically significant differences at $p < 0.05$ level of Duncan's Multiple Range test.

2.3. Screening Optimum Concentrations of Plant Growth Regulators for Embryogenic Callus Induction

Pomegranate embryogenic calli were then induced using the calli we obtained from the last step. In general, one key step for embryogenic callus induction is to screen suitable culture conditions, particularly the concentration of exogenous plant growth regulators. Here, we used MS medium as the basal medium and supplemented it with 6-BA and NAA at different concentrations (0.5, 1.0, and 1.5 mg/L for both types of plant growth regulators) to obtain the optimum combination of plant growth regulators for pomegranate embryogenic callus induction. Each combination was examined with 30 biological replicates. After culturing for 30 days, the growth state of the embryogenic calli was recorded and photographed using a Leica M205 (Leica Microsystems Inc., Buffalo Grove, IL, USA), and the induction rate of embryogenic calli was measured as follows: embryogenic callus induction rate = number of embryogenic calli induced from explants/number of inoculated explants × 100%. Statistical differences among 9 combinations of different 6-BA and NAA concentrations (detailed in Table 2) were measured by two-way ANOVA using SPSS software. Multiple comparison procedure was performed by the post hoc DMRT.

Table 2. Induction rate and growth status of embryogenic calli induced by the media with different types and concentrations of plant growth regulators.

Medium	Plant Growth Regulator (mg/L)		Embryogenic Callus Induction Rate (%)	Callus Growth State
	6-BA	NAA		
B1	0.5	0.5	14.44 ± 1.92^{e}	First yellow, slowly brown, grew slowly, not dense
B2	0.5	1.0	74.44 ± 5.09^{a}	Light yellow to light green, grew fast, not dense
B3	0.5	1.5	0 ± 0^{f}	Brown, did not grow well
B4	1.0	0.5	35.56 ± 5.09^{c}	Light yellow, partially brown, modest growth, modestly dense
B5	1.0	1.0	61.11 ± 8.39^{b}	White, first grew fast and then grew slowly, dense
B6	1.0	1.5	25.56 ± 5.09^{d}	Yellow, slow grew, not dense
B7	1.5	0.5	7.78 ± 1.92^{ef}	White, grew slowly, dense
B8	1.5	1.0	43.33 ± 6.67^{c}	Milky white, grew modestly, not dense
B9	1.5	1.5	2.22 ± 3.85^{f}	First white, slowly brown, almost did not grow

Notes: Different lowercase letters upward indicate statistically significant differences at $p < 0.05$ level of Duncan's Multiple Range test.

2.4. Evaluation of Different Explants on Embryogenic Callus Induction

We further compared the performance of using stem segments and hypocotyls as explants on embryogenic callus induction. The hypocotyls were from pomegranate seedlings that germinated and grew for 7 days. All the explants were inoculated onto MS medium supplemented with 1.0 mg/L NAA and 0.5 mg/L 6-BA, according to the screening results (detailed in Section 3.2). For each type of explant, the experiments were repeated 105 times, with explants inoculated on 15 Petri dishes and each dish containing 7 replicates. The embryogenic callus induction rate was then calculated by the equation described above. The difference in induction of embryogenic calli between the two types of explants was evaluated by one-way ANOVA in SPSS software.

2.5. Evaluation of Different Subculture Cycles on Proliferation of Embryogenic Calli

We also examined the impacts of the different subculture cycles on the proliferation of embryogenic calli. Well-grown embryogenic calli were selected and inoculated on MS medium with 0.5 mg/L 6-BA and 1.0 mg/L NAA to produce new embryogenic calli. The embryogenic calli were weighed immediately after inoculation to serve as the fresh weight of the initial calli and were then weighed and recorded every 2 days until the fresh weight no longer increased. Origin 2021 (OriginLab Corp., Northampton, MA, USA) was used to fit the growth curve.

2.6. Screening Optimum Concentrations of Plant Growth Regulators for Somatic Embryogenesis

The embryogenic calli of good growth status were transferred to the culture medium for somatic embryo induction. Similar to embryogenic calli, the induction of somatic embryos is also sensitive to the concentrations of plant growth regulators in the medium. Thus, in this study, we tested the impacts of different 6-BA and NAA concentrations on somatic embryogenesis. Specifically, the MS medium was supplemented with 0.1, 0.5, and 1.0 mg/L 6-BA and NAA, and the embryogenic calli were then inoculated. Each culture condition was examined with 75 biological replicates. The growth status of the somatic embryos was recorded and photographed using a Leica M205. The somatic embryo induction rate was computed after 35 days of growth with the following formula: somatic embryo induction rate = number of somatic embryos induced from calli/number of calli inoculated × 100%. Two-way ANOVA was implemented to quantify the statistically significant level among the nine combinations of different 6-BA and NAA concentrations (detailed in Table 3), and DMRT was employed to measure the pairwise differences between culture conditions.

Table 3. Induction rate and growth status of somatic embryos induced by the media with different concentrations of plant growth regulators.

Medium	Plant Growth Regulators (mg/L)		Somatic Embryo Induction Percentage (%)	Number of Somatic Embryoids Formed from One Callus Explant
	6-BA	NAA		
C1	0.1	0.1	0 ± 0^{e}	0 ± 0^{e}
C2	0.1	0.5	9.33 ± 2.31^{e}	1.17 ± 0.29^{d}
C3	0.1	1.0	9.33 ± 4.62^{e}	1.67 ± 0.33^{c}
C4	0.5	0.1	20.00 ± 4.00^{d}	1.85 ± 0.13^{c}
C5	0.5	0.5	78.67 ± 6.11^{a}	2.83 ± 0.15^{a}
C6	0.5	1.0	57.33 ± 6.11^{b}	2.26 ± 0.41^{b}
C7	1.0	0.1	22.67 ± 4.62^{d}	1.74 ± 0.05^{c}
C8	1.0	0.5	50.67 ± 6.11^{b}	1.80 ± 0.17^{c}
C9	1.0	1.0	40.00 ± 10.58^{c}	1.45 ± 0.17^{cd}

Notes: Different lowercase letters upward indicate statistically significant differences at $p < 0.05$ level of Duncan's Multiple Range test.

2.7. Rooting and Regeneration of Somatic Embryo Seedlings

The somatic embryo plantlets were inoculated on WPM supplemented with 0.6 mg/L IBA, 5.5 g/L agar, and 25 g/L sucrose (rooting medium) for rooting and growth. The somatic embryo seedlings were washed and transplanted onto the seedling medium in a 6.5 × 6.5 cm nutrient bowl for cultivation. The seedling medium was a mixture of nutritive soil and vermiculite in a ratio of 1:1. Before use, the seedling medium was disinfected at a high temperature and was sprayed with 50 mL 0.8 mg/L IBA. After the transplantation, the nutrient bowl was covered by plastic film, and the medium was sprayed with water every morning and evening to keep the substrate moist. Air humidity was kept above 50%. Every 5 days, the seedings were sprayed with Hoagland's nutrient solution until they were viable (~25 days). After taking roots in the medium, the somatic embryo seedlings were washed and transplanted onto the seedling medium in a 6.5 × 6.5 cm nutrient bowl for cultivation. The seedling medium was a mixture of nutritive soil and vermiculite in a ratio of 1:1. Before use, the seedling medium was disinfected at a high temperature, and was sprayed with 50 mL 0.8 mg/L IBA. After the transplantation, the nutrient bowl was covered by plastic films, and the medium was sprayed with water every morning and evening to keep the substrate moist. Air humidity was kept above 50%. EFor every 5 days, the seedings were was sprayed with Hoagland's nutrient solution until they were viable (~25 days). After that, the root-bearing seedlings were transplanted to the natural environment, and their growth status was recorded.

3. Results

3.1. Effects of Different Combinations of Media and Plant Growth Regulators on Callus Induction

Table 1 shows the induction percentage and growth status of calli formed from pomegranate stem segments growing under different combinations of media and plant growth regulators. For all the growth scenarios, more than 60% of the explants were successfully induced into calli. The explants grown on WPM supplemented with 6-BA and IBA (medium A2) exhibited the highest induction percentage of calli (81.11 ± 5.09%), which was significantly higher than that in other scenarios ($p < 0.05$). However, on this medium, the calli were light brown and dense and were prone to browning and necrosis over time. In contrast, the calli on the WPM with IBA and NAA (medium A3) and on the MS medium with IBA and NAA (medium A6) exhibited the best growth status in that they were mostly gray-white and not dense but had a relatively low induction rate. Taken together, these results suggested that MS medium supplemented with 1.0 mg/L 6-BA and 1.0 mg/L NAA (medium A4) was the optimal culture condition for callus induction from stem explants and was, therefore, used for subsequent experiments.

3.2. Effects of Different Concentrations of Plant Growth Regulators on Embryogenic Callus Induction

We then tested the impacts of different 6-BA and NAA concentrations on embryogenic callus induction and found that both the induction rate and growth status of embryogenic calli were sensitive to changes in the concentrations of both plant growth regulators (Figure 1; Table 2). Either too high or too low a concentration of NAA in the medium led to a dramatic decrease in the induction rate. On the medium with 1.5 mg/L NAA (media B3, B6 and B9), less than 30% of the explants were induced into embryogenic calli, and they were yellow to brown and did not grow well. In contrast, the medium with 0.5 or 1 mg/L 6-BA was more conducive to the induction of pomegranate embryogenic calli than that with 1.5 mg/L 6-BA. Among the nine combinations, the medium supplemented with 0.5 mg/L 6-BA and 1 mg/L NAA (B2) was most suitable for the induction of pomegranate embryogenic calli, where the induction rate (74.44 ± 5.09%) was significantly higher than all the other culture conditions ($p < 0.05$), and the produced embryogenic calli were light yellow to light green, fast-growing and not dense (Figures 1 and 2; Table 2). Thus, this medium was used for inducing embryogenic calli from pomegranate stem segments.

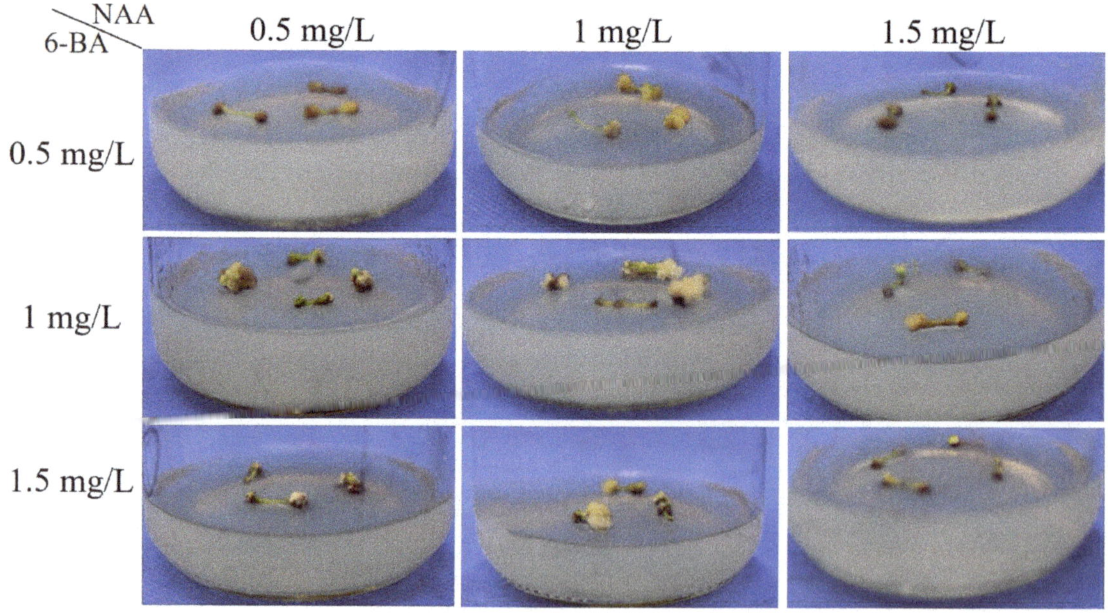

Figure 1. Pomegranate embryogenic callus induced by the media with different concentrations of plant growth regulators.

3.3. Effect of Different Explants on Embryogenic Callus Induction

We also examined the performance of different explants on embryogenic callus induction. When using hypocotyl explants, 72.38 ± 4.36% of the inoculated explants were successfully induced into embryogenic calli. However, after 37 days of culture, the growth of induced calli slowed and some calli began to turn brown and necrose (Figure 3A). In contrast, the induction using stem segments performed better than that using hypocotyls, with a significantly higher induction rate of embryogenic calli (80.00 ± 4.95%; $p < 0.05$). In addition, no obvious stagnation was observed in their growth rate on the 35th day (Figure 3B); thus, stem segments were more suitable than hypocotyl explants for pomegranate somatic embryogenesis.

Figure 2. Pomegranate embryogenic calli induced using medium B2 (see Table 2 for details). Red arrow points to a magnified shot of the induced embryogenic calli.

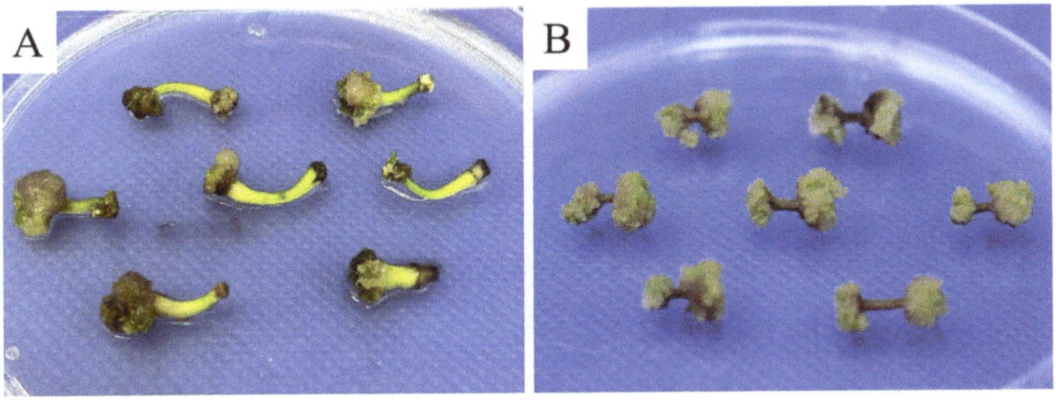

Figure 3. Pomegranate embryogenic calli induced by hypocotyl (**A**) and stem explants (**B**).

3.4. Effect of Subculture Cycle on Embryogenic Callus Proliferation

The obtained well-grown embryogenic calli were transferred to fresh B2 medium to maintain embryogenicity and proliferation. Then, we drew the growth curve according to the measurement data from the fresh weight of the embryogenic calli (Figure 4). The growth curve showed that the fresh weight of the embryogenic calli continuously increased over time. The callus proliferation rate increased after 15 days of inoculation and entered the rapid growth period at the 20th to 30th days. During this period, the color of the calli was light green. After 30 days, the growth rate began to slow down, and the tissue gradually degenerated, the color gradually turned yellow, and the granular texture softened. From the 40th to 55th days, almost all calli died of browning. Therefore, a suitable subculture cycle for embryogenic calli was determined to be 30–35 days to effectively maintain callus quality.

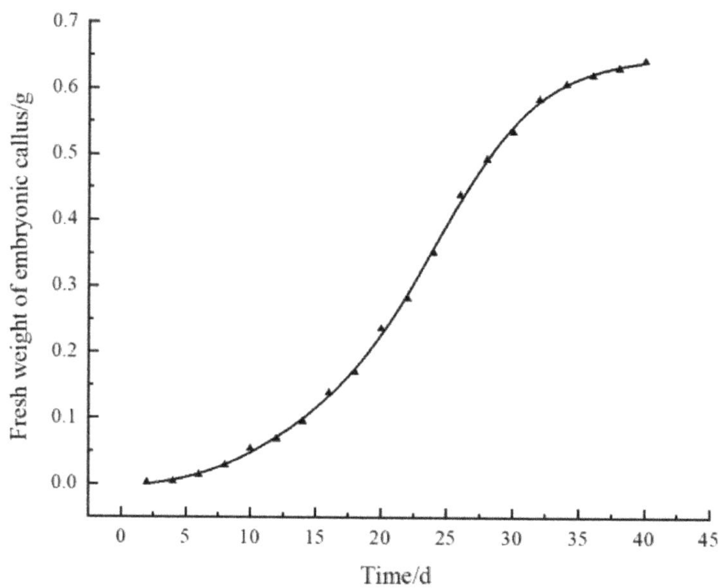

Figure 4. Growth curve of pomegranate embryonic calli.

3.5. Effects of Different Concentrations of 6-BA and NAA on Somatic Embryo Induction

In addition to embryogenic calli, 6-BA and NAA concentrations in the MS medium were also found to substantially affect the induction of somatic embryos. Among all the tested culture conditions, only C5, C6, and C8 exhibited somatic embryo induction rates higher than 50% (Table 3). Of them, the MS medium with 0.5 mg/L 6-BA and 0.5 mg/L NAA (medium C5) was the most suitable for somatic embryo induction, with 78.67 ± 6.11% of the calli induced into somatic embryos after being inoculated for 25 days, which was a significantly higher rate than in the others ($p < 0.05$). Meanwhile, granular proembryonic masses were observed on the surface of the embryogenic calli after 50 days of culture (Figure 5B). As the culture time was extended, the embryos, in turn, appeared globular and heart-shaped/torpedo-shaped in form (Figure 5C–E), and eventually developed into cotyledon embryos by approximately 50 days (Figure 5F). Under the stereomicroscope, we observed plumules at one end of the cotyledon embryos and the radicles at the other end that were connected to the callus (Figure 6). In addition, we also observed the apparent structure of the vascular bundle in the cotyledon, which was independent of the embryogenic calli.

Figure 5. Embryogenic calli and different somatic embryo stages of pomegranate. (**A**) Calli induced by stem explants. (**B**) Embryogenic calli (**C**) Globular embryo. (**D**) Extended globular embryo. (**E**) Heart-shaped/torpedo-shaped embryo. (**F**) Cotyledon embryo. In each of the panel (**B–F**), the induced somatic embryos was marked by red arrow.

3.6. Rooting and Regeneration of Somatic Embryo Seedlings

The embryogenic calli were inoculated on the medium C5 for 45 days, and the mature somatic embryos developed complete seedlings (Figure 7A,B). The somatic embryo seedlings were transferred to the natural environment to form regenerated plants (Figure 7C,D). During this process, radicle formation is a key step that determines whether the production of somatic embryo seedlings can be industrialized and commercialized. On the rooting medium, all the somatic embryo seedlings grew normally and generated healthy roots (Figure 7C,D). Then, these root-bearing seedlings were developed into healthy and vigorous plants in the natural environment (Figure 7E,F).

Figure 6. Schematic (**A**) and microscopic views (**B**) of pomegranate somatic embryogenesis. SE, somatic embryo; EC, embryonic callus.

Figure 7. Somatic embryo seedlings of pomegranate. (**A,B**) Mature somatic embryos with buds. (**C**) Somatic embryo seedlings inoculated onto rooting medium. (**D**) Roots of somatic embryo seedlings in culture medium. (**E,F**) Somatic embryo seedlings after seedling refinement and transplantation.

4. Discussion

Somatic embryogenesis faces many difficulties and problems in woody plants that are not encountered with herbaceous plants. In this study, we first established a protocol for the effective induction of somatic embryogenesis in pomegranate using stem explants. In this system, we employed an indirect organogenesis strategy for somatic embryogenesis induction, that involved first generating embryogenic calli from dedifferentiated pomegranate explant cells as an intermediate stage and then inducing somatic embryos from the produced calli. The results showed that with our system, we successfully induced pomegranate embryogenic calli from stem segments (Figures 2 and 3). Under the optimal combination of plant growth regulators, the granular proembryo mass, as well as somatic embryos at various stages, was then elicited at the surface of the calli, and the mature cotyledonary somatic embryos could develop into healthy somatic seedlings (Figures 5 and 7). After the somatic embryo seedlings were transferred to the rooting medium, all the pomegranate seedlings developed healthy roots within 20 days, indicating their good rooting ability both in vivo and in vitro (Figure 7).

During this process, obtaining high-quality embryogenic calli is a vital step [12]. Here, we first explored the effects of different combinations of two types of media, WPM and MS, and three types of plant growth regulators on pomegranate callus induction. The results showed that the induction rate of calli when using MS medium was higher than that when using WPM. This may be because of their differences in the composition of nutrients and elements [28]. In addition to the basal medium, plant growth regulators also play a critical role in regulating plant somatic embryogenesis. For pomegranate, exogenous auxin and cytokinin are generally supplemented to the culture medium to improve embryogenesis induction and shoot proliferation [25,29,30]. In this study, we assessed the performance of the combination of different types of plant growth regulators 6-BA, NAA, and IBA in pomegranate somatic embryogenesis and showed that supplementing the MS medium with 6-BA and NAA achieved the best balance between callus induction rate and growth status (Table 1). NAA is a synthetic auxin that is usually used for eliciting plant rooting and vegetative propagation from stem and leaf cuttings [18,19], and 6-BA is a plant growth regulator of the cytokinin family that can stimulate plant growth and development. However, for both types of plant growth regulators, high concentrations in the medium were found to have adverse impacts on the callus induction and growth. In our system, the appropriate 6-BA concentration for induction of pomegranate embryogenic calli and somatic embryos is 0.5 mg/L, and the optimum NAA concentration is 1.0 and 0.5 mg/L, respectively, where the induction rates of embryogenic calli and somatic embryos were ~74% and 79%, respectively (Tables 2 and 3). This is because auxins at high concentrations are toxic to plants, while a high cytokinin concentration inhibits auxin polar transport and, thus, inhibits somatic embryogenesis [21,30,31]. However, different combinations of plant growth regulators are required for in vitro propagation of pomegranate when using different explants [25]. For instance, Kantharajah et al. [32] showed that MS medium supplemented with 1.0 mg/L 6-BA and 0.4 mg/L NAA gave rise to the best callus initiation and growth from nodal explants, while the medium containing 1.0 mg/L 6-BA achieved the best performance for leaf explant-derived callus culture. These results indicate that the choice of plant growth regulators used in a nutrient medium and their ratio (auxin/cytokinin) are largely dependent on the morphogenic potentials of the pomegranate explants employed for in vitro culture.

In general, all types of plant tissues with strong meristematic ability can be used as explants for tissue culture, but different explants have different abilities in forming calli [33]. Jaidka and Mehra [34] demonstrated the success of using pomegranate stem explants in inducing calli when cultured in MS medium containing 4.0 mg/L NAA, 2.0 mg/L kinetin, and 15% coconut water, indicating the feasibility of using stem explants for pomegranate somatic embryogenesis. In addition, Yang and Ludders [35] reported organogenesis in pomegranate initiated from stem explants. In the current study, we first induced pomegranate somatic embryogenesis using stem segments as explants and showed better performance using stem explants than using hypocotyls in producing pomegranate calli in terms of both induction rate and the growth status of embryogenic calli. Although pomegranate calli and somatic embryos can also be generated using other types of explants, such as leaf, shoot tip, nodal segment, and cotyledonary tissue, these explants exhibit their own shortcomings: shoot tips are not able to provide enough materials, and the materials from leaves and leafstalk are prone to browning, which leads to tissue culture failure [25]. Comparatively, stem segments are more easily assessable, and the embryogenic calli formed by stem segments did not show obvious browning and grew well. These advantages make stem explants more suitable for initial culture, especially for industrial and commercial production.

The callus subculture cycle also plays a key role in eliciting plant somatic embryos, and an appropriate number of subculture cycles can ensure the good growth status of embryogenic calli [36]. The pomegranate embryogenic calli inoculated for 20–30 days entered the rapid growth period, during which the callus proliferation rate was the fastest, and its color was brightly light green. The calli turned brown after 40 days of culture, and almost all died after 55 days, indicating that they had lost their capacity for somatic embryogenesis. Therefore, taking 30–35 days as the subculture cycle is more suitable for long-term subculture of pomegranate calli and maintenance of somatic embryogenesis.

5. Conclusions

In this study, we established an effective system for somatic embryogenesis and plant regeneration in pomegranate using stem segments as explants. In our system, the MS medium containing 1.0 mg/L 6-BA and 1.0 mg/L NAA achieved the best performance in pomegranate callus induction, and the MS medium with 0.5 mg/L 6-BA and 1.0 and 0.5 mg/L NAA was the optimal condition for the induction of embryogenic calli and somatic embryos, where the induction rates of calli, embryogenic calli, and somatic embryos were ~72%, 74%, and 79%, respectively. Good somatic embryo induction indicates the effectiveness and stability of our system. In addition, compared to the existing systems, the stem explant employed in our protocol is more easily acquired and suitable for widespread commercial production. Taken together, our system overcomes the recalcitrant nature of pomegranate and provides a solution to mass somatic embryo induction and plant regeneration of pomegranate. With the help of this somatic embryogenesis protocol, we aim to build a genetic transformation system for future bioengineering improvement of pomegranate with favorable agronomic traits.

Author Contributions: Conceptualization, G.C. and J.Q.; validation, J.W. (Jingting Wang) and J.Q.; formal analysis, J.W. (Jingting Wang); investigation, J.W. (Jingting Wang), X.X., J.T., J.W. (Jun Wang), W.Z., M.Q. and J.L.; writing—original draft preparation, J.W. (Jingting Wang), X.X., Y.Y. and J.Q.; writing—review and editing, X.X., Y.Y., G.Q. and J.Q.; visualization, J.W.(Jingting Wang) and J.Q; supervision, G.C., Y.Y. and J.Q.; project administration, G.C., Y.Y. and J.Q.; funding acquisition, J.Q., Y.Y. and G.Q. All authors have read and agreed to the published version of the manuscript.

Funding: This research was funded by Anhui Province Key Research and Development Plan Project, grant number 202204c06020062; Project of Industry and School and Research Institution, grant number AKZY2022110; Anhui Province Natural Sciences Fund, grant number 2023AH040281; National Natural Science Foundation of China, grant number 32201420; Natural Science Foundation of Anhui Province, grant number 2308085MC95.

Data Availability Statement: No new data were created or analyzed in this study.

Conflicts of Interest: The authors declare no conflict of interest.

References

1. Caruso, A.; Barbarossa, A.; Tassone, A.; Ceramella, J.; Carocci, A.; Catalano, A.; Basile, G.; Fazio, A.; Iacopetta, D.; Franchini, C.; et al. Pomegranate: Nutraceutical with Promising Benefits on Human Health. *Appl. Sci.* **2020**, *10*, 6915. [CrossRef]
2. Hu, L.; Zhang, X.; Ni, H.; Yuan, F.; Zhang, S. Identification and Functional Analysis of CAD Gene Family in Pomegranate (*Punica granatum*). *Genes* **2023**, *14*, 26. [CrossRef] [PubMed]

3. Liu, C.; Li, J.; Qin, G. Genome-wide distribution of simple sequence repeats in pomegranate and their application to the analysis of genetic diversity. *Tree Genet. Genom.* **2020**, *16*, 36. [CrossRef]
4. Xia, X.; Fan, M.; Liu, Y.; Chang, X.; Wang, J.; Qian, J.; Yang, Y. Genome-wide alternative polyadenylation dynamics underlying plant growth retardant-induced dwarfing of pomegranate. *Front. Plant Sci.* **2023**, *14*, 1189456. [CrossRef] [PubMed]
5. Tao, J.; Chen, S.; Qin, C.; Li, Q.; Cai, J.; Sun, C.; Wang, W.; Weng, Y. Somatic embryogenesis in mature zygotic embryos of Picea pungens. *Sci. Rep.* **2021**, *11*, 19072. [CrossRef] [PubMed]
6. Sánchez-Romero, C. Somatic embryogenesis in *Olea* spp. *Plant Cell Tissue Organ Cult. (PCTOC)* **2019**, *138*, 403–426. [CrossRef]
7. Park, Y.-S. Conifer Somatic Embryogenesis and Multi-Varietal Forestry. In *Challenges and Opportunities for the World's Forests in the 21st Century*; Fenning, T., Ed.; Springer: Dordrecht, The Netherlands, 2014; pp. 425–439. [CrossRef]
8. Corredoira, E.; Ballester, A.; Ibarra, M.; Vieitez, A.M. Induction of somatic embryogenesis in explants of shoot cultures established from adult *Eucalyptus globulus* and *E. saligna* × E. maidenii trees. *Tree Physiol.* **2015**, *35*, 678–690. [CrossRef]
9. Pais, M.S. Somatic Embryogenesis Induction in Woody Species: The Future After OMICs Data Assessment. *Front. Plant Sci.* **2019**, *10*, 240. [CrossRef]
10. Lowe, K.; Wu, E.; Wang, N.; Hoerster, G.; Hastings, C.; Cho, M.-J.; Scelonge, C.; Lenderts, B.; Chamberlin, M.; Cushatt, J.; et al. Morphogenic Regulators Baby boom and Wuschel Improve Monocot Transformation. *Plant Cell* **2016**, *28*, 1998–2015. [CrossRef]
11. Altpeter, F.; Springer, N.M.; Bartley, L.E.; Blechl, A.E.; Brutnell, T.P.; Citovsky, V.; Conrad, L.J.; Gelvin, S.B.; Jackson, D.P.; Kausch, A.P.; et al. Advancing Crop Transformation in the Era of Genome Editing. *Plant Cell* **2016**, *28*, 1510–1520. [CrossRef]
12. Liu, Y.; Wei, C.; Wang, H.; Ma, X.; Shen, H.; Yang, L. Indirect somatic embryogenesis and regeneration of Fraxinus mandshurica plants via callus tissue. *J. For. Res.* **2021**, *32*, 1613–1625. [CrossRef]
13. Yan, R.; Sun, Y.; Sun, H. Current status and future perspectives of somatic embryogenesis in Lilium. *Plant Cell Tissue Organ Cult. (PCTOC)* **2020**, *143*, 229–240. [CrossRef]
14. Guan, Y.; Li, S.-G.; Fan, X.-F.; Su, Z.-H. Application of Somatic Embryogenesis in Woody Plants. *Front. Plant Sci.* **2016**, *7*, 938. [CrossRef]
15. Baskaran, P.; Van Staden, J. Plant regeneration via somatic embryogenesis in Drimia robusta. *Plant Cell Tissue Organ Cult. (PCTOC)* **2014**, *119*, 281–288. [CrossRef]
16. Khan, T.; Reddy, V.S.; Leelavathi, S. High-frequency regeneration via somatic embryogenesis of an elite recalcitrant cotton genotype (*Gossypium hirsutum* L.) and efficient Agrobacterium-mediated transformation. *Plant Cell Tissue Organ Cult. (PCTOC)* **2010**, *101*, 323–330. [CrossRef]
17. Lu, L.; Holt, A.; Chen, X.; Liu, Y.; Knauer, S.; Tucker, E.J.; Sarkar, A.K.; Hao, Z.; Roodbarkelari, F.; Shi, J.; et al. miR394 enhances WUSCHEL-induced somatic embryogenesis in *Arabidopsis thaliana*. *New Phytol.* **2023**, *238*, 1059–1072. [CrossRef] [PubMed]
18. Syeed, R.; Mujib, A.; Malik, M.Q.; Gulzar, B.; Zafar, N.; Mamgain, J.; Ejaz, B. Direct somatic embryogenesis and flow cytometric assessment of ploidy stability in regenerants of *Caladium* × *hortulanum* 'Fancy'. *J. Appl. Genet.* **2022**, *63*, 199–211. [CrossRef]
19. Xiong, Y.; Wei, Z.; Yu, X.; Pang, J.; Zhang, T.; Wu, K.; Ren, H.; Jian, S.; Teixeira da Silva, J.A.; Ma, G. Shoot proliferation, embryogenic callus induction, and plant regeneration in *Lepturus repens* (G. Forst.) R. Br. *Vitr. Cell. Dev. Biol. Plant* **2021**, *57*, 1031–1039. [CrossRef]
20. Wang, Y.; Chen, F.; Wang, Y.; Li, X.; Liang, H. Efficient Somatic Embryogenesis and Plant Regeneration from Immature Embryos of Tapiscia sinensis Oliv., an Endemic and Endangered Species in China. *HortScience Horts* **2014**, *49*, 1558–1562. [CrossRef]
21. Bernula, D.; Benkő, P.; Kaszler, N.; Domonkos, I.; Valkai, I.; Szőllősi, R.; Ferenc, G.; Ayaydin, F.; Fehér, A.; Gémes, K. Timely removal of exogenous cytokinin and the prevention of auxin transport from the shoot to the root affect the regeneration potential of Arabidopsis roots. *Plant Cell Tissue Organ Cult. (PCTOC)* **2020**, *140*, 327–339. [CrossRef]
22. Binte Mostafiz, S.; Wagiran, A. Efficient Callus Induction and Regeneration in Selected Indica Rice. *Agronomy* **2018**, *8*, 77. [CrossRef]
23. Ming, N.J.; Binte Mostafiz, S.; Johon, N.S.; Abdullah Zulkifli, N.S.; Wagiran, A. Combination of Plant Growth Regulators, Maltose, and Partial Desiccation Treatment Enhance Somatic Embryogenesis in Selected Malaysian Rice Cultivar. *Plants* **2019**, *8*, 144. [CrossRef] [PubMed]
24. Dai, C.-W.; Yan, Y.-Y.; Liu, Y.-M.; Liu, Y.-M.; Deng, Y.-W.; Yao, H.-Y. The regeneration of *Acer rubrum* L. "October Glory" through embryonic callus. *BMC Plant Biol.* **2020**, *20*, 309. [CrossRef] [PubMed]
25. Teixeira da Silva, J.A.; Rana, T.S.; Narzary, D.; Verma, N.; Meshram, D.T.; Ranade, S.A. Pomegranate biology and biotechnology: A review. *Sci. Hortic.* **2013**, *160*, 85–107. [CrossRef]
26. Qian, J.; Wang, N.; Ren, W.; Zhang, R.; Hong, X.; Chen, L.; Zhang, K.; Shu, Y.; Hu, N.; Yang, Y. Molecular Dissection Unveiling Dwarfing Effects of Plant Growth Retardants on Pomegranate. *Front. Plant Sci.* **2022**, *13*, 866193. [CrossRef] [PubMed]
27. Shin, U.; Chandra, R.; Kang, H. In vitro and Ex vitro Propagations of *Astilboides tabularis* (Hemsl.) Engl. as a Rare and Endangered Species. *Hortic. J.* **2019**, *6*, 261.
28. Hazrati, R.; Zare, N.; Asghari-Zakaria, R.; Sheikhzadeh, P.; Johari-Ahar, M. Factors affecting the growth, antioxidant potential, and secondary metabolites production in hazel callus cultures. *AMB Express* **2022**, *12*, 109. [CrossRef]
29. Gaj, M.D. Factors Influencing Somatic Embryogenesis Induction and Plant Regeneration with Particular Reference to *Arabidopsis thaliana* (L.) Heynh. *Plant Growth Regul.* **2004**, *43*, 27–47. [CrossRef]
30. Fehér, A. Callus, Dedifferentiation, Totipotency, Somatic Embryogenesis: What These Terms Mean in the Era of Molecular Plant Biology? *Front. Plant Sci.* **2019**, *10*, 536. [CrossRef]

31. Liu, Y.; Dong, Q.; Kita, D.; Huang, J.-B.; Liu, G.; Wu, X.; Zhu, X.; Cheung, A.Y.; Wu, H.-M.; Tao, L.-Z. RopGEF1 Plays a Critical Role in Polar Auxin Transport in Early Development. *Plant Physiol.* **2017**, *175*, 157–171. [CrossRef]
32. Kantharajah, A.; Dewitz, I.; Jabbari, S. The effect of media, plant growth regulators and source of explants on in vitro culture of pomegranate (*Punica granatum* L.). *Erwerbsobstbau* **1998**, *40*, 54–58.
33. Vinterhalter, B.; Mitić, N.; Vinterhalter, D.; Uzelac, B.; Krstić-Milošević, D. Somatic embryogenesis and in vitro shoot propagation of Gentianautriculosa. *Biologia* **2016**, *71*, 139–148. [CrossRef]
34. Jaidka, K.; Mehra, P.N. Morphogenesis in *Punica granatum* (pomegranate). *Can. J. Bot.* **1986**, *64*, 1644–1653. [CrossRef]
35. Yang, Z.; Ludders, P. Organogenesis of *Punica granatum* L. var. nana. *Angew. Bot.* **1993**, *67*, 151–156.
36. Ren, Y.; Yu, X.; Xing, H.; Tretyakova, I.N.; Nosov, A.M.; Yang, L.; Shen, H. Interaction of Subculture Cycle, Hormone Ratio, and Carbon Source Regulates Embryonic Differentiation of Somatic Cells in *Pinus koraiensis*. *Forests* **2022**, *13*, 1557. [CrossRef]

Disclaimer/Publisher's Note: The statements, opinions and data contained in all publications are solely those of the individual author(s) and contributor(s) and not of MDPI and/or the editor(s). MDPI and/or the editor(s) disclaim responsibility for any injury to people or property resulting from any ideas, methods, instructions or products referred to in the content.

horticulturae

MDPI

Article

Decoding the Genomic Landscape of Pomegranate: A Genome-Wide Analysis of Transposable Elements and Their Structural Proximity to Functional Genes

Samuel Simoni †, Gabriele Usai †, Alberto Vangelisti, Marco Castellacci, Tommaso Giordani, Lucia Natali, Flavia Mascagni * and Andrea Cavallini

Department of Agriculture, Food and Environment (DAFE), University of Pisa, Via del Borghetto 80, 56124 Pisa, Italy; samuel.simoni@agr.unipi.it (S.S.); gabriele.usai@agr.unipi.it (G.U.); alberto.vangelisti@unipi.it (A.V.); marco.castellacci@phd.unipi.it (M.C.); tommaso.giordani@unipi.it (T.G.); lucia.natali@unipi.it (L.N.); andrea.cavallini@unipi.it (A.C.)

* Correspondence: flavia.mascagni@unipi.it

† These authors contributed equally to this work.

Abstract: Transposable elements (TEs) significantly drive dynamic changes that characterize genome evolution. However, understanding the variability associated with TE insertions among different cultivars remains challenging. The pomegranate (*Punica granatum* L.) has yet to be extensively studied regarding the roles of TEs in the diversification of cultivars. Herein, we explored the genome distribution of TEs and its potential functional implications among four pomegranate cultivars, 'Bhagwa', 'Dabenzi', 'Taishanhong' and 'Tunisia', whose genome sequences are available. A total of 8404 full-length TEs were isolated. The content of TEs varied among the cultivars, ranging from 41.67% of 'Taishanhong' to 52.45% of 'Bhagwa'. In all cultivars, the *Gypsy* superfamily of retrotransposons accounted for a larger genome proportion than the *Copia* superfamily. Seventy-three full-length TEs were found at the same genomic loci in all four cultivars. By contrast, 947, 297, 311, and 874 TEs were found exclusively in 'Bhagwa', 'Dabenzi', 'Taishanhong', and 'Tunisia' cultivars, respectively. Phylogenetic clustering based on the presence of TE insertions in specific loci reflected the geographic origins of the cultivars. The insertion time profiles of LTR-REs were studied in the four cultivars. Shared elements across the four cultivars exhibited, on average, a more ancient insertion date than those exclusive to three, two, or one cultivars. The majority of TEs were located within 1000 bp from the nearest gene. This localization was observed for 57% of DNA TEs and 55% of long-terminal repeat retrotransposons (LTR-RE). More than 10% of TEs resulted inserted within genes. Concerning DNA TEs, 3.91% of insertions occurred in introns, while 2.42% occurred in exons. As to LTR-REs, 4% of insertions occurred in exons and 1.98% in introns. Functional analysis of the genes lying close to TEs was performed to infer if differences in TE insertion can affect the fruit quality. Two TE insertions were found close to two genes encoding 4-coumarate--CoA ligase, an enzyme involved in the phenylpropanoid pathway. Moreover, a *TIR/Mariner* element was found within the exon of a gene encoding anthocyanidin reductase in the 'Tunisia' genotype, crucial in the biosynthesis of flavan-3-ols and proanthocyanidins, strictly correlated with the nutraceutical properties of pomegranate. Although functional and metabolomic studies are essential to elucidate the consequences of TE insertions, these results contribute to advancing our comprehension of the role of TEs in pomegranate genomics, providing insights for crop breeding.

Keywords: *Punica granatum* L.; transposable element insertions; DNA transposons; retrotransposons; genome evolution

1. Introduction

The pomegranate stands out as an economically important tree species due to the nutraceutical attributes of its fruits and finds widespread consumption in various forms, including as fresh fruit, juice, wine, and medicinal products [1,2]. Globally, it is cultivated across approximately 0.55 million hectares, yielding a total production of about 6.5 million tonnes, and it is considered an important crop in semi-arid tropical areas [3]. Pomegranate has a short history of extensive breeding and artificial selection, generally followed by vegetative propagation [4], with well-defined and maintained cultivars. The genome is relatively small (328 Mb), and currently, high-quality genomes of four cultivars of different origins have been sequenced at the chromosome or scaffold level.

The first pomegranate genome assembly for the Chinese cultivar 'Dabenzi', based on short-read sequencing, was released by Qin [5]. This assembly unveiled a whole-genome duplication event specific to the Myrtales lineage, which occurred in the common ancestor before the pomegranate and eucalyptus diverged. Subsequently, Yuan [6] released the genome of the Chinese cultivar 'Taishanhong', resolving the previously disputed taxonomic classification of the *Punica* genus and reclassifying it within the Lythraceae family. In 2020, Luo [4] published a high-quality draft genome sequence (based on long-read sequencing) for the soft-seeded 'Tunisia' cultivar. They also performed resequencing of 26 genetically diverse pomegranate varieties, varying in terms of seed hardness and geographical distribution, to elucidate the genetic distinctions between soft-seeded and hard-seeded cultivars. Finally, in 2022, the genome of the most diffused Indian cultivar, 'Bhagawa', was sequenced and assembled [7], which showed high syntenic relationships between 'Bhagawa' and 'Dabenzi'.

Genome sequences allowed to clarify many aspects of pomegranate metabolism and development. Putative genes involved in anthocyanin and punicalagin (ellagitannins unique to pomegranate) metabolism were identified [5]. They reported that the *INNER NO OUTER* (*INO*) gene was under positive selection and likely played a role in developing the fleshy outer layer of the seed coat, the edible part of the pomegranate fruit. Yuan [6] used the genome sequence to clarify ellagitannin-based compound biosynthesis, the evolution of the anthocyanin biosynthetic pathway, and the peculiar ovule development processes in pomegranates. Luo [4] identified loci encoding *SUC8-like* and *SUC6*, involved in sucrose allocation, transport, and seed hardness. Sowjanya [7] identified important genes for resistance/susceptibility to major diseases and pests, such as bacterial blight, *Ceratocystis* wilt, and fruit-sucking moths. Despite these advances, there is still an important lack of information concerning an essential fraction of the genome—the one constituted by repeated sequences. Pomegranate transposable elements (TEs) remain only partially characterized, presenting an intriguing area for further exploration, also concerning intraspecific variability.

TEs are DNA sequences capable of autonomously moving within the genome through specific transposition mechanisms. TEs are classified into two principal classes: retrotransposons (REs), or Class I TEs, and DNA TEs, also known as Class II TEs. Both these classes comprise autonomous and non-autonomous elements, depending on the presence or absence of specific open reading frames encoding transposon-related proteins. DNA TEs use a "cut-and-paste" mechanism for transposition, while REs use a "copy-and-paste" replication mechanism that necessitates an intermediate RNA molecule [8] and implies the proliferation of the element.

It is widely acknowledged that TEs constitute the majority of plant genomes. For instance, the sunflower genome (*Helianthus annuus*) is composed of over 81% TEs [9], bread wheat (*Triticum aestivum*) exhibits TEs for over 85% of its genome [10] and, similarly, around 85% of the maize genome (*Zea mays*) is composed of TEs [11].

In plants, the most abundant TEs are REs, especially those characterized by long-terminal repeat (LTR) sequences. The two flanking LTRs vary in size from a few hundred base pairs to over 10 kb, and complete autonomous elements include coding segments with two open reading frames (ORFs) for element replication and integration within the host genome. The two ORFs consist of "*gag*", which encodes a virus-like particle structural protein, and "*pol*", which encodes a polyprotein with protease, reverse transcriptase, RNaseH, and integrase enzyme domains [12]. LTR-REs of higher plants are separated into the *Copia* and *Gypsy* superfamilies, differing in the position of the integrase domain within the polyprotein [13]. The two major superfamilies are further categorized into distinct evolutionary lineages, primarily based on the sequence similarities within their coding regions. Notably, in most plant species among *Gypsy* REs, the *Chromovirus* lineage is prevalent, including *Galadriel, Tekay, Reina,* and *CRM* sublineages, which are characterized by a chromodomain at the 3′ end of the coding sequence. In particular, *Chromovirus/CRM* elements are mainly located in the centromeres and in the pericentromeric regions, where they probably play a structural role [14–17] participating in plant centromere evolution [18]. Conversely, the non-*Chromovirus Gypsy* lineages, including *Athila, Tat, Ogre,* and *Retand* sublineages, lack the chromodomain. As for *Copia* REs, they encompass key lineages like *Ale, Ivana, Ikeros, Tork, Alesia, Angela, Bianca, SIRE,* and *TAR* [17].

As for DNA TEs, these sequences consist of a transposase gene flanked by two terminal inverted repeats (TIRs). DNA TEs are classified into different families depending on their coding sequence, TIRs, and/or TSDs. Some of the best-known subclass I families include *Tc1/mariner, PIF/Harbinger, hAT, Mutator, Merlin, Transib, P, piggyBac,* and *CACTA* [19]. *Helitron* and *Maverick* TEs belong to subclass II, as they use a different transposition and insertion mechanism [19–22]. Among the non-autonomous DNA TEs, the miniature inverted-repeat transposable elements (MITEs) are the most common in many eukaryotes, especially in plants [23].

TEs can rapidly increase in number, driving their proliferation. Conversely, the TE component of a genome can also reduce its abundance through mechanisms like unequal homologous and illegitimate recombination that produces the so-called "solo-LTRs" [24–26].

The activity of TEs can lead to a broad spectrum of alterations of genome structure, gene expression, and functionality [27]. These alterations can have detrimental, neutral, or even advantageous effects to the host [28]. For instance, TEs can promote chromosome rearrangements by facilitating unequal homologous recombination between sites located at a distance from each other, either within the same chromosome or across different chromosomes [29] leading to gene deletion, translocation, and inversion. TEs can also participate in the formation of novel regulatory networks and the creation of new genes through processes like exon shuffling and the mechanism of exaptation [20,30].

Even more significantly, TEs can be inserted within or in proximity to a gene. The insertion within the coding region of a gene can modify gene functionality or give rise to new splicing patterns, resulting in mutations and alterations in the encoded protein [22,31,32]. It is also well-documented that TEs are often situated within less than 2 kb either upstream or downstream from the genes themselves [33]. These modifications can lead to putative phenotypic variations, as observed in sunflowers [9], orange tree [34], and four cucurbit species [35]. Particularly, TE integration into regulatory regions, influencing promoter functionality and cis-regulatory mechanisms, can result in abnormal gene expression, even in response to different stimuli [36–39]. In certain grape varieties, the loss of pigmentation can be attributed to the insertion of a RE into the promoter region of a MYB transcription factor gene [40]. Similarly, the insertion of a RE into the upstream promoter region of the apple MdMYB1-1 gene is associated with the red fruit skin phenotype [41].

This work aimed to characterize the repetitive component of the pomegranate genome, making a comparative analysis of the abundance and evolutionary dynamics of TEs in the four sequenced genomes. Moreover, the insertion of TEs in proximity or within genes was also assessed at a genome-wide level in the four cultivars to hypothesize the possible functional implications of TE activity in *P. granatum* genetic diversity.

2. Materials and Methods

2.1. Collection of Sequence Data for the Four Pomegranate Cultivars

The collection of four genome assemblies belonging to different *P. granatum* L. cultivars were downloaded from the National Center for Biotechnology Information (NCBI, https://www.ncbi.nlm.nih.gov/datasets/genome; accessed on 3 July 2023). The cultivar 'Dabenzi' (Bioproject PRJNA360679), 'Taishanhong' (Bioproject PRJNA355913), 'Tunisia' (Bioproject

PRJNA565884), and 'Bhagwa' (Bioproject PRJNA445950) were used for all the subsequent analyses [4–7].

2.2. Collection and Abundance Estimation of Full-Length Transposable Elements

The four pomegranate genomes were scanned for Class I and II TEs using EDTA v1.9.3 [42]. EDTA implementing a combination of LTR_FINDER v1.06 [43], LTRharvest v1.5.10 [44], and LTR_retriever v2.5 [45] was used for the search of LTR-REs. Generic Repeat Finder v1.0.2 [46] and TIR-Learner v1.18 [47] were used for the identification of DNA MITE and TIR elements, respectively. HelitronScanner v1.1 [48] was used for searching the *Helitron* elements. All the program parameters were automatically set, as reported in the default pipeline [42], and only full-length TEs were retained for analysis. For the lineage-level classification of LTR-REs, the elements were subjected to domain-based annotation using DANTE v1.1.8, accessible on the RepeatExplorer2 Galaxy-based website (https://repeatexplorer-elixir.cerit-sc.cz/galaxy/; accessed on 17 July 2023). The annotation was carried out with default settings, using the REXdb database of transposable element protein domains [17] and applying a BLOSUM80 scoring matrix. Protein matches were subsequently filtered based on their significance, following the parameters provided by the platform. For abundance estimation, the libraries of LTR-REs, MITEs, TIRs, and *Helitrons* obtained using EDTA were merged and used to mask the whole four genomes using RepeatMasker v4.1.5 [49] with the following parameters: -no_is, -nolow, -X.

2.3. Identification of Shared Transposable Element Insertion Sites and Phylogenetic Analysis

To determine the position of the full-length TEs across the four genome assemblies, i.e., to identify TEs inserted at genomic loci that are common or not across the four cultivars, we exploited the flanking regions of the elements themselves. All the previously obtained libraries of full-length LTR-REs, MITEs, TIRs, and *Helitrons* were used for this analysis. Each element was extracted from the corresponding genome assembly with 1000 bp extended downstream and upstream. This procedure was carried out using the "getfasta" function within BEDTools v2.30.0 [50]. Subsequently, all the extended TEs were subjected to clustering using CD-HIT v4.7 with the "s" parameter set to 0.9 [51]. Full-length elements at the same locus in all four genome assemblies were grouped into a single cluster, resulting in a cluster with four elements. For instances where an element occurred at the same locus in three out of four genome assemblies, these were grouped into a single cluster, and so forth. Elements exclusive to a single genome were isolated into separate clusters, each with one element. Ambiguous clusters were manually curated.

Data on the presence/absence of TEs were used to evaluate the phylogenetic relationships among the four pomegranate cultivars. These data were transformed into a matrix dataset and utilized to conduct a hierarchical clustering analysis using the UPGMA method. The analysis was executed with the R package "pvclust" v2.2-0, supported by 10,000 bootstrap replications [52]. A graphical representation of the data was produced using the "ggplot2" R package v3.4.1 [53].

2.4. Localization of Shared Transposable Element Insertion Sites in Genes or Their Proximity

To determine the positional relationship between TEs and protein-coding genes across the four genome assemblies, we extracted 1000 bp upstream and downstream of each full-length TE inserted using BEDTools. The extracted sequences were joined and aligned on the 'Tunisia' transcriptome using BLAST tool v2.6.0+ by a blastn search [54], enabling us to identify TE insertion sites compared to genes. If the entire joined sequence aligned to a transcript, indicated a TE inserted into exon. Conversely: (i) if one end of the joined sequence aligned to the transcript indicated the TE position within an intron or in an intergenic region; (ii) if both ends of the joined sequence aligned to the transcript, yet with a non-overlapping internal portion, indicated TE positioning within an intron. Lastly, we identified TEs in proximity to genes by comparing genome coordinates of protein-coding genes with those of TEs in all four genome assemblies, within a maximum distance of

1000 bp upstream or downstream of the genes using BEDTools "intersect" function. We conducted a 2-way ANOVA to assess the primary sources of data variation attributed to both cultivars and TE insertions. The statistical analysis was carried out with GraphPad PRISM v9.0.0 (GraphPad Software, Inc., La Jolla, CA, USA).

2.5. Profiling the Insertion Time of Full-Length LTR-Retrotransposons

The insertion time of different LTR-RE lineages was assessed by computing the distributions of pairwise divergence comparisons of the 5′- and 3′-LTRs. LTR pairwise alignments were calculated using the "stretcher" tool of the EMBOSS v6.6.0.0 suite, applying the Kimura two-parameter model of sequence evolution [55]. Distance matrices were generated using the "distmat" tool within the same suite [56]. To estimate the insertion times of lineages with at least ten full-length LTR-REs in the four genome assemblies, a mutation rate of 4.72×10^{-9}, i.e., two-fold the rate calculated for synonymous substitutions in gene sequences in *Populus trichocarpa* [57] was used. This adjustment accounts for the fact that LTR-REs accumulate mutations at twice the rate of gene sequences [58]. Peaks in frequency distribution were interpreted as transposition burst events, where lower divergence values suggested recent proliferation [59]. Insertion times of LTR-REs among the four pomegranate genotypes and their genomic locations were tested with ANOVA, followed by post-hoc analyses using Tukey's method. Outlier values were automatically removed from analysis by the software, while separate tests were performed for the *Gypsy* and *Copia* superfamilies. Finally, Statistical analysis was carried out using GraphPad PRISM, with a graphical representation of the data generated by "ggplot2" R package.

2.6. Functional Analysis of Genes in Proximity to or Interrupted by Transposable Elements

To infer the impact of TEs on gene function, we analysed the Gene Ontology (GO) functional annotations of genes lying nearby or interrupted by TEs. The GO terms were derived from the annotated 'Tunisia' genome [4]. For the GO enrichment analysis on genes in proximity to or interrupted by TEs compared to the entire transcriptome, we utilized Blast2GO v5.2.5, employing Fisher's exact test (p-value < 0.05) [60]. Subsequently, KEGG Orthology (KO) id codes of corresponding genes were submitted to KEGG for pathway network analysis (Kyoto Encyclopaedia of Genes and Genomes) [61]. Subsequently, REVIGO was used to remove redundant GO terms with the parameter "tiny similarity" [62].

3. Results

3.1. Collection and Estimation of Abundance of Full-Length Transposable Elements

The genome assemblies of the four available pomegranate cultivars, namely 'Bhagwa', 'Dabenzi', 'Taishanhong', and 'Tunisia', were scrutinized to isolate full-length TEs belonging to both Class I and Class II. Overall, we identified a total of 8404 TEs (Table 1, Supplementary Data S1–S4). The highest number of elements was found in the 'Bhagwa' genome, with a total of 2511. A similar amount was retrieved in the 'Tunisia' genome, with a total of 2465 elements. The analyses of the 'Taishanhong' and 'Dabenzi' genomes returned 1822 and 1606 elements, respectively.

Regarding Class I elements, the *Copia* lineages identified were *Ale*, *Alesia*, *Angela*, *Ikeros*, *Ivana*, *TAR*, and *Tork*. The *Ale* lineage was abundant in all four genome assemblies (Table 1), predominating in the 'Taishanhong' and 'Dabenzi' genomes. However, in the 'Tunisia' and 'Bhagwa' genomes, the *Angela* lineage was the most abundant. Interestingly, *Angela* elements were present in significantly fewer copies in the 'Taishanhong' and 'Dabenzi' genomes. Another notable difference can be observed concerning the *Tork* lineage, which was highly represented in the 'Tunisia' and 'Bhagwa' genomes but less abundant in the 'Taishanhong' and 'Dabenzi' genomes.

As for the *Gypsy* superfamily, the lineages identified in the four genome assemblies were *Chromovirus*, including the four sublineages *CRM*, *Galadriel*, *Reina*, and *Tekay*, and non-*Chromovirus*, including *Athila* and *Tat/Ogre*. Most identified elements belonged to the *Chromovirus/CRM* lineage. A considerable disparity in the number of non-*Chromovirus/Tat/Ogre* elements was also observed by comparing 'Taishanhong' and 'Dabenzi' to the 'Bhagwa' and 'Tunisia' genomes, with the latter showing a much higher amount.

Table 1. Number (nr) of transposable elements identified in each pomegranate genome assemblies.

Order	Superfamily	Lineage	Tunisia (nr)	Bhagwa (nr)	Taishanhong (nr)	Dabenzi (nr)
Class I (Retrotransposons)	*Copia*	*Ale*	179	179	148	132
		Alesia	1	1	1	1
		Angela	229	230	70	20
		Ikeros	10	10	10	7
		Ivana	53	52	32	32
		TAR	66	65	30	29
		Tork	148	143	71	23
	Gypsy	*Chromovirus/CRM*	190	229	90	46
		Chromovirus/Galadriel	14	13	12	8
		Chromovirus/Reina	24	24	19	19
		Chromovirus/Tekay	6	8	1	1
		Non-*Chromovirus/Athila*	56	52	23	8
		Non-*Chromovirus/Tat/Ogre*	59	58	8	1
	Unknown		121	138	63	57
	LINE		1	1	nd	nd
	Pararetrovirus		nd	nd	1	nd
Class II (DNA Transposons)	*TIR*	*hAT*	127	110	111	110
		CACTA	141	171	148	142
		PIF/Harbinger	28	37	27	29
		Mutator	393	374	368	356
		Tc1/Mariner	19	17	13	13
	MITE	*hAT*	88	90	84	78
		CACTA	15	15	12	16
		PIF/Harbinger	16	12	15	14
		Mutator	96	96	85	90
		Tc1/Mariner	1	2	4	5
	Helitron	*Helitron*	378	373	371	366
Unknown			6	11	5	3
Total			2465	2511	1822	1606

Concerning Class II TEs, the number of full-length elements in the four genome assemblies was comparable (Table 1). In particular, *hAT*, *CACTA*, *PIF/Harbinger*, *Mutator*, *Tc1/Mariner* (for both TIR and MITE superfamilies), and *Helitron* elements were identified. *Mutator* elements, considering TIR and MITE superfamilies, were the most abundant in the four pomegranate genome assemblies. The least abundant were the *Tc1/Mariner* elements. Noteworthy, a relatively large number of *Helitron* elements were detected to a similar frequency in all four genome assemblies.

The abundance of TEs was evaluated across the four genotypes by masking each genome assembly with TE libraries. Overall, TE abundance resulted highly variable, ranging from 41.67 to 52.45% of 'Taishanhong' and 'Bhagwa', respectively.

The total content of LTR-REs was higher in 'Bhagwa' and lower in the 'Taishanhong' genome. The overall abundance of *Gypsy* was approximately two-fold greater than *Copia* regarding the 'Tunisia', 'Bhagwa', and 'Dabenzi' genomes. In the case of 'Taishanhong', the difference in abundance of the two LTR-RE superfamilies is reduced.

Among the *Copia* LTR-REs, *Angela* was the most abundant lineage (above 1.79%), except in the 'Taishanhong' genome, where *Ale* was the most represented (1.79%). On the contrary, *Ale* was the second most abundant lineage among 'Dabenzi', 'Tunisia', and 'Bhagwa' (ranging from 1.44 to 1.62%). The lineage non-*Chromovirus/Tat/Ogre*, which belongs to the *Gypsy* superfamily, was the most abundant LTR-RE in all four pomegranate genomes (Table 2).

Table 2. Abundance of transposable elements of the four pomegranate genome assemblies, specified for each order, superfamily, and lineage; %: refers to the percentage of genomic abundance.

Order	Superfamily	Lineage	Tunisia (%)	Bhagwa (%)	Taishanhong (%)	Dabenzi (%)
Class I (Retrotransposons)	*Copia*	*Ale*	1.72	1.61	1.79	1.80
		Alesia	0.01	0.01	0.01	0.01
		Angela	2.26	2.08	1.49	1.90
		Ikeros	0.15	0.13	0.15	0.16
		Ivana	0.33	0.32	0.33	0.36
		TAR	0.44	0.41	0.34	0.4
		Tork	0.72	0.7	0.56	0.65
		Total	5.63	5.26	4.67	5.28
	Gypsy	*Chromovirus/CRM*	3.05	3.22	1.65	2.15
		Chromovirus/Galadriel	0.08	0.07	0.08	0.08
		Chromovirus/Reina	0.12	0.11	0.13	0.13
		Chromovirus/Tekay	0.48	0.62	0.18	0.26
		Non-*Chromovirus/Athila*	0.48	0.45	0.36	0.4
		Non-*Chromovirus/Tat/Ogre*	8.88	8.98	4.98	6.59
		Total	13.09	13.45	7.38	9.61
	Total *Copia/Gypsy*		18.72	18.71	12.05	14.89
	Unknown		11.88	15.32	7.42	9.0
	LINE		0.05	0.05	0.06	0.05
	pararetrovirus		0.08	0.07	0.08	0.08
Class II (DNA Transposons)	*TIR*	*hAT*	1.32	1.08	1.5	1.23
		CACTA	3.13	2.93	3.71	3.32
		PIF/Harbinger	1.26	1.19	1.42	1.35
		Mutator	5.61	5.27	6.27	5.99
		Tc1/Mariner	0.38	0.36	0.43	0.41
	MITE	*hAT*	0.24	0.27	0.36	0.3
		CACTA	0.04	0.04	0.05	0.05
		PIF/Harbinger	0.07	0.07	0.11	0.08
		Mutator	0.42	0.39	0.55	0.46
		Tc1/Mariner	0.02	0.02	0.05	0.02
	Helitron	*Helitron*	6.9	6.68	7.61	7.45
Total			50.12	52.45	41.67	44.68

3.2. Identification and Phylogenetic Analysis of Shared Transposable Element Insertion Sites

In relation to the TEs identified and annotated in the four pomegranate cultivars, we determined if each TE position was maintained across the four genome assemblies through a clustering approach. The analysis produced 5025 clusters, each composed of one to four elements (Supplementary Table S1), according to whether an element was exclusive to a single genome or shared across multiple genomes. The clusters were categorized to represent the number of shared TE insertions for every genotype combination in relation to the element class (Table 3, Supplementary Table S2).

In total, we identified 73 TEs at the same genomic loci in all four genome assemblies, comprising 23 REs and 50 DNA TEs. Regarding elements shared by three genotypes, the 'Dabenzi', 'Taishanhong', and 'Tunisia' assemblies presented the highest number, totalling 211 shared TEs (38 REs and 173 DNA TEs). The three genotypes with the fewest shared elements were 'Bhagwa', 'Dabenzi', and 'Tunisia', with a total of 81 TEs (17 REs and 64 DNA TEs).

Among the four genome assemblies, 'Bhagwa' possessed the highest number of exclusive elements, totalling 947 (comprising 621 REs and 326 DNA TEs), closely followed by 'Tunisia' with 878 exclusive elements (including 554 REs and 324 DNA TEs).

Nevertheless, it is important to highlight that the failure to identify certain TEs in specific loci may depend on the accuracy of the assembly and the sequencing technologies used, potentially over-rating the differences among genomes.

Table 3. Number of shared transposable element insertions identified in the four pomegranate genome assemblies. Each genotype combination is reported. The number of insertions refers to total transposable elements, REs, and DNA TEs.

Genotype Combination	Number of Insertion Sites (nr)		
	Total Transposable Elements	Retrotransposons	DNA Transposons
Bhagwa–Dabenzi–Taishanhong–Tunisia	73	23	50
Bhagwa–Dabenzi–Taishanhong	200	38	162
Bhagwa–Dabenzi–Tunisia	81	17	64
Bhagwa–Taishanhong–Tunisia	125	63	62
Dabenzi–Taishanhong–Tunisia	211	38	173
Bhagwa–Dabenzi	310	68	242
Bhagwa–Taishanhong	362	125	237
Bhagwa–Tunisia	402	252	150
Dabenzi–Taishanhong	137	30	107
Dabenzi–Tunisia	293	68	225
Taishanhong–Tunisia	396	141	255
Bhagwa	947	621	326
Dabenzi	297	105	192
Taishanhong	313	124	189
Tunisia	878	554	324

The presence/absence data of TEs in the four pomegranate genomes were used to investigate the relationship among cultivars (Figure 1, Supplementary Table S3). The resulting dendrogram showed that 'Taishanhong' and 'Dabenzi' genomes exhibit a closer phylogenetic relationship to each other compared to 'Tunisia' and 'Bhagwa'. This relationship probably reflects the geographical origins of these cultivars, where 'Taishanhong' and 'Dabenzi' were of Chinese origin, and 'Tunisia' and 'Bhagwa' originated from Tunisia and India, respectively.

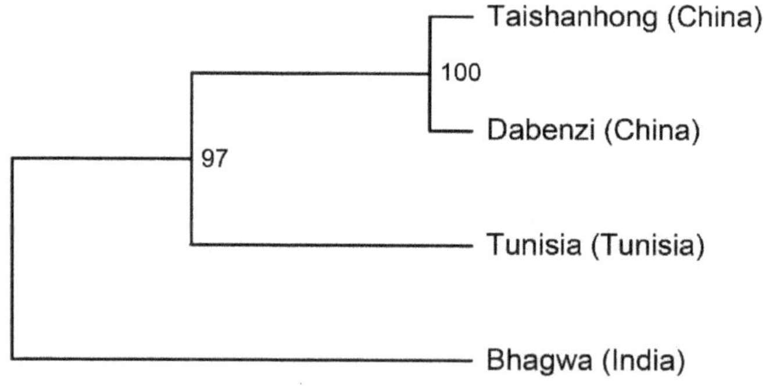

Figure 1. Dendrogram resulting from hierarchical clustering analysis using the presence/absence data of transposable elements in the four pomegranate genome assemblies, along with information about their geographic origins (shown in parenthesis). Bootstrap resampling values are indicated at the nodes.

3.3. *Profiling the Insertion Time of Full-Length LTR-Retrotransposons*

The proliferation time profiles of the full-length LTR-REs were inferred in the four pomegranate genome assemblies by measuring pairwise distances between the LTRs of each element, based on the principle that the two LTR sequences of a RE are identical immediately after the insertion event and then accumulate mutations over time (see Section 2). Although LTR-RE age calculation based on this assumption is subjected to errors due to casualties in mutation events, this method still appears to be the most useful for inferring RE proliferation dynamics [55]. In pomegranate, this analysis showed the proliferation of *Copia* and *Gypsy* REs in the last 40 million years (Figures 2 and 3).

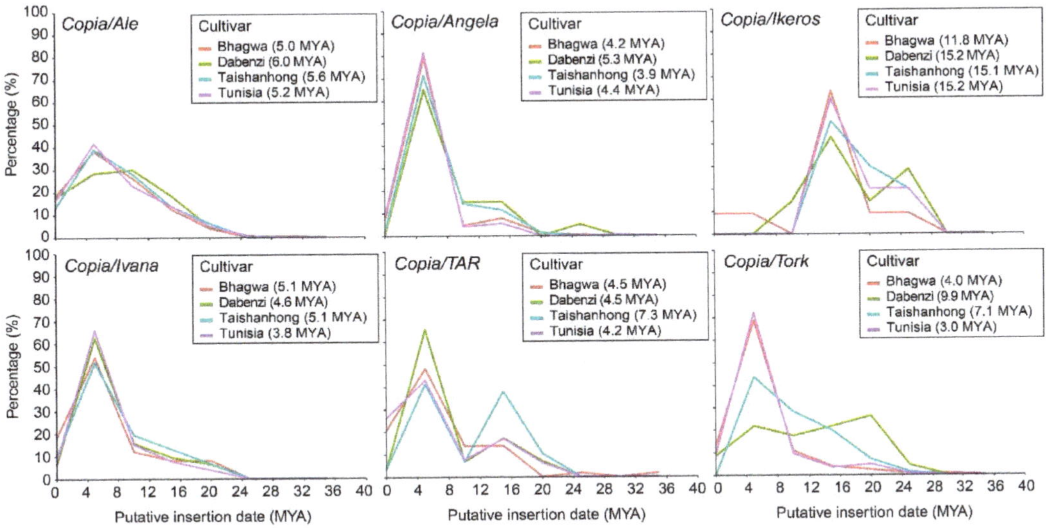

Figure 2. Insertion time of six retrotransposon lineages belonging to the *Copia* superfamily in the four pomegranate genome assemblies. Each cultivar is indicated with a different colour. The average insertion time (million years ago = MYA) for each cultivar is reported in parentheses.

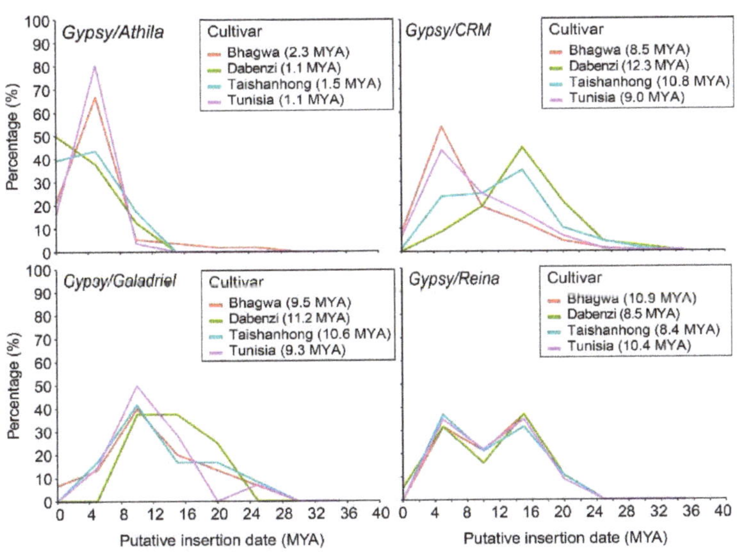

Figure 3. Insertion time of four retrotransposon lineages belonging to the *Gypsy* superfamily in the four pomegranate genome assemblies. Each cultivar is indicated with a different colour. The average insertion time (million years ago = MYA) for each cultivar is reported in parentheses.

The cultivars presented similar putative TE insertion time profiles, with differences specific to each LTR-RE lineage. Most of the lineages of the *Copia* superfamily showed a proliferation peak about six million years ago (MYA) (Figure 2), except for elements belonging to lineages *TAR* and *Tork* that showed older proliferation peaks in 'Taishanhong' and 'Dabenzi', respectively. The *Ikeros* lineage presented a more ancient proliferation peak at 16 MYA in the four cultivars. The *Tork* elements of the 'Tunisia' cultivar were identified as the youngest (average insertion time of 3 MY), while the *Ikeros* elements of the 'Dabenzi' and 'Tunisia' cultivars resulted the oldest (average insertion time of 15.2 MY).

As regards the *Gypsy* superfamily, the lineages generally showed a different proliferation activity compared to *Copia* (Figure 3). The *Athila* lineage showed a proliferation peak at 5 MYA in all cultivars except for 'Dabenzi', in which this lineage appears to be still proliferating. The lineage *Galadriel* displayed a proliferation peak at 10 MYA, whereas the *Reina* lineage showed a pattern with two proliferation peaks, one at 5 MYA and one at 15 MYA in all genotypes. *CRM* lineage exhibited the oldest proliferation peak at 16 MYA in 'Taishanhong' and 'Dabenzi' cultivars, whereas the transposition burst in 'Bhagwa' and 'Tunisia' is observed at 5 MYA. The *Athila* elements of the 'Dabenzi' and 'Tunisia' cultivars were identified as the youngest (average insertion time of 1.1 MY), while the *CRM* elements of the 'Dabenzi' cultivar were the oldest (average insertion time of 12.3 MY).

The putative insertion times of the LTR-REs were also analysed in relation to the presence of the same element in the same locus in four, three, or two genotypes or to its presence in one specific genotype (Figure 4). Overall, the LTR-REs shared in the same genomic loci across all four pomegranate genome assemblies had a higher average insertion date than elements shared between three or two cultivars or specific to one cultivar. In brief, the more elements are shared at the same locus among cultivars, the older their average insertion date is.

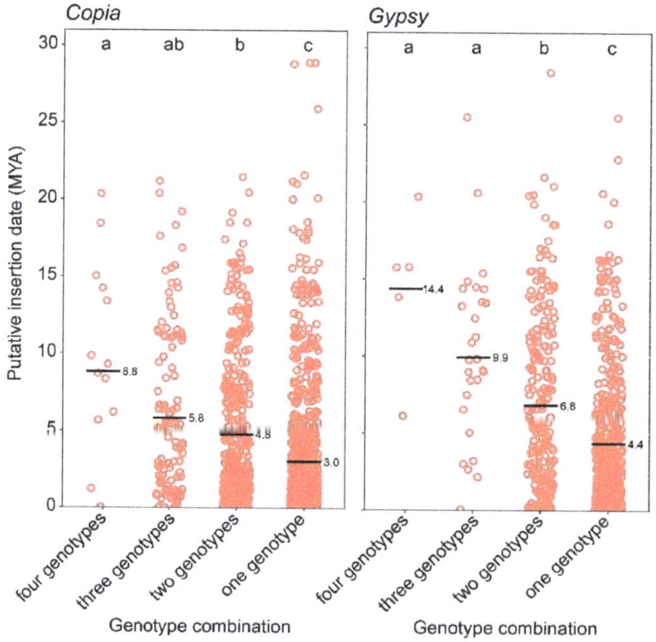

Figure 4. Putative insertion times of LTR-REs are subdivided into four groups based on their presence in the same locus in four, three, or two genotypes or specific to one genotype. Data for the *Copia* and *Gypsy* superfamilies are presented separately. The black bar represents each genotype combination's average LTR-RE insertion time (million years ago = MYA). Significant differences for each group of measurements are indicated by letters a, b, and c: groups with the same letter are not significantly different (p-value < 0.05) according to Tukey's test.

3.4. Localization of Shared TE Insertion Sites in Genes or Their Proximity

To identify TEs inserted in proximity or within gene regions (either exons or introns), 2000 bp-long sequences retrieved for each full-length TE in the four genomes (joining 1000 bp upstream and 1000 bp downstream sequences) were aligned against the 'Tunisia' transcriptome (see Section 2). The alignments between entire joined sequences and gene transcripts indicated insertions into exons, while alignments of only one or both ends of the joined sequences to the transcripts indicated insertions in the introns or intergenic regions. Also, TE insertions in the proximity of genes were identified by comparing the genome coordinates of protein-coding genes with those of TEs and retaining all full-length elements lying within 1000 bp upstream or downstream of the coding portion of a gene.

Considering all the insertion sites identified in the four pomegranate genome assemblies for the instances where TEs are shared among, it was observed that most TEs were located near genes (within 1000 bp). This localization was consistent for DNA TEs and REs, with approximately 57% and 55% of insertion sites, respectively (Figure 5). Insertion sites far from genes (i.e., distance more than 1000 bp) represented approximately 36% of DNA TE and 38% of RE insertions. Insertions within gene exons and introns were rare for both DNA TEs and REs.

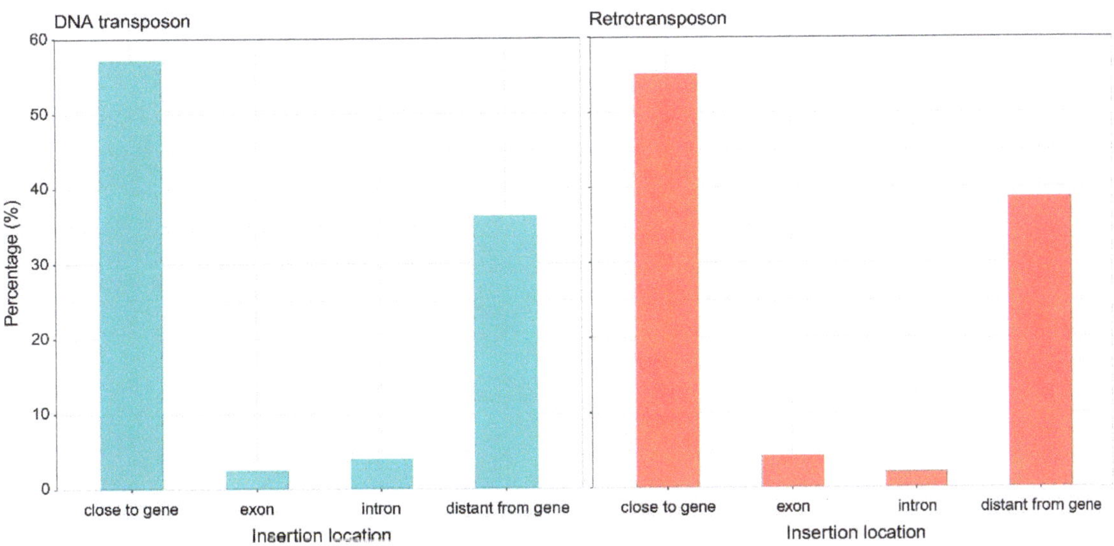

Figure 5. Distribution of total transposable element insertion sites in the four pomegranate genome assemblies. The percentage of the insertion sites is relative to transposable element classes and insertion location.

Detailed results of the distribution of TE insertions among all genotype combinations are reported in Supplementary Figure S1. The complete list of the genes showing TE insertions can be found in Supplementary Table S4.

The number of TEs in the proximity of genes ranged from 874 in 'Dabenzi' to 1278 in 'Tunisia'. The number of TEs within exons in 'Bhagwa' and 'Tunisia' genomes was higher that of 'Dabenzi' and 'Taishanhong'. The highest number of intronic TE insertions was found in the 'Bhagwa' genome assembly. In terms of data variation, the TE insertion location contributed the most to the variation (95.76%) compared to the variation provided by cultivars (Table 4).

Table 4. Two-factorial analysis of variance (ANOVA) for the insertion of transposable elements and pomegranate cultivars. ns: not significant; ***: p-value < 0.001.

Cultivar	TE Insertion Location		
	Close to Gene (nr)	**Exon (nr)**	**Intron (nr)**
'Bhagwa'	1264	66	69
'Dabenzi'	874	40	50
'Taishanhong'	968	39	58
'Tunisia'	1278	61	63
Source of variation	**Percentage of variation (%)**	**Significance**	
Cultivar	1.07	ns	
TE insertion location	95.76	***	

The temporal insertion profile of LTR-REs in relation to their insertion locations was also explored (Supplementary Figure S2). This analysis showed no significant differences between the groups in both superfamilies. In the *Copia* superfamily, the average insertion ages varied from a minimum of 3.3 MYA for elements inserted into exons to a maximum of 4.2 MYA for those distant from genes. In the *Gypsy* superfamily, insertion ages ranged from 2.6 MYA for elements inserted in introns to a maximum of 6.2 MYA for those distant from genes.

3.5. Functional Analysis of Genes in Proximity to or Interrupted by Transposable Elements

The potential impact of TE insertions on the function of genes lying in proximity to the element or interrupted by the element was explored by functionally annotating these genes using Gene Ontology (GO) and KEGG enrichment analyses. The GO and KEGG codes of analysed genes can be found in Supplementary Table S6. GO enrichment analysis (Figure 6) showed that the most recurrent GO terms of the genes in proximity of at least one TE were 'tetrapyrrole binding' (GO:0046906), 'heme binding' (GO:0020037), 'oxidoreductase activity, acting on paired donors, with incorporation or reduction of molecular oxygen' (GO:0016705), and 'iron ion binding' (GO:0005506) (Figure 6a). The most abundant enriched GO terms associated with genes interrupted by a TE in the introns were 'catalytic activity' (GO:0003824), 'hydrolase activity' (GO:0016787), and 'ATP binding' (GO:0005524) (Figure 6b). Similarly, GO terms like 'carbohydrate derivative binding' (GO:0097367), 'heterocyclic compound binding' (GO:1901363), 'adenyl ribonucleotide binding' (GO:0032559), and 'anion binding' (GO:0043168), were the most represented for genes interrupted by a TE in the exons (Figure 6c).

KEGG analysis was performed to analyse the genes of the phenylpropanoid pathway (Table 5) that is crucial for producing polyphenolic compounds, the main secondary metabolites in pomegranate, including flavonoids, anthocyanins, and tannins that are of value for pomegranate fruits [2]. Some genes of the phenylpropanoid pathway were found to be located in the proximity of at least one TE or interrupted by one TE, suggesting that TE insertions might change the regulation of the metabolism of these compounds, contributing to biodiversity between cultivars (Table 5). Overall, we found genes encoding two *flavonoid 3′-monooxygenase* (*F3′H*) and two *4-coumarate--CoA ligase* (*4CL*) located in the proximity of TEs. Furthermore, genes encoding an *anthocyanidin reductase* (*ANR*) and two *peroxidases* (*POD*) were found to be interrupted by a TE in the exonic region. No cases of TE insertion in intronic regions were identified.

For both *F3′H* genes, the TE proximal to the gene coexisted at the same genomic locus across all four pomegranate genome assemblies (Table 5). In both instances, the element belonged to lineages of the *Copia* superfamily; specifically, one was an *Ale* element, and the other was an *Ivana* element. Concerning the two *4CL* genes, one exhibits a TE insertion at a shared genomic locus between 'Bhagwa' and 'Taishanhong'. In this case, the element belonged to the *Chromovirus/CRM* lineage. The other *4CL* gene was interrupted by an element belonging to the *Helitron* class shared among 'Dabenzi', 'Taishanhong', and 'Tunisia' cultivars.

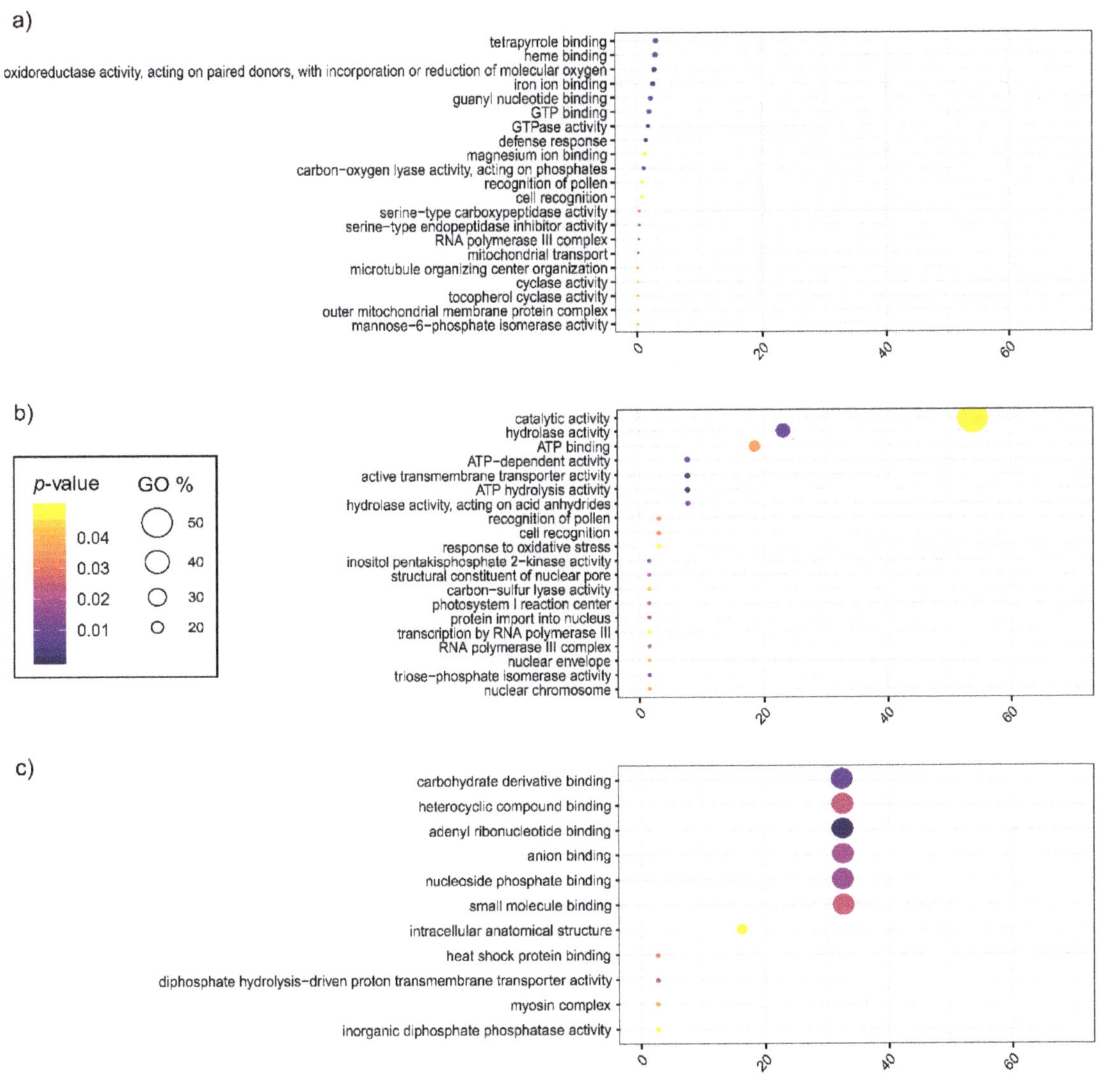

Figure 6. Distribution of genes in the proximity or interrupted by a transposable element in Gene Ontology classes. The intensity of the colour (yellow to purple) indicates significance (p-value < 0.05). The size of the circle indicates the percentage of the Gene Ontology class. (**a**) Genes in the proximity of transposable elements; (**b**) Genes interrupted by a transposable element in the introns; (**c**) Genes interrupted by a transposable element in the exons.

Concerning cases where a TE is inserted within exonic gene regions, we observed an *ANR* gene, interrupted exclusively in the 'Tunisia' cultivar and belonging to the *TIR/Mariner* family (Table 5). Finally, the two *POD* genes were interrupted by two different *Chromovirus/Reina* elements. The first POD gene was interrupted in 'Bhagwa' and 'Dabenzi' genotypes, the second in 'Taishanhong' and 'Tunisia' cultivars.

Table 5. Phenylpropanoid-related genes located in proximity or interrupted by transposable element insertion in the four pomegranate genome assemblies. The table provides details for each gene, including insertion location, family/lineage of the inserted element, gene ID, gene name, and the genotype combination sharing the element at the same genomic locus. Genotype names are abbreviated as follows: Bh = 'Bhagwa', Da = 'Dabenzi', Ta = 'Taishanhong', Tu = 'Tunisia'.

Insertion Location	Transposable Element Family/Lineage	Gene ID	Gene Name	Gene Code	Genotype Combination
Close to gene	*Copia/Ale*	XM_031520924.1	*flavonoid 3′-monooxygenase*	*F3′H*	Bh I Da I Ta I Tu
	Copia/Ivana	XM_031528957.1	*flavonoid 3′-monooxygenase*	*F3′H*	Bh I Da I Ta I Tu
	Gypsy/Chromovirus/CRM	XM_031526933.1	*4-coumarate--CoA ligase*	*4CL*	Bh I Ta
	Helitron	XM_031516428.1	*4-coumarate--CoA ligase*	*4CL*	Da I Ta I Tu
Exon	*TIR/Mariner*	XM_031530037.1	*anthocyanidin reductase*	*ANR*	Tu
	Gypsy/Chromovirus/Reina	XM_031520605.1	*peroxidase*	*POD*	Bh I Da
	Gypsy/Chromovirus/Reina	XM_031520605.1	*peroxidase*	*POD*	Ta I Tu

4. Discussion

Our work provides a comprehensive characterization of full-length TEs in the genome of *P. granatum* through a comparative analysis of the genome assemblies of four pomegranate cultivars (i.e., 'Bhagwa', 'Dabenzi', 'Taishanhong', and 'Tunisia'), focusing on the intraspecies variability of TE insertion loci and its possible functional implications.

The content of TEs varied among the four cultivars, ranging from 41.67 to 52.45%, in a proportion similar to that found in other small-sized genomes such as apple [63], pear [64], fig [65], and blackberry [66]. Most of the repeat component of the pomegranate genome is composed of LTR-REs. The common occurrence of these elements in the fraction of repeated sequences is a widespread characteristic of higher plant genomes, where REs account for one of the major forces driving genome size evolution [67–69]. Based on TEs abundance, the Chinese cultivars 'Taishanhong' and 'Dabenzi' differ from 'Tunisia' and 'Bhagwa' due to the lower abundance of the LTR-REs. In all four cultivars, *Gypsy* accounted for a larger proportion than *Copia* elements, confirming what is generally observed in the Angiosperms, with valuable exceptions, such as pear, date palm, and banana [70]. Notably, the *Gypsy* lineage *Tat/Ogre* was the most abundant in all four pomegranate genomes, as observed in pea [71] and 23 plant genomes belonging to the *Fabeae* tribe [72], indicating the importance of this lineage in determining the genome size evolution of pomegranate. The *Copia* superfamily abundance ranged between 4.67 to 5.63% in 'Taishanhong' and 'Tunisia', respectively. Similar results were observed in pomegranate by Qin [5] and Yuan [6], accounting for 4.8% and 5.87%, respectively. Concerning the *Copia* superfamily, except for 'Taishanhong', the *Angela* elements were the most frequent in all cultivars, followed by *Ale*, as observed in *Stevia rebaudiana* [73] and grape [74].

Our analysis identified only 73 full-length TEs shared across all four pomegranate genome assemblies. Conversely, the four genotypes showed a high number of elements uniquely present in one genotype. For example, 'Bhagwa' exhibited the highest number of exclusive elements (947), followed by 'Tunisia' (874) (Table 3). Hierarchical clustering based on TE presence/absence reflected a closer phylogenetic relationship between 'Taishanhong' and 'Dabenzi' cultivars, in line with their shared Chinese origin, distinct from 'Tunisia' and 'Bhagwa', which originated in Tunisia and India, respectively (Figure 1).

The number of TEs shared among the four genotypes can be underestimated because of genome misassembling. However, it is also plausible that in many cases TEs have been subjected to rearrangements and mutations so that the same full-length element could not be found at the same locus in all cultivars. It is possible that in some loci only TE remnants are maintained, which cannot be recognized by the bioinformatic tools used for identifying full-length TEs. The large number of TEs unique to single genotypes may also suggest that TE proliferation and/or insertions have occurred following the divergence of these genotypes, or that many full-length LTR-REs present in the progenitor have experienced TE removal by unequal recombination or by DNA loss [75].

The full-length LTR-REs were further characterized by their insertion time profiles, which evidenced transposition bursts, presumably associated with plant evolution [76]. The insertion time profiles for different LTR-RE lineages were similar among cultivars, although in some cases different transposition peaks were displayed (Figures 2 and 3). The *Copia* superfamily showed that the insertions of the isolated full-length elements were relatively recent, with a transposition peak at 4–6 million years ago, except for the *Ikeros* lineage, where the transposition burst was around 15 million years ago.

Among *Gypsy* LTR-REs, the insertion times revealed a more ancient transposition burst of *Chromovirus/Galadriel* and *Chromovirus/Reina* lineages in all four cultivars compared to non-*Chromovirus/Athila*. Interestingly, in the cultivar 'Dabenzi' the *Athila* lineage has not yet reached the peak of proliferation. The transposition burst characterizing *Chromovirus/CRM* lineage occurred more recently in 'Tunisia' and 'Bhagwa' compared to the Chinese cultivars.

Relating the putative insertion dates of LTR-REs to the presence of the element at the same locus in one, two, three, or four cultivars, indicated that shared elements of both *Copia* and *Gypsy* superfamilies exhibited more ancient average insertion dates than those exclusive to individual cultivars (Figure 4). This coherence is logical, as the presence of shared elements among multiple genotypes should imply that their replication and insertion occurred before the divergence of these genotypes. The trend according to which the more an element is shared between the cultivars, the older its insertion, is generally statistically significant. However, in some cases, even full-length elements found only in one, two, or three cultivars, exhibit insertion ages older than the average (Figure 4). This could indeed suggest that very ancient TEs have either been lost from one or more genotypes after their separation or that these TEs have undergone rearrangements that prevented their recognition by the bioinformatic tools used for their identification.

Regarding full-length TE insertion sites, the majority were located within 1000 bp of the encoding portion of a gene (Figure 5). Overall, our results might be influenced by the identification of full-length TE itself. To identify the full-length RE the sequence must exhibit a conserved sequence, and a higher level of conservation may be more favored for TEs located in gene-rich regions that are less exposed to purifying selection. On the other hand, the tendency of TEs to lie near genes has already been observed especially in TE-rich species [28]. This tendency was observed for both DNA TEs (57%) and REs (55%). Less than 5% of full-length TEs and LTR-REs were found interrupting gene exons or introns, suggesting the occurrence of purifying selection against the insertion in the coding portions of genes, as expected because of the potentially negative effect of TE insertion for the gene functionality.

In several plant species, including tomato, soybean, melon, orange, sunflower, and others, functionally relevant TE insertions in the proximity of genes have been well-documented (reviewed by Fambrini [33]). The insertion of a TE near a gene can change its proximal promoter sequence, with possible consequences on the regulation of gene activity [28]; moreover, the inserted TE can modulate the expression rate of a close gene by inducing epigenetic modifications along the chromosomal locus [77].

Our data indicate that TE insertions occurred in the proximity of genes regardless of their function as determined by GO analysis, although some GO (for example those related to binding) resulted overrepresented, also when considering genes interrupted in their exonic portion by a TE (Figure 6). It is noteworthy that, among genes showing proximity to full-length TEs or interrupted in their transcribed portion by a full-length TE, some are involved in the phenylpropanoid biosynthetic pathway.

The pomegranate fruit, celebrated for its health benefits attributed to antioxidant polyphenolic compounds, such as flavonols, flavonoids, hydrolyzable tannins (ellagitannins), gallagic acid, punicalin, anthocyanins, and proanthocyanidins, has received considerable attention [78–82]. Among these secondary metabolites, anthocyanins are one of the most important flavonoids that contribute to the colour of fruits [83], and the content of these compounds was also characterised in the four cultivars whose genome is available. In particular, anthocyanin biosynthesis and the accumulation in ripe fruits occur earlier in 'Tunisia' than in 'Dabenzi' [84]. 'Taishanhong' displays bright red fruits at the ripe stage [6], boasting high total anthocyanin concentration [85]. Similarly, 'Bhagwa', the most widespread Indian cultivar, is distinguished by its high anthocyanin content [7].

Our results showed events of TE insertions close to two genes encoding 4-*coumarate--CoA ligase* (4*CL*), a pivotal enzyme in the phenylpropanoid pathway directing precursors toward various phenylpropanoids [86]. Notably, one of these insertions was observed in two genotypes, 'Bhagwa' and 'Taishanhong', of Indian and Chinese origin, respectively, suggesting an ancient, pre-divergence origin for this insertion. The other 4*CL* gene is shared among three genotypes, i.e., 'Dabenzi', 'Taishanhong', and 'Tunisia', indicating that the inserted TE was lost in the fourth genotype ('Bhagwa') or that TE insertion occurred after the divergence of the 'Bhagwa' genotype from the common ancestor of the other three genotypes.

Finally, a gene encoding *anthocyanidin reductase* (*ANR*) was found to be disrupted by a *TIR/Mariner* element inserted within the exon. *ANR* is pivotal in the biosynthesis of flavan-3-ols and proanthocyanidins (PAs) [87]; the significant presence of ellagitannins and anthocyanins in pomegranates, primarily in the form of flavan-3-ol monomers and dimers, enhances the nutraceutical properties of pomegranate juice, showing superior bioavailability compared to larger oligomers and polymers [82]. PAs, as condensed tannins, are usually associated with plant astringency and the darkening of fruit skin upon exposure to air. Increased ANR activity could potentially enhance astringency in plant tissues, like fruit skins and seeds. Interestingly, this insertion was only found in the Tunisian genotype. This could suggest a recent mobilization event exclusive of this genotype.

Changes in the phenylpropanoid phenotype have been induced by insertional mutagenesis in *Arabidopsis thaliana* [88]. Our data show that such insertional mutagenesis has occurred naturally in *P. granatum*, and such TE insertions can have induced changes in the phenylpropanoid profile of the pomegranate fruit, affecting nutraceutical properties of pomegranate juice.

In recent years, the availability of genomic resources, even for minor crops like pomegranate, has clarified important aspects related to the structure of the plant genome and potential functional aspects. Despite being challenging, characterizing the repetitive fraction and assessing the variability linked to TE abundance and insertions across different cultivars proves pivotal.

Undoubtedly, the profound impact of transposable elements on genome evolution is widely acknowledged, and this study represents an initial foray into comprehending their functional dynamics in pomegranate. Nevertheless, the functional influence of TEs in pomegranate, which extends beyond their proximity to genes, necessitates targeted functional analyses coupled with in-depth metabolomic studies. Exploring potential candidate targets through screening and evaluating the phenotypic effects of specific TE insertions will unravel the functional repercussions of TE activity. Overall, these elements can generate new genetic variants and be exploited as molecular markers to select plants with specific traits or facilitate genetic mapping, with potential implications for pomegranate breeding and crop improvement.

Supplementary Materials: The following supporting information can be downloaded at: https://www.mdpi.com/article/10.3390/horticulturae10020111/s1, Figure S1: Distribution of TEs among all genotype combinations, their insertion location and annotation. The genotype combination is shown above. The names of the genotypes were abbreviated as follows: Bh = 'Bhagwa', Da = 'Dabenzi', Ta = 'Taishanhong', Tu = 'Tunisia'; Figure S2: Retrotransposon insertion time based on the insertion locations in the four pomegranate cultivars. Data for the *Copia* and *Gypsy* superfamilies are presented. The black bar represents each genotype combination's average retrotransposon insertion time (in MYA). No significant differences were identified according to Tukey's test. Table S1: ST_1; Table S2: ST_2; Table S3: ST_3; Table S4: ST_4; Table S5: ST_5. Data S1: Transposable element prediction in 'Bhagwa' genome assembly; Data S2: Transposable element prediction in 'Dabenzi' genome assembly; Data S3: Transposable element prediction in 'Taishanhong' genome assembly; Data S4: Transposable element prediction in 'Tunisia' genome assembly.

Author Contributions: L.N., T.G., F.M. and A.C. planned and designed the project. S.S., G.U., A.V. and M.C. performed the computational analysis. S.S. and G.U. wrote the manuscript with contributions from all authors. All authors have read and agreed to the published version of the manuscript.

Funding: University of Pisa: Project "Plantomics".

Data Availability Statement: Publicly available datasets were analyzed in this study. This data can be found here: https://www.ncbi.nlm.nih.gov/ (accessed on 3 July 2023).

Conflicts of Interest: The authors declare no conflicts of interest.

References

1. Patel, C.; Dadhaniya, P.; Hingorani, L.; Soni, M.G. Safety assessment of pomegranate fruit extract: Acute and subchronic toxicity studies. *Food Chem. Toxicol.* **2008**, *46*, 2728–2735. [CrossRef] [PubMed]
2. Johanningsmeier, S.D.; Harris, G.K. Pomegranate as a functional food and nutraceutical source. *Annu. Rev. Food Sci. Technol.* **2011**, *2*, 181–201. [CrossRef]
3. Sarkhosh, A.; Yavari, A.M.; Zamani, Z. (Eds.) *The Pomegranate: Botany, Production and Uses*; CABI: Wallingford, UK, 2020.
4. Luo, X.; Li, H.; Wu, Z.; Yao, W.; Zhao, P.; Cao, D.; Yu, H.; Li, K.; Poudel, K.; Zhao, D.; et al. The pomegranate (*Punica granatum* L.) draft genome dissects genetic divergence between soft-and hard-seeded cultivars. *Plant Biotechnol. J.* **2020**, *18*, 955–968. [CrossRef]
5. Qin, G.; Xu, C.; Ming, R.; Tang, H.; Guyot, R.; Kramer, E.M.; Hu, Y.; Yi, X.; Qi, Y.; Xu, X. The pomegranate (*Punica granatum* L.) genome and the genomics of punicalagin biosynthesis. *Plant J.* **2017**, *91*, 1108–1128. [CrossRef] [PubMed]
6. Yuan, Z.; Fang, Y.; Zhang, T.; Fei, Z.; Han, F.; Liu, C.; Liu, M.; Xiao, W.; Zhang, W.; Wu, S.; et al. The pomegranate (*Punica granatum* L.) genome provides insights into fruit quality and ovule developmental biology. *Plant Biotechnol. J.* **2018**, *16*, 1363–1374. [CrossRef] [PubMed]
7. Sowjanya, P.R.; Shilpa, P.; Patil, G.P.; Babu, D.K.; Sharma, J.; Sangnure, V.R.; Mundewadikar, D.M.; Natarajan, P.; Marathe, A.R.; Reddy, U.K.; et al. Reference quality genome sequence of Indian pomegranate cv.'Bhagawa' (*Punica granatum* L.). *Front. Plant Sci.* **2022**, *13*, 947164. [CrossRef] [PubMed]
8. Finnegan, D.J. Eukaryotic transposable elements and genome evolution. *Trends Genet.* **1989**, *5*, 103–107. [CrossRef]
9. Mascagni, F.; Barghini, E.; Giordani, T.; Rieseberg, L.H.; Cavallini, A.; Natali, L. Repetitive DNA and plant domestication: Variation in copy number and proximity to genes of LTR-retrotransposons among wild and cultivated sunflower (*Helianthus annuus*) genotypes. *Genome Biol. Evol.* **2015**, *7*, 3368–3382. [CrossRef]
10. Wicker, T.; Gundlach, H.; Spannagl, M.; Uauy, C.; Borrill, P.; Ramirez-Gonzalez, R.H.; De Oliveira, R. Impact of transposable elements on genome structure and evolution in bread wheat. *Genome Biol.* **2018**, *19*, 103. [CrossRef]
11. Jiao, Y.; Peluso, P.; Shi, J.; Liang, T.; Stitzer, M.C.; Wang, B.; Campbell, M.S.; Stein, J.C.; Wei, X.; Chin, C.-S.; et al. Improved maize reference genome with single-molecule technologies. *Nature* **2017**, *546*, 524–527. [CrossRef]
12. Kumar, A.; Bennetzen, J.L. Plant retrotransposons. *Annu. Rev. Genet.* **1999**, *33*, 479–532. [CrossRef] [PubMed]
13. Gifford, R.J.; Blomberg, J.; Coffin, J.M.; Fan, H.; Heidmann, T.; Mayer, J.; Stoye, J.; Tristem, M.; Johnson, W.E. Nomenclature for endogenous retrovirus (ERV) loci. *Retrovirology* **2018**, *15*, 59. [CrossRef] [PubMed]
14. Sharma, A.; Presting, G.G. Centromeric retrotransposon lineages predate the maize/rice divergence and differ in abundance and activity. *Mol. Genet. Genom.* **2008**, *279*, 133–147. [CrossRef]
15. Gong, Z.; Wu, Y.; Koblížková, A.; Torres, G.A.; Wang, K.; Iovene, M.; Neumann, P.; Zhang, W.; Novák, P.; Buell, C.R.; et al. Repeatless and repeat-based centromeres in potato: Implications for centromere evolution. *Plant Cell* **2012**, *24*, 3559–3574. [CrossRef] [PubMed]
16. Su, H.; Liu, Y.; Liu, Y.-X.; Lv, Z.; Li, H.; Xie, S.; Gao, Z.; Pang, J.; Wang, X.-J.; Lai, J.; et al. Dynamic chromatin changes associated with de novo centromere formation in maize euchromatin. *Plant J.* **2016**, *88*, 854–866. [CrossRef] [PubMed]
17. Neumann, P.; Novák, P.; Hoštáková, N.; Macas, J. Systematic survey of plant LTR-retrotransposons elucidates phylogenetic relationships of their polyprotein domains and provides a reference for element classification. *Mob. DNA* **2019**, *10*, 1. [CrossRef]
18. Neumann, P.; Navrátilová, A.; Koblížková, A.; Kejnovský, E.; Hřibová, E.; Hobza, R.; Widmer, A.; Doležel, J.; Macas, J. Plant centromeric retrotransposons: A structural and cytogenetic perspective. *Mob. DNA* **2011**, *2*, 4. [CrossRef]
19. Muñoz-López, M.; García-Pérez, J.L. DNA transposons: Nature and applications in genomics. *Curr. Genom.* **2010**, *11*, 115–128. [CrossRef]
20. Morgante, M.; Brunner, S.; Pea, G.; Fengler, K.; Zuccolo, A.; Rafalski, A. Gene duplication and exon shuffling by helitron-like transposons generate intraspecies diversity in maize. *Nat. Genet.* **2005**, *37*, 997–1002. [CrossRef]
21. Wicker, T.; Sabot, F.; Hua-Van, A.; Bennetzen, J.L.; Capy, P.; Chalhoub, B.; Flavell, A.; Leroy, P.; Morgante, M.; Panaud, O.; et al. A unified classification system for eukaryotic transposable elements. *Nat. Rev. Genet.* **2007**, *8*, 973–982. [CrossRef]
22. Bourque, G.; Burns, K.H.; Gehring, M.; Gorbunova, V.; Seluanov, A.; Hammell, M.; Imbeault, M.; Izsvák, Z.; Levin, H.L.; Macfarlan, T.S.; et al. Ten things you should know about transposable elements. *Genome Biol.* **2018**, *19*, 199. [CrossRef] [PubMed]
23. Viviani, A.; Ventimiglia, M.; Fambrini, M.; Vangelisti, A.; Mascagni, F.; Pugliesi, C.; Usai, G. Impact of transposable elements on the evolution of complex living systems and their epigenetic control. *Biosystems* **2021**, *210*, 104566. [CrossRef] [PubMed]

24. Devos, K.M.; Brown, J.K.; Bennetzen, J.L. Genome size reduction through illegitimate recombination counteracts genome expansion in Arabidopsis. *Genome Res.* **2002**, *12*, 1075–1079. [CrossRef] [PubMed]
25. Vitte, C.; Panaud, O. Formation of solo-LTRs through unequal homologous recombination counterbalances amplifications of LTR retrotransposons in rice *Oryza sativa* L. *Mol. Biol. Evol.* **2003**, *20*, 528–540. [CrossRef] [PubMed]
26. Wang, Q.; Dooner, H.K. Remarkable variation in maize genome structure inferred from haplotype diversity at the bz locus. *Proc. Natl. Acad. Sci. USA* **2006**, *103*, 17644–17649. [CrossRef] [PubMed]
27. Oliver, K.R.; McComb, J.A.; Greene, W.K. Transposable elements: Powerful contributors to angiosperm evolution and diversity. *Genome Biol. Evol.* **2013**, *5*, 1886–1901. [CrossRef]
28. Lisch, D. How important are transposons for plant evolution? *Nat. Rev. Genet.* **2013**, *14*, 49–61. [CrossRef]
29. Pinosio, S.; Giacomello, S.; Faivre-Rampant, P.; Taylor, G.; Jorge, V.; Le Paslier, M.C.; Morgante, M. Characterization of the poplar pan-genome by genome-wide identification of structural variation. *Mol. Biol. Evol.* **2016**, *33*, 2706–2719. [CrossRef]
30. Ventimiglia, M.; Marturano, G.; Vangelisti, A.; Usai, G.; Simoni, S.; Cavallini, A.; Giordani, T.; Natali, L.; Zuccolo, A.; Mascagni, F. Genome wide identification and characterization of exapted transposable elements in the large genome of sunflower (*Helianthus annuus* L.). *Plant J.* **2023**, *113*, 734–748. [CrossRef]
31. Hirsch, C.D.; Springer, N.M. Transposable element influences on gene expression in plants. *Biochim. Biophys. Acta Gene Regul. Mech.* **2017**, *1860*, 157–165. [CrossRef]
32. Drongitis, D.; Aniello, F.; Fucci, L.; Donizetti, A. Roles of transposable elements in the different layers of gene expression regulation. *Int. J. Mol. Sci.* **2019**, *20*, 5755. [CrossRef] [PubMed]
33. Fambrini, M.; Usai, G.; Vangelisti, A.; Mascagni, F.; Pugliesi, C. The plastic genome: The impact of transposable elements on gene functionality and genomic structural variations. *Genesis* **2020**, *58*, e23399. [CrossRef] [PubMed]
34. Wang, L.; Huang, Y.; Liu, Z.; He, J.; Jiang, X.; He, F.; Lu, Z.; Yang, S.; Chen, P.; Yu, H.; et al. Somatic variations led to the selection of acidic and acidless orange cultivars. *Nat. Plants* **2021**, *7*, 954–965. [CrossRef] [PubMed]
35. Liu, H.N.; Pei, M.S.; Ampomah-Dwamena, C.; He, G.Q.; Wei, T.L.; Shi, Q.F.; Yu, Y.H.; Guo, D.L. Genome-wide characterization of long terminal repeat retrotransposons provides insights into trait evolution of four cucurbit species. *Funct. Integr. Genom.* **2023**, *23*, 218. [CrossRef] [PubMed]
36. Hollister, J.D.; Smith, L.M.; Guo, Y.L.; Ott, F.; Weigel, D.; Gaut, B.S. Transposable elements and small RNAs contribute to gene expression divergence between *Arabidopsis thaliana* and *Arabidopsis lyrata*. *Proc. Natl. Acad. Sci. USA* **2011**, *108*, 2322–2327. [CrossRef] [PubMed]
37. Dubin, M.J.; Scheid, O.M.; Becker, C. Transposons: A blessing curse. *Curr. Opin. Plant Biol.* **2018**, *42*, 23–29. [CrossRef]
38. Schrader, L.; Schmitz, J. The impact of transposable elements in adaptive evolution. *Mol. Ecol.* **2019**, *28*, 1537–1549. [CrossRef]
39. Vangelisti, A.; Simoni, S.; Usai, G.; Ventimiglia, M.; Natali, L.; Cavallini, A.; Mascagni, F.; Giordani, T. LTR-retrotransposon dynamics in common fig (*Ficus carica* L.) genome. *BMC Plant Biol.* **2021**, *21*, 221. [CrossRef]
40. Kobayashi, S.; Goto-Yamamoto, N.; Hirochika, H. Retrotransposon-induced mutations in grape skin color. *Science* **2004**, *304*, 982. [CrossRef]
41. Zhang, L.; Hu, J.; Han, X.; Li, J.; Gao, Y.; Richards, C.M.; Zhang, C.; Tian, Y.; Liu, G.; Gul, H.; et al. A high-quality apple genome assembly reveals the association of a retrotransposon and red fruit colour. *Nat. Commun.* **2019**, *10*, 1494. [CrossRef]
42. Ou, S.; Su, W.; Liao, Y.; Chougule, K.; Agda, J.R.; Hellinga, A.J.; Lugo, C.S.B.; Elliott, T.A.; Ware, D.; Peterson, T.; et al. Benchmarking transposable element annotation methods for creation of a streamlined, comprehensive pipeline. *Genome Biol.* **2019**, *20*, 275. [CrossRef] [PubMed]
43. Xu, Z.; Wang, H. LTR_FINDER: An efficient tool for the prediction of full-length LTR retrotransposons. *Nucleic Acids Res.* **2007**, *35*, W265–W268. [CrossRef] [PubMed]
44. Ellinghaus, D.; Kurtz, S.; Willhoeft, U. *LTRharvest*, an efficient and flexible software for de novo detection of LTR retrotransposons. *BMC Bioinform.* **2008**, *9*, 18. [CrossRef] [PubMed]
45. Ou, S.; Jiang, N. LTR_retriever: A highly accurate and sensitive program for identification of long terminal repeat retrotransposons. *Plant Physiol.* **2018**, *176*, 1410–1422. [CrossRef] [PubMed]
46. Shi, J.; Liang, C. Generic repeat finder: A high-sensitivity tool for genome-wide de novo repeat detection. *Plant Physiol.* **2019**, *180*, 1803–1815. [CrossRef]
47. Su, W.; Gu, X.; Peterson, T. TIR-Learner, a new ensemble method for TIR transposable element annotation, provides evidence for abundant new transposable elements in the maize genome. *Mol. Plant* **2019**, *12*, 447–460. [CrossRef]
48. Xiong, W.; He, L.; Lai, J.; Dooner, H.K.; Du, C. HelitronScanner uncovers a large overlooked cache of Helitron transposons in many plant genomes. *Proc. Natl. Acad. Sci. USA* **2014**, *111*, 10263–10268. [CrossRef]
49. Smit, A.F.A.; Hubley, R.; Green, P. RepeatMasker Open-4.0. 2013–2015. Available online: http://www.repeatmasker.org (accessed on 3 July 2023).
50. Quinlan, A.R.; Hall, I.M. BEDTools: A flexible suite of utilities for comparing genomic features. *Bioinformatics* **2010**, *26*, 841–842. [CrossRef]
51. Li, W.; Godzik, A. Cd-hit: A fast program for clustering and comparing large sets of protein or nucleotide sequences. *Bioinformatics* **2006**, *22*, 1658–1659. [CrossRef]
52. Suzuki, R.; Shimodaira, H. Pvclust: An R package for assessing the uncertainty in hierarchical clustering. *Bioinformatics* **2006**, *22*, 1540–1542. [CrossRef]

53. Wickham, H. Data analysis. In *ggplot2. Use R!* Springer: Cham, Switzerland, 2016. [CrossRef]
54. Altschul, S.F.; Gish, W.; Miller, W.; Myers, E.W.; Lipman, D.J. Basic local alignment search tool. *J. Mol. Biol.* **1990**, *215*, 403–410. [CrossRef] [PubMed]
55. Kimura, M.A. simple method for estimating evolutionary rates of base substitutions through comparative studies of nucleotide sequences. *J. Mol. Evol.* **1980**, *16*, 111–120. [CrossRef] [PubMed]
56. Rice, P.; Longden, I.; Bleasby, A. EMBOSS: The European molecular biology open software suite. *Trends Genet.* **2000**, *16*, 276–277. [CrossRef] [PubMed]
57. Mascagni, F.; Usai, G.; Natali, L.; Cavallini, A.; Giordani, T. A comparison of methods for LTR-retrotransposon insertion time profiling in the *Populus trichocarpa* genome. *Caryologia* **2018**, *71*, 85–92. [CrossRef]
58. SanMiguel, P.; Gaut, B.S.; Tikhonov, A.; Nakajima, Y.; Bennetzen, J.L. The paleontology of intergene retrotransposons of maize. *Nat. Genet.* **1998**, *20*, 43–45. [CrossRef] [PubMed]
59. Usai, G.; Mascagni, F.; Natali, L.; Giordani, T.; Cavallini, A. Comparative genome-wide analysis of repetitive DNA in the genus *Populus* L. *Tree Genet. Genomes* **2017**, *13*, 96. [CrossRef]
60. Conesa, A.; Götz, S.; García-Gómez, J.M.; Terol, J.; Talón, M.; Robles, M. Blast2GO: A universal tool for annotation, visualization and analysis in functional genomics research. *Bioinformatics* **2005**, *21*, 3674–3676. [CrossRef] [PubMed]
61. Kanehisa, M.; Goto, S. KEGG: Kyoto encyclopedia of genes and genomes. *Nucleic Acids Res.* **2000**, *28*, 27–30. [CrossRef]
62. Supek, F.; Bošnjak, M.; Škunca, N.; Šmuc, T. REVIGO summarizes and visualizes long lists of gene ontology terms. *PLoS ONE* **2011**, *6*, 7. [CrossRef]
63. Velasco, R.; Zharkikh, A.; Affourtit, J.; Dhingra, A.; Cestaro, A.; Kalyanaraman, A.; Fontana, P.; Bhatnagar, S.K.; Troggio, M.; Pruss, D.; et al. The genome of the domesticated apple (*Malus* × *domestica* Borkh.). *Nat. Genet.* **2010**, *42*, 833–839. [CrossRef]
64. Wu, J.; Wang, Z.; Shi, Z.; Zhang, S.; Ming, R.; Zhu, S.; Khan, M.A.; Tao, S.; Korban, S.S.; Wang, H.; et al. The genome of the pear (*Pyrus bretschneideri* Rehd.). *Genome Res.* **2013**, *23*, 396–408. [CrossRef] [PubMed]
65. Usai, G.; Mascagni, F.; Giordani, T.; Vangelisti, A.; Bosi, E.; Zuccolo, A.; Ceccarelli, M.; King, R.; Hassani-Pak, K.; Liceth, S.Z.; et al. Epigenetic patterns within the haplotype phased fig (*Ficus carica* L.) genome. *Plant J.* **2020**, *102*, 600–614. [CrossRef] [PubMed]
66. Brůna, T.; Aryal, R.; Dudchenko, O.; Sargent, D.J.; Mead, D.; Buti, M.; Cavallini, A.; Hytönen, T.; Andrés, J.; Pham, M.; et al. A chromosome-length genome assembly and annotation of blackberry (*Rubus argutus*, cv. "Hillquist"). *G3* **2023**, *13*, jkac289. [CrossRef] [PubMed]
67. Neumann, P.; Koblížková, A.; Navrátilová, A.; Macas, J. Significant expansion of *Vicia pannonica* genome size mediated by amplification of a single type of giant retroelement. *Genetics* **2006**, *173*, 1047–1056. [CrossRef] [PubMed]
68. Christelová, P.; Valárik, M.; Hřibová, E.; De Langhe, E.; Doležel, J. A multi gene sequence-based phylogeny of the Musaceae (banana) family. *BMC Evol. Biol.* **2011**, *11*, 130. [CrossRef]
69. Tenaillon, M.I.; Hufford, M.B.; Gaut, B.S.; Ross-Ibarra, J. Genome size and transposable element content as determined by high-throughput sequencing in maize and *Zea luxurians*. *Genome Biol. Evol.* **2011**, *3*, 219–229. [CrossRef]
70. Vitte, C.; Fustier, M.A.; Alix, K.; Tenaillon, M.I. The bright side of transposons in crop evolution. *Brief. Funct. Genom.* **2014**, *13*, 276–295. [CrossRef]
71. Kreplak, J.; Madoui, M.A.; Cápal, P.; Novák, P.; Labadie, K.; Aubert, G.; Burstin, J. A reference genome for pea provides insight into legume genome evolution. *Nat. Genet.* **2019**, *51*, 1411–1422. [CrossRef]
72. Macas, J.; Novák, P.; Pellicer, J.; Čížková, J.; Koblížková, A.; Neumann, P.; Leitch, I.J. In depth characterization of repetitive DNA in 23 plant genomes reveals sources of genome size variation in the legume tribe Fabeae. *PLoS ONE* **2015**, *10*, e0143424. [CrossRef]
73. Simoni, S.; Clemente, C.; Usai, G.; Vangelisti, A.; Natali, L.; Tavarini, S.; Angelini, C.L.; Cavallini, A.; Mascagni, F.; Giordani, T. Characterisation of LTR-retrotransposons of *Stevia rebaudiana* and their use for the analysis of genetic variability. *Int. J. Mol. Sci.* **2022**, *23*, 6220. [CrossRef]
74. He, G.Q.; Jin, H.Y.; Cheng, Y.Z.; Yu, Y.H.; Guo, D.L. Characterization of genome-wide long terminal repeat retrotransposons provide insights into trait evolution of four grapevine species. *J. Syst. Evol.* **2023**, *61*, 414–427. [CrossRef]
75. Usai, G.; Mascagni, F.; Vangelisti, A.; Giordani, T.; Ceccarelli, M.; Cavallini, A.; Natali, L. Interspecific hybridisation and LTR-retrotransposon mobilisation-related structural variation in plants: A case study. *Genomics* **2020**, *112*, 1611–1621. [CrossRef] [PubMed]
76. Zhang, Q.J.; Gao, L.Z. Rapid and recent evolution of LTR retrotransposons drives rice genome evolution during the speciation of AA-genome *Oryza* species. *G3* **2017**, *7*, 1875–1885. [CrossRef] [PubMed]
77. Arnaud, P.; Goubely, C.; Pelissier, T.; Deragon, J.M. SINE retroposons can be used in vivo as nucleation centers for de novo methylation. *Mol. Cell Biol.* **2000**, *20*, 3434–3441. [CrossRef] [PubMed]
78. Gil, M.I.; Tomás-Barberán, F.A.; Hess-Pierce, B.; Holcroft, D.M.; Kader, A.A. Antioxidant activity of pomegranate juice and its relationship with phenolic composition and processing. *J. Agric. Food Chem.* **2000**, *48*, 4581–4589. [CrossRef] [PubMed]
79. Longtin, R. The pomegranate: Nature's power fruit? *J. Natl. Cancer Inst.* **2003**, *95*, 346–348. [CrossRef] [PubMed]
80. Tzulker, R.; Glazer, I.; Bar-Ilan, I.; Holland, D.; Aviram, M.; Amir, R. Antioxidant activity, polyphenol content, and related compounds in different fruit juices and homogenates prepared from 29 different pomegranate accessions. *J. Agric. Food Chem.* **2007**, *55*, 9559–9570. [CrossRef] [PubMed]
81. Sreekumar, S.; Sithul, H.; Muraleedharan, P.; Azeez, J.M.; Sreeharshan, S. Pomegranate fruit as a rich source of biologically active compounds. *Biomed. Res. Int.* **2014**, *2014*, 686921. [CrossRef]

82. Díaz-Mula, H.M.; Tomás-Barberán, F.A.; García-Villalba, R. Pomegranate fruit and juice (cv. Mollar), rich in ellagitannins and anthocyanins, also provide a significant content of a wide range of proanthocyanidins. *J. Agric. Food Chem.* **2019**, *67*, 9160–9167. [CrossRef]
83. Horbowicz, M.; Kosson, R.; Grzesiuk, A.; Dębski, H. Anthocyanins of fruits and vegetables-their occurrence, analysis and role in human nutrition. *J. Fruit Ornam. Plant Res.* **2008**, *68*, 5–22. [CrossRef]
84. Zhao, J.; Qi, X.; Li, J.; Cao, Z.; Liu, X.; Yu, Q.; Qin, G. Metabolic Profiles of Pomegranate Juices during Fruit Development and the Redirection of Flavonoid Metabolism. *Horticulturae* **2023**, *9*, 881. [CrossRef]
85. Zhu, F.; Yuan, Z.; Zhao, X.; Yin, Y.; Feng, L. Composition and contents of anthocyanins in different pomegranate cultivars. *Acta Hortic* **2015**, *1089*, 35–41.
86. Lavhale, S.G.; Kalunke, R.M.; Giri, A.P. Structural, functional and evolutionary diversity of 4-coumarate-CoA ligase in plants. *Planta* **2018**, *248*, 1063–1078. [CrossRef] [PubMed]
87. Zhao, L.; Jiang, X.L.; Qian, Y.M.; Wang, P.Q.; Xie, D.Y.; Gao, L.P.; Xia, T. Metabolic characterization of the anthocyanidin reductase pathway involved in the biosynthesis of Flavan-3-ols in elite Shuchazao Tea (*Camellia sinensis*) cultivar in the field. *Molecules* **2017**, *22*, 2241. [CrossRef]
88. Wisman, E.; Hartmann, U.; Sagasser, M.; Baumann, E.; Palme, K.; Hahlbrock, K.; Saedler, H.; Weisshaar, B. Knock-out mutants from an En-1 mutagenized *Arabidopsis thaliana* population generate phenylpropanoid biosynthesis phenotypes. *Proc. Natl. Acad. Sci. USA* **1998**, *95*, 12432–12437. [CrossRef]

Disclaimer/Publisher's Note: The statements, opinions and data contained in all publications are solely those of the individual author(s) and contributor(s) and not of MDPI and/or the editor(s). MDPI and/or the editor(s) disclaim responsibility for any injury to people or property resulting from any ideas, methods, instructions or products referred to in the content.

horticulturae

MDPI

Article

Cytogenetics Study of Four Edible and Ornamental *Zingiber* Species (Zingiberaceae) from Thailand

Piyaporn Saensouk [1,2], Surapon Saensouk [2,3,*], Rattanavalee Senavongse [4], Duangkamol Maensiri [5] and Phetlasy Souladeth [6]

1 Department of Biology, Faculty of Science, Mahasarakham University, Kantarawichai, Maha Sarakham 44150, Thailand; pcornukaempferia@yahoo.com
2 Diversity of Family Zingibeaceae and Vascular Plant of Its Applications Research Unit, Mahasarakham University, Kantarawichai District, Maha Sarakham 44150, Thailand
3 Walai Rukhavej Botanical Research Institute, Mahasarakham University, Kantarawichai District, Maha Sarakham 44150, Thailand
4 School of Crop Production Technology, Institute of Agricultural Technology, Suranaree University of Technology, Nakhon Ratchasima 30000, Thailand; araceaefamily@gmail.com
5 School of Biology, Institute of Science, Suranaree University of Technology, Nakhon Ratchasima 30000, Thailand; duangkamol@sut.ac.th
6 Faculty of Forest Science, National University of Laos, Vientiane P.O. Box 7322, Laos; p.souladeth@nuol.edu.la
* Correspondence: surapon.s@msu.ac.th

Abstract: A cytological study was carried out on four *Zingiber* species from Thailand, namely, *Z. chrysostachys*, *Z. isanense*, *Z. junceum*, and *Z. niveum*, which are edible and beautiful ornamental plants. They all have somatic chromosomal numbers of $2n = 22$. This research contributes to karyological knowledge regarding this species. The somatic chromosomal counts of *Z. niveum* and *Z. isanense* are reported for the first time, as are the NFs of all species, which were all discovered to be 44. All four edible and ornamental species had their karyotypes: 16m + 6sm for *Z. chrysostachys*, 4m + 18sm for *Z. isanense*, 12m + 10sm for *Z. junceum*, and 14m + 4sm + 4st for *Z. niveum*. The dominant characteristics of these four *Zingiber* species are as follows: *Z. chrysostachys* has yellow bracts, pale yellow flowers, and a red labellum with white dots; *Z. isanensis* has red-brown bracts, white flowers, and a white labellum; *Z. junceum* has green bracts, yellow flowers, and a yellow labellum; and *Z. niveum* has white bracts, yellow flowers, and a yellow labellum. Additionally, principal component analysis (PCA) of the karyotype formula was used to divide the four *Zingiber* species into two groups via various points using the chromosome indexes (CIs): *Z. niveum* (D) with *Z. chrysostachys* (A), and *Z. junceum* (C) with *Z. isanensis* (B). This finding implies that, while being in the same stage, the CIs of these four *Zingiber* species can be used to distinguish them, revealing their resemblance at unique stages and close relationship. Accordingly, the chromosomal structure, karyotype formulae, and CIs can be used to distinguish these four edibles and ornamental *Zingiber* species from Thailand.

Keywords: cytogenetics; chromosome number; karyotype; *Zingiber*; Zingiberaceae

1. Introduction

The Zingiberaceae family, also referred to as the ginger family, consists of a vast range of flowering plants that are found across tropical and subtropical regions globally, covering Africa, the Americas, and Asia. There are around 57 recognized groups and more than 1900 distinct types of species [1]. Zingiberaceae plants exhibit adaptability to several ecological circumstances, although they are primarily distributed in tropical rainforests and damp habitats. They flourish in tropical and subtropical environments characterized by high temperatures and moisture, where they can benefit from ample precipitation and adequate shelter. However, certain species can also be found in arid regions exposed to direct sunshine. This family consists of both terrestrial and epiphytic species. The Zingiberaceae family is well-known for its exceptional biodiversity, encompassing a variety of plants that are both economically useful and of great importance [2]. These plants exhibit a diverse array of colors, forms, and sizes. Furthermore, they have been employed for diverse applications, and the understanding of their usage has also been transmitted through generations in human societies, specifically in Southeast Asia [1,2]. Thailand has an extensive variety of Zingiberaceae species, making it one of the countries with the highest abundance and diversity of this plant family. Thailand harbors around 29 recognized groups and more than 400 species of Zingiberaceae [1–17]. Thailand's tropical climate and diversified biosphere provide ideal conditions for the flourishing and spread of Zingiberaceae species. Plants belonging to the ginger family have diverse applications, serving as food and spice sources, medicinal plants, decorative flowers, and cultural symbols. They are involved in traditions, such as offering flowers to monks, and tourism, like the Krachiew flower fields in Chaiyaphum province. Additionally, they are used in the production of cosmetics, dyes, and essential oil extracts and in spas [2]. Thailand is highly significant for understanding the wide variety and range of evolutionary tendencies within this intriguing plant family [6].

The large genus *Zingiber* belongs to the Zingiberaceae family [2]. *Zingiber* has been found in China, India, Australia, Southeast Asia, Papua New Guinea, and China [17–19]. The medicinal herbs *Zingiber officinale* (or ginger), *Z. montanum*, and *Z. zerumbet*, among others, are important [16]. There are a total of 56 Thai *Zingiber* species, consistent with 60 taxa [17]. *Zingiber*, as a result, belongs to one of the most diverse genera in Thailand [17]. There are many uses for the plants in this genus. Many *Zingiber* species' rhizomes are often utilized in Thai food as sauces or other components. Other species are also used as food, medicinal plants, ornamental plants, commercial plants, and ritual plants [6]. *Zingiber corallinum*, *Z. myoga*, *Z. officinale*, *Z. striolatum*, and *Z. zerumbet* have been the focus of studies on biological and pharmacological activity, but a vast number of phytochemical constituents have since been discovered that are thought to represent genus features [18], suggesting that many more species in this genus should possess the same application potential.

The study of the chromosomes of plants in the ginger family mostly involves studying their chromosome numbers [20–26]. As for the study of karyotypes and ideograms, it was found that there are very few studies [23–25]. This may be due to the small size of chromosomes in the plant family, and Evaluating the distribution of chromosomes is a difficult assignment. This could explain the low number of studies conducted on the chromosomes of plants within the ginger family [13–15]. Moreover, it has been found that there are very few chromosome studies on the *Zingiber* genus [26–30].

Notwithstanding, previous cytogenetic studies on *Zingiber* showed that *Zingiber chrysostachys* [22], *Z. corallinum* [11], *Z. junceum* [23], *Z. ligulatum* [25], *Z. mekongense* [20], *Z. mioga* [28], *Z. macrostachyum* [25], *Z. montanum* [10,29], *Z. officinale* [15,25,29], *Z. ottensii* [22,29], *Z. parishii* subsp. *phuphanense* [25], *Z. purpureum* [22,25], *Z. roseum* [25], *Z. rubens* [23], *Z. wightianum* [25], *Z.* aff. *wrayi* [15], *Z. zerumbet* [22,23,25], have a chromosomal count of $2n = 22$.

Additionally, karyotype studies have reported the formulae of a few *Zingiber* species, i.e., *Zingiber montanum* (2a + 18m + 2sm) [10], *Zingiber officinale* [31], and *Z. ligulatum* (10m + 12sm), as well as *Z. parishii* subsp. *phuphanense* (16m + 6sm) [25]. *Zingiber* genus ideogram investigations have been limited to three species: *Zingiber ligulatum* [25], *Z. montanum* [10], and *Z. parishii* subsp. *phuphanense* [25].

Research team discovered four species of *Zingiber*, namely, *Z. chrysostachys*, *Z. isanensis*, *Z. junceum*, and *Z. niveum*, while collecting samples from the Zingiberaceae family in the northeastern part of Thailand. Additionally, villagers utilized these plants in a variety of ways; for example, young inflorescences, leaves, and rhizomes can all be eaten as vegetables or used in local food, including as ingredients in local soups and salads. The rhizomes, pseudostems, and young inflorescences of the four *Zingiber* species mentioned above were traded at the local market. Furthermore, villagers cultivated them in their gardens for ornamental use, as the inflorescences and pseudostems of all the plants are attractive [2,6,16]. Therefore, each of the four *Zingiber* plants is widely sought after for both food and ornamental uses and can be readily purchased in the regions where they are native. These forest plants were discovered to be extensively cultivated and propagated within the community for the purpose of harvesting flowers and rhizomes, which are then sold in the local market or sent to other provinces. These four species of *Zingiber* plants are economically significant at the community level in Northeastern Thailand and have been integral to the local people's way of life since ancient times [16,17,23].

Moreover, they are consumed locally. *Zingiber junceum* and *Z. isanense* are listed as least-concern (LC) species on the IUCN Red List [19]; however, *Zingiber chrysostachys* and *Z. niveum* are listed as endangered (EN). All four species are classified as rare in Thailand according to the World Checklist of Selected Plant Families [1]. Researchers are interested in researching the chromosomal information of these plants due to the aforementioned reasons. The purpose of this research is to provide fundamental data for the subsequent advancement of these plants as widely consumed indigenous food crops. The goal of this research is to advocate for the cultivation of ornamental plants and flowers as profitable agricultural commodities. Hence, knowledge of chromosome information holds great significance for future advancements.

Several experts have recognized the significance of chromosomal information in plant systematics and evolution. Chromosome morphology can help us comprehend taxonomic relationships at the general and subgenomic levels. Chromosomal morphologies and structures serve as the basis for taxonomy [20,21]. Previous investigations have clarified the chromosomal structures of very few *Zingiber* species [22–24], although there is still a lack of knowledge of karyology. The chromosomal numbers of these four edibles and ornamental *Zingiber* species, namely, *Z. chrysostachys*, *Z. isanensis*, *Z. junceum*, and *Z. niveum*, have been recorded in a few publications. However, their karyological aspects have never been studied before. As a result, the purpose of this study was to provide more information regarding their chromosome numbers, the Fundamental Number (NF: number of chromosome arms), and karyotype forms.

2. Materials and Methods

2.1. Sample Collection

Zingiber chrysostachys (coll. No. Saensouk 3900), *Z. niveum* (coll. No. Saensouk 3901), *Z. isanense* (coll. No. Saensouk 3902), and *Z. junceum* (coll. No. Saensouk 3903) were obtained from natural settings in different provinces in the northeastern part of Thailand. *Zingiber chrysostachys* was discovered in a villager's garden (a deciduous dipterocarp forest) in Nakhon Phanom Province, which is located in the upper region of the northeastern part from Thailand. *Z. niveum* and *Z. isanense* were collected from a deciduous dipterocarp forest (a villager's garden) in Ubon Ratchathani Province, which is located in the lower region of Northeast Thailand. *Z. junceum* was obtained from a villager's garden (which was like a deciduous dipterocarp forest) in Beung Kan Province, located in the upper region of Northeast Thailand. They were kept alive by being grown in a nursery at Mahasarakham University in Thailand's Maha Sarakham Province. Figure 1 depicts the blooms of the species investigated in this study.

2.2. Chromosome Number and Karyotype Study

Root tips from four Zingiber species, which are both edible and ornamental, were prepared with paradichlorobenzene at 4 °C for 6 h before being fixed in ethanol-acetic acid (3:1, *v:v*) at room temperature for 30 min. The roots were kept at 4 °C in case they needed to be used later. The samples were rinsed in distilled water, hydrolyzed in 1 M of HCl for 5 min at 60 °C, and then washed again in distilled water. They were dyed with 2% aceto-orcein and examined using the squash technique. Observations were made using a light microscope (Zeiss Axiostar Plus: Carl Zeiss Light Microscopy, Göttingen, Germany) operated at 400× magnification [25,26]. The metaphase plates with well-individualized chromosomes were imaged. The chromosomes were counted, and the karyotype formulas were determined from measurements of the metaphase chromosomes apparent in photomicrographs. The chromosome morphology was described using the terminology used [22,24–35].

Figure 1. Flowers of edible and ornamental *Zingiber* species in this study: (**a**) *Zingiber chrysostachys*, (**b**) *Z. isanense*, (**c**) *Z. junceum*, and (**d**) *Z. niveum*. Scale bars = 1 cm.

2.3. Statistical Analysis

The average karyotype of all four *Zingiber* species, which includes measurements of the short arm chromosome, long arm chromosome, total chromosome arms, relative length, and centromeric index, was used as a representation of all the results. The average was calculated by taking into account all the variables, along with one standard deviation (SD). We used calculations, following the methods defined in references [28–31], to evaluate the similarity among the four *Zingiber* species. Our study focused on the karyotype formula data and the presence or absence of distinguishing features related to somatic chromosomes. We took into account a somatic chromosomal number of $2n = 22$. In addition, we assigned a binary value of 1 or 0 to indicate the morphological traits of the four *Zingiber* species that were analyzed. A distance matrix was used in the dendrogram (UPGMA) (unweighted pair group method with arithmetic mean) [32–34], and principal component analysis (PCA) [22,34–39] was performed using SPSS version 29.

3. Results

The *Zingiber* (family: Zingiberaceae) species *Z. chrysostachys*, *Z. isanense*, *Z. junceum*, and *Z. niveum*, obtained from several provinces in Thailand for this study (Figure 1), were identified as rare species in Thailand. Two species are considered endangered (*Zingiber chrysostachys* and *Z. niveum*), while two are considered least-threatened (*Zingiber isanense* and *Z. junceum*). Figure 2 shows photomicrographs of the chromosomes from the four species. Table 1 summarizes the somatic chromosomal counts corresponding to the root tips of four rare *Zingiber* species. For each species, chromosomal counts were performed on ten metaphase plates. All the species' chromosomal counts were discovered to be $2n = 22$. Figures 3 and 4 show the species' karyotypes and ideograms, respectively.

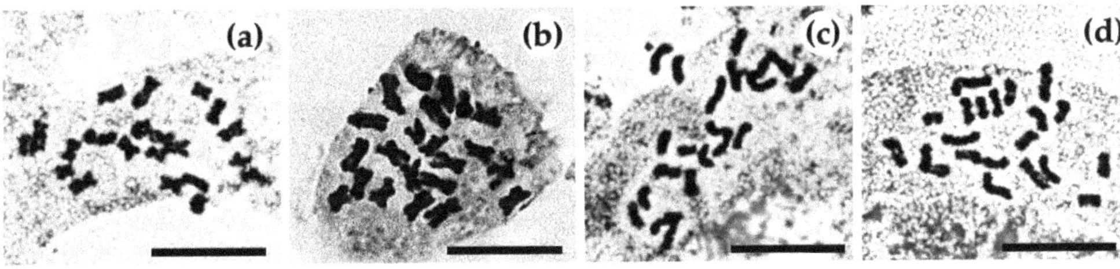

Figure 2. Photomicrographs of somatic metaphase plate $2n = 22$: (**a**) *Zingiber chrysostachys*, (**b**) *Z. isanense*, (**c**) *Z. junceum*, and (**d**) *Z. niveum*. Scale bars = 10 μm.

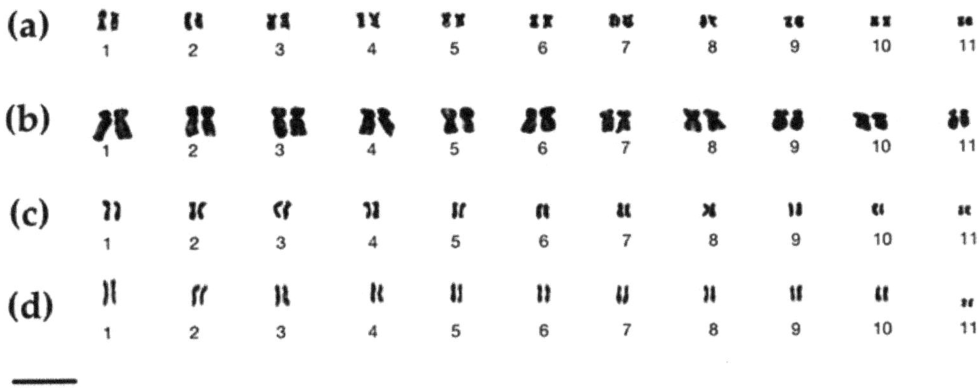

Figure 3. Karyotypes of (**a**) *Zingiber chrysostachys*, (**b**) *Z. isanense*, (**c**) *Z. junceum*, and (**d**) *Z. niveum*. Scale bars = 10 μm.

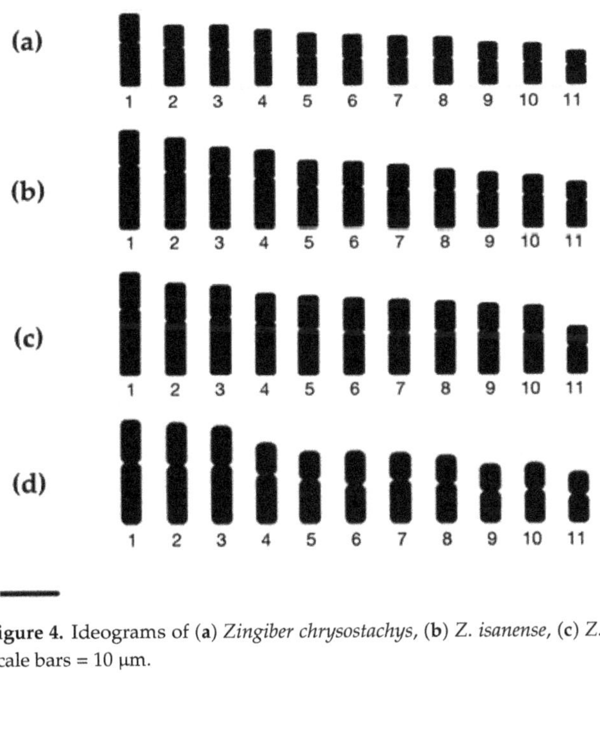

Figure 4. Ideograms of (**a**) *Zingiber chrysostachys*, (**b**) *Z. isanense*, (**c**) *Z. junceum*, and (**d**) *Z. niveum*. Scale bars = 10 μm.

Table 1. The cytogenetics with conservation status, dominant characteristics, and traditional uses of the four species in the genus *Zingiber* analyzed in the present study.

Species	*n*	2*n*	NF	Karyotype Formula (Symmetry)	Conservation Status (Based on WCSP/IUCN 2022)	Dominant Characteristics	Traditional Uses	Location (Provinces)	References
Zingiber chrysostachys	-	22	44 *	16m + 6sm * (Symmetry karyotype)	Rare species/ Endangered	yellow bracts, pale yellow flowers, and a red labellum with white dots	- Young inflorescences and young leaves are used as vegetables or local food. - Ornamental plants	Thailand (Nakhon Phanom)	Present study
	11	-	-	-	-			Malaysia	[22]
	-	22	-	-	-			Thailand (Sakon Nakhon)	[23]
Z. isanense	-	22 *	44 *	2m + 20sm * (Symmetry karyotype)	Rare species/ Least Concern	red-brown bracts, white flowers, and a white labellum	- Young leaves are used as vegetables or local food. - Ornamental plants	Thailand (Ubon ratchathani)	Present study *
Z. junceum	-	22	44 *	12m + 10sm * (Symmetry karyotype)	Rare species/ Least Concern	green bracts, yellow flowers, and a yellow labellum	- Young inflorescences, young leaves, and young rhizomes are used as vegetables or local food. - Ornamental plants	Thailand (Bueng Kan)	Present study
	-	22	-	-	-			Thailand (Sakon Nakhon)	[23]
Z. niveum	-	22 *	44 *	14m + 4sm + 4st * (Asymmetry karyotype)	Rare species/ Endangered	white bracts, yellow flowers, and a yellow labellum	- Young inflorescences, young leaves, and young rhizomes are used as vegetables or local food. - Ornamental plants	Thailand (Ubon ratchathani)	Present study *

* = First time report.

The most important distinguishing characteristics of these four *Zingiber* species, which can be used for both food and ornamentation, are as follows: *Z. chrysostachys* has yellow bracts, pale yellow flowers, and a red labellum adorned with white dots. *Z. isanensis* has red-brown bracts, white flowers, and a white labellum. *Z. junceum* features green bracts, yellow flowers, and a yellow labellum. Lastly, *Z. niveum* presents white bracts, yellow flowers, and a yellow labellum.

In the field survey, we recorded the conventional utilization of all four edible and ornamental *Zingiber* species. The young inflorescences and young leaves of all species are utilized as edible vegetables or local foods. Additionally, young rhizomes of two specific species (*Z. junceum* and *Z. niveum*) were eaten as fresh vegetables or as part of the local cuisine. All of these species are utilized for ornamental purposes, namely, as plants for one's home.

These four species, which can be eaten or used for ornamentation, have a highly beneficial arrangement of chromosomes in each cell. Consequently, in this study, specific counting of chromosomes could be conducted by using intact plant root parts and employing treatments such as chromosomal identifying and technological addition, resulting in good karyotype and ideogram data.

Zingiber chrysostachys has a somatic chromosomal number of $2n = 22$ and a fundamental number (NF) of 44 (Figure 2a) (Table 1). The karyotype of *Zingiber chrysostachys* contained eight pairs of metacentric chromosomes and three pairs of submetacentric chromosomes. The karyotype formula was 16m + 6sm, indicating that the karyotype was symmetrical (Table 1, Figure 3a). Because of the arm ratio, it had a symmetrical karyotype. The short arm length was 0.68 ± 0.04 to 1.40 ± 0.09 μm, the long arm length was 0.98 ± 0.06 to 2.17 ± 0.65 μm, and the total chromosomal length was 1.66 ± 0.11 to 3.57 ± 0.73 μm. The centromeric index (CI) ranged from 0.54 ± 0.06 to 0.64 ± 0.10, and the relative chromosomal length (RL) ranged from 5.91 ± 0.04 to 12.72 ± 0.04% (Table 2). The CI and the average length of chromosomes were used to arrange the chromosomes into an ideogram with size decreasing from left to right (Figure 4a).

Table 2. Mean lengths of short-arm chromosomes (Ls), long-arm chromosomes (Ll), total chromosome arm (LT), relative length (RL), and centromeric index (CI) of four *Zingiber* species obtained from 10 metaphase plates.

			***Zingiber chrysostachys* (2*n* = 22)**			
Chro. Pair	**Ls ± SD (μm)**	**Ll ± SD (μm)**	**LT ± SD (μm)**	**RL (%)**	**CI**	**Chromosome Type**
1.	1.40 ± 0.09	2.17 ± 0.65	3.57 ± 0.73	12.72 ± 0.04	0.61 ± 0.08	Submetacentric
2.	1.09 ± 0.07	1.92 ± 0.13	3.02 ± 0.21	10.75 ± 0.06	0.64 ± 0.10	Submetacentric
3.	1.20 ± 0.08	1.81 ± 0.12	3.01 ± 0.20	10.74 ± 0.08	0.60 ± 0.06	Submetacentric
4.	1.18 ± 0.07	1.61 ± 0.12	2.78 ± 0.18	9.92 ± 0.06	0.58 ± 0.04	Metacentric
5.	1.10 ± 0.06	1.49 ± 0.10	2.59 ± 0.16	9.24 ± 0.04	0.57 ± 0.08	Metacentric
6.	1.15 ± 0.07	1.36 ± 0.09	2.51 ± 0.16	8.94 ± 0.08	0.54 ± 0.06	Metacentric
7.	1.03 ± 0.06	1.40 ± 0.10	2.43 ± 0.16	8.68 ± 0.08	0.58 ± 0.04	Metacentric
8.	1.00 ± 0.06	1.36 ± 0.08	2.36 ± 0.14	8.41 ± 0.04	0.58 ± 0.04	Metacentric
9.	0.91 ± 0.06	1.18 ± 0.08	2.08 ± 0.14	7.43 ± 0.04	0.56 ± 0.06	Metacentric
10.	0.91 ± 0.06	1.13 ± 0.08	2.04 ± 0.13	7.27 ± 0.08	0.56 ± 0.06	Metacentric
11.	0.68 ± 0.04	0.98 ± 0.06	1.66 ± 0.11	5.91 ± 0.04	0.59 ± 0.08	Metacentric
			***Zingiber isanensis* (2*n* = 22)**			
Chro. Pair	**Ls ± SD (μm)**	**Ll ± SD (μm)**	**LT ± SD (μm)**	**RL (%)**	**CI**	**Chromosome Type**
1.	1.37 ± 0.05	2.45 ± 0.01	3.83 ± 0.06	13.00 ± 0.02	0.64 ± 0.01	Submetacentric
2.	1.29 ± 0.02	2.22 ± 0.01	3.51 ± 0.02	12.00 ± 0.01	0.63 ± 0.01	Submetacentric
3.	1.12 ± 0.02	2.03 ± 0.01	3.15 ± 0.03	11.00 ± 0.01	0.64 ± 0.01	Submetacentric
4.	1.07 ± 0.03	1.96 ± 0.01	3.03 ± 0.04	10.00 ± 0.01	0.65 ± 0.01	Submetacentric
5.	1.04 ± 0.03	1.58 ± 0.01	2.62 ± 0.03	9.00 ± 0.01	0.59 ± 0.01	Metacentric
6.	0.98 ± 0.01	1.54 ± 0.01	2.52 ± 0.01	9.00 ± 0.01	0.61 ± 0.01	Submetacentric
7.	0.94 ± 0.02	1.50 ± 0.01	2.44 ± 0.02	8.00 ± 0.01	0.61 ± 0.01	Submetacentric

Table 2. *Cont.*

Zingiber isanensis (2n = 22)						
Chro. Pair	**Ls ± SD (µm)**	**Ll ± SD (µm)**	**LT ± SD (µm)**	**RL (%)**	**CI**	**Chromosome Type**
8.	0.83 ± 0.01	1.44 ± 0.01	2.26 ± 0.01	8.00 ± 0.01	0.64 ± 0.01	Metacentric
9.	0.77 ± 0.01	1.39 ± 0.01	2.16 ± 0.01	7.00 ± 0.01	0.64 ± 0.01	Submetacentric
10.	0.75 ± 0.01	1.28 ± 0.01	2.04 ± 0.01	7.00 ± 0.01	0.63 ± 0.01	Submetacentric
11.	0.68 ± 0.01	1.08 ± 0.01	1.76 ± 0.01	6.00 ± 0.01	0.61 ± 0.01	Submetacentric
Zingiber junceum (2n = 22)						
Chro. Pair	**Ls ± SD (µm)**	**Ll ± SD (µm)**	**LT ± SD (µm)**	**RL (%)**	**CI**	**Chromosome Type**
1.	1.24 ± 0.08	2.22 ± 0.65	3.46 ± 0.73	12.00 ± 0.16	0.64 ± 0.13	Submetacentric
2.	1.27 ± 0.07	1.83 ± 0.13	3.10 ± 0.20	10.76 ± 0.15	0.59 ± 0.12	Metacentric
3.	1.06 ± 0.07	1.95 ± 0.13	3.01 ± 0.19	10.45 ± 0.13	0.65 ± 0.12	Submetacentric
4.	1.00 ± 0.06	1.76 ± 0.12	2.76 ± 0.18	9.58 ± 0.14	0.64 ± 0.13	Submetacentric
5.	1.12 ± 0.06	1.54 ± 0.10	2.66 ± 0.16	9.21 ± 0.13	0.58 ± 0.12	Metacentric
6.	1.14 ± 0.07	1.44 ± 0.09	2.58 ± 0.16	8.95 ± 0.12	0.56 ± 0.12	Metacentric
7.	1.13 ± 0.06	1.40 ± 0.10	2.54 ± 0.16	8.80 ± 0.12	0.55 ± 0.10	Metacentric
8.	1.00 ± 0.06	1.45 ± 0.09	2.46 ± 0.14	8.51 ± 0.10	0.59 ± 0.10	Metacentric
9.	0.93 ± 0.05	1.45 ± 0.09	2.38 ± 0.14	8.26 ± 0.10	0.61 ± 0.09	Submetacentric
10.	0.89 ± 0.05	1.39 ± 0.08	2.28 ± 0.13	7.92 ± 0.09	0.61 ± 0.09	Submetacentric
11.	0.66 ± 0.04	0.95 ± 0.06	1.61 ± 0.10	5.57 ± 0.08	0.59 ± 0.10	Metacentric
Zingiber niveum (2n = 22)						
Chro. Pair	**Ls ± SD (µm)**	**Ll ± SD (µm)**	**LT ± SD (µm)**	**RL (%)**	**CI**	**Chromosome Type**
1.	1.65 ± 0.10	2.33 ± 0.68	3.98 ± 0.78	12.50 ± 0.20	0.58 ± 0.10	Metacentric
2.	1.61 ± 0.09	2.24 ± 0.15	3.85 ± 0.24	12.08 ± 0.12	0.58 ± 0.10	Metacentric
3.	1.47 ± 0.09	2.25 ± 0.14	3.72 ± 0.23	11.66 ± 0.09	0.61 ± 0.12	Subtalocentric
4.	1.19 ± 0.08	1.88 ± 0.13	3.07 ± 0.21	9.63 ± 0.12	0.61 ± 0.12	Subtalocentric
5.	1.09 ± 0.07	1.67 ± 0.11	2.76 ± 0.18	8.68 ± 0.09	0.60 ± 0.10	Submetacentric
6.	1.29 ± 0.08	1.47 ± 0.10	2.76 ± 0.18	8.66 ± 0.10	0.53 ± 0.09	Metacentric
7.	1.05 ± 0.07	1.63 ± 0.11	2.68 ± 0.18	8.42 ± 0.09	0.61 ± 0.12	Submetacentric
8.	1.06 ± 0.06	1.51 ± 0.09	2.56 ± 0.16	8.05 ± 0.08	0.59 ± 0.09	Metacentric
9.	1.03 ± 0.07	1.22 ± 0.09	2.25 ± 0.15	7.06 ± 0.06	0.54 ± 0.08	Metacentric
10.	1.02 ± 0.06	1.26 ± 0.08	2.28 ± 0.15	7.16 ± 0.09	0.55 ± 0.09	Metacentric
11.	0.86 ± 0.06	1.09 ± 0.07	1.94 ± 0.13	6.10 ± 0.12	0.56 ± 0.010	Metacentric

Zingiber isanense has $2n = 22$ somatic chromosomes and an NF = 44 (Figure 2b, Table 1). There were two pairs of metacentric chromosomes and nine pairs of submetacentric chromosomes in the corresponding karyotype. Its arm ratio and karyotype formula of 4m + 18sm place it in the symmetrical karyotype group (Figure 3b). The short arm length was 0.68 ± 0.01 to 1.37 ± 0.05 µm, the long arm length was 1.08 ± 0.01 to 2.45 ± 0.01 µm, and the total chromosomal length was 1.76 ± 0.01 to 3.83 ± 0.06 µm. The RL ranged from 6.00 ± 0.01 to 13.00 ± 0.02%, while the CI ranged from 0.59 ± 0.01 to 0.65 ± 0.01, as shown in Table 2. Figure 4b shows the ideogram that was created.

Zingiber junceum has $2n = 22$ chromosomes and an NF = 44 (Figure 2c) (Table 1). Six pairs of metacentric and five pairs of submetacentric chromosomes were discovered. Because of the formula, namely, 12m + 10sm, and the arm ratio of the chromosomes, the karyotype seemed to be symmetrical. The short arm length was 0.66 ± 0.04 to 1.24 ± 0.08 µm, the long arm length was 0.95 ± 0.06 to 2.22 ± 0.65 µm, and the total chromosomal length was 1.61 ± 0.10 to 3.46 ± 0.73 µm. Table 2 shows that the RL ranged from 5.57 ± 0.08 to 12.00 ± 0.16% and that the CI ranged from 0.59 ± 0.10 to 0.64 ± 0.13 (Table 2). Figure 4c shows the ideogram that was created.

The somatic chromosomal number of *Zingiber niveum*, reported herein for the first time, is $2n = 22$, with an NF = 44 (Figure 2d, Table 1). There were seven pairs of metacentric chromosomes, four pairs of submetacentric chromosomes, and four pairs of subtelocentric chromosomes found. 14m + 4sm + 4st was the karyotype formula. The arm ratio revealed that it was a asymmetrical karyotype (Table 1, Figure 3d). The short arm length was 0.86 ± 0.06 to 1.65 ± 0.10 µm, the long arm length was 1.09 ± 0.07 to 2.33 ± 0.68 µm, and the total chromosomal length was 1.94 ± 0.13 to 3.98 ± 0.78 µm. The RL ranged from 6.10 ± 0.12 to 12.50 ± 0.20, and the CI varied from 0.56 ± 0.10 to 0.58 ± 0.10 (Table 2). The ideogram of *Zingiber niveum* is depicted in Figure 4d.

3.1. The Similarity Index of Four Rare Zingiber Species

The results of a UPMG cluster analysis, which examined the similarity among four species of *Zingiber*, included *Zingiber chrysostachys*, *Z. isanense*, *Z. junceum*, and *Z. niveum*. This analysis was based on their karyotype formulas, which represent the chromosome characteristics and 30 morphological characteristics of each species. The UPMG cluster analysis produced a dendrogram, a diagram showing the relationships among these species based on their similarities in terms of chromosome characteristics and morphology. The dendrogram revealed that *Zingiber junceum* and *Z. niveum* are the most similar to each other in terms of their karyotype formulas and morphologies; these two species have similar flowers and inflorescences. Following them, *Z. chrysostachys* showed a moderate level of similarity, while *Z. isanense* exhibited the least similarity to the other species (Figure 5).

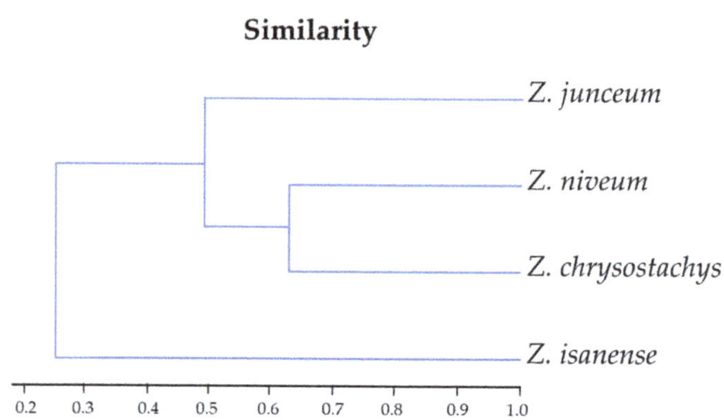

Figure 5. UPGM cluster analysis dendrogram of four rare *Zingiber* species based on the distant matrix of karyotype formulars, similarity index, and 30 morphological characteristics with a cophenetic correlation = 0.9686.

The UPMG cluster analysis provided valuable insights into the similarities among four distinct species of *Zingiber*: *Z. chrysostachys*, *Z. isanense*, *Z. junceum*, and *Z. niveum*. The analysis yielded karyotype formulas, similarity indices, and 30 morphological characteristics. The cophenetic correlation, which is 0.9686, indicates a high level of similarity among the four *Zingiber* species, with a 96.86% similarity rate. Crucially, this analysis utilized karyotype formulas as the basis for comparison, offering a glimpse into the unique chromosome characteristics of each species according to their morphologies.

3.2. The PCA Score Plot of CI

The PCA score plot of *Zingiber chrysostachys*, *Z. isanensis*, *Z. junceum*, and *Z. niveum*, based on CIs, showed that component 1 accounts for 60.25% of the total variance, while component 2 accounts for 37.64%. All four species are in the same stage, which means that their CIs are similar to each other. The CIs can be used to differentiate between different points. This presentation divides the four species into two groups: *Zingiber niveum* (D) with

Z. chrysostachys (A), and *Z. junceum* (C) with *Z. isanensis* (B). This indicates that the CIs of each group are closely related (Figure 6). This result suggests that the CIs of these four rare *Zingiber* species can be employed to differentiate them even if they are in the same stage, signifying their similarity yet distinct stages while remaining closely related.

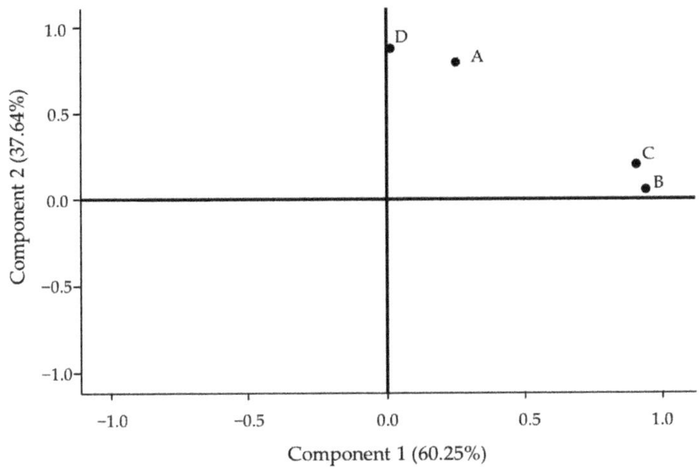

Figure 6. PCA score plot of (A) *Zingiber chrysostachys*, (B) *Z. isanense*, (C) *Z. junceum*, and (D) *Z. niveum* based on CI.

4. Discussion

According to the Plants of the World Online database [17], four edible and ornamental species of *Zingiber* are rare in Thailand: *Zingiber chrysostachys*, *Z. isanensis*, *Z. junceum*, and *Z. niveum*. Furthermore, two species are classified as endangered (*Zingiber chrysostachys* and *Z. niveum*), and two as being of least concern (*Zingiber isanensis* and *Z. junceum*), in the International Union for Conservation of Nature and Natural Resources database [16].

Zingiber chrysostachys, *Z. isanensis*, *Z. junceum*, and *Z. niveum* are four rare edible and ornamental *Zingiber* species discovered in Thailand. They all have the same somatic chromosomal number, $2n = 22$ [23,24]. Similarly, it was determined that this species' haploid chromosome number in Malaysia was $n = 11$ [23]. The somatic chromosomal counts of *Zingiber niveum* and *Z. isanense* are reported herein for the first time. We determined the somatic chromosomal numbers of *Zingiber chrysostachys* and *Z. junceum* to be $2n = 22$ [23] (Table 1). The NFs of these four *Zingiber* species were determined to be 44, which aligns with the research results reported by Saensouk and Saensouk [25]. The arm ratio analysis revealed that individuals of three species (*Z chrysostachys*, *Z. isanensis*, and *Z. junceum*) had a karyotype that showed symmetrical characteristics, which corresponds to the findings published by Saensouk and Saensouk [25]. On the other hand, *Z. niveum* had an asymmetrical karyotype. Nevertheless, the NFs, karyotypes, and ideograms of all four species have been determined for the first time. Therefore, while the chromosome number was consistent across all four species, there were differences in the CIs, RL (%), karyotypes, and ideograms among the species. These differences can be attributed to environmental factors such as the plant's location, soil composition, water availability, air quality, etc.

The chromosomal structures and karyotype formulae of these four *Zingiber* species can be used to distinguish them. Researchers can distinguish plants by analyzing their chromosomal and morphological characteristics [36]. The total chromosomal length ranged from 3.46 ± 0.73 μm (*Z. junceum*) to 3.98 ± 0.78 μm (*Z. niveum*), which is shorter than the measurements reported in a previous work by Saensouk and Saensouk [25]. This difference in length can be attributed to various environmental conditions, age, or periods of development. The RL content varied from 12.00 ± 0.16% (*Z. junceum*) to 13.00 ± 0.02% (*Z. isanensis*), which differs from the earlier findings reported by Saensouk and Saensouk [25]. This difference could be attributed to variations in environmental circumstances, age, or developmental stages. Furthermore, the CI ranged from 0.65 ± 0.01 (*Z. isanense*) to 0.58 ± 0.10 (*Z. niveum*), which contrasts with the previous results documented by Saensouk and Saensouk [25]. This disparity can be ascribed to differences in environmental conditions, age, or phases of development.

Karyological data can be employed in cytotaxonomy or karyosystematics since they can represent the genetic links between the species being studied [37]. These features are considered advantageous due to the fact that they are not influenced by environmental circumstances, age, or developmental stages [38]. This research has found that the chromosomal numbers of all four *Zingiber* species are the same. Nonetheless, banding patterns in karyotypes with identical morphologies can differ significantly [39]. Banding patterns in karyotypes with identical shapes can, however, vary dramatically [39].

The PCA score plot of *Zingiber chrysostachys*, *Z. isanense*, *Z. junceum*, and *Z. niveum*, based on CI, showed that all four rare species are in the same stage, which means that their CIs are similar to each other. However, this study based on CIs could be divided into two groups, namely, *Zingiber niveum* (D) with *Z. chrysostachys* (A) and *Z. junceum* (C) with *Z. isanense* (B), which is consistent with [17], who reported the morphological features of leaf blades of *Z. niveum* (D) and *Z. chrysostachys* (A) were reported as oblong to lanceolate. On the other hand, the second group, consisting of *Z. junceum* (C) and *Z. isanensis* (B), showed narrow lanceolate leaf blade morphology.

Beyond the four rare edible plants belonging to the *Zingiber* genus, four had distinct karyotype formulas and chromosome structures. However, not all evidence can be sued to differentiate between these four species. The CI is capable of distinguishing between the four species, notwithstanding the fact that the CI values seem to be aligned in the same direction. Additionally, the leaf blade morphologies of these four species can serve as a characteristic with which to differentiate them. According to the findings of this study, it is feasible to categorize all four species of *Zingiber* using the PCA of the karyotype formula, the CI, and the morphological characteristics of the leaf blade.

Therefore, these data will help in supporting the development of plant breeding for commercial purposes in the future and are expected to generate more income for the community.

5. Conclusions

Cytological research was conducted on four species of *Zingiber* (*Z. chrysostachys*, *Z. isanensis*, *Z. junceum*, and *Z. niveum*) that are edible and ornamental, with the aim of obtaining more knowledge of these species. Their somatic chromosomal numbers are all $2n = 22$. Our knowledge of these species' karyology has been increased by this investigation. For the first time, both the somatic chromosomal counts of *Zingiber niveum* and *Z. isanense*—in addition to the NFs of all species, all of which were found to be 44—were reported. The karyotypes for all four of these endangered species were made available: 16m + 6sm for *Zingiber chrysostachys*, 4m + 18sm for *Z. isanense*, 12m + 10sm for *Z. junceum*, and 14m + 4sm + 4st for *Z. niveum*. The karyotypes are symmetrical for each individual. Chromosome architecture and karyotype formulae were used to distinguish four rare *Zingiber* species from Thailand.

The CIs can be used to categorize these into various points. The four species were divided into two groups in this presentation: *Zingiber niveum* (D) and *Z. chrysostachys* (A), and *Z. junceum* (C) and *Z. isanensis* (B). This suggests a close relationship between the CIs of each group (Figure 5). This finding implies that, while in the same stage, the CIs of these four rare *Zingiber* species can be used to distinguish them, indicating their resemblance at unique stages while still being closely related.

The information obtained from this study helps promote the conservation of plants in the areas that are their habitats, which will lead to the conservation of these rare plants and their habitats. Consequently, the genetics of these plants persist, allowing for their propagation and development into ornamental plants with economic value, which could potentially boost a community's income.

Author Contributions: P.S. (Piyaporn Saensouk) and S.S., methodology; S.S., formal analysis; R.S., data curation; P.S. (Piyaporn Saensouk), S.S., R.S., D.M. and P.S. (Phetlasy Souladeth), writing—original draft preparation; S.S., writing—review and editing; S.S., funding acquisition. All authors have read and agreed to the published version of the manuscript.

Funding: This research was funded by Mahasarakham University.

Data Availability Statement: All data produced and examined are available in this article.

Acknowledgments: This research was financially supported by Mahasarakham University. We are grateful to the Walai Rukhavej Botanical Research Institute, Mahasarakham University, Diversity of Family Zingibeaceae and Vascular Plant of Its Applications Research Unit, and Mahasarakham University, for their facilities during this study. I would like to thank Jolyon Dodgson for conducting language editing and providing suggestions regarding how to improve the manuscript.

Conflicts of Interest: The authors declare no conflicts of interest.

References

1. POWO. Plant of the World Online, Facilitated by the Royal Botanic Gardens, Kew. Available online: http://www.plantsoftheworldonline.org/ (accessed on 12 February 2023).
2. Larsen, K.; Larsen, S.S. *Ginger of Thailand*; Queen Sirikit Botanic Garden, The Botanical Garden Organization: Chiang Mai, Thailand, 2006.
3. Kress, W.J.; Prince, L.M.; Williams, J.K. The phylogeny and a new classifcation of the gingers (Zingiberaceae): Evidence from molecular data. *Am. J. Bot.* **2002**, *89*, 1682–1696. [CrossRef] [PubMed]
4. Mood, J.D.; Ardiyani, M.; Veldkamp, J.F.; Mandáková, T.; Prince, L.M.; de Boer, H.J. Nomenclatural changes in Zingiberaceae: Haplochorema is reduced to *Boesenbergia*. *Gard. Bull. Singap.* **2020**, *72*, 77–95. [CrossRef] [PubMed]
5. Aung, M.M.; Tanaka, N.; Miyake, N. Two gingers, *Zingiber orbiculatum* and *Z. flavomaculosum* (Zingiberaceae), newly recorded from Myanmar. *Bull. Natl. Mus. Nat. Sci.* **2015**, *41*, 107–112.
6. Saensouk, S.; Saensouk, P.; Pasorn, P.; Chantaranothai, P. Diversity, Traditional Uses and New Record of Zingiberaceae in Nam Nao National Park, Petchabun Province, Thailand. *Agric Nat. Resour.* **2016**, *50*, 445–453.
7. Triboun, T.; Chantaranothai, P.; Larsen, K. Taxonomic changes regarding three species of *Zingiber* (Zingiberaceae) from Thailand. *Acta Phytotax. Sin.* **2007**, *45*, 403–404.
8. Ikeda, H.; Nam, B.M.; Yamamoto, N.; Funakoshi, H.; Takano, A.; Im, H.T. Chromosome number of myoga ginger (*Zingiber mioga*: Zingiberaceae) in Korea. *Korean J. Plant Taxon.* **2021**, *51*, 100–102. [CrossRef]
9. Khumkratok, S.; Boontiang, K.; Chutichudet, P.; Pramaul, P. Geographical distributions and ecology of ornamental *Curcuma* (Zingiberaceae) in Northeastern Thailand. *Pak. J. Biol. Sci.* **2012**, *15*, 929–939. [CrossRef]
10. Saensouk, P.; Saensouk, S.; Phechphakdee, T.; Ragsasilp, A. Cytogenetic study in seven species of Zingiberaceae family from Bueng Kan Province, Thailand. *Biodiversitas* **2023**, *24*, 68–77. [CrossRef]
11. Chen, Z.Y. Evolutionary patterns in cytology and pollen structure of Asian Zingiberaceae. In *Tropical Forests*; Holm-Nielsen, B., Nielsen, I.C., Balslev, H., Eds., Academic Press Limited: New York, NY, USA, 1989; pp. 185–191.
12. Lin, Y.-C.; Chao, C.-T.; Chang, C.-Y.; Tseng, Y.-H. Taxonomic revision of *Zingiber* (Zingiberaceae) of Taiwan. *Eur. J. Taxon.* **2022**, *839*, 74–102. [CrossRef]
13. Eksomtramage, L.; Boontum, K. Chromosome counts of Zingiberaceae. *Songklanakarin J. Sci. Technol.* **1995**, *17*, 291–297.
14. Eksomtramage, L.; Sirirugsa, P.; Sawangchote, P.; Jornead, S.; Saknimit, T.; Leeratiwong, C. Chromosome numbers of some monocot species from Ton-Nga-Chang Wildlife Sanctuary, Southern Thailand. *Thai For. Bull. (Bot.)* **2001**, *29*, 63–71.
15. Eksomtramage, L.; Sirirugsa, P.; Jivanit, P.; Maknoi, C. Chromosome counts of some Zingiberaceous species from Thailand. *Songklanakarin J. Sci. Technol.* **2002**, *24*, 311–319.
16. Ragsasilp, A.; Saensouk, P.; Saensouk, S. Ginger family from Bueng Kan Province, Thailand: Diversity, conservation status, and traditional uses. *Biodiversitas* **2022**, *23*, 2739–2752. [CrossRef]
17. Triboun, P.; Larsen, K.; Chantaranothai, P. A key to the genus *Zingiber* (Zingiberaceae) in Thailand with descriptions of 10 new taxa. *Thai J. Bot.* **2003**, *6*, 53–77.
18. Deng, M.; Yun, X.; Ren, S.; Qing, Z.; Luo, F. Plants of the Genus *Zingiber*: A review of their ethnomedicine, phytochemistry and pharmacology. *Molecules* **2022**, *27*, 2826. [CrossRef] [PubMed]
19. IUCN. *The IUCN Red List of Threatened Species*; Version 2022-2; International Union for Conservation of Nature and Natural Resources: Gland, Switzerland, 2023.
20. Chaiyasut, K. *Cytogenetics and Cytotaxonomy of the Family Zephyranthes*; Department of Botany, Faculty of Science, Chulalongkorn University: Bangkok, Thailand, 1989.
21. Stebbins, G.L. *Chromosomal Evolution in Higher Plants*; Addison-Wesley Pub. Co.: San Francisco, CA, USA, 1971.

22. Beltran, I.C.; Kiew, K.Y. Cytotaxonomic studies in the Zingiberaceae. *Notes R. Bot. Gard. Edinb.* **1984**, *41*, 541–559.
23. Saensouk, S.; Chantaranothai, P. The Family Zingiberaceae in Phu Phan National Park. In Proceedings of the 3rd Symposium on the Family Zingiberaceae, Khon Kaen, Thailand, 7–12 July 2002; pp. 16–25.
24. Saenprom, K.; Saensouk, S.; Saensouk, P.; Senakun, C. Karyomorphological analysis of four species of Zingiberaceae from Thailand. *Nucleus* **2018**, *61*, 111–120. [CrossRef]
25. Saensouk, S.; Saensouk, P. New report on karyotype and ideogram of two *Zingiber* species, *Z. ligulatum* and *Z. parishii* subsp. *phuphanense* from Thailand. *Nucleus* **2021**, *64*, 115–121. [CrossRef]
26. Bhadra, S.; Bandyopadhyay, M. New chromosome number counts and karyotype analyses in three important genera of Zingiberaceae. *Nucleus* **2016**, *59*, 35–40. [CrossRef]
27. Levan, A.; Fredya, K.; Sandberg, A.A. Nomenclature for centromeric position on chromosome. *Hereditas* **1964**, *52*, 201–220. [CrossRef]
28. Morinaga, T.; Fukushina, E.; Kanui, T.; Tamasaki, Y. Chromosome numbers of cultivated plants. *Bot. Mag.* **1929**, *43*, 589–594. [CrossRef]
29. Saensouk, S.; Saensouk, P. Chromosome number of some Zingiberaceous in Thailand. *KKU Res. J.* **2004**, *9*, 3–9.
30. Hong, D.Y.; Zhang, S.Z. Observations on chromosomes of some plants from western Sichuan. *Cathaya* **1990**, *2*, 191–197.
31. Das, A.B.; Rai, S.; Das, P. Estimation of 4C DNA and karyotype analysis in ginger (*Zingiber officinale* Rosc.) II. *Cytologia* **1998**, *63*, 133–139. [CrossRef]
32. Senavongse, R.; Saensouk, S.; Saensouk, P. Karyological study in three native species of genus *Alocasia* (Araceae) in the northeast of Thailand. *Nucleus* **2020**, *63*, 81–85. [CrossRef]
33. Hammer, Ø.; Harper, D.A.T.; Ryan, P.D. PAST: Paleontological Statistics Software Package for Education and Data Analysis. *Palaeontol. Electron.* **2001**, *4*, 1–9.
34. Sengthong, A.; Saensouk, S.; Saensouk, P.; Souladeth, P. Cytogenetic Study of Five Varieties of *Callisia repens* (Jacq.) L. (Commelinaceae) from Laos. *Horticulturae* **2023**, *9*, 1050. [CrossRef]
35. Bro, R.; Smilde, A.K. Principal component analysis. *Anal. Methods* **2014**, *6*, 2812–2831. [CrossRef]
36. Santhosh, B. *Cytological and Palynological Studies on the Family Apocynaceae*; Department of Botany, University of Kerala: Thiruvananthapuram, India, 1999.
37. Dobigny, G.; Ducroz, J.F.; Robinson, T.J.; Volobouev, V. Cytogenetics and Cladistics. *Syst. Biol.* **2004**, *53*, 470–484. [CrossRef]
38. Guerra, M. Chromosome numbers in plant taxonomy: Concepts and implications. *Cytogenet. Genome Res.* **2008**, *120*, 339–350. [CrossRef]
39. Dobigny, G.; Aniskin, V.; Volobouev, V. Explosive chromosome evolution and speciation in the gerbil genus *Taterillus* (Rodentia, Gerbillinae): A case of two new cryptic species. *Cytogenet. Cell Genet.* **2002**, *96*, 117–124. [CrossRef] [PubMed]

Disclaimer/Publisher's Note: The statements, opinions and data contained in all publications are solely those of the individual author(s) and contributor(s) and not of MDPI and/or the editor(s). MDPI and/or the editor(s) disclaim responsibility for any injury to people or property resulting from any ideas, methods, instructions or products referred to in the content.

horticulturae

MDPI

Article

NGS-Based Multi-Allelic InDel Genotyping and Fingerprinting Facilitate Genetic Discrimination in Grapevine (*Vitis vinifera* L.)

Guiying Jia [1,2,†], Na Zhang [1,†], Yingxia Yang [1,3], Qingdong Jin [3], Jianfu Jiang [4], Hong Zhang [1,2], Yutong Guo [1,2], Qian Wang [1], He Zhang [1], Jianjin Wu [5], Rui Chen [1], Jianquan Huang [1,*] and Mingjie Lyu [1,*]

1 State Key Laboratory of Vegetable Biobreeding, Tianjin Academy of Agricultural Sciences, Tianjin 300192, China; jgying99@163.com (G.J.); zhna200@126.com (N.Z.); yingxiayang@126.com (Y.Y.); zhhong724@163.com (H.Z.); guo_yutong@mail.nankai.edu.cn (Y.G.); wangqian881001@126.com (Q.W.); zhanghe1969@126.com (H.Z.); chenrui_aglab@126.com (R.C.)
2 College of Life Sciences, Nankai University, Tianjin 300071, China
3 National Key Laboratory of Crop Genetic Improvement, National Center of Oil Crop Improvement, College of Plant Science and Technology, Huazhong Agricultural University, Wuhan 430070, China; jinqingdong0202@webmail.hzau.edu.cn
4 Zhengzhou Fruit Research Institute, Chinese Academy of Agricultural Sciences, Zhengzhou 450009, China; jiangjianfu@caas.cn
5 Tianjin Agricultural Development Service Center, Tianjin 300061, China; wjj142@126.com
* Correspondence: huangjianquan200@126.com (J.H.); lvmingjie_good@163.com (M.L.)
† These authors contributed equally to this work.

Abstract: Molecular markers play a crucial role in marker-assisted breeding and varietal identification. However, the application of insertion/deletion markers (InDels) in grapevines has been limited by the low throughput and separability of gel electrophoresis. To developed effective InDel markers for grapevines, this study reports a novel, effective and high-throughput pipeline for InDel marker development and identification. After rigorous filtering, 11 polymorphic multi-allelic InDel markers were selected. These markers were then used to perform genetic identification of 123 elite grape cultivars using agarose gel electrophoresis and next-generation sequencing (NGS). The polymorphism rate of the InDel markers identified by gels was 37.92%, while the NGS-based results demonstrated a higher polymorphism rate of 61.12%. Finally, the NGS-based fingerprints successfully distinguished 122 grape varieties (99.19%), surpassing the gels, which could distinguish 116 grape varieties (94.31%). Specifically, we constructed phylogenetic trees based on the genotyping results from both gels and NGS. The population structure revealed by the NGS-based markers displayed three primary clusters, consisting of the patterns of the evolutionary divergence and geographical origin of the grapevines. Our work provides an efficient workflow for multi-allelic InDel marker development and practical tools for the genetic discrimination of grape cultivars.

Keywords: grapevine; multi-allelic InDels; next-generation sequencing; fingerprinting; population structure

1. Introduction

The grapevine (*V. vinifera* L.) is one of the earliest domesticated fruit plants, characterized by the absence of genetic barriers within the genus [1]. Over the past thousands of years, grapevine introduction, breeding and intensive trade have blurred genetic relationships [2]. Moreover, the genetic diversity of grapes gradually decreases while domestication progresses [3–5]. Currently, synonyms and homonyms are prevalent in the grape market, which has negative effects on breeding activities and breeders' rights [6]. Therefore, economical, efficient and accurate variety identification methods are urgently needed to improve breeding efficiency and protect the rights of grape breeders.

DNA molecular markers are crucial for grape varietal identification [7]. Among them, simple sequence repeat (SSR) and amplified fragment length polymorphism (AFLP) are widely used in grapevine identification. The use of 20 SSR markers helped identify three genetic populations among 1378 varieties [8]. The genetic variability in 'Sangiovese', 'Sanforte' and 'Montepulciano' grapes was analyzed using multiple molecular markers such as SSR and AFLP [9]. However, the application of SSR markers is hindered by the low throughput and intensive data processing [10].

With the development of sequencing technology, a large number of single nucleotide polymorphism (SNP) and InDel markers were identified. SNP and InDel have become the most promising markers in genetic research and molecular marker-assisted breeding due to their wide distribution in genomes and suitability for high-throughput genotyping [11,12]. Emanuelli et al. [13] employed SNP markers to verify European grapevine cultivars, while Wang et al. [14] used SNP markers to distinguish major grape cultivars in China. Generally, InDels are defined as short insertion or deletion of up to 50 nucleotides at a single locus. Compared to SNPs, InDel markers can be easily detected using gel electrophoresis. In recent years, InDel markers have been applied as fingerprints in various crops, such as rice [15], maize [16], peach [17], apple [18], tomato [19] and cucumber [20]. However, as of now, few InDel markers have been developed for grape-variety fingerprinting.

Multi-allelic InDels refer to variations caused by multiple different sizes of insertion or deletion at one allele within a population. Multi-allelic InDels are common in plants [21]. The frequent selfing and hybridization in plants increase the diversity of alleles, and the relatively high rate of genetic mutations promotes the formation of multi-allelic InDels [22]. Gel electrophoresis has a relatively low resolution and may not be sensitive enough to separate different fragments of multi-allelic InDels [23]. The NGS allows the accurate identification of multi-allelic InDels and facilitates data processing. However, the application of NGS is limited by the high costs and lack of professional bioinformatics analysis tools.

This study established a practical workflow for multi-allelic InDel selection and identification. A total of 11 polymorphic InDel primers were selected from a variant database constructed using the resequencing data of 499 grapevine lines. A total of 123 core germplasms were collected and genotyped using gel electrophoresis and NGS. The polymorphism rates and discriminability of different genotyping strategies highlight the advantages of NGS-based methods. The fingerprints constructed by multi-allelic InDels in this study provide valuable tools for the genetic discrimination of grapes.

2. Materials and Methods

2.1. Materials

Preliminary screening of InDel markers was conducted using 499 whole-genome resequencing data from the public databases (Table S1). In August 2022, 123 young grapevine leaf samples were collected at the Institute of Forestry and Pomology, Tianjin Academy of Agricultural Sciences (Table S2). These samples exhibit a diverse genetic background and a wide range of phenotypic variations. Healthy and young leaves were collected under cool and dry conditions in the early morning. The collected leaf samples were homogenized with a 4 mm steel ball in a Retsch MM 400 Mixer Mill after chilling in liquid nitrogen. The total genomic DNA of all samples was extracted using the Plant Genomic Extraction Kit (Tiangen Biotech, Beijing, China). Subsequently, DNA concentration and purity were assessed using a NanoDrop™ 2000 spectrophotometer (Thermo Fisher Scientific, Waltham, MA, USA). Qualified DNA samples were then stored at −20 °C.

2.2. Screening of InDels Markers

Initially, grape resequencing data were downloaded to construct a variant database, followed by quality control of the raw data. Fastp software (v.0.23.4) was used for data quality assessment and preprocessing of the sequencing data [24]. Subsequently, the pre-processed sequencing data were aligned to the reference genome using the Burrows–Wheeler aligner (BWA, v.0.7.17) to generate bam files [25]. SAMtools (v.1.14) was utilized for sorting, indexing, filtering and statistical analysis of these files [26]. Following this, the Genome Analysis Toolkit (GATK, v.4.0.10.0) was employed to detect InDels, and programming was utilized to generate a final Variant Call Format (VCF) file, containing information about variations such as InDel locations, types, quality and their distribution across different samples [27].

High-quality polymorphic InDel sites were obtained by further filtering using the following criteria: (i) InDels are not on repeat regions in the genome, (ii) multi-allelic InDels, and (iii) the size of an InDel is more than 20 bp. Additionally, the length of the PCR product of the InDel was required to be within the range of 150–300 bp to enable separation by agarose gel electrophoresis. Genetic diversity parameters such as polymorphic information content (PIC), heterozygosity (Het), minor allele frequency (MAF) and gene diversity index (GDI) were calculated using the Python 3.12.4 script (https://github.com/Lvmingjie/indel-seq, accessed on 16 May 2024).

2.3. InDel Primers Design and Selection

Primer design was carried out using Primer3 software [28]. The 100 bp upstream and downstream sequences of each InDel were extracted. The primer design parameters were set as follows: (1) GC content less than 60%; (2) primer size between 18 bp and 22 bp; (3) melting temperature (Tm) between 55 °C and 62 °C. Primer sequences that amplify unique loci were selected using e-pcr software (v.2.3.12) [29]. Subsequently, the genotype distribution data were analyzed, and 24 primer pairs exhibiting an even genotype distribution were chosen for further experiments.

2.4. InDel Genotyping

A total of 24 pairs of synthesized primers were screened using PCR amplification and agarose gel electrophoresis (Table S3). Based on the gel results, the core markers were selected based on the following principles: multiple and clear bands, insertion or deletion sizes over 20 bp and tendency to be located on different chromosomes. Finally, eleven core primers were chosen for the subsequent amplification of DNA from 123 grape samples (Table S3). PCR amplification was conducted in a 25 μL reaction volume containing 2 μL DNA (50 ng/μL), 0.25 μL forward primer (10 μM), 0.25 μL reverse primer (10 μM), 12.5 μL 2x Taq Master Mix and 10 μL ddH_2O. The amplification process was carried out using an Applied Biosystems Veriti Thermal Cycler with the following conditions: 98 °C for 5 min, 35 cycles at 95 °C for 20 s, 56 °C for 20 s, and 72 °C for 20 s, followed by a final extension at 72 °C for 10 min and a concluding step at 4 °C to halt the reaction. Subsequently, 5 μL of the PCR products were separated by 2% agarose gel electrophoresis, and band information was visualized under UV light. The bands on the agarose gel electrophoresis profiles were recorded sequentially from smallest to largest as "1" "2" "3", with the absence of a band recorded as "0". For example, a single band was noted as "1_1", and double bands were noted as "1_2". This information was used to conduct further genetic analysis.

Since gel electrophoresis was not sufficient to fully discriminate the genotyping results, sequencing was used for further genotyping in this study. Eleven adapter-equipped core primers and universal barcode primers were used for two rounds of amplification. The universal barcode primers and PCR system used in this experiment were based on the system described in the study by Liu et al. [30]. Products from the second round of amplification, originating from distinct individual plants, were then mixed in equal amounts and subjected to 150 bp paired-end NGS. The sequences of each target site were decoded using a custom script. To avoid and exclude sequencing errors, we set filtering requirements: total reads per well 1000; each genotype $\geq$ 100; Top1 percentage of genotypes per well $\geq$ 30%; Top2/Top1 percentage of genotypes per well $\geq$ 30%. This confirms the presence of the genotype in the sample, thereby improving the accuracy of the sequencing results.

2.5. Phylogenetic Analyses

The phylogenetic tree was constructed using FastTree (v2.1.11) [31] and illustrated with FigTree (v1.4.4) (http://tree.bio.ed.ac.uk/software/figtree/, accessed on 10 December 2023). Principal Component Analysis (PCA) was performed using GCTA (v1.940) software [32] and PCA diagrams were generated using the R 4.4.1 language (https://github.com/Lvmingjie/indel-seq/MM_InDel, accessed on 16 May 2024).

3. Results

3.1. Screening of InDels Markers

To select a set of stable and polymorphic multi-allelic InDel markers, we developed a rigorous filtering pipeline (Figure 1A). A total of 126,493,127 InDels were identified using 499 grape resequencing data (Figure 1B). Subsequently, 80 high-quality and highly polymorphic markers were retained based on the following criteria (Figure 1C; Table S2): (1) the number of InDel alleles ≥ 4; (2) InDel lengths between 20 bp and 100 bp; (3) the size of each of the two types ≥ 20 cp; (4) no repeats within 100 bp upstream and downstream of the InDel site; (5) missing rate $\leq 10\%$; and (6) maximum genotype frequency $\leq 50\%$. Based on the genotype frequencies observed in 499 samples, 24 InDels were selected, with uniformly distributed genotype frequencies across 19 chromosomes. Subsequently, based on the PCR and agarose gel electrophoresis results of 24 InDels, a total of 11 primers, which yielded clear and multiple amplified bands, were selected as the core markers (Figure 1D).

3.2. Genotyping of 123 Grape Lines Using Agarose Gel Electrophoresis

To investigate the genetic discriminability of the 11 core markers. We collected 123 elite grape cultivars with diverse ecotypes and phenotypes in China. PCR and agarose gel electrophoresis were used to analyze the 11 core multi-allelic InDels in the grape cultivars, resulting in a total of 1353 distinct bands on agarose gel electrophoresis, indicating an average polymorphism rate of 37.92% (Figure S1; Table S4). Based on the genotyping results, GDI, Het, MAF and PIC values were calculated for the 11 InDels (Figure 2). The PIC values of the InDels ranged from 0.305 to 0.566, with a mean of 0.407. Most (82%) fell within the range of 0.3 to 0.5. The Het values ranged from 0.359 to 0.611, with a mean of 0.488. The average MAF was 0.306, ranging from 0.210 to 0.411. The average GDI was 0.4882, ranging from 0.359 to 0.611. These results indicate that the 11 candidate multi-allelic InDels exhibited a high level of polymorphism, making them optimal markers for grape fingerprinting.

3.3. NGS-Based Genotyping of 123 Grape Lines

In addition to agarose gel electrophoresis, NGS was performed on all 123 grape samples for further analysis. The genotyping outcomes derived from sequencing can provide richer information on InDels (Table S5). NGS revealed a higher average polymorphism rate of 61.12%. Based on the genotyping results, GDI, Het, MAF and PIC values were calculated for the 11 InDels (Figure 2). The mean values for PIC, GDI and Het of the 11 markers were 0.860, 0.806 and 0.647, respectively, which were significantly higher than those revealed by the gels.

3.4. Discrimination and Fingerprints of 123 Grape Lines

The InDel fingerprints were constructed based on the gels and NGS. Based on the fingerprinting data, heatmaps were used to illustrate the unique genotypes of grapes through distinct colors (Figure 3A,B). Two sets of genotype data from 123 accessions were analyzed to assess the identification efficiency. The NGS-based fingerprints successfully discriminated 122 accessions, achieving an identification efficiency of 99.18% (Figure 3D). The only exception was observed for 'Maple Leaf Grapes' and 'NO.8', both of which are round leaf grapes. The results based on agarose gel electrophoresis only distinguished 116 grape varieties. These results proved that the NGS method is more efficient than agarose gel electrophoresis in the genetic discrimination of grape varieties (Figure 3C).

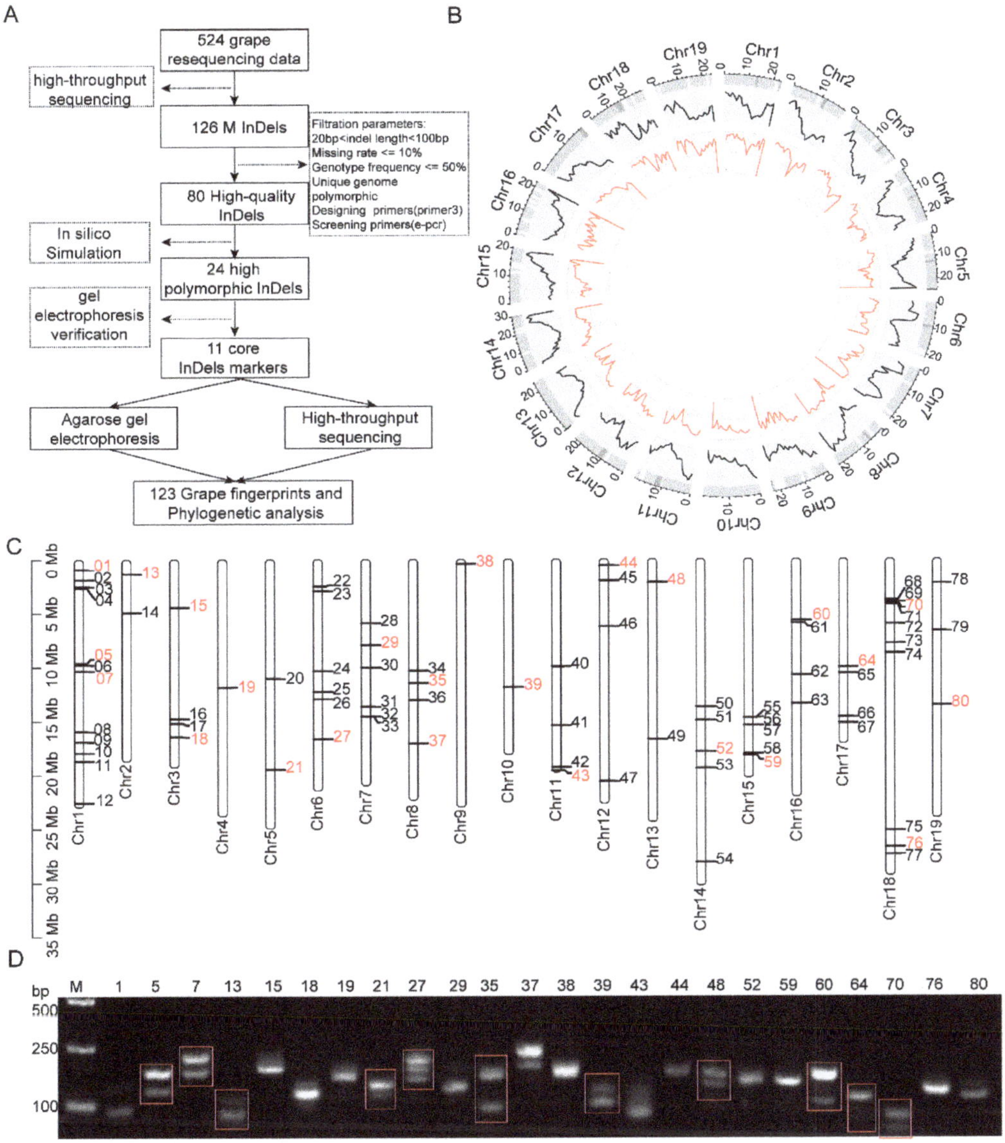

Figure 1. Development of multi-allelic InDel markers in grapevine. (**A**) Pipeline for grapevine fingerprint database construction based on multi-allelic InDel markers. (**B**) Gene (black line plot in the outer track) and InDel (red line plot in the inner track) density derived from 499 grape resequencing data across 19 chromosomes. (**C**) Physical location of 80 high-quality multi-allelic InDels. A total of 24 highly polymorphic primers were labeled in red style. (**D**) Agarose gel electrophoresis identification of 24 polymorphic InDels selected from the 80 high-quality markers. A total of 11 markers in the red boxes were selected for further analysis.

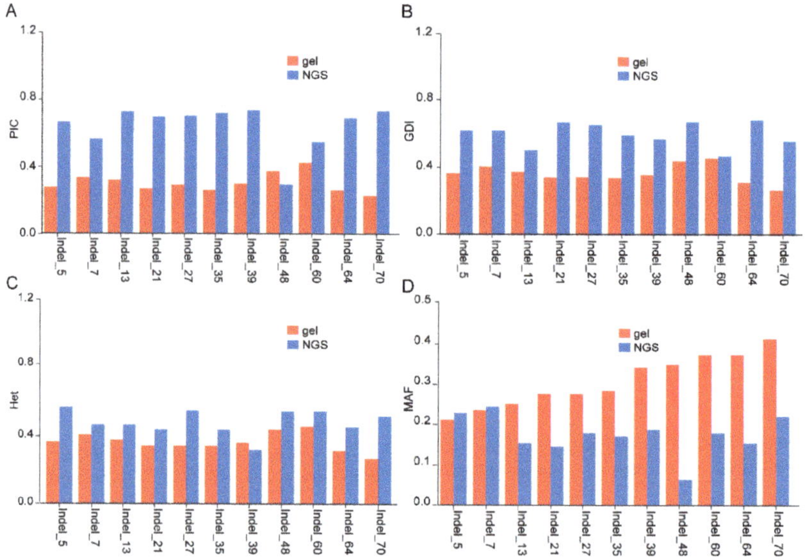

Figure 2. Genetic information of 11 core multi-allelic InDels. (**A**) Polymorphic information content (PIC) of 11 core markers in 123 grape cultivars. (**B**) Gene diversity index (GDI) of 11 core markers in 123 grape cultivars. (**C**) Heterozygosity (Het) of 11 core markers in 123 grape cultivars. (**D**) Minor allele frequency (MAF) of 11 core markers in 123 grape cultivars.

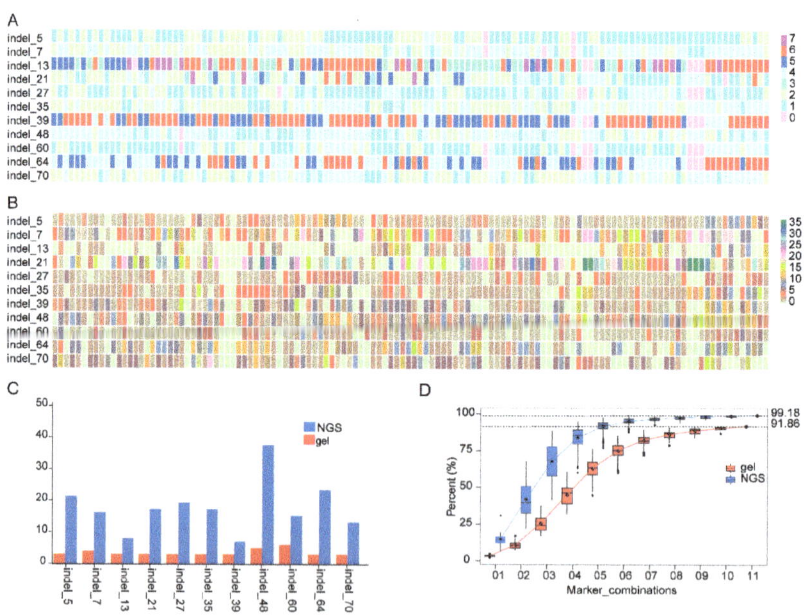

Figure 3. Fingerprinting of 123 grape cultivars. (**A**) The fingerprints of 123 grape cultivars based on agarose gel electrophoresis results. (**B**) The fingerprints of 123 grape cultivars based on NGS. (**C**) The number of genotypes identified by agarose gel electrophoresis and NGS. The y-axis was the genotype number of markers revealed by NGS and gel. (**D**) The discernibility of different combinations of 11 multi-allelic InDel markers for 123 grape accessions.

3.5. Phylogenetic Analyses of 123 Grapes

Phylogenetic trees of 123 grape varieties were constructed based on the genotyping data of gels and NGS. The NGS-based hierarchical cluster successfully classified the 123 grape samples into three distinct populations (Pop−1, Pop−2, Pop−3). Interestingly, in Pop−1(ID 1–19), 71% of the grapes were rootstock, such as 'cr1', 'cr2', and '5BB♀'. 'SP137', 'ST0998', and 'Zhuosexiang16' were rootstock. Pop−2 (ID 20–64) included 45 species, most of which were *V. vinifera* × *V. labrusca* hybrids, 18 were *V. vinifera*, such as '1103♀' and 'maple leaf grape', and three interspecific hybrids within American populations. Among the 59 Pop−3 varieties (ID 65–123), 73% consisted of *V. vinifera*, while 24% were hybrids of *V. vinifera* and *V. labrusca*. Cultivars with relatively close relationships clustered together, such as '5BB', '101−14', 'Hongfushi' and 'Izu_Nishiki', 'Tianshan" and 'Shaoxing_1' (Figure 4C). The clustering diagram indicates that multi-allelic InDel markers can significantly reflect the genetic variation within grapevine populations and exhibit strong varietal differentiation ability. Compared to the gel-based results, the clustering tree constructed based on NGS provides a more comprehensive and accurate differentiation of grape populations (Figure 4A). In addition, the dispersion in PCA subgroups of NGS-based genotyping was clearer than that of gel-based, which coincided with the results of the hierarchical clustering (Figure 4B,D).

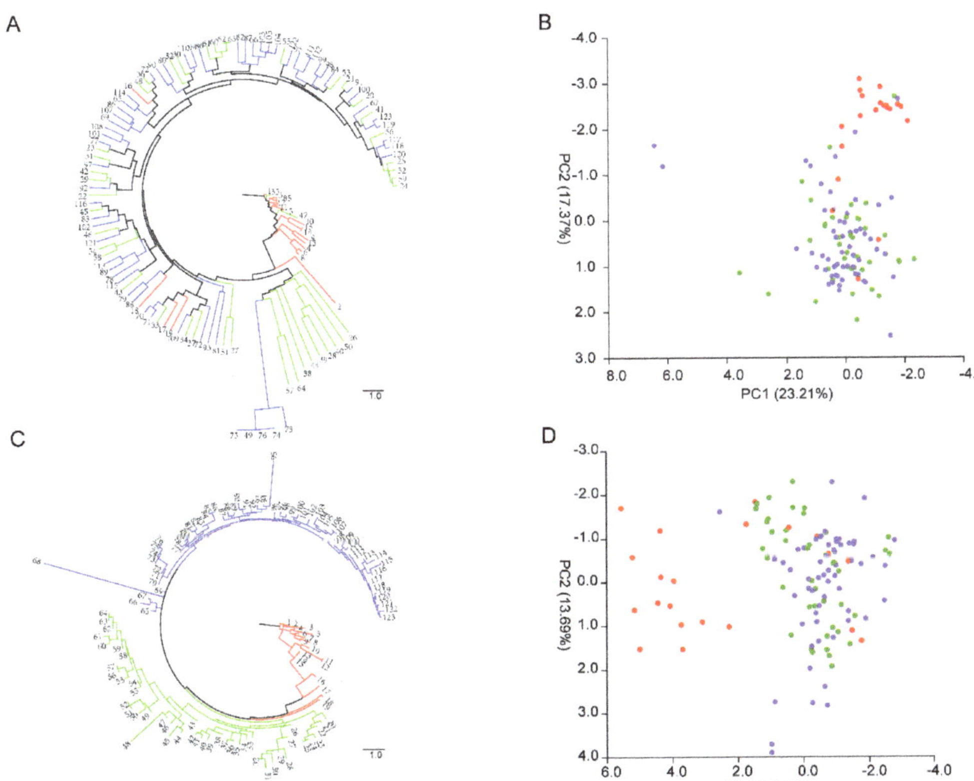

Figure 4. Population structure analysis of 123 grape cultivars. (**A**) The phylogenetic tree of 123 grape cultivars based on agarose gel electrophoresis results. (**B**) PCA analysis based on agarose gel electrophoresis results. (**C**) The phylogenetic tree of 123 grape cultivars based on the NGS results. (**D**) PCA analysis based on the NGS results. All colors were marked according to the NGS results: Pop−1 (red), Pop−2 (green) and Pop−3 (purple).

4. Discussion

The grape, an important economic fruit crop with a storied history of cultivation and diverse varieties, is ranked as one of the world's most extensively cultivated fruit crops [33]. The cultivation of *V. vinifera* extends across diverse geographical regions, facilitating frequent exchanges of cultivars and common interspecific hybridization, which complicates the taxonomy and identification of varieties [34]. InDel molecular markers, known for their polymorphism and genetic stability, are crucial tools for crop genetic improvement, varietal identification and functional genomics research. Their application spans population genetic analysis, molecular breeding and medical diagnostics, demonstrating their potential to augment genetic improvement and germplasm innovation across various species [35–38]. Despite the successful application of InDel markers in several crops and organisms, a dedicated InDel-based fingerprinting system for grapevines remains elusive. The establishment of an efficient and precise fingerprinting platform is imperative for the identification of grapevine varieties and population genetic analysis.

The rapid development of NGS has provided unprecedented opportunities for genomic research. Although the cost of NGS technology is continuously declining, the cost of high-throughput sequencing is still high, especially in large-scale sequencing projects. Multiplex sequencing is a practical approach to address the issue. By pooling the DNA or RNA of multiple samples, this approach significantly reduces experimental costs, reagent expenses and manpower resources. The Hi-TOM platform is an online platform for the sequencing of multiple samples and multiple target sites. The Hi-TOM strategy markedly conserves time and cost, proving to be more economical and expedient, especially when applied to a large number of samples or loci. To ensure the quality of multiplex sequencing, the demultiplexed method and InDel genotyping strategy are important. By fixing the bridge sequences and barcoding primers, the Hi-TOM tool has high reliability and sensitivity in tracking various mutations, especially complex chimeric mutations, frequently induced by genome editing. Consequently, multiplex sequencing based on the Hi-TOM sequencing platform was chosen for the identification of NGS-based indel marker in this study. To eliminate sequencing errors and enhance the identification accuracy of InDel markers, we developed a comprehensive filtration method for multi-allelic InDels based on several parameters, such as the read depth, heterozygosity, missing rate in population.

Multi-allelic InDels, characterized by the multiple insertions or deletions in single locus, exhibit higher polymorphism than traditional SNPs or InDels. However, the resolution of gel electrophoresis may not be able to identify all events of InDels. The high sensitivity of NGS enables the accurate detection of small insertion or deletion events, making it possible to identify more genotypes of an InDel. In this study, we developed a pipeline, including data quality control, genotype phasing and variety fingerprinting, for the analysis of NGS-based multi-allelic InDels. The NGS-based identification recalled all genotypes obtained from gel electrophoresis and demonstrated higher discrimination power.

Using 11 multi-allelic InDel markers, we performed genotyping of 123 grape cultivars, including Eurasian species, American species and European and American hybrids. The grapes exhibited a high degree of Het, with values ranging from 0.429 to 0.762, and a mean value of 0.647. These values were higher than those reported in previous studies [39]. Due to the polymorphic nature of InDel markers, their PIC values are relatively high, ranging from 0.398 to 0.981. The average PIC value in this study was 0.86, which is higher than the PIC value of 0.38 for SNP markers [40] and 0.33 for broccoli [41], indicating a higher rate of polymorphism of 11 multi-allelic InDels. The average GDI value was 0.806, which is higher than the value of 0.271 observed in winter wheat [42]. Among the InDel markers, 81.82% had GDI values ranging from 0.7 to 1, indicating the higher level of GDI of the core markers. The 11 NGS-based core markers used in this study could discriminate 122 grape varieties, resulting in an identification efficiency of 99.18%. The remaining unseparated 'Maple Leaf grapes' and 'No. 8' were round-leafed grapes. This study suggests the superiority of NGS over agarose gel electrophoresis in genetically discriminating grape varieties.

5. Conclusions

Molecular markers and fingerprints are crucial in the marker-assisted breeding and varietal identification of grapevines. In this study, we developed a novel, effective and high-throughput pipeline for multi-allelic-InDel selection and identification. Using this pipeline, we developed 11 high polymorphic multi-allelic InDel markers based on 499 resequencing data for the genetic discrimination of grapevines. We performed genetic identification of 123 grape cultivars using agarose gel electrophoresis and NGS. The NGS-based InDel markers processed a higher rate of polymorphism and showed a superior performance in the genetic discrimination of grapes compared to the gel approach. Our work provides a practical workflow for multi-allelic InDel marker development and valuable tools for the genetic discrimination and marker-assisted breeding in grapes.

Supplementary Materials: The following supporting information can be downloaded at: https://www.mdpi.com/article/10.3390/horticulturae10070752/s1, Figure S1: Agarose gel electrophoresis results of 123 grape samples using 11 core multi-allelic InDel markers; Table S1: Grape resequencing data used in this study; Table S2: 123 grape cultivars used in this study; Table S3: Information for 80 high-quality multi-allelic InDel primers; Table S4: Genotyping results for 123 cultivars based on agarose gel electrophoresis; Table S5: Genotyping results for 123 cultivars based on NGS.

Author Contributions: Conceptualization, M.L. and J.H.; methodology, N.Z. and M.L.; software, G.J.; validation, G.J., M.L., Y.Y., H.Z. (Hong Zhang), Y.G., Q.W. and R.C.; formal analysis, G.J.; investigation, M.L.; resources, J.H., Q.J., J.J., H.Z. (He Zhang) and J.W.; data curation, G.J.; writing—original draft preparation, G.J.; writing—review and editing, G.J. and M.L.; visualization, G.J.; supervision, M.L.; project administration, N.Z.; funding acquisition, J.H. All authors have read and agreed to the published version of the manuscript.

Funding: This work was supported by the Natural Science Foundation of Tianjin (23JCYBJC00770 to M.L., 22JCYBJC00190 to Y.Y.), the National Natural Science Foundation of China (32302579 to Y.Y.), the China Agriculture Research System of MOF and MARA (CARS-29-2 to J.H.), the Innovative Research of Seed Industry of Tianjin Academy of Agricultural Science (2022ZYCX014 to J.H.) and the Innovative Research and Experimental Projects for Young Researchers of Tianjin Academy of Agricultural Science (2021002 to N.Z.).

Data Availability Statement: The raw sequence data reported in this paper have been deposited in the Genome Sequence Archive in the National Genomics Data Center of China (GSA: CRA016408) that are publicly accessible at https://ngdc.cncb.ac.cn/gsa (accessed on 11 May 2024). The analysis scripts for this study are deposited in the online repository at https://github.com/Lvmingjie/indel-seq (accessed on 16 May 2024).

Conflicts of Interest: The authors declare that they have no conflicts of interest.

References

1. Dong, Y.; Duan, S.; Xia, Q.; Liang, Z.; Dong, X.; Margaryan, K.; Musayev, M.; Goryslavets, S.; Zdunić, G.; Bert, P.-F.; et al. Dual domestications and origin of traits in grapevine evolution. *Science* **2023**, *379*, 892–901. [CrossRef] [PubMed]
2. Grassi, F.; Gabriella, D.L. Back to the Origins: Background and Perspectives of Grapevine Domestication. *Int. J. Mol. Sci.* **2021**, *22*, 4518. [CrossRef] [PubMed]
3. Sabir, A.; Tangolar, S.; Büyükalaca, S.; Kafkas, S. Ampelographic and molecular diversity among grapevine (*Vitis* spp.) cultivars. *Czech J. Genet. Plant Breed.* **2009**, *45*, 160–168.
4. Zhou, Y.; Massonnet, M.; Sanjak, J.S.; Cantu, D.; Gaut, B.S. Evolutionary genomics ofgrape (*Vitis vinifera* ssp. *vinifera*) domestication. *Proc. Natl. Acad. Sci. USA* **2017**, *114*, 11715–11720.
5. Xiao, H.; Liu, Z.; Wang, N.; Long, Q.; Cao, S.; Huang, G.; Liu, W.; Peng, Y.; Riaz, S.; Walker, A.M.; et al. Adaptive and maladaptive introgression in grapevine domestication. *Proc. Natl. Acad. Sci. USA* **2023**, *120*, e2222041120. [CrossRef]
6. Martínez, M.C.; Boso, S.; Gago, P.; Muñoz-Organero, G.; De Andrés, M.T.; Gaforio, L.; Cabello, F.; Santiago, J.L. Value of two Spanish live grapevine collections in the resolution of synonyms, homonyms and naming errors. *Aust. J. Grape Wine R.* **2018**, *24*, 430–438. [CrossRef]
7. Tessier, C.; David, J.; This, P.; Boursiquot, J.M.; Charrier, A. Optimization of the choice of molecular markers for varietal identification in *Vitis vinifera* L. *Theor. Appl. Genet.* **1999**, *98*, 171–177. [CrossRef]
8. Shen, X.; Guo, W.; Zhu, X.; Yuan, Y.; Yu, J.Z.; Kohel, R.J.; Zhang, T. Molecular mapping of QTLs for fiber qualities in three diverse lines in Upland cotton using SSR markers. *Mol. Breeding* **2005**, *15*, 169–181. [CrossRef]

9. Zombardo, A.; Meneghetti, S.; Morreale, G.; Calò, A.; Costacurta, A.; Storchi, P. Study of inter-and intra-varietal genetic variability in grapevine cultivars. *Plants* **2022**, *11*, 397. [CrossRef] [PubMed]
10. Taheri, S.; Abdullah, T.L.; Yusop, M.R.; Hanafi, M.M.; Sahebi, M.; Azizi, P.; Shamshiri, R.R. Mining and development of novel SSR markers using next generation sequencing (NGS) data in plants. *Molecules.* **2018**, *23*, 399. [CrossRef]
11. Lv, Y.; Liu, Y.; Zhao, H. mInDel: A high-throughput and efficient pipeline for genome-wide InDel marker development. *BMC Genom.* **2016**, *17*, 290. [CrossRef]
12. Rimbert, H.; Darrier, B.; Navarro, J.; Kitt, J.; Choulet, F.; Leveugle, M.; Duarte, J.; Rivière, N.; Eversole, K. High throughput SNP discovery and genotyping in hexaploid wheat. *PLoS ONE* **2018**, *13*, e0186329. [CrossRef]
13. Emanuelli, F.; Lorenzi, S.; Grzeskowiak, L.; Catalano, V.; Stefanini, M.; Troggio, M.; Myles, S.; Martinez-Zapater, J.M.; Zyprian, E.; Moreira, F.M.; et al. Genetic diversity and population structure assessed by SSR and SNP markers in a large germplasm collection of grape. *BMC Plant Biol.* **2013**, *13*, 39. [CrossRef] [PubMed]
14. Wang, F.; Fan, X.; Zhang, Y.; Sun, L.; Liu, C.; Jiang, J. Establishment and application of an SNP molecular identification system for grape cultivars. *J. Integr. Agric.* **2022**, *21*, 1044–1057. [CrossRef]
15. Lü, Y.; Cui, X.; Li, R.; Huang, P.; Zong, J.; Yao, D.; Li, G.; Zhang, D.; Yuan, Z. Development of genome-wide insertion/deletion markers in rice based on graphic pipeline platform. *J. Integr. Plant Biol.* **2015**, *57*, 980–991. [CrossRef]
16. Liu, Z.; Zhao, Y.; Zhang, Y.; Xu, L.; Zhou, L.; Yang, W.; Zhao, H.; Zhao, J.; Wang, F. Development of Omni InDel and supporting database for maize. *Front. Plant Sci.* **2023**, *14*, 1216505. [CrossRef]
17. Guan, L.; Cao, K.; Li, Y.; Guo, J.; Xu, Q.; Wang, L. Detection and application of genome-wide variations in peach for association and genetic relationship analysis. *BMC Genet.* **2019**, *20*, 101. [CrossRef] [PubMed]
18. Wang, X.; Shen, F.; Gao, Y.; Wang, K.; Chen, R.; Luo, J.; Yang, L.; Zhang, X.; Qiu, C.; Li, W.; et al. Application of genome-wide insertion/deletion markers on genetic structure analysis and identity signature of Malus accessions. *BMC Plant Biol.* **2000**, *20*, 540. [CrossRef]
19. Liu, X.; Geng, X.; Zhang, H.; Shen, H.; Yang, W. Association and genetic identification of loci for four fruit traits in tomato using InDel markers. *Front. Plant Sci.* **2017**, *8*, 1269. [CrossRef] [PubMed]
20. Shen, D.; Liu, B.; Qiu, Y.; Zhang, X.; Zhang, Z.; Wang, H.; Li, X.; Li, S. Development and application of cucumber InDel markers based on genome re-sequencing. *Plant Genet. Res.* **2013**, *14*, 278–283.
21. Liang, S.; Lin, F.; Qian, Y.; Zhang, T.; Wu, Y.; Qi, Y.; Ren, S.; Ruan, L.; Zhao, H. A cost-effective barcode system for maize genetic discrimination based on bi-allelic InDel markers. *Plant Methods* **2020**, *16*, 101. [CrossRef] [PubMed]
22. This, P.; Lacombe, T.; Thomas, M.R. Historical origins and genetic diversity of wine grapes. *Trends Genet.* **2006**, *22*, 511–519. [CrossRef]
23. Chung, H.Y.; Won, S.Y.; Kim, Y.K.; Kim, J.S. Development of the chloroplast genome-based InDel markers in Niitaka (*Pyrus pyrifolia*) and its application. *Plant Biotechnol. Rep.* **2019**, *13*, 51–61. [CrossRef]
24. Chen, S.; Zhou, Y.; Chen, Y.; Gu, J. Fastp: An ultra-fast all-in-one FASTQ preprocessor. *Bioinformatics* **2018**, *34*, i884–i890. [CrossRef]
25. Li, H.; Durbin, R. Fast and accurate short read alignment with burrows-wheeler transform. *Bioinformatics* **2009**, *25*, 1754–1760. [CrossRef] [PubMed]
26. Li, H.; Handsaker, B.; Wysoker, A.; Fennell, T.; Ruan, J.; Homer, N.; Marth, G.; Abecasis, G.; Durbin, R. The sequence alignment/map format and SAMtools. *Bioinformatics* **2009**, *25*, 2078–2079. [CrossRef] [PubMed]
27. McKenna, A.; Hanna, M.; Banks, E.; Sivachenko, A.; Cibulskis, K.; Kernytsky, A.; Garimella, K.; Altshuler, D.; Gabriel, S.; Daly, M.; et al. The genome analysis toolkit: A MapReduce framework for analyzing next-generation DNA sequencing data. *Genome Res.* **2010**, *20*, 1297–1303. [CrossRef]
28. Rozen, S.; Skaletsky, H. Primer3 on the WWW for general users and for biologist programmers. In *Bioinformatics Methods and Protocols*; Humana Press: Totowa, NJ, USA, 1999; pp. 365–386.
29. Schuler, G.D. Sequence mapping by electronic PCR. *Genome Res.* **1997**, *7*, 541–550. [CrossRef] [PubMed]
30. Liu, Q.; Wang, C.; Jiao, X.; Zhang, H.; Song, L.; Li, Y.; Gao, C.; Wang, K. Hi-TOM: A platform for high-throughput tracking of mutations induced by CRISPR/Cas systems. *Sci. China Life Sci.* **2019**, *62*, 1–7. [CrossRef] [PubMed]
31. Price, M.N.; Dehal, P.S.; Arkin, A.P. FastTree 2–approximately maximum-likelihood trees for large alignments. *PLoS ONE* **2010**, *5*, e9490. [CrossRef] [PubMed]
32. Yang, J.; Lee, S.H.; Goddard, M.E.; Visscher, P.M. GCTA: A tool for genome-wide complex trait analysis. *Am. J. Hum. Genet.* **2011**, *88*, 76–82. [CrossRef]
33. Liang, Z.; Duan, S.; Sheng, J.; Zhu, S.; Ni, X.; Shao, J.; Liu, C.; Nick, P.; Du, F.; Fan, P.; et al. Whole-genome resequencing of 472 *Vitis* accessions for grapevine diversity and demographic history analyses. *Nat. Commun.* **2019**, *10*, 1190. [CrossRef]
34. Myles, S.; Boyko, A.R.; Owens, C.L.; Brown, P.J.; Grassi, F.; Aradhya, M.K.; Prins, B.; Reynolds, A.; Chia, J.-M.; Ware, D.; et al. Genetic structure and domestication history of the grape. *Proc. Natl. Acad. Sci. USA* **2011**, *108*, 3530–3535. [CrossRef]
35. Kvikstad, E.M.; Tyekucheva, S.; Chiaromonte, F.; Makova, K.D. A macaque's-eye view of human insertions and deletions: Differences in mechanisms. *PLoS Comput. Biol.* **2007**, *3*, e176. [CrossRef] [PubMed]
36. Jain, A.; Roorkiwal, M.; Kale, S.; Garg, V.; Yadala, R.; Varshney, R.K. InDel markers: An extended marker resource for molecular breeding in chickpea. *PLoS ONE* **2019**, *14*, e0213999. [CrossRef] [PubMed]
37. Adedze, Y.M.N.; Lu, X.; Xia, Y.; Sun, Q.; Nchongboh, C.G.; Alam, M.A.; Liu, M.; Yang, X.; Zhang, W.; Deng, Z.; et al. Agarose-resolvable InDel markers based on whole genome re-sequencing in cucumber. *Sci. Rep.* **2021**, *11*, 3872. [CrossRef]

38. Pan, G.; Li, Z.; Huang, S.; Tao, J.; Shi, Y.; Chen, A.; Li, J.; Tang, H.; Chang, L.; Deng, Y.; et al. Genome-wide development of insertion-deletion (InDel) markers for Cannabis and its uses in genetic structure analysis of Chinese germplasm and sex-linked marker identification. *BMC Genom.* **2021**, *22*, 595. [CrossRef] [PubMed]
39. Lijavetzky, D.; Cabezas, J.; Ibáñez, A.; Rodríguez, V.; Martínez-Zapater, J.M. High throughput SNP discovery and genotyping in grapevine (*Vitis vinifera* L.) by combining a re-sequencing approach and SNPlex technology. *BMC Genom.* **2007**, *8*, 424. [CrossRef] [PubMed]
40. Yang, Y.; Lyu, M.; Liu, J.; Wu, J.; Wang, Q.; Xie, T.; Li, H.; Chen, R.; Sun, D.; Yang, Y.; et al. Construction of an SNP fingerprinting database and population genetic analysis of 329 cauliflower cultivars. *BMC Plant Biol.* **2022**, *22*, 522. [CrossRef]
41. Shen, Y.; Wang, J.; Shaw, R.K.; Yu, H.; Sheng, X.; Zhao, Z.; Li, S.; Gu, H. Development of GBTS and KASP panels for genetic diversity, population structure, and fingerprinting of a large collection of broccoli (*Brassica oleracea* L. var. *italica*) in China. *Front. Plant Sci.* **2021**, *12*, 655254. [CrossRef]
42. Eltaher, S.; Sallam, A.; Belamkar, V.; Emara, H.A.; Nower, A.A.; Salem, K.F.M.; Poland, J.; Baenziger, P.S. Genetic diversity and population structure of F3: 6 Nebraska winter wheat genotypes using genotyping-by-sequencing. *Front. Genet.* **2018**, *9*, 76. [CrossRef]

Disclaimer/Publisher's Note: The statements, opinions and data contained in all publications are solely those of the individual author(s) and contributor(s) and not of MDPI and/or the editor(s). MDPI and/or the editor(s) disclaim responsibility for any injury to people or property resulting from any ideas, methods, instructions or products referred to in the content.

horticulturae

MDPI

Article

Unified Assembly of Chloroplast Genomes: A Comparative Study of Grapes Representing Global Geographic Diversity

Yue Song [1,†], Lujia Wang [1,†], Lipeng Zhang [2], Junpeng Li [1], Yuanxu Teng [2], Zhen Zhang [1], Yuanyuan Xu [1], Dongying Fan [1], Juan He [1] and Chao Ma [1,*]

1 Shanghai Collaborative Innovation Center of Agri-Seeds, School of Agriculture and Biology, Shanghai Jiao Tong University, Shanghai 200240, China; wanglj1028@sjtu.edu.cn (L.W.); dirklee41@sjtu.edu.cn (J.L.); zhangzhen_1217@sjtu.edu.cn (Z.Z.); xyy19971004@sjtu.edu.cn (Y.X.); juan.he@sjtu.edu.cn (J.H.)

2 Key Laboratory of Special Fruits and Vegetables Cultivation Physiology and Germplasm Resources Utilization of Xinjiang Production and Construction Corps, Department of Horticulture, Agricultural College of Shihezi University, Shihezi 832003, China

* Correspondence: chaoma2015@sjtu.edu.cn

† These authors contributed equally to this work and should be considered co-first authors.

Abstract: The genus *Vitis*, known for its economically important fruit—grape—is divided into three geographical groups, American, East Asian, and Eurasian, along with a hybrid group. However, previous studies on grape phylogeny using chloroplast genomes have been hindered by limited sample sizes and inconsistent methodologies, resulting in inaccuracies. In this study, we employed the GetOrganelle software with consistent parameters to assemble the chloroplast genomes of 21 grape cultivars, ensuring comprehensive representation across four distinct groups. A comparative analysis of the 21 grape cultivars revealed structural variation, showing chloroplast genome sizes ranging from 160,813 bp to 161,275 bp. In 21 *Vitis* cultivars, genome annotation revealed 134 to 136 genes, comprising 89 to 91 protein-coding genes (PCGs), 37 tRNAs, and 8 rRNAs. Our observations have pinpointed specific occurrences of contraction and expansion phenomena at the interfaces between inverted repeat (IR) regions and single-copy (SC) regions, particularly in the vicinity of the *rpl2*, *ycf1*, *ndhF*, and *trnN* genes. Meanwhile, a total of 193 to 198 SSRs were identified in chloroplast genomes. The diversification pattern of chloroplast genomes exhibited strong concordance with the phylogenetic relationships of the *Euvitis* subgenera. Phylogenetic analysis based on conserved chloroplast genome strongly clustered the grape varieties according to their geographical origins. In conclusion, these findings enhance our understanding of chloroplast genome variation in *Vitis* populations and have important implications for cultivar selection, breeding, and conservation efforts.

Keywords: chloroplast genome; grape; phylogenetic analysis

1. Introduction

Belonging to the *Vitaceae* family, grapes are a crucial crop with significant global importance due to their widespread cultivation and economic value [1]. The genus *Vitis* is divided into two subgenera, *Euvitis* and *Muscadinia*, with all economically important wine, table, and raisin grapes belonging to *Euvitis* [2,3]. The *Euvitis* subgenus encompasses approximately 70 species, comprising over 23,000 cultivated varieties, predominantly distributed across the Northern Hemisphere [4]. Geographically, the *Euvitis* subgenus can be categorized into three distinct groups, the American group, the East Asian group, and the Eurasian group, with an additional category encompassing hybrids derived from these, collectively termed the Hybrid group [5]. Currently, the dominant grape species under cultivation largely comprise *Vitis vinifera* cultivars, originating within the Eurasian group, and are primarily utilized for the purpose of winemaking [6]. However, hybrids have gained significant traction in certain regions, notably for table grape cultivation [7]. In light of recent efforts to explore the diversity of grape genetic resources and to develop a plethora of high-quality hybrid cultivars, there is a paucity of comprehensive studies elucidating the evolution of *Vitis* [8].

The chloroplast, a pivotal organelle in plants, serves not only as the site of photosynthesis but also as a hub for a myriad of other metabolic pathways [9,10]. Characteristically, the chloroplast contains its own genome, which is distinct from the nuclear genome, and it typically lacks recombination, following a uniparental mode of inheritance [11]. The chloroplast genome is rich in a large number of diverse gene loci and non-coding regions, replete with substantial DNA sequence data, and it has been lauded as a potent instrument for phylogenetic analysis [12,13]. This facilitates the advancement of species identification methodologies and the formulation of conservation strategies. The complete chloroplast genome is thus regarded as a foundational resource for unraveling phylogenetic evolution and for the accurate classification of plant relatives.

Advancements in next-generation sequencing technology have made it relatively easy to obtain the complete chloroplast genome of plants, such as grapes [14]. Most angiosperm chloroplast genomes exhibit a highly conserved structure, typically organized into a quadripartite arrangement consisting of a large single-copy (LSC) region, a small single-copy (SSC) region, and a pair of inverted repeats (IRs) [15–17]. Generally, angiosperm chloroplast genomes encode between 110 and 130 genes, with genome sizes ranging from 120 to 160 kb [18]. Variation in genome size is primarily attributed to the expansion, contraction, or even loss of IR regions [19]. In the vast majority of angiosperms, the chloroplast genome is maternally inherited.

Chloroplast genomes have proven to be invaluable tools for understanding the phylogenetic relationships among different plant taxa. For instance, studies on rice have utilized complete chloroplast genome sequences to elucidate the evolutionary links between cultivated rice and its wild relatives, providing insights for crop improvement and conservation strategies [20,21]. In the legume family, the unique structural features of chloroplast genomes, including extensive rearrangements and inversions, have been shown to play a significant role in phylogenetic studies, offering critical insights into the domestication of crops such as soybean and various species of beans [22–24]. Research on citrus chloroplast genomes has confirmed that phylogenetic analyses can uncover common ancestors among related genera and provide information on selective pressures influencing gene evolution, as observed in the *matK* and *ndhF* genes [25,26]. These findings underscore the broad applicability of chloroplast genomics in reconstructing evolutionary histories and understanding the adaptive mechanisms of diverse plant lineages.

Over the last two decades, researchers have extensively explored the phylogenetic relationships within the *Vitaceae* family and the *Vitis* genus. In a foundational molecular study, Soejima and Wen (2006) reconstructed the phylogeny of *Vitaceae*, proposing five major clades [27]. Despite their contribution, the limited sampling of both taxa and molecular markers hindered the resolution of deeper evolutionary relationships within the family. With the development of next-generation sequencing (NGS) technologies, Zhang et al. undertook a comprehensive genomic study of chloroplast and mitochondrial sequences from 27 species [28]. Their findings further resolved the deep phylogenetic structure of *Vitaceae*, strongly supporting the division of the family into five distinct clades. This work underscores the value and reliability of chloroplast genome analysis in constructing phylogenetic frameworks.

Numerous studies have employed chloroplast genomes to investigate the phylogenetic relationships of *Vitaceae* and its evolutionary position within the rosid clade and angiosperms [29,30]. However, phylogenetic research specifically addressing species within the *Vitis* genus using chloroplast genome data is still insufficient. Comparative analyses of *Vitaceae* chloroplast genomes remain incomplete, with some studies lacking coverage of all four major *Vitis* groups, which restricts their capacity to fully represent the diversity of *Vitis* species. Furthermore, in the realm of comparative chloroplast genomics, disparities in the choice of reference genomes and assembly methodologies employed hinder the direct comparability of assembled chloroplast sequences. The primary objective of this study is to provide a comprehensive, unified analysis of the chloroplast genomes of *Vitis* species, addressing gaps in previous research that have either lacked complete representation of the major *Vitis* groups or employed inconsistent methods for genome assembly. Specifically, we aim to assemble high-quality, consistent chloroplast genomes across all four major *Vitis* groups, compare their structural variation, and construct a phylogenetic tree that accurately reflects the evolutionary relationships within the *Vitis* genus. This will help to overcome the limitations of previous studies by offering a standardized reference for chloroplast genome analysis, ensuring greater comparability and consistency in future research. In the present study, we amassed, assembled, and annotated the genome sequencing data of 21 grapevine cultivars, spanning four distinct clusters, leveraging a unified reference genome and assembly protocol. Subsequently, we conducted a detailed comparative genomic analysis of the chloroplast genomes of the selected species, employing a variety of bioinformatic tools to assess sequence variation, structural differences, and phylogenetic relationships, aiming to decipher the phylogenetic trajectories of the select species within this diverse assortment.

2. Materials and Methods

2.1. Collection of Raw Data for Grapes Sequencing

Our investigation hinged upon a broad spectrum of genome sequences, encompassing those newly acquired within our laboratory's confines and those previously reported by other academic research entities. The raw sequencing data for these grapes are publicly accessible via the NCBI Sequence Read Archive (SRA) database or the National Genomics Data Center (NGDC) database, with specific accession numbers detailed in Table S1.

2.2. Genome Assembly and Annotation

In this study, we processed a dataset of 21 samples. First, we performed a rigorous quality control step by assessing the raw data using FastQC (version 0.12.1) to identify any potential issues [31]. Subsequently, Trimmomatic (version 0.39) was employed with default parameters to meticulously filter out adapters and low-quality reads, ensuring a clean dataset for downstream analysis [32]. In our efforts, we utilized GetOrganelle (version 1.7.1) to execute a highly efficient de novo assembly [33], applying the following parameters: the chloroplast type was "embplant_pt", the assembly process was iterated 10 times, and we specified a k-mer size of 121 for the SPAdes assembly algorithm. The circular plastid graphs generated were validated using Bandage (version 0.8.1), ensuring the quality and authenticity of the assembled sequences [34]. Annotation of the plastomes from the 21 *Vitis* cultivars was carried out with the aid of the GeSeq (https://chlorobox.mpimp-golm.mpg.de/geseq.html, accessed on 3 September 2024) tool, adopting *Vitis vinifera* (GenBank accession NC 007957) as the guiding reference [35]. To visually provide a portrayal of the 21 genome maps of *Vitis* cultivars, we employed OGDRAW (https://chlorobox.mpimp-golm.mpg.de/OGDraw.html, accessed on 3 September 2024), offering a detailed visualization that captures the intricacies of the genomic structure [36].

2.3. Comparative Analysis of Chloroplast Genomes

To elucidate the extent of plastome divergence across 21 *Vitis* cultivars, we conducted a comparative genomic analysis utilizing the mVISTA (https://genome.lbl.gov/vista/mvista/submit.shtml, accessed on 5 September 2024), online tool in the default mode, with *Vitis vinifera* as our reference species [37]. Additionally, we aligned the complete chloroplast genome sequences using the MAFFT (https://mafft.cbrc.jp/alignment/server/, accessed on 5 September 2024)online server, configuring the parameters to "Same as input" for UPPERCASE/lowercase and output order, and selecting "Adjust direction according to the first sequence" for accurate alignment [38]. For a nuanced assessment of nucleotide diversity (π), we employed DnaSP (version 6), utilizing a window length of 600 bp and a step size of 200 bp, offering a fine-grained view of genetic diversity within the species [39]. The visualization of the boundaries of the inverted repeat (IR) regions was achieved through the use of IRScope (version 0.1.R) [40]. To identify dispersed repeats, encompassing forward (F), palindromic (P), reverse (R), and complementary (C) variants, we utilized REPuter (https://bibiserv.cebitec.uni-bielefeld.de/reputer, accessed on 5 September 2024), configuring the maximum computed repeats threshold at 50 bp and the minimal repeat size at 30 bp, while allowing for a Hamming distance of 3 bp, ensuring a rigorous analysis framework [41]. Furthermore, we utilized MISA (https://webblast.ipk-gatersleben.de/misa/, accessed on 5 September 2024) to detect simple sequence repeats (SSRs) across *Vitaceae* species, setting detection thresholds at 12 for mononucleotide, 6 for dinucleotide, 5 for trinucleotide, 3 for tetranucleotide, pentanucleotide, and 2 for hexanucleotide SSRs, thereby ensuring a thorough examination of repetitive genomic motifs [42].

2.4. Phylogenetic Tree Construction

In our analysis, we constructed a phylogenetic tree leveraging a dataset that integrates 21 chloroplast (cp) genome sequences previously employed in this manuscript, in conjunction with 27 additional sequences obtained from the SRA database. The selection of outgroups was a critical step in our analysis; we opted for *Parthenocissus heptaphylla* and *Parthenocissus tricuspidata* to serve in this capacity, providing a comparative baseline that facilitated the elucidation of the evolutionary relationships within the *Vitis* lineage.

2.5. Statistical Analysis

Statistical analysis of the data was performed using Microsoft Excel (2016). The bar charts were created in Microsoft Excel. Heatmap visualizations for the relative synonymous codon usage (RSCU) values was generated using TBtools software, 2.136 (https://github.com/CJ-Chen/TBtools, accessed on 13 September 2024).

3. Result

3.1. Chloroplast Genome Features of Vitis Species

The complete chloroplast genomes of 21 species in the *Vitis* were successfully assembled and a consistent pattern in the overall structure, gene sequences, orientation, and GC content was identified (Figure S1). The lengths of the 21 *Vitis* plastomes ranged between 160,813 bp (*V. heyneana*) and 161,275 bp (*V. rotundifolia*). In each case, the cp genomes featured the standard arrangement of four connected sections: a pair of inverted repeats (IRs) spanning 26,354 to 26,411 bp, separated by a large single-copy (LSC) region from 89,033 to 89,464 bp and a small single-copy (SSC) region ranging from 18,941 to 19,075 bp. The IR regions exhibited a higher GC content (42.92–42.96%) than the LSC region (35.25–35.35%) and SSC region (31.52–31.69%). In 21 *Vitis* cultivars, genome annotation revealed a total of 134 to 136 genes, including 89 to 91 protein-coding genes (PCGs), 37 tRNAs, and 8 rRNAs (Table S2). The LSC region contained 84 to 86 genes, in contrast to the SSC region, which housed only 11 to 12 genes. The IRa and IRb regions harbored 17 and 16 genes, respectively, with an additional 5 to 6 genes located at the junctions between these regions, spanning four connection points (Table S1).

A Python-generated graphical map of circular genomes was constructed to examine sequence variations among the 21 chloroplast genomes in *Vitis* (Figure 1). This map provides an in-depth look at both conserved and variable regions of the plastomes. The GC skew is visualized by red lines, revealing strand bias often related to replication origins or selective pressures. The GC content, represented by black lines, highlights regions that may be influenced by different evolutionary constraints, with intergenic regions showing greater variation in both GC content and skew compared to the more conserved coding regions.

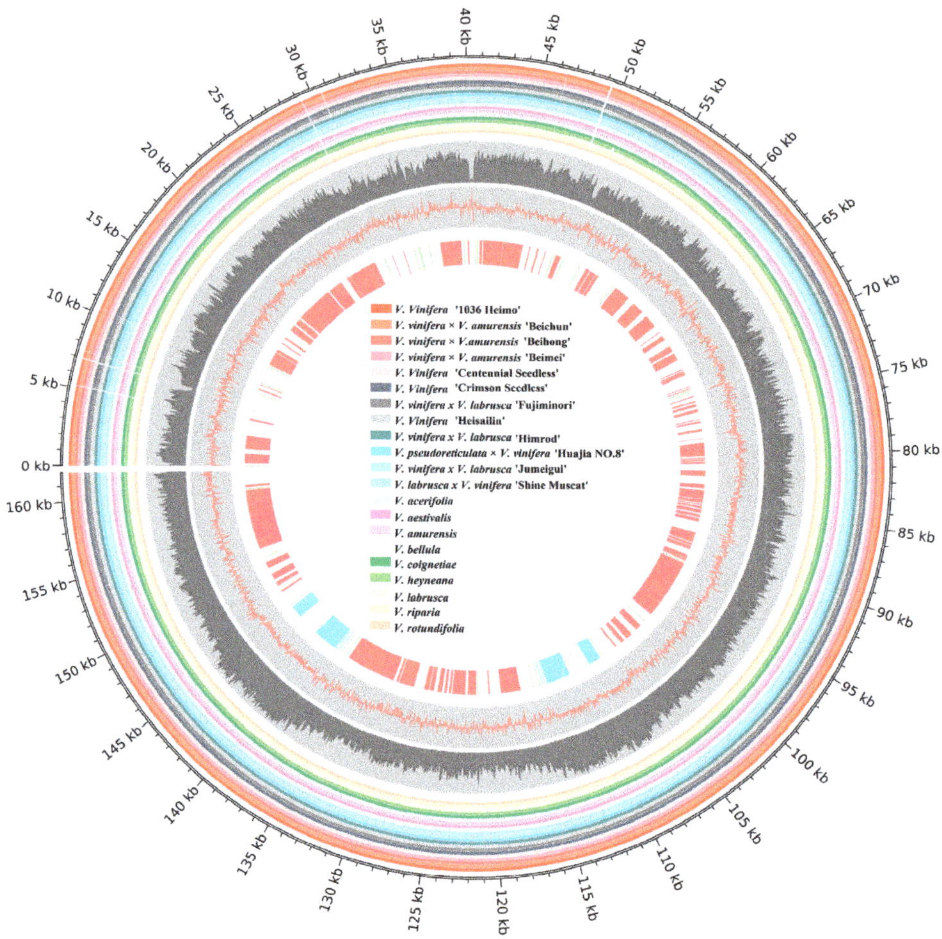

Figure 1. Graphical map of circular genomes, providing the overall visualization of the 21 *Vitis* plastomes. A circos diagram, crafted through Python, was employed to visually represent the organization of the 21 *Vitis* plastomes. The innermost level shows the arrangement of CDS genes (blue), rRNA genes (red), and tRNA genes (green) along the DNA strands, highlighting the functional elements and their distribution within the plastid genomes. GC Skew: Moving outward, the diagram shows the GC skew with red lines, indicating the difference in guanine (G) and cytosine (C) content between the forward and reverse DNA strands. GC Content: Next, the GC content is shown as black lines, providing a quantitative measure of the genomic content of guanine and cytosine bases. BLAST Analysis: The intermediate rings show the results of BLAST analyses of the plastome sequences. Conserved regions across the grape varieties are color-coded according to the specific varieties, with each variety's corresponding color indicated in the center of the diagram. Variable loci are highlighted in white, signifying regions where significant differences exist among the 21 *Vitis* plastomes. The outermost ring denotes the plastid genome size in kilobase pairs.

In the plastid genomes of *Vitis* cultivars, encompassing a repertoire of 89–91 protein-coding genes (PCGs), the initiation codons of *rps19*, *psbC*, and *psbL* exhibit non-canonical configurations. Specifically, GTG replaces the standard ATG as the start codon in both *rps19* and *psbC*, while ACG assumes the role of the initiation signal for *psbL*.

Our analysis of various grape species revealed a high degree of concordance in relative synonymous codon usage (RSCU) values. Particularly, the RSCU values for TGG (1), GGC (0.335–0.336), CAG (0.444–0.447), and CAA (1.553–1.556) varied within narrow margins. The uniformity in codon usage suggests a common genetic strategy or physiological constraint that may have shaped the evolution of grapevine genomes. Otherwise, AGA (1.839–1.855), TTA (1.818–1.849), and GCT (1.824–1.848) displayed high RSCU values. In contrast, AGC (0.324–0.333), CGC (0.326–0.333), and GGC (0.335–0.336) exhibited low RSCU values. In the genetic code, 31 codons have an RSCU over 1, with the exception of AUG, the initiator of translation, and UUG, which terminates invariably in the nucleotides Adenine or Uracil (Figure 2 and Table S3).

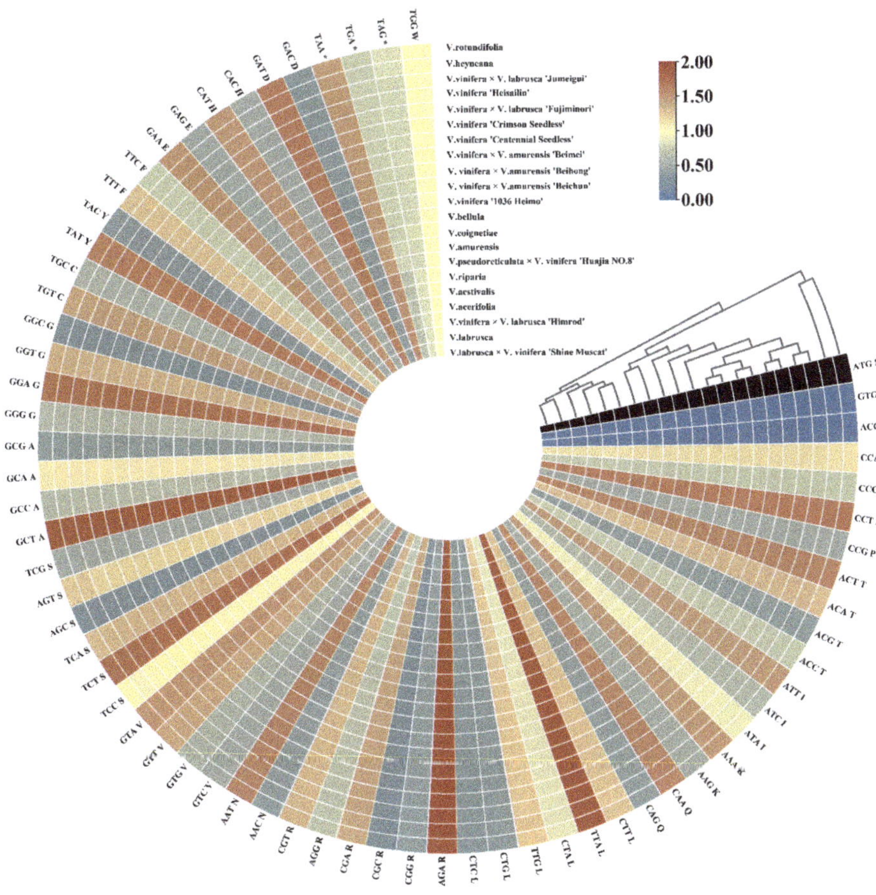

Figure 2. Relative synonymous codon usage (RSCU) values in the 21 grape chloroplast genomes. * Red represents higher RSCU values, while blue indicates lower RSCU values.

A quantitative analysis of simple sequence repeats (SSRs) across multiple grape cultivars was conducted, and their abundances were evaluated. The number of SSRs in the 21 chloroplast genomes examined ranged from 193 to 198, as shown in Figure 3A. *V. bellula*, *V. vinifera* '1036 Heimo', and *V. rotundifolia* had the highest SSR counts (198), while *V. acerifolia* exhibited the lowest (193). The most common SSR type identified was the hexa-nucleotide (P6), with the majority of SSRs ranging from 12 to 15 bp in length, playing a crucial role in the development of molecular markers for varietal identification (Tables S4 and S5).

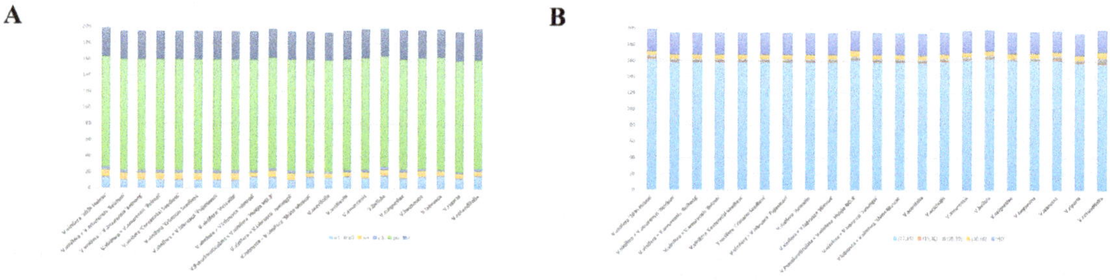

Figure 3. Simple sequence repeats (SSRs) in the 21 *Vitis* chloroplast genomes. (**A**) A bar chart illustrating the distribution of different SSR types across the 21 *Vitis* plastomes, with p1 through p6 representing mono-, di-, tri-, tetra-, penta-, and hexanucleotide repeats, respectively, and 'c' denoting compound SSRs. (**B**) A bar chart displaying the distribution of SSRs based on size.

Based on REPuter software, we identified between 40 and 46 long repetitive sequences (Figure 4 and Table S6). '1036 Heimo' and 'Huajia NO.8' had the fewest (40), while 'Himrod' had the most (46). Additionally, there were notable differences in the counts of P, F, R, and C types. Typically, the P, F, R, and C types occurred 21–23, 17–18, 2–4, and 0–1 time. Notably, this includes 18 occurrences of the P type in *V. rotundifolia*, 16 and 19 of the F type in *V. pseudoreticulata* × *V. vinifera* 'Huajia NO.8' and *V. labrusca* × *V. vinifera* 'Shine Muscat', 5 of the R type in *V. rotundifolia*, and 2 of the C type in *V. vinifera* × *V. labrusca* 'Himrod'.

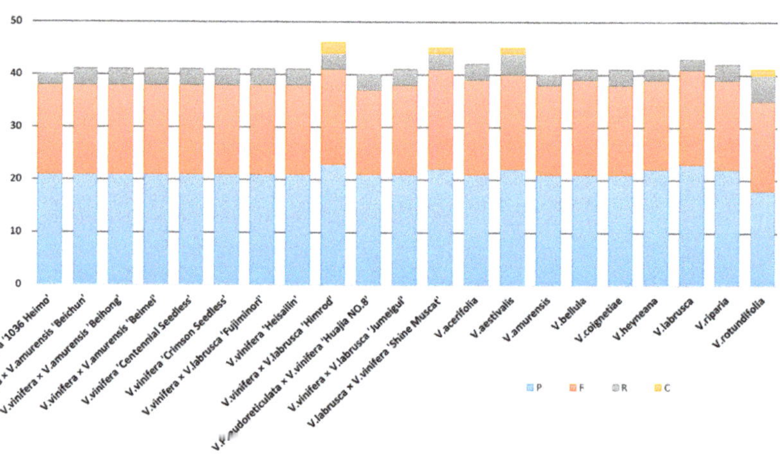

Figure 4. Distribution of four types of long repetitive sequences. Long repetitive sequences in the 21 *Vitis* plastomes, categorized as forward (F), reverse (R), complement (C), and palindromic (P) types.

3.2. The Comparison of Chloroplast Genome Sequences

Visualization of the IR/LSC and IR/SSC boundary locations highlights several key gene positions (Figure 5). For instance, the *rps19* and *rpl2* genes consistently appear near the LSC/IRb boundary in all plastomes, with *rps19* spanning 233 bp in the LSC region, except in *V. rotundifolia*, where it measures 234 bp. Likewise, *rpl2* is positioned 114–116 bp from the LSC/IRb boundary, except for *V. rotundifolia*, where it is 124 bp. The *ycf1* and *ndhF* genes are located at the IRb/SSC junction. With the exception of *V. rotundifolia*, *V. riparia*, and *V. coignetiae*, the positions of *ycf1* and *ndhF* follow two distinct patterns across the other cultivars. Similarly, these two genes were identified at the IRa/SSC boundary of most chloroplast genomes. Similarly, we found *rps19* and *trnH* genes at the IRa/LSC boundary. This suggests that these genes are conserved in the chloroplast genome.

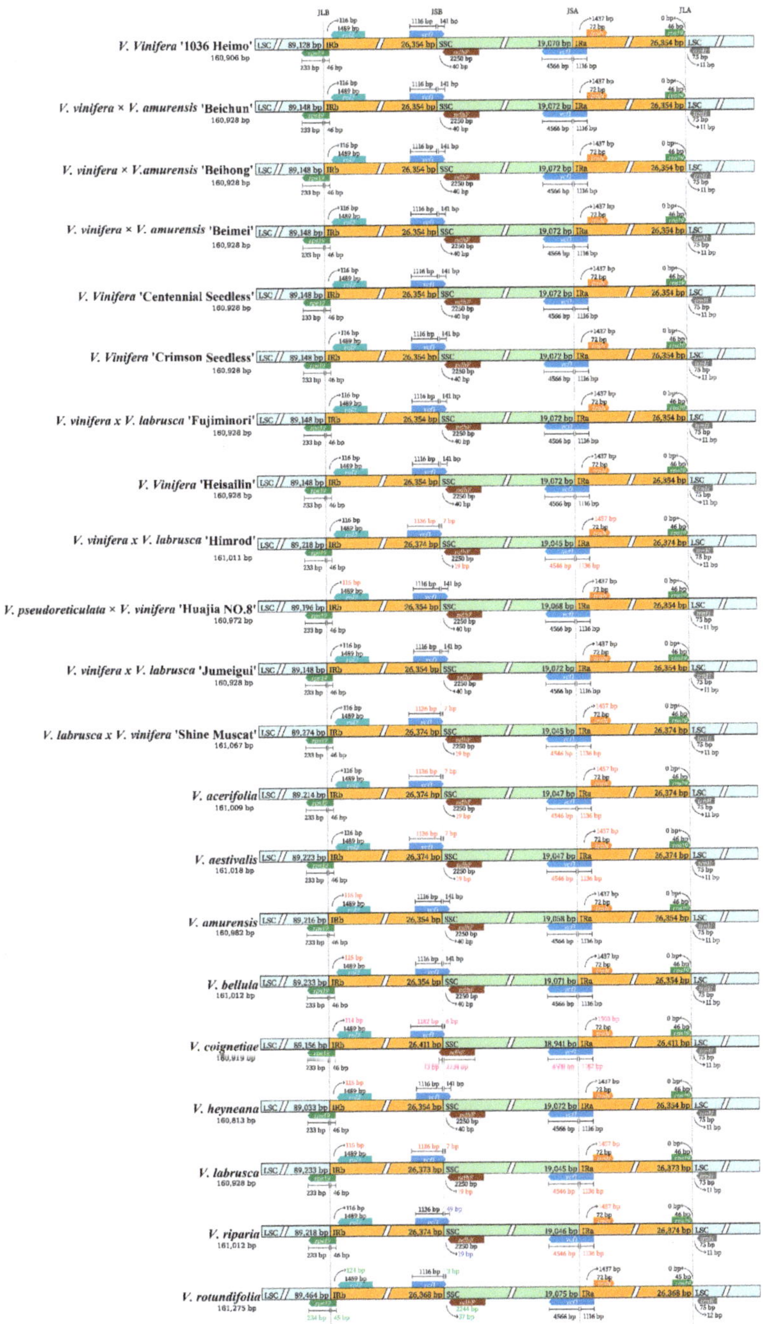

Figure 5. Comparison of the junctions between the LSC/SSC and IR regions among the 21 *Vitis* chloroplast genomes, using cyan for LSC, orange for IRa and IRb, green for SSC.

Utilizing the Mauve (version 2.4.0), software, we assessed the collinearity of 21 *Vitis* cultivars at the chloroplast genome level, as depicted in Figure 6. Our analysis reveals no instances of inversion or rearrangement within the chloroplast genomes, indicating a high degree of conservation in the genomic architecture across these varieties. However, sequence alignment identified 11 mutation hotspots; further functional annotation highlighted three protein-coding genes (*petB*, *ycf1*, and *rpl32*) with Pi > 0.005, and eight intergenic and non-coding regions (*trnK—rps16*, *rps16—trnQ*, *trnS—trnG*, *psbZ—trnG*, *ndhC—trnV*, *accD—psaI*, *ndhF—rpl32*, *trnL—ccsA*) with Pi > 0.005 (Figure 7). Further examination of the highly variable genes revealed that most were linked to photosynthesis, involving subunits of NADH-dehydrogenase (*ndhC* and *ndhF*), photosystem I (*psaI*), and photosystem II (*psbZ*). This suggests that the photosynthetic system is regulated by these genes.

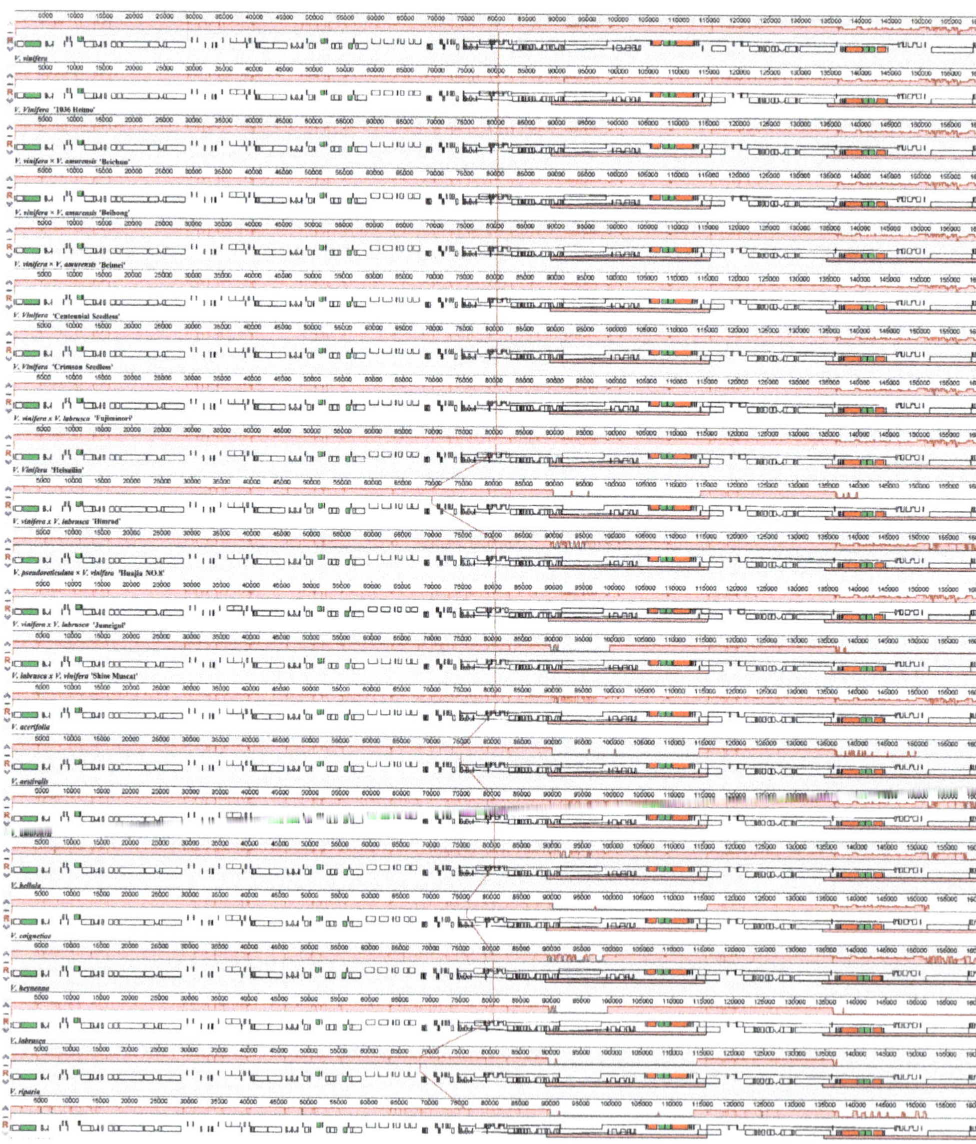

Figure 6. The Mauve alignment of 21 chloroplast genomes from *Vitis*. The reference genome utilized in this analysis is *Vitis Vinifera*.

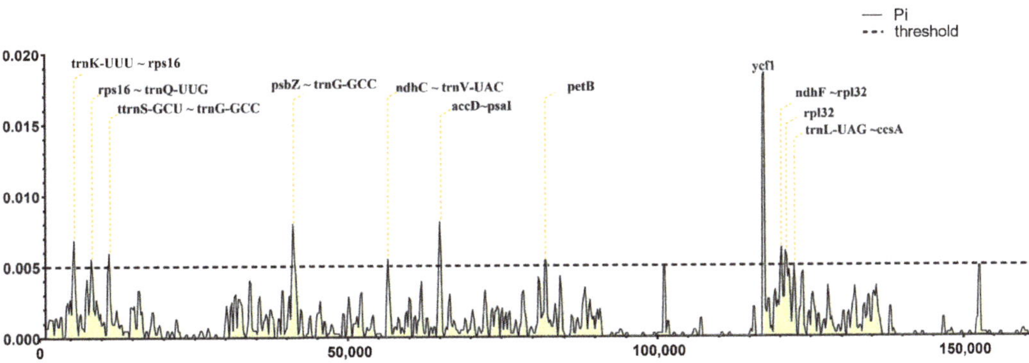

Figure 7. The Mauve alignment of 21 chloroplast genomes from *Vitis*. The reference genome utilized in this analysis is *Vitis* Vinifera.

Employing the annotated genome sequence of *Vitis vinifera* (NC_007957) as a benchmark, an mVISTA analysis was conducted to identify regions of divergence within the multiple alignments of 21 *Vitis* chloroplast genomes (Figure 8). The analysis revealed that the highest levels of genetic variation were predominantly localized in the non-coding sequence (CNS) regions, including intervals such as between *rps16* and *trnQ—UUG*, *trnS—GCU* and *trnG—GCC*, *trnC—GCA* and *petN*, and *psbZ* and *trnG—GCC*, among others. Conversely, the untranslated regions and the majority of the coding sequence (CDS) regions exhibited a high degree of conservation, with only minor sequence variations observed in select genes, such as *atpA*. This suggests that while functional regions critical for essential processes are highly conserved, the non-coding regions are thought to be less constrained by selective pressures, allowing for higher variability that may reflect evolutionary processes.

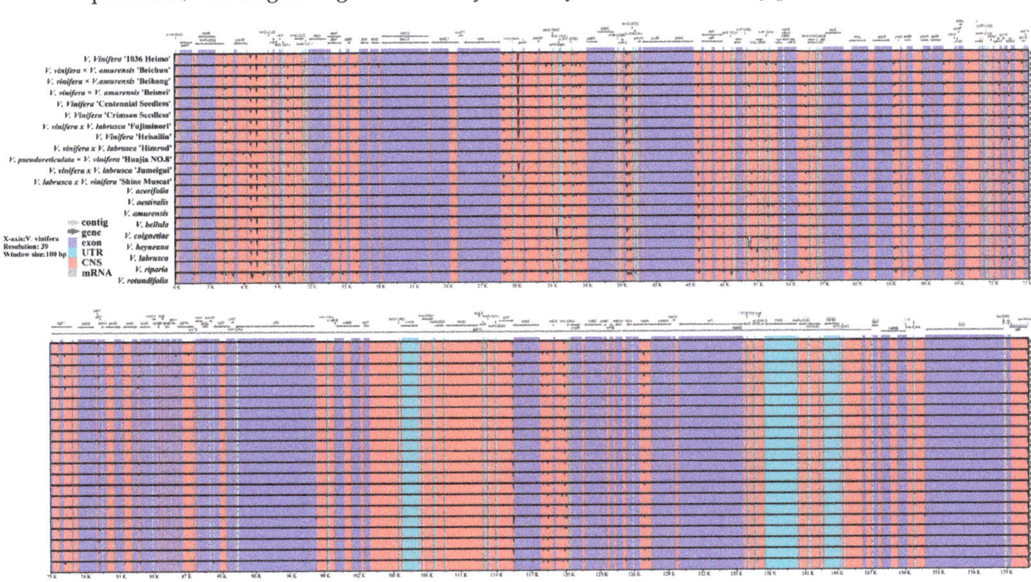

Figure 8. mVISTA alignment for chloroplast genomes. Illustrated is an alignment of complete chloroplast genomes from 21 *Vitis* species, with *Vitis vinifera* serving as the reference. Gray arrows indicate gene direction, dark blue areas denote exons, light-blue signifies untranslated regions (tRNA and rRNA), and pink shows non-coding sequences (CNS). Sequence identity is depicted on the vertical scale, spanning 50% to 100%.

3.3. Phylogenetic Analyses

To elucidate the evolutionary relationships among *Vitis* cultivars, we conducted a maximum likelihood phylogenetic analysis of 48 complete chloroplast genome sequences using IQ-TREE2 (version 2.0.7) software (Figure 9). This dataset includes representative genomes from various species, ensuring both the comprehensiveness and representativeness of the analysis. In the resulting phylogenetic trees, distinct colors were used to differentiate the various clades within the *Vitis* genus, clearly illustrating interspecies evolutionary relationships. To ensure precise rooting of the phylogenetic tree, two outgroup species from the genus *Parthenocissus* were incorporated, highlighted in gray. This strategy improved the confidence of the phylogenetic inference and provided robust support for rooting *Vitis*. The results revealed that all *Vitis* cultivars formed a well-supported monophyletic group, further validating the hypothesis of *Vitis* as a distinct evolutionary lineage. Two primary branches within the genus *Vitis*, corresponding to the subgenera *Vitis* and *Muscadinia* (marked in purple), received strong bootstrap support. To further explore the evolutionary relationships within the subgenus *Euvitis*, a more detailed analysis was conducted, revealing two major geographical clades, the Eurasian clade and the American group, with the latter marked in blue. This division correlates closely with geographic distribution, underscoring the significant role of regional flora in species evolution. Additionally, the Eurasian clade was subdivided into an East Asian group (red) and a Eurasian group (green), reflecting the radiation and adaptive divergence of grape species across Eurasia. These findings not only offer fresh insights into the phylogeny and taxonomy of *Vitis*, but also provide a strong theoretical foundation for understanding the evolutionary diversity within this genus. Furthermore, the distinct branches in the phylogenetic tree clarify the global distribution patterns and evolutionary relationships of *Vitis* cultivars, offering key insights into their origin, migration routes, and environmental adaptation.

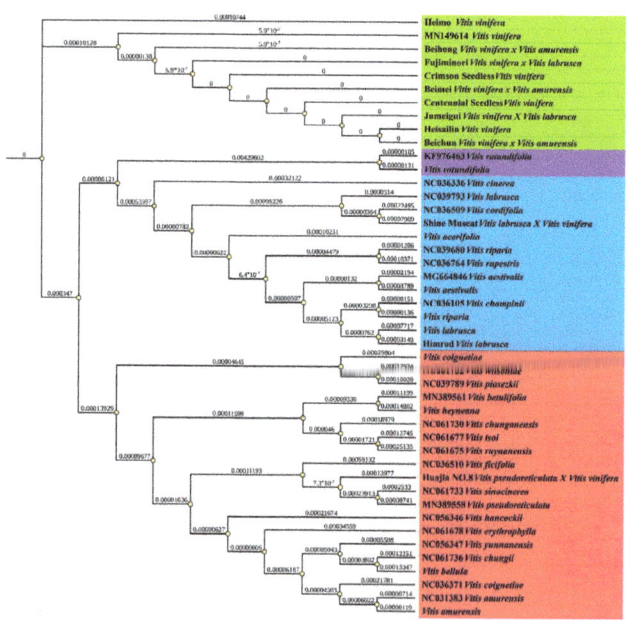

Figure 9. Phylogenetic tree of 48 *Vitis* cultivars. A maximum likelihood (ML) phylogenetic tree of the complete chloroplast genomes was constructed using *Parthenocissus* as the outgroup. Cultivars were color-coded based on their taxonomic status and geographic groups, with the outgroup in gray, subgenus *Muscadinia* in purple, and the three geographic groups of subgenus *Euvitis*—American, East Asian, and Eurasian—colored blue, green, and red, respectively. The numbers at the nodes of the phylogenetic tree represent branch lengths.

4. Discussion

4.1. The Architecture Features of the Chloroplast Genome

The chloroplast genome has long been recognized as a key genetic tool for reconstructing the evolutionary histories of plant species, including grapes. Its well-documented circular DNA structure, comprising a large single-copy (LSC) region, a small single-copy (SSC) region, and two inverted repeat (IR) regions, mirrors the structure found in previous studies of *Vitis* chloroplast genomes [15–17]. In this analysis of 21 grape varieties, the number of annotated genes ranged from 134 to 136, with the differences mainly due to gene copy number variation, while no unique, variety-specific genes were identified. This high level of gene conservation is consistent with earlier work on grape and other plant chloroplast genomes, which have shown limited variability at the gene level [43]. It shows that grape chloroplast genomes shared a high degree of similarity in length, GC content, codon usage, and distribution of SSR loci and repetitive sequences [44]. While most coding sequences (CDSs) and untranslated regions were conserved, minor variations were found in certain genes such as *atpA*. *V. bellula*, and *V. vinifera*. The '1036 Heimo' are distinguished by their high SSR counts (198), whereas *V. acerifolia* has the lowest (193) [45,46]. Among the SSRs, hexa-nucleotides (P6) are the most abundant. It is speculated that P6 type SSRs play a significant role in the dynamics of chloroplast genomes (Figure 2, Table S3) [47]. The variations observed in specific genes and SSRs, such as those in atpA, could play a role in the species' ability to adapt to specific environmental conditions. For instance, certain SSR loci might be involved in modulating gene expression in response to abiotic stressors like temperature extremes or drought, potentially contributing to the ecological success of specific varieties in distinct geographical regions. Further functional studies are needed to explore how these variations might influence the adaptability of these grape species. These findings, like those in previous studies, open new avenues for developing molecular markers and genetic tools to explore the evolutionary and taxonomic relationships within the *Vitis* genus.

4.2. The Chloroplast Genome Comparison

The expansion and contraction of the inverted repeat (IR) and single-copy (SC) boundary regions are regarded as pivotal evolutionary mechanisms that contribute to chloroplast genome size variation [48]. In this study, as shown in Figure 5, a comparative analysis of IR/SC junction positions across *Vitis* plastomes revealed substantial diversity. *V. rotundifolia* presented a unique boundary configuration, setting it apart from the other 20 *Vitis* cultivars. Generally, closely related species are expected to display similar responses to environmental stimuli, leading to modest changes in their IR boundaries, involving a small gene repertoire. The variation observed in the IR/SC junctions in this study aligns with the major phylogenetic clades within *Vitis*, suggesting that these boundary modifications may reflect deeper evolutionary splits, further corroborating trends seen in previous chloroplast genome studies [43,49].

4.3. The Phylogeny Revealed by Chloroplast

Phylogenetic relationships delineated through chloroplast genome analysis carry significant weight in elucidating the evolutionary lineage of plant species. The progression in genome sequencing and assembly technologies has offered a new dimension to phylogenetic investigations of the *Vitis* genus, capitalizing on the conserved and maternally inherited attributes of plant chloroplast genomes. Accordingly, a phylogenetic tree utilizing the complete chloroplast genome sequence was constructed. The phylogenetic structure generated by IQtree showed high support values for branches corresponding to grape cultivars from different geographic origins, whereas lower bootstrap values were observed within cultivar groups from the same geographic region. The 48 chloroplast genomes were broadly segregated into five major clades: an outgroup, *V. rotundifolia*, American group, East Asian group, and Eurasian group. The phylogenetically close relationship among Himrod, Shine Muscat, Beimei, and Jumeigui grapes further validates the maternal inheritance trait. Unlike prior research that reported a confusing intermingling between the subgenus *Muscadinia* and subgenus *Euvitis* based on chloroplast rbcL and UTL sequences, the phylogenetic tree constructed herein affirms a clear demarcation of *Vitis* into subgenus *Euvitis* and *Muscadinia* [50].

5. Conclusions

An in-depth comparative study of the chloroplast genome structure across 21 *Vitis* cultivars has uncovered valuable genomic resources, with the high degree of sequence conservation providing invaluable insights into this subfamily and its evolutionary lineage. Concurrently, the identification of 19 hotspot regions of mutation, comprising nine protein-coding genes—namely *atpF, rpoC2, rps18, psbC, atpB, rbcL, rps20, ycf1*, and *ycf15*—and 10 non-coding regions, including *trnK—rps16, rps16—trnQ, trnE—trnT, psbZ—trnG, ndhC—trnV, accD—psaI, ycf2—trnL, ndhF—rpl32, ccsA—ndhD*, and *trnL—ycf2*—highlights the potential of these photosynthesis-related variation sites as specific DNA barcodes. Additionally, a maximum likelihood phylogenetic analysis of 48 complete chloroplast genomes revealed that all *Vitis* cultivars formed a well-supported monophyletic group, confirming *Vitis* as a distinct evolutionary lineage. Two primary branches within the genus were identified, corresponding to the subgenera *Euvitis* and *Muscadinia*. Within the subgenus *Vitis*, further analysis showed two major geographical clades: the Eurasian and American groups. The Eurasian clade was further subdivided into an East Asian and a Eurasian group, highlighting the influence of geography on species evolution. These findings provide new insights into the evolutionary diversity, global distribution patterns, and adaptive differentiation of *Vitis* species.

Supplementary Materials: The following supporting information can be downloaded at https://www.mdpi.com/article/10.3390/horticulturae10111218/s1, Figure S1: The chloroplast maps of 21 grape cultivars were linearized by unfolding the circular genomes at a consistent incision point for comparative analysis. Genes transcribed in the reverse and forward directions are displayed above and below the line, respectively; Table S1: Summary of major characteristics of the 21 *Vitis* chloroplast genomes; Table S2: List of protein-coding genes present in Vitis chloroplast genome; Table S3: Relative synonymous codon usage (RSCU) values in the 21 grape chloroplast genomes; Table S4 Types and numbers of SSRs detected in 21 *Vitis* cultivars; Table S5 Length of SSRs in the 21 *Vitis* cultivars; Table S6: The number of four long repeat types in the 21 *Vitis* cultivars.

Author Contributions: Conceptualization, Y.S., L.W. and C.M.; Data curation, Y.S., L.W., L.Z., J.L. and Y.T.; Funding acquisition, C.M.; Investigation, Y.S., Z.Z. and D.F.; Methodology, Y.S., L.W.; Project administration, C.M.; Resources, Y.X.; Software, Y.S. and L.Z.; Supervision, L.Z., J.H. and C.M.; Validation, Y.S., L.W.; Visualization, Y.S., L.W.; Writing—original draft, Y.S., L.W.; Writing—review and editing, L.Z., J.L., J.H. and C.M. All authors have read and agreed to the published version of the manuscript.

Funding: This work was supported by the Agricultural Breeding Project of Ningxia Hui Autonomous Region (NXNYYZ202101), the National Natural Science Foundation of China (Grant No. 32341041, 32372673), and the earmarked fund for CARS-29.

Data Availability Statement: The data used to support the findings of this study are available from the corresponding author. The basic materials and data of this study were available from the corresponding author upon request.

Conflicts of Interest: The authors declare no conflicts of interest.

References

1. Wen, J.; Lu, L.; Nie, Z.; Liu, X.; Zhang, N.; Ickert-Bond, S.; Gerrath, J.; Manchester, S.R.; Boggan, J.; Chen, Z. A new phylogenetic tribal classification of the grape family (*Vitaceae*). *J. Syst. Evol.* **2018**, *56*, 262–272. [CrossRef]
2. Goto-Yamamoto, N.; Sawler, J.; Myles, S. Genetic Analysis of East Asian Grape Cultivars Suggests Hybridization with Wild *Vitis*. *PLoS ONE* **2015**, *10*, e0140841. [CrossRef] [PubMed]
3. Wang, X.; Xie, X.; Chen, N.; Wang, H.; Li, H. Study on current status and climatic characteristics of wine regions in China. *Vitis* **2018**, *57*, 9–16.

4. Zecca, G.; Abbott, J.R.; Sun, W.-B.; Spada, A.; Sala, F.; Grassi, F. The timing and the mode of evolution of wild grapes (*Vitis*). *Mol. Phylogenet. Evol.* **2012**, *62*, 736–747. [CrossRef]
5. Gerrath, J.; Posluszny, U.; Melville, L. *Taming the Wild Grape: Botany and Horticulture in the Vitaceae*; Springer: Berlin/Heidelberg, Germany, 2015.
6. Hussain, S.Z.; Naseer, B.; Qadri, T.; Fatima, T.; Bhat, T.A. Grapes (*Vitis vinifera*)—Morphology, Taxonomy, Composition and Health Benefits. In *Fruits Grown in Highland Regions of the Himalayas: Nutritional and Health Benefits*; Hussain, S.Z., Naseer, B., Qadri, T., Fatima, T., Bhat, T.A., Eds.; Springer International Publishing: Cham, Switzerland, 2021; pp. 103–115.
7. Morales-Cruz, A.; Aguirre-Liguori, J.A.; Zhou, Y.; Minio, A.; Riaz, S.; Walker, A.M.; Cantu, D.; Gaut, B.S. Introgression among North American wild grapes (*Vitis*) fuels biotic and abiotic adaptation. *Genome Biol.* **2021**, *22*, 1–27. [CrossRef] [PubMed]
8. Ma, Z.; Nie, Z.; Liu, X.; Tian, J.; Zhou, Y.; Zimmer, E.; Wen, J. Phylogenetic relationships, hybridization events, and drivers of diversification of East Asian wild grapes as revealed by phylogenomic analyses. *J. Syst. Evol.* **2023**, *61*, 273–283. [CrossRef]
9. Neuhaus, H.E.; Emes, M.J. Nonphotosynthetic Metabolism in Plastids. *Annu. Rev. Plant Biol.* **2000**, *51*, 111–140. [CrossRef]
10. Daniell, H.; Lin, C.S.; Yu, M.; Chang, W.J. Chloroplast genomes: Diversity, evolution, and applications in genetic engineering. *Genome Biol.* **2016**, *17*, 134. [CrossRef]
11. Bock, R.; Knoop, V. Genomics of Chloroplasts and Mitochondria. In *Advances in Photosynthesis and Respiration*; Springer: Berlin/Heidelberg, Germany, 2012.
12. Dong, W.; Liu, J.; Yu, J.; Wang, L.; Zhou, S. Highly Variable Chloroplast Markers for Evaluating Plant Phylogeny at Low Taxonomic Levels and for DNA Barcoding. *PLoS ONE* **2012**, *7*, e35071. [CrossRef]
13. Hurst, G.D.D.; Jiggins, F.M. Problems with mitochondrial DNA as a marker in population, phylogeographic and phy-logenetic studies: The effects of inherited symbionts. *Proc. R. Soc. B Biol. Sci.* **2005**, *272*, 1525–1534. [CrossRef]
14. Cronn, R.; Liston, A.; Parks, M.; Gernandt, D.S.; Shen, R.; Mockler, T. Multiplex sequencing of plant chloroplast genomes using Solexa sequencing-by-synthesis technology. *Nucleic Acids Res.* **2008**, *36*, e122. [CrossRef] [PubMed]
15. Ma, C.; Fu, P.; Wang, L.; Zhao, L.; Xu, W.; Zhang, C.; Wang, S.; Lu, J.; Song, S. The complete chloroplast genome sequence of *Vitis pseudoreticulata*. *Mitochondrial DNA Part B* **2019**, *4*, 3630–3631. [CrossRef] [PubMed]
16. Zhang, L.; Guo, D.; Liu, M.; Dou, F.; Ren, Y.; Li, D.; Wang, S.; Wang, L.; Ma, C.; Zha, Q. The complete chloroplast genome sequence of *Vitis vinifera* × *Vitis labrusca* 'Shenhua'. *Mitochondrial DNA B Resour.* **2021**, *6*, 166–167. [CrossRef] [PubMed]
17. Guo, D.; Liu, M.; Dou, F.; Ren, Y.; Li, D.; Wang, S.; Wang, L.; Ma, C.; Zha, Q.; Su, L.; et al. The complete chloroplast genome sequence of *Vitis vinifera* Muscat Hamburg. *Mitochondrial DNA B Resour.* **2019**, *5*, 117–118. [CrossRef]
18. Palmer, J.D. Comparative Organization of Chloroplast Genomes. *Annu. Rev. Genet.* **1985**, *19*, 325–354. [CrossRef] [PubMed]
19. Ma, J.; Yang, B.; Zhu, W.; Sun, L.; Tian, J.; Wang, X. The complete chloroplast genome sequence of *Mahonia bealei* (Berberidaceae) reveals a significant expansion of the inverted repeat and phylogenetic relationship with other angiosperms. *Gene* **2013**, *528*, 120–131. [CrossRef]
20. Waters, D.L.E.; Nock, C.J.; Ishikawa, R.; Rice, N.; Henry, R.J. Chloroplast genome sequence confirms distinctness of Australian and Asian wild rice. *Ecol. Evol.* **2012**, *2*, 211–217. [CrossRef]
21. Sotowa, M.; Ootsuka, K.; Kobayashi, Y.; Hao, Y.; Tanaka, K.; Ichitani, K.; Flowers, J.M.; Purugganan, M.D.; Nakamura, I.; Sato, Y.-I.; et al. Molecular relationships between Australian annual wild rice, Oryza meridionalis, and two related perennial forms. *Rice* **2013**, *6*, 26. [CrossRef]
22. Kazakoff, S.H.; Imelfort, M.; Edwards, D.; Koehorst, J.; Biswas, B.; Batley, J.; Scott, P.T.; Gresshoff, P.M. Capturing the Biofuel Wellhead and Powerhouse: The Chloroplast and Mitochondrial Genomes of the Leguminous Feedstock Tree *Pongamia pinnata*. *PLoS ONE* **2012**, *7*, e51687. [CrossRef]
23. Martin, G.E.; Rousseau-Gueutin, M.; Cordonnier, S.; Lima, O.; Michon-Coudouel, S.; Naquin, D.; de Carvalho, J.F.; Aïnouche, M.; Salmon, A.; Aïnouche, A. The first complete chloroplast genome of the Genistoid legume *Lupinus luteus*: Evidence for a novel major lineage-specific rearrangement and new insights regarding plastome evolution in the legume family. *Ann. Bot.* **2014**, *113*, 1197–1210. [CrossRef]
24. Schwarz, E.N.; Ruhlman, T.A.; Sabir, J.S.; Hajrah, N.H.; Alharbi, N.S.; Al-Malki, A.L.; Bailey, C.D.; Jansen, R.K. Plastid genome sequences of legumes reveal parallel inversions and multiple losses of rps16 in papil-ionoids. *J. Syst. Evol.* **2015**, *53*, 458–468. [CrossRef]
25. Carbonell-Caballero, J.; Alonso, R.; Ibañez, V.; Terol, J.; Talon, M.; Dopazo, J. A Phylogenetic Analysis of 34 Chloroplast Genomes Elucidates the Relationships between Wild and Domestic Species within the Genus *Citrus*. *Mol. Biol. Evol.* **2015**, *32*, 2015–2035. [CrossRef] [PubMed]
26. Caspermeyer, J. Most Comprehensive Study to Date Reveals Evolutionary History of Citrus. *Mol. Biol. Evol.* **2015**, *32*, 2217–2218. [CrossRef]
27. Soejima, A.; Wen, J. Phylogenetic analysis of the grape family (*Vitaceae*) based on three chloroplast markers. *Am. J. Bot.* **2006**, *93*, 278–287. [CrossRef]
28. Zhang, N.; Wen, J.; Zimmer, E.A. Congruent Deep Relationships in the Grape Family (*Vitaceae*) Based on Sequences of Chloroplast Genomes and Mitochondrial Genes via Genome Skimming. *PLoS ONE* **2015**, *10*, e0144701. [CrossRef] [PubMed]
29. Jansen, R.K.; Kaittanis, C.; Saski, C.; Lee, S.-B.; Tomkins, J.; Alverson, A.J.; Daniell, H. Phylogenetic analyses of *Vitis* (*Vitaceae*) based on complete chloroplast genome sequences: Effects of taxon sampling and phylogenetic methods on resolving relationships among rosids. *BMC Evol. Biol.* **2006**, *6*, 32. [CrossRef] [PubMed]

30. Zhang, L.; Meng, Y.; Wang, D.; He, G.-H.; Zhang, J.-M.; Wen, J.; Nie, Z.-L. Plastid genome data provide new insights into the dynamic evolution of the tribe Ampelopsideae (*Vitaceae*). *BMC Genom.* **2024**, *25*, 247. [CrossRef]
31. Andrews, S. FastQC: A Quality Control Tool for High Throughput Sequence Data. 2012. Available online: http://www.bioinformatics.babraham.ac.uk/projects/fastqc (accessed on 3 September 2024).
32. Bolger, A.M.; Lohse, M.; Usadel, B. Trimmomatic: A flexible trimmer for Illumina sequence data. *Bioinformatics* **2014**, *30*, 2114–2120. [CrossRef]
33. Jin, J.-J.; Yu, W.-B.; Yang, J.-B.; Song, Y.; Depamphilis, C.W.; Yi, T.-S.; Li, D.-Z. GetOrganelle: A fast and versatile toolkit for accurate de novo assembly of organelle genomes. *Genome Biol.* **2020**, *21*, 241. [CrossRef]
34. Wick, R.R.; Schultz, M.B.; Zobel, J.; Holt, K.E. Bandage: Interactive visualization of de novo genome assemblies. *Bioinformatics* **2015**, *31*, 3350–3352. [CrossRef]
35. Tillich, M.; Lehwark, P.; Pellizzer, T.; Ulbricht-Jones, E.S.; Fischer, A.; Bock, R.; Greiner, S. GeSeq—versatile and accurate annotation of organelle genomes. *Nucleic Acids Res.* **2017**, *45*, W6–W11. [CrossRef] [PubMed]
36. Greiner, S.; Lehwark, P.; Bock, R. OrganellarGenomeDRAW (OGDRAW) version 1.3.1: Expanded toolkit for the graphical visualization of organellar genomes. *Nucleic Acids Res.* **2019**, *47*, W59–W64. [CrossRef] [PubMed]
37. Frazer, K.A.; Pachter, L.; Poliakov, A.; Rubin, E.M.; Dubchak, I. VISTA: Computational tools for comparative genomics. *Nucleic Acids Res.* **2004**, *32*, W273–W279. [CrossRef]
38. Katoh, K.; Standley, D.M. MAFFT Multiple Sequence Alignment Software Version 7: Improvements in Performance and Usability. *Mol. Biol. Evol.* **2013**, *30*, 772–780. [CrossRef]
39. Rozas, J.; Ferrer-Mata, A.; Sánchez-DelBarrio, J.C.; Guirao-Rico, S.; Librado, P.; Ramos-Onsins, S.E.; Sánchez-Gracia, A. DnaSP 6: DNA Sequence Polymorphism Analysis of Large Data Sets. *Mol. Biol. Evol.* **2017**, *34*, 3299–3302. [CrossRef]
40. Amiryousefi, A.; Hyvönen, J.; Poczai, P. IRscope: An online program to visualize the junction sites of chloroplast genomes. *Bioinformatics* **2018**, *34*, 3030–3031. [CrossRef] [PubMed]
41. Kurtz, S.; Choudhuri, J.V.; Ohlebusch, E.; Schleiermacher, C.; Stoye, J.; Giegerich, R. REPuter: The manifold applications of repeat analysis on a genomic scale. *Nucleic Acids Res.* **2001**, *29*, 4633–4642. [CrossRef]
42. Beier, S.; Thiel, T.; Münch, T.; Scholz, U.; Mascher, M. MISA-web: A web server for microsatellite prediction. *Bioinformatics* **2017**, *33*, 112833. [CrossRef]
43. Zhang, L.; Song, Y.; Li, J.; Liu, J.; Zhang, Z.; Xu, Y.; Fan, D.; Liu, M.; Ren, Y.; He, J.; et al. Identification, comparative and phylogenetic analysis of eight *Vitis* species based on the chloroplast genome revealed their contribution to heat tolerance in grapevines. *Sci. Hortic.* **2024**, *327*, 12833. [CrossRef]
44. Kim, J.E.; Kim, K.M.; Kim, Y.S.; Chung, G.Y.; Che, S.H.; Na, C.S. Chloroplast Genomes of *Vitis flexuosa* and *Vitis amurensis*: Molecular Structure, Phylogenetic, and Comparative Analyses for Wild Plant Conservation. *Genes* **2024**, *15*, 761. [CrossRef]
45. Bhatt, B.S.; Awasthi, M.; George, B.; Singh, A.K. Comparative analysis of microsatellites in chloroplast genomes of lower and higher plants. *Curr. Genet.* **2015**, *61*, 665–677. [CrossRef]
46. Zhu, M.; Feng, P.; Ping, J.; Li, J.; Su, Y.; Wang, T. Phylogenetic significance of the characteristics of simple sequence repeats at the genus level based on the complete chloroplast genome sequences of Cyatheaceae. *Ecol. Evol.* **2021**, *11*, 14327–14340. [CrossRef] [PubMed]
47. Cipriani, G.; Marrazzo, M.T.; Di Gaspero, G.; Pfeiffer, A.; Morgante, M.; Testolin, R. A set of microsatellite markers with long core repeat optimized for grape (*Vitis* spp.) genotyping. *BMC Plant Biol.* **2008**, *8*, 127. [CrossRef] [PubMed]
48. Yan, M.; Dong, S.; Gong, Q.; Xu, Q.; Ge, Y. Comparative chloroplast genome analysis of four *Polygonatum* species insights into DNA barcoding, evolution, and phylogeny. *Sci. Rep.* **2023**, *13*, 16495. [CrossRef]
49. Wen, J.; Harris, A.J.; Kalburgi, Y.; Zhang, N.; Xu, Y.; Zheng, W. Chloroplast phylogenomics of the New World grape species (*Vitis*, *Vitaceae*). *J. Syst. Evol.* **2018**, *56*, 297–308. [CrossRef]
50. Pelsy, F. Untranslated leader region polymorphism of Tvv1, a retrotransposon family, is a novel marker useful for analyzing genetic diversity and relatedness in the genus *Vitis*. *Theor. Appl. Genet.* **2007**, *116*, 15–27. [CrossRef]

Disclaimer/Publisher's Note: The statements, opinions and data contained in all publications are solely those of the individual author(s) and contributor(s) and not of MDPI and/or the editor(s). MDPI and/or the editor(s) disclaim responsibility for any injury to people or property resulting from any ideas, methods, instructions or products referred to in the content.

horticulturae

MDPI

Article

Genetic Diversity and Population Structure of *Prunus persica* Cultivars Revealed by Genotyping-by-Sequencing (GBS)

Ekaterina Vodiasova [1,*], Artem Pronozin [2], Irina Rozanova [1], Valentina Tsiupka [1], Gennady Vasiliev [2], Yuri Plugatar [1], Sergey Dolgov [1,3] and Anatoly Smykov [1]

1 Federal State Funded Institution of Science "The Labor Red Banner Order Nikita Botanical Gardens—National Scientific Center of the RAS", Nikita, 298648 Yalta, Russia; fermoza@gmail.com (I.R.); valentina.brailko@yandex.com (V.T.); selectfruit@yandex.ru (A.S.)

2 Institute of Cytology and Genetics, Siberian Branch of the Russian Academy of Sciences, 630090 Novosibirsk, Russia; pronozinartem95@gmail.com (A.P.)

3 Branch of Shemyakin and Ovchinnikov Institute of Bioorganic Chemistry, 142290 Puschino, Russia

* Correspondence: eavodiasova@gmail.com

Abstract: Peach (*Prunus persica* (L.)) is one of the major commercial stone fruit crops. A genetic analysis of peach collections around the world is essential for effective breeding programmes, and the development of genomic and marker-assisted selection. This study focuses on research on peach collection at the Nikita Botanical Garden and aims to identify single-nucleotide polymorphisms (SNPs) at the genome level and analyse the genetic diversity, population structure, and the linkage disequilibrium (LD) pattern among 161 cultivars and hybrids. A total of 288,784 SNPs were identified using the genotyping-by-sequencing (GBS) approach and, after filtering, 7803 high-quality SNPs were used in the analyses. The 161 accessions were clustered into two groups using principal component analyses (PCoA) and seven populations by ADMIXTURE v.1.3 software, which was confirmed using phylogenetic analyses. The distribution of the genotypes within subpopulations reflected any fruit-related traits. A low level of genetic diversity and medium linkage disequilibrium was detected in peach cultivars. The observed heterozygosity was lower than expected and varied from 0.11 to 0.22 in genotypes with different origins. Our results based on 7803 SNPs were compared with those based on 12 microsatellite markers and differences in clustering, observed heterozygosity, and phylogeny were identified. This highlights the need to analyse collections using whole-genome approaches.

Keywords: peach; LD patterns; population genetics; GBS; SNP

1. Introduction

Prunus persica (L.) Batsch (peach) is a perennial fruit crop with ornamental and edible varieties [1]. It is one of the most commercially important [2] and the third most cultivated fruit crop in temperate climates, after apple and pear [3]. Peach is grown in more than 60 countries worldwide [4]. FAO data [5] show that its plantations cover more than 165 million hectares. Today *P. persica* is cultivated almost everywhere on all continents except Greenland, northern regions of Europe, and some regions in central Africa.

As a species, peach originated about 2.5 million years ago in southwestern China. The oldest peach stones were found in the Kuahuqiao (8000–7000 BC) [6]. According to different hypotheses, peach domestication began 4000–7500 years ago in China [6,7]. The crop spread to Europe about 2000 years ago, arriving via ancient trade routes through

Persia [8,9]. Peach was introduced to the Americas by Spanish and Portuguese settlers, with the first breeding programmes appearing there in the late 18th century [8,10,11]. With the discovery of Mendel's laws, there was a breakthrough in breeding programmes. A total of 1092 cultivars of *P. persica* were registered from 1991 to 2001 [12]. A collection of peach accessions is being developed and studied around the world (Table 1).

Table 1. Some peach collections in different countries.

Country	Collection	Accessions Number	Reference
China	Six major peach breeding institutes	>600	[13]
USA	University of Florida	168	[14]
	U.S. Department of Agriculture (USDA)	112	[15]
Spain	Agrifood Research and Technology Centre of Aragon (CITA) Experimental Station of Aula Dei (EEAD)-CSIC	287	[16]
Turkey	Ataturk Central Horticultural Research Institute	≈50	[17,18]
Brazil	Embrapa	900	[19]
Russia	Nikita Botanical Garden North Caucasian Federal Scientific Center of Horticulture, Viticulture, and Winemaking Russian Research Institute of Floriculture and Subtropical Crops	>600	[20–22]
India	Patharchatta of G. B. Pant University of Agriculture and Technology ICARNational Bureau of Plant Genetic Resources	41	[23]
Italy	MAS.PES Germplasm Bank	123	[24]
France	INRA, National Institute for Agricultural research	28	[25]

One of the largest collections of stone fruits, especially peaches, in Russia is the collection of the Nikita Botanical Garden, which has existed for over 200 years. There are unique collections of peach, nectarine, apricot, almond, plum, cherry, and sweet cherry. The *P. persica* collection has 624 specimens, including 150 specimens of nectarine and ornamental peach (local varieties and selective forms of local breeding, and varieties from North America, Southern Europe, Central Asia, and the Caucasus) [26].

The first studies analysing peach populations started in the 1980s and were based on protein markers [27,28]; in the 1990s, molecular genetic methods such as restriction fragment length polymorphism (RFLP) [29] and Random Amplified Polymorphic DNA (RAPD) appeared [30]. To date, different SSR markers (simple sequence repeats or microsatellites) have been used in most of the studies on peach population structure, where the number of SSR markers used to study populations varies from 15 SSR markers [24] to 50 [31]; the most-used series of microsatellites are BPPCT, UDP, CPPCT, and Pchgms [32–36].

With the development of next-generation sequencing (NGS) and the publication of the complete peach genome [37], studies of genetic diversity using single-nucleotide polymorphisms (SNPs) of different peach collections are becoming an extremely relevant task, as such data could be the basis for the search for SNPs associated with valuable agronomic traits. A genome-wide association study (GWAS) compares the genomes (whole or part) from many different cultivars to find the SNPs associated with a particular phenotype. Therefore, genotyping peach collections using SNP instead of SSR markers could allow for the detection of genome sites associated with agronomic traits.

Two approaches are used to identify SNPs in peach populations: array (9K SNP v1 peach array, 18K SNP v2 peach array) [16,38,39] and genotyping by sequencing (GBS) [40]. At present, such studies have been carried out for 287 accessions from two Spanish peach germplasm collections (CITA and CSIC) [16,41,42], 183 accessions from MAS.PES Italy collection [43], 161 accessions from the Hexi Corridor and the Tarim Basin, Northwest China [44], 195 accessions from China, the USA, and Italy [45], 417 accessions from China [1,46,47],

220 peach genotypes from Brazilian peach breeding germplasm [48], and 1576 peach accessions from collections from Spain, Italy, France and China [38]. These collections do not contain varieties of peach, which are present in the NBG collections. These studies on a large number of varieties revealed some SNPs associated with the following agronomic traits: maturation, fruit hairiness, fruit shape, flesh colour, texture, flesh colour around the stone, fruit weight and soluble solid content, olecranon-type traits, seed characteristics (kernel taste), pollen fertility traits, flower characteristics, resistance to diseases, and resistance to abiotic factors, as chilling and drought [16,38,41–43,45–56].

However, none of the Russian peach collections have been studied using NGS methods. The most recent study of the peach population structure of the Nikita Botanical Garden collection was conducted on 85 peach cultivars using SSR [26].

This study is devoted to genotyping 161 peach accessions from the NBG collection using GBS. This approach allowed us to describe genetic diversity, population structure, and LD decay in more detail. Understanding the genetic diversity of each cultivar is the basis for the improvement and breeding of peach varieties. Moreover, it is needed for the next GWAS.

2. Materials and Methods

2.1. Plant Material

A total of 161 cultivars and hybrids of *P. persica* from the Nikita Botanical Garden collection were investigated. Among the 161 accessions, 105 are local breeds and 35 are from the USA. The rest of the 21 cultivars are from Armenia, Azerbaijan, Canada, France, Italy, Moldova, Uzbekistan, Romania, Russia, Serbia, Spain, and Ukraine. The total list of some cultivar features, such as their origin, fruit texture (melting/non-melting), flesh colour (yellow, white, cream), and flesh adhesion (freestone, semi-free and clingstone) is presented in the Supplementary Materials.

2.2. DNA Extraction, Library Preparation, and Sequencing

Genomic DNA was extracted from fresh young leaves using GeneJET Plant Genomic DNA Purification Kit (ThermoScientific, Waltham, MA, USA), following the manufacturer's protocol. The DNA samples were quality-tested and quantified using a NanoPhotometer N60 (Implen, Westlake Village, CA, USA) and electrophoresis on 1.5% agarose gel. The DNA was diluted to 20 ng/µL concentration and used for the next library preparation.

High-quality genomic DNA (200 ng) from individual samples was digested in 20 µL with the Msp I and Pst I restriction enzymes with NEBNext Cut Smart (New England Biolabs, Ipswich, MA, USA) buffer, followed by adapter ligation and DNA purification with KAPA Pure Beads (Roche, Basel, Switzerland). Each DNA sample was barcoded and amplified by PCR (15 cycles). PCR products were purified again. Finally, library concentration was normalised to 4 nM. Libraries were sequenced using NextSeq550 Mid Output Kit v2.5 (150 Cycles) with single-end 150 bp reads on the Illumina NextSeq550 platform.

2.3. Read Mapping and SNP Calling

To analyse the resulting libraries, we used the bioinformatic pipeline GBS-DP [57] (Figure 1).

The pipeline consists of three main steps: (1) data preprocessing, (2) polymorphism identification, and (3) genetic diversity analysis. Data preprocessing includes FastQC v0.12.1, FastP v0.23.4 [58], and MultiQC v1.17 [59] programmes that perform quality controls of raw reads and remove adaptors. The filtered reads for each sample were aligned with the reference genome *Prunus persica* version NCBIv2 [37] downloaded from Ensembl plants [60]. For this purpose, we used bwa mem v0.7.17 [61]. Based on the mapping results, the reading depth was calculated using Samtools depth v1.6 [62] and coverage across the entire genome was ensured using bedtools genomecov v2.30.0. Variant calling was performed with samtools mpileup in combination with bcftools call [63]; both were run using the default options. The resulting SNP calls were filtered by MAPQ $\geq$ 30, missing rate > 0.25, and MAF > 0.01 in order to obtain high-quality SNPs.

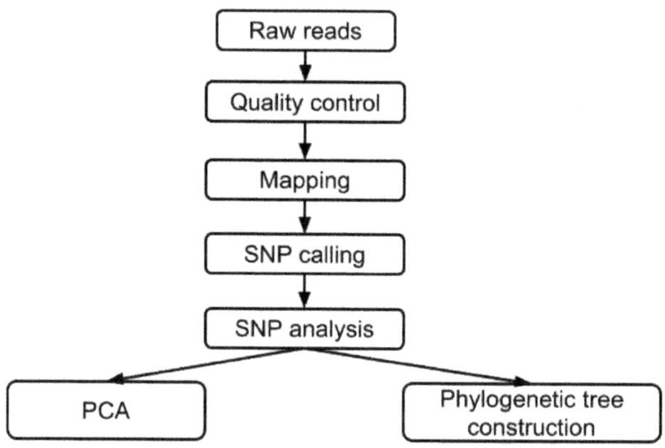

Figure 1. The diagram of the GBS-DP bioinformatics pipeline.

2.4. *Clustering and Phylogenetic Tree Construction*

The filtered files in VCF format were converted into the Genomic Data Structure (GDS), with the R package SeqArray v1.46.0 [64], which allowed for the RAM to be reduced. Linkage disequilibrium (LD) decay was calculated and visualised using PopLDdecay v3.43 [65] software with default parameters. In the resulting file, the removal of redundant SNPs based on linkage disequilibrium (LD) information (cutoff threshold 0.2) was performed by means of SNPRelate v3.13 [64].

For filtered polymorphisms, a principal component analysis and phylogenetic tree construction were carried out, also using the R package v4.2.3 SNPRelate. The phylogenetic tree was constructed with the IQ-TREE v.2.2.2.6 programme [66]. It applied a maximum likelihood algorithm based on multiple alignments of the sequences under study to construct a phylogenetic tree. The bootstrap was chosen to be equal to 1000. The phylogenetic tree was drawn and visualised using iTOL v7 [67].

2.5. *Determination of Population Structure, Statistical Analyses, and LD*

The population structure was determined using a filtered SNP of all individuals that were used for PCA and phylogenetic tree construction. For this purpose, we used ADMIXTURE v1.3 software [68] in combination with R package LEA [69]. The estimation of clusters (K) was performed in ten replications using a K of 1 to 15. The population substructure analysis was performed on an SNP dataset of individuals from 15 growing regions populations. We used cross-validation error rates to determine the optimal K value [68]. Our data were visualised using the LEA package v3.18.0.

Observed heterozygosity (Ho), expected heterozygosity (He), and inbreeding coefficient (F_{IS}) at each locus for each population were obtained using the VCFtools v0.1.16 programme [63]. F-statistics (F_{ST}) between populations were obtained using PLINK v1.90b6.12 [70]. For F_{ST} analysis, we used the following population combinations: NBG vs. USSR, NBG vs. Europe, and NBG vs. USA.

Linkage disequilibrium (LD) decay was calculated and visualised by using PopLDdecay v3.43 [65] software with parameters 'MaxDist' = 100 kb and 3000 kb. We performed an LD analysis of the whole set and of clusters that we determined based on the phylogenetic tree analysis.

3. Results

3.1. SNP Identification

All GBS data for the 161 peach accessions were of good quality (Q20 ≥ 89%, Q30 ≥ 78%) and the GC content was 45–49%. The total sequencing data volume was 43 Gb, with an average of 3.18 million single-end 100 bp reads per accession.

On average, 90% of reads from each GBS library aligned with the reference genome, but some accessions had a mapping rate of 75% (Figure 2a). The average sequencing depth for the GBS reads was 6–7.5× and for over 25% of the samples, the sequencing reads covered 5–8% of the peach reference genome (Figure 2b,c).

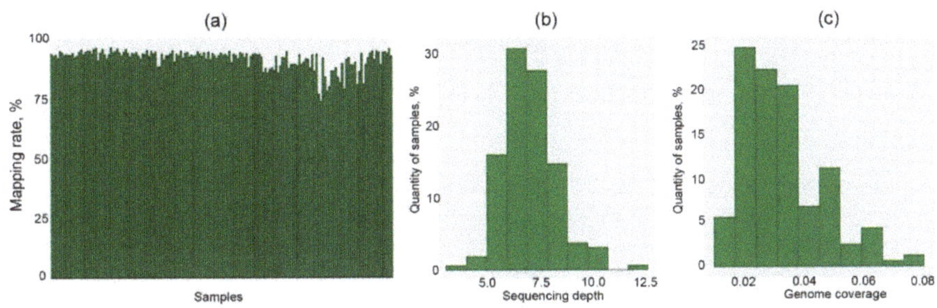

Figure 2. GBS quality. (**a**) The mapping rate of each sample. (**b**) The sequencing depth. (**c**) The peach reference genome coverage.

Out of 288,784 polymorphic sites, an average SNP density of 546 SNPs per 100 kb was detected after the quality filter. A total of 25,863 indels were revealed. More transitions than transversions were identified for 262,921 SNPs, and the ratio of transition/transversion was 1.31. The number of SNPs for each cultivar was estimated and to range from 3093 to 21,989 (Figure 3).

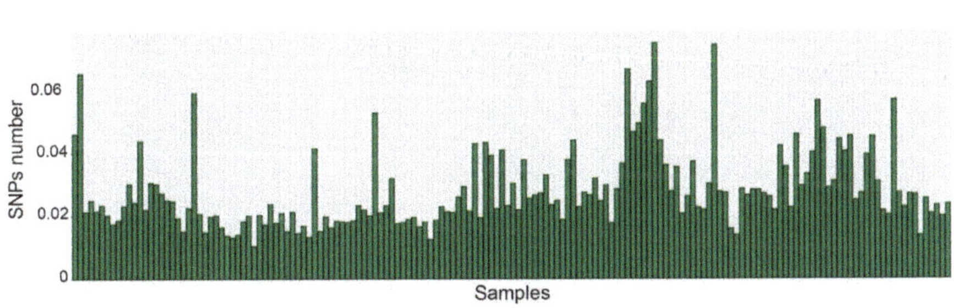

Figure 3. The number of SNPs for each cultivar.

All 288,784 SNPs were mapped across eight major scaffolds corresponding to peach chromosomes (Figure 4). SNPs were widely and equally distributed over eight chromosomes except for the top end of chromosome 2, where the observed density of the SNPs was three times higher than that of the other chromosomes. The most SNPs were observed on chromosome 1 (47,828 SNPs) and the fewest on chromosome 5 (21,348 SNPs), which is correlated with chromosome length.

All detected SNPs were filtered with MAF > 0.01; a missing rate of 25% and 7803 high-quality SNPs were obtained, which corresponds to an average of 34.4 SNPs/Mb (peach whole-genome size is 227,411,381 bp). These 7803 SNPs were used for subsequent analysis.

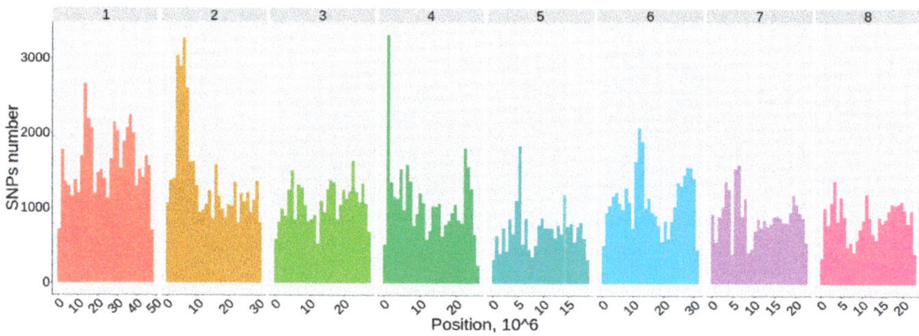

Figure 4. SNP distribution across the eight peach chromosomes.

3.2. Population Structure

The population structure of the 161 peach cultivars from different countries was analysed. According to three methods of cluster estimation (CV, deltaK, and L (K)), the most probable number of populations at K = 7 was detected and used to describe the population structure using ADMIXTURE analysis (Figure 5a).

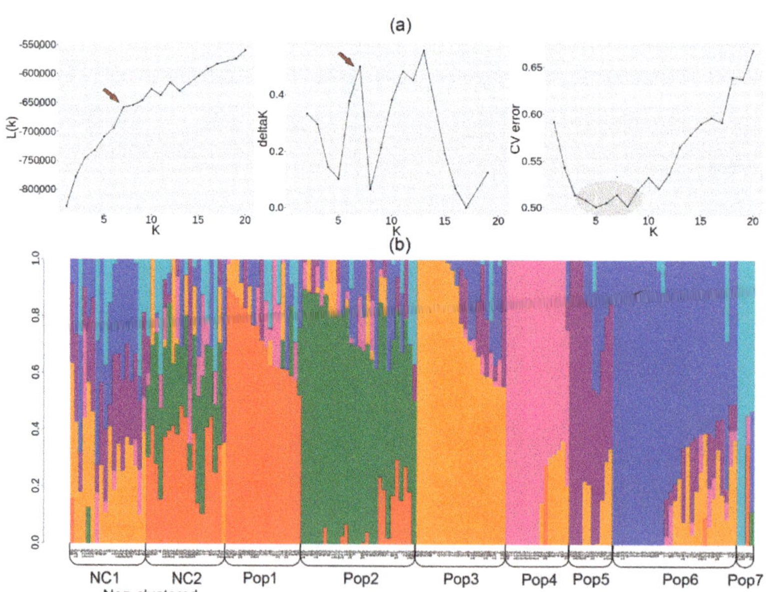

Figure 5. Genome-wide SNP-based population genetic structure of 161 *Prunus persica* cultivars with different origins from the NBG collection. (**a**) The estimation of the population number. The arrows indicate the best value of K (**b**) Results of an ADMIXTURE analysis with K = 7. The different colours represent different peach populations.

Two values of the member threshold were considered when analysing the population structure. All genotypes with a member threshold < 0.5 were identified as non-clustered (NC) and grouped separately. The remaining accessions were divided into seven populations according to the maximum estimated membership of the inferred K groups (member threshold > 0.5). Within each population, all accessions were divided into two groups. The population membership for accessions with a score higher than 0.75 was considered reliable, while accessions with a membership value of 0.5–0.75 could be considered as partial admixture. The 124 peach cultivars were divided into seven populations, and 37 accessions were assigned to a non-clustered group (Figure 5b). The estimated population membership of the inferred K groups for each cultivar is presented in the Supplementary Materials.

The two populations Pop2 (27 genotypes) and Pop6 (29 genotypes) were the most abundant. These populations included peach with a different flesh colour (yellow or white) and a variable stone adhesion (freestone, semi-free, clingstone). Some populations were contained only peaches with melting flesh (Pop2, Pop3, Pop4), whereas Pop5 (10 genotypes) and Pop7 (4 genotypes) consisted of melting-flesh peaches as well as those characterised as non-melting. No significant correlation was observed between population grouping and the fruit characteristics of the accessions.

The population structure was also obtained using the principal component analysis (PCA) for the 7803 SNPs. All 161 accessions were annotated based on the results of the ADMIXTURE when K = 7 (Figure 6).

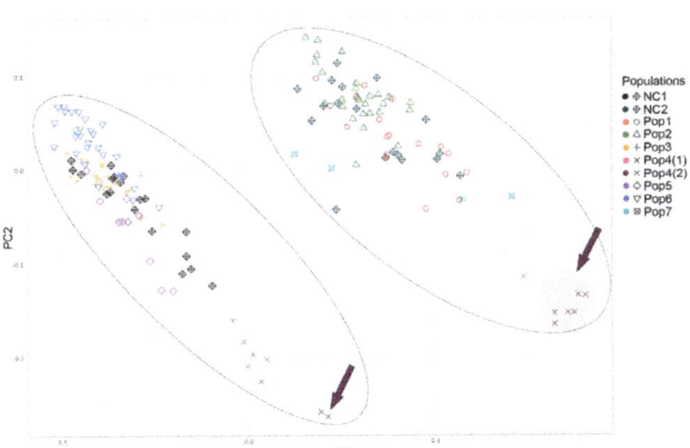

Figure 6. Principal component analysis (PCoA) of 161 peach accessions based on 7803 SNPs. Colours denote populations obtained via ADMIXTURE. Arrow points to the cultivar "Sovetskiy" and its radio mutants.

PCoA revealed two main clusters that corresponded to different populations. Cluster C1 included Pop3, Pop5, and Pop6, whereas C2 consisted of Pop1 and Pop2. The admixture group was divided into NC1 and NC2, which were separated according to the main clusters. Also, Pop4 was divided between C1 and C2, where genotypes formed distinct groups inside each cluster. This population includes the cultivar "Sovetskiy" and its radio mutants (marked with arrows and named Pop4 (2) in Figure 6). The variety "Sovetskiy" and one radio mutant were included in cluster C1, while the other radio mutants of this variety formed a separate group in cluster C2. The PCoA also revealed no dependence between the main clusters and the origin of the accessions or ecology–geographic groups.

3.3. Genetic Diversity Analysis

Genetic variability was analysed for varieties with different origins: NBG, Europe, and the USA. Varieties obtained under the USSR breeding programmes were analysed separately (Table 2). The two main clusters resulting from PCoA were also analysed separately. The differences between clusters are smaller than those between varieties of different origins. Observed heterozygosity varied from 0.11 (NGB selection) to 0.22 (Europe); inbreeding coefficient varied from 0.38 to 0.55. In each sample, the observed heterozygosity was less than He, with an inbreeding coefficient > 0. The maximum level of F_{IS} was observed in varieties from NBG and other USSR selection programmes. On average, the studied peach varieties are characterised by a low heterozygosity (Ho = 0.11) and a high level of inbreeding (F_{IS} = 0.51).

Table 2. Population genetic diversity statistics: observed heterozygosity (Ho), expected heterozygosity (He), and inbreeding coefficient (F_{IS}).

Country/Breeder	N	Ho	He	F_{IS}
NBG	114	0.11	0.24	0.51
USSR selection programme	22	0.14	0.32	0.55
USA	36	0.13	0.24	0.46
Europe	6	0.22	0.36	0.38
Cluster 1	87	0.12	0.22	0.43
Cluster 2	74	0.11	0.25	0.54
Total	161	0.11	0.22	0.51

The genetic differentiation between cultivars with different origins was tested using an F_{ST} statistic estimated from pairwise analysis. Pairwise F_{ST} values were 0.0003 (between NBG and USSR varieties), 0.02 (between NBG and Europe varieties), and 0.029 (between NBG and USA varieties).

3.4. Phylogeny of 161 Peach Cultivars

The phylogenetic analysis based on 7803 SNPs revealed three main phylogroups: A, B and C (Figure 7).

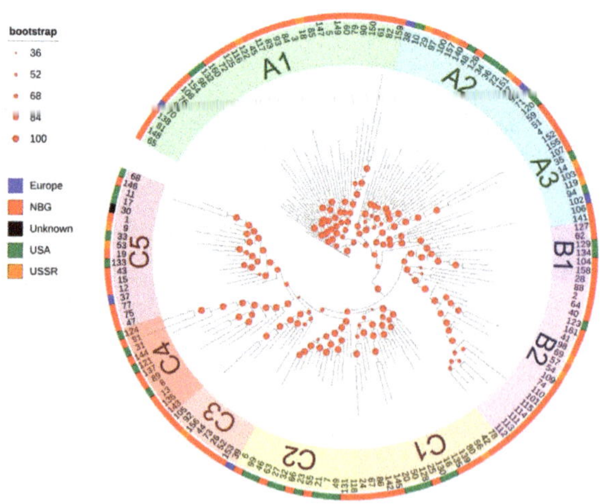

Figure 7. Phylogenetic tree of the 161 peach cultivars. The coloured boxes indicate the origins of the varieties. The red dots denote the value of the bootstrap support.

The phylogeny results correlated with those of the PCoA clustering, but not completely. Phylogroup A corresponds to cluster C2; phylogroup C corresponds to C1. Phylogroup B represents a separate branch with high level of bootstrap support and consists of genotypes belonging to both cluster C1 and cluster C2. Each phylogroup is subdivided into subgroups and reflects the populations identified by ADMIXTURE analysis. Subgroup A1 mainly contains genotypes from the Pop2 population and a few admixture genotypes. Subgroup A2 includes genotypes from the Pop1 population and a few genotypes from the Pop2 population, with a membership score of 0.5–0.75. Subgroup A3 includes all genotypes from Pop7 and non-clustered populations. Genotypes from the Pop5 and Pop4 populations, respectively, represent subgroups B1 and B2. In phylogroup C, three subgroups, C1, C3, and C4, revealed a considerable admixture, while C2 consists of the genotypes from Pop3 and subgroup C5 consists of genotypes from Pop6. Thus, phylogenetic analysis revealed a stronger differentiation of the isolated populations than the PCoA analysis. We can identify phylogenetic subgroups with a high level of bootstrap support, represented predominantly by genotypes from one population (A1, A2, B1, B2, C2, or C5), while genotypes with membership scores < 0.5 represent other phylogenetic subgroups. A correlation between phylogenetic clades and the place of origin of the varieties was also not found. This could be explained by the fact that the same parental forms may be used during hybridisation in breeding centres in different countries or when cultivars with different origins are used as parental forms. All this leads to a lack of correlation between the origin of varieties and phylogenetic clades.

3.5. Linkage Disequilibrium

The LD estimates (measured as r^2) and the extent of LD decay were calculated in the two clusters obtained with ADMIXTURE (C1 and C2), as well as in all 161 peach genotypes (Figure 8).

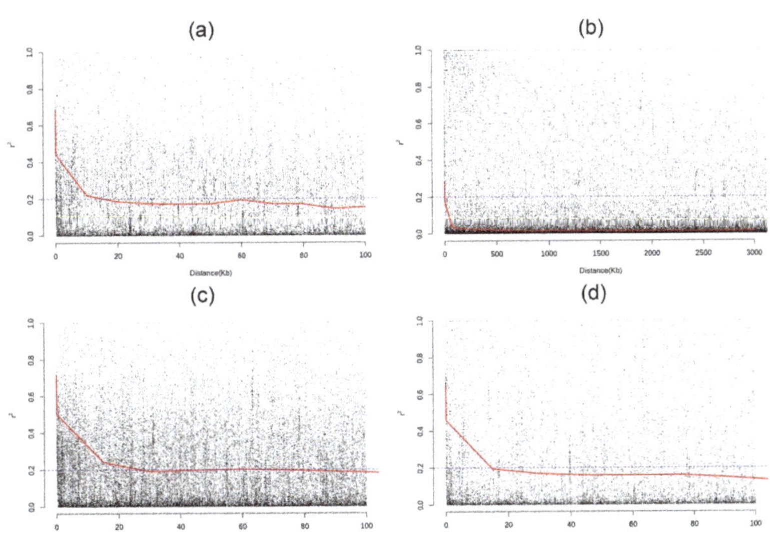

Figure 8. Linkage disequilibrium measures (r^2) using the physical distance between pairs of 7803 SNP markers. (**a**) LD decay for all 161 peach accessions until 100 Kb. (**b**) LD decay for all 161 peach accessions until 3000 Kb. (**c**) LD decay for genotypes from cluster C1. (**d**) LD decay for genotypes from cluster C2.

The average value in the whole set was 0.1, the average value in C1 was 0.45, and the average value in C2—0.41. The r^2 was lower than 0.2 at around 14 Kb in all accessions. Different LD patterns were observed in the two clusters (Figure 8c,d). In cluster C1, r^2 decreased below 0.02, at approximately 29 Kb, decreasing two times more than in cluster C2 (around 15 Kb).

4. Discussion

Peach breeding is a long and labour-intensive process. One of the main tasks is to analyse collections using molecular methods (SSR or ISSR markers), assessing genetic variability and population structure and determining the degree of the relationship between different varieties. At the same time, with the development of whole-genome sequencing, genomic selection is beginning to develop, improving the efficiency of breeding programmes. This requires the use of whole-genomic approaches to study different collections instead of traditional microsatellite markers. The Nikita Botanical Garden peach collection was previously partially (85 peach accessions only) studied using 12 SSRs [26]. Thus, this work is the first to investigate the genetic diversity and population structure of 161 peach cultivars from the NBG collection on a genome-wide scale. Our results revealed some differences when examining the collection using SSR markers and GBS, which are discussed further below.

Due to the small size of the peach genome and the sufficient genome coverage (5–8%), 288,784 SNPs were obtained, of which only 7803 SNP markers were retained after filtering. This is explained by the fact that GBS is a sequencing approach that results in a high missing rate and a high minor allele frequency. The number of SNPs obtained for further analysis is comparable to other studies, where the number of SNPs before filtering ranged from 9998 [40] to 4,063,377 [46]. Despite the high loss of SNPs after filtering (missing rate > 0.25, MAF > 0.01), which is typical for GBS data, this approach allows for more data to be obtained compared to the 9K SNP v1 peach array and the 18K SNP v2 peach array, where the number of possible SNPs is limited [16,38]. It has also been shown that the number of filtered SNPs can be increased by increasing the number of accessions [40].

An analysis of the distribution of 288,784 SNPs across all peach chromosomes showed the highest density of SNPs at the top of chromosome 2. The same data were obtained when comparing *P. persica* and *P. ferganensis* based on 953,357 SNPs [37]. This is probably explained by the fact that these species have a higher density of genes on chromosome 2, encoding receptor proteins from the family of conserved nucleotide-binding leucine-rich proteins (R-proteins), which are involved in immunity and have greater genetic variability [71]. No other regions of the genome with an increased SNP density were identified.

The analysis of genetic variability based on SNPs showed a lower mean value of the observed heterozygosity (Ho = 0.11) in the NBG peach collection compared to the previous study based on SSR (Ho = 0.31) [26]. A lower value of observed heterozygosity using SNPs compared to SSR markers was also reported in other studies of peach collections in Europe, Brazil, and China [16,38,48]. Such differences in the estimation of heterozygosity using SNP and SSR markers are explained by the bi-allelic nature of SNPs, in contrast to the high number of SSR alleles and could differ [72].

Differences were found between peach accessions from the NBG collection and other peach genotype datasets that were studied. The lower value of observed heterozygosity in the NBG peach collection compared to other studies, e.g., the analysis of the Brazilian collection (Ho = 0.29) [48] or analysis of 1580 peach genotypes from different countries including China (Ho = 0.28–0.33) [38], is explained by the specificity of the collections that were investigated. It is known that peach populations in China are characterised by higher genetic diversity compared to other countries. A limited number of peach cultivars from

China were introduced to Europe and then to America, and this led to the use of a limited number of founders in breeding programmes and resulted in reduced genetic variability and high linkage disequilibrium (LD). The peach collection of the Nikita Botanical Garden largely contains varieties from the USA and Europe, which make up a large part of the collection and are often used as parental forms. At the same time, the NBG collection contains a poor representation of varieties from China. This may explain the low value of Ho = 0.11–0.22. A similar situation was observed in peach populations from Spain, where Ho ranges from 0.11 to 0.24, which is comparable to the results for the NBG collection. The LD estimates in this study also differ from those previously reported for peach [48], where the LD decay dropped to below 0.2 within a distance of about 38 Kb, as opposed to the distance of about 15 Kb observed in the present study. Moreover, other studies have shown a correlation between population stratification and geographical origin [38,73,74] or melting/non-melting flesh [16,31,38,48,73]. In this study, we did not find any correlation between these characteristics and the population structure.

Such differences can be explained by the different sets of genotypes analysed or the use of other restriction enzymes. All GBS studies on peach collections were carried out using ApeKI restriction enzymes, whereas we used double-digest GBS with PstI/MspI, resulting in a different set of SNPs. This approach has, so far, only been used for Prunus rootstock [75]. This hypothesis about the effect of different restriction enzymes used for GBS needs to be tested.

The PCoA and phylogenetic analysis showed the presence of two main groups (clusters C1 and C2, corresponding to phylogroups A and C) and a third group containing genotypes of the variety 'Sovetskiy' and its radio mutants (Figures 6 and 7). The variability within this group of radio mutants is comparable to the variability between different varieties. Similar results were obtained in the study of radio mutants of two chrysanthemum cultivars [76] and require further study. In previous work using 12 SSRs, PCoA did not show such clustering [26], probably due to the small number of markers used. A similar situation, where PCoA based on SSRs does not detect clusters detected by SNPs, occurs in other plants [77].

Differences between SSRs and SNPs were also found in population structure analyses (using STRUCTURE and ADMIXTURE, respectively). In this paper, we determined the best value for the number of populations K = 7, which correlates well with the results of the phylogenetic analysis. Within the phylogroups identified, a subdivision into subgroups corresponding to either distinct populations (subgroups A1, A2, B1, B2, C2, and C5) or mixed genotypes (subgroups A3, C1, C2, and C4) was observed (Figure 7). Previously, only four groups were identified based on microsatellite loci [26]. One group, separated from other accessions studied by PCoA, corresponds to subgroup B2. Other groups separated by SSR do not agree with our data based on SNPs. For example, the varieties 'Asmik' (subgroup C4), 'Jerseyglo' (subgroup C1), 'Kievskij Samyj Rannij' (subgroup C4), and 'Zheltoplodnyj Rannij' (subgroup A2) were erroneously assigned to one group. Such differences in the analyses of the population structure of the peach collection are explained either using molecular markers with a lower resolution (12 SSR markers vs. 7803 SNPs) or by a limited set of genotypes (85 accessions vs. 161) [26]. Therefore, the findings support the need to analyse collections using whole-genome approaches and SNP markers.

The study showed that the NBG peach collection had the lowest heterozygosity compared to the other collections studied due to the peculiarities of the breeding programmes and the limited number of parental forms. No relationship was found between the selected populations and phenotypic characteristics or geographical origin. For the first time, it was shown that the variability within a group of peach radio mutants was comparable to the variability between different cultivars.

Moreover, this study of the NBG collection using the GBS approach identified 7803 SNPs that can be used to further search for associations with agronomic traits. It should be noted that peach genotypes of the NBG selection are not contained in other collections and have not been investigated previously, so the information obtained in this study may be useful to other breeding centres. The results obtained from this study of the genetic variability and population structure in NBG peach collection could be used to select a set for GWAS analysis and a validation set.

Supplementary Materials: The following supporting information can be downloaded at: https://www.mdpi.com/article/10.3390/horticulturae11020189/s1, Table S1: Characteristics of the studied peach cultivars.

Author Contributions: Conceptualization, E.V.; methodology, V.T., G.V. and I.R.; software, A.P.; formal analysis, E.V. and A.P.; investigation, A.P. and V.T.; resources, A.S. and Y.P.; writing—original draft preparation, E.V., I.R., V.T. and A.P.; writing—review and editing, E.V. and S.D.; visualisation, A.P. and E.V.; supervision, E.V. All authors have read and agreed to the published version of the manuscript.

Funding: The study is supported by the Kurchatov Genomic Centre of the NBG–NSC (075-15-2019-1670).

Data Availability Statement: Data is contained within the article or Supplementary Material.

Acknowledgments: The GBS was carried out in ICG core facility «Center for Genome Studies» (Novosibirsk, Russia).

Conflicts of Interest: The authors declare no conflicts of interest.

References

1. Cao, K.; Zheng, Z.; Wang, L.; Liu, X.; Zhu, G.; Fang, W.; Cheng, S.; Zeng, P.; Chen, C.; Wang, X.; et al. Comparative population genomics reveals the domestication history of the peach, *Prunus persica*, and human influences on perennial fruit crops. *Genome Biol.* **2014**, *15*, 415. [CrossRef] [PubMed]
2. Sabbadini, S.; Ricci, A.; Limera, C.; Baldoni, D.; Capriotti, L.; Mezzetti, B. Factors affecting the regeneration, via organogenesis, and the selection of transgenic calli in the peach rootstock Hansen 536 (*Prunus persica* × *Prunus amygdalus*) to express an RNAi construct against PPV virus. *Plants* **2019**, *8*, 178. [CrossRef] [PubMed]
3. Bassi, D.; Mignani, I.; Spinardi, A.; Tura, D. Chapter 23—PEACH (*Prunus persica* (L.) Batsch). In *Nutritional Composition of Fruit Cultivars*; Simmonds, M.S.J., Preedy, V.R., Eds.; Academic Press: San Diego, CA, USA, 2016; pp. 535–571, ISBN 978-0-12-408117-8.
4. Smykov, A.V.; Mesyats, N.V. State analysis of horticulture and peach culture in the world. *Plant Biol. Hortic. Theory Innov.* **2020**, *155*, 130–137. [CrossRef]
5. FAOSTAT. 2024. Available online: http://www.fao.org/faostat (accessed on 25 October 2024).
6. Zheng, Y.; Crawford, G.W.; Chen, X. Archaeological evidence for peach (*Prunus persica*) cultivation and domestication in China. *PLoS ONE* **2014**, *9*, e106595. [CrossRef]
7. Yu, Y.; Fu, J.; Xu, Y.; Zhang, J.; Ren, F.; Zhao, H.; Tian, S.; Guo, W.; Tu, X.; Zhao, J.; et al. Genome re-sequencing reveals the evolutionary history of peach fruit edibility. *Nat. Commun.* **2018**, *9*, 5404. [CrossRef]
8. Hesse, C.O. Peaches. In *Advances in Fruit Breeding*; Purdue University Press: West Lafayette, IN, USA, 1975; pp. 285–335.
9. Byrne, D.H.; Bassols, M.; Bassi, D.; Piagnani, M.; Gasic, K.; Reighard, G.; Moreno, M.; Pérez, S. Peach. In *Fruit Breeding*; Springer Science: Boston, MA, USA, 2012; pp. 505–569.
10. Scorza, R.; Okie, W.R. Peaches (*Prunus*). *Acta Hortic.* **1991**, *290*, 177–234. [CrossRef]
11. Faust, M.; Timon, B. Origin and Dissemination of Peach. In *Horticultural Reviews*; John Wiley & Sons Inc.: Hoboken, NJ, USA, 1995; pp. 331–379.
12. Strada, G.D.; Fideghelli, C. Le cultivar de drupacee introdottee del 1991 al 2001. *l'Inf. Agrar.* **2003**, *59*, 65–70.
13. Li, Y.; Wang, L. Genetic resources, breeding programs in China, and gene mining of peach: A review. *Hortic. Plant J.* **2020**, *6*, 205–215. [CrossRef]

14. Chavez, D.J.; Beckman, T.G.; Werner, D.J.; Chaparro, J.X. Genetic diversity in peach [*Prunus persica* (L.) Batsch] at the University of Florida: Past, present and future. *Tree Genet. Genomes* **2014**, *10*, 1399–1417. [CrossRef]
15. Chen, C.; Okie, W.R. Population Structure and Phylogeny of Some US Peach Cultivars. *J. Am. Soc. Hortic. Sci.* **2022**, *147*, 1–6. [CrossRef]
16. Mas-Gomez, J.; Cantin, C.M.; Moreno, M.A.; Martinez-Garcia, P.J. Genetic Diversity and Genome-Wide Association Study of Morphological and Quality Traits in Peach Using Two Spanish Peach Germplasm Collections. *Front. Plant Sci.* **2022**, *13*, 854770. [CrossRef]
17. Demirel, S.; Pehluvan, M.; Aslantaş, R. Evaluation of Genetic Diversity and Population Structure of Peach (*Prunus persica* L.) Genotypes Using Inter-Simple Sequence Repeat (ISSR) Markers. *Genet. Resour. Crop Evol.* **2024**, *71*, 1301–1312. [CrossRef]
18. Eroğlu, Z.Ö.; Mısırlı, A.; Küden, A. The cross-breeding performances of some peach varieties. *Yuz. Yıl Univ. J. Agric. Sci.* **2016**, *26*, 89–97.
19. Correa, E.R.; Nardino, M.; Barros, W.S.; Raseira, M.D.C.B. Genetic progress of the peach breeding program of embrapa over 16 years. *Crop Breed. Appl. Biotechnol.* **2019**, *19*, 319–328. [CrossRef]
20. Smykov, A.V.; Komar-Temnaya, L.D.; Gorina, V.M.; Khokhlov, S.Y.; Shoferistov, E.P.; Latsko, T.A.; Shishkina, E.L.; Fedorova, O.S.; Tsiupka, S.Y.; Korzin, V.V.; et al. *Atlas of Fruit Crops Varieties of the Nikita Botanical Gardens Collection*; Plugatar, Y.V., Ed.; ARIAL: Simferopol, Ukraine, 2018; pp. 5–212.
21. Eremin, V.G.; Eremin, G.V. Genetic collections of stone fruit crops and their use to accelerate the breeding process. *Sci. Proc. N. Cauc. Fed. Sci. Cent. Hortic. Vitic. Winemak.* **2020**, *30*, 15–24. [CrossRef]
22. Smagin, N.E.; Tsymbalova, A.A. Promising peach varieties in the collection of VNIYATSISK. *Subtrop. Ornam. Hortic.* **2020**, *72*, 53–58. [CrossRef]
23. Kumar, R.; Dimri, D.C.; Karki, K.; Rai, K.M.; Singh, N.K.; Shivran, J.S.; Bharti, S. SSR marker based profiling and population structure analysis in peach (*Prunus persica*) germplasm. *Indian J. Agric. Sci.* **2023**, *93*, 1080–1085. [CrossRef]
24. Linge, C.D.S.; Pacheco, I.; Rossini, L.; Bassi, D.; Foschi, S.; Chietera, G.; Biffani, S.; Lama, M. Genetic Variability and Population Structure of Peach Accessions from MAS. PES Germplasm Bank. *Acta Hort.* **2015**, *1084*, 233–240. [CrossRef]
25. Parveaud, C.E.; Gomez, C.; Libourel, G.; Warlop, F.; Mercier, V. Assessment of disease susceptibility and fruit quality of 28 peach cultivars. *GRAB INRA* **2012**, 201–208. Available online: https://www.cabidigitallibrary.org/doi/pdf/10.5555/20133110086 (accessed on 30 January 2025).
26. Trifonova, A.A.; Boris, K.V.; Mesyats, N.V.; Tsiupka, V.A.; Smykov, A.V.; Mitrofanova, I.V. Genetic diversity of peach cultivars from the collection of the Nikita Botanical Garden based on SSR markers. *Plants* **2021**, *10*, 2609. [CrossRef]
27. Carter, G.E., Jr.; Brock, M.M. Identification of peach cultivars through protein analysis. *HortScience* **1980**, *15*, 292–293. [CrossRef]
28. Arulsekar, S.; Parfitt, D.E.; Beres, W.; Hansche, P.E. Genetics of malate dehydrogenase isozymes in the peach. *J. Hered.* **1986**, *77*, 49–51. [CrossRef]
29. Belthoff, L.E.; Ballard, R.; Abbott, A.; Baird, W.V.; Morgens, P.; Callahan, A.; Scorza, R.; Monet, R. Development of a saturated linkage map of *Prunus persica* using molecular based marker systems. *Acta Hortic.* **1993**, *336*, 51–56. [CrossRef]
30. Chaparro, J.X.; Conner, P.J.; Beckman, T.G. 'GulfAtlas' peach. *HortScience* **2014**, *49*, 1093–1094. [CrossRef]
31. Aranzana, M.J.; Abbassi, E.K.; Howad, W.; Arus, P. Genetic variation, population structure and linkage disequilibrium in peach commercial varieties. *BMC Genet.* **2010**, *11*, 69. [CrossRef]
32. Dirlewanger, E.; Cosson, P.; Tavaud, M.; Aranzana, M.J.; Poizat, C.; Zanetto, A.; Arus, P.; Laigret, F. Development of microsatellite markers in peach [*Prunus persica* (L.) Batsch] and their use in genetic diversity analysis in peach and sweet cherry (*Prunus avium* L.). *Theor. Appl. Genet.* **2002**, *105*, 127–138. [CrossRef]
33. Dirlewanger, E.; Graziano, E.; Joobeur, T.; Garriga-Calderé, F.; Cosson, P.; Howad, W.; Arús, P. Comparative mapping and marker-assisted selection in *Rosaceae* fruit crops. *Proc. Natl. Acad. Sci. USA* **2004**, *101*, 9891–9896. [CrossRef]
34. Aranzana, M.J.; Garcia-Mas, J.; Carbo, J.; Arus, P. Development and variability analysis of microsatellite markers in peach. *Plant Breed.* **2002**, *121*, 87–92. [CrossRef]
35. Sosinski, B.; Gannavarapu, M.; Hager, L.D.; Beck, L.E.; King, G.J.; Ryder, C.D.; Rajapakse, S.; Baird, W.V.; Ballard, R.E.; Abbott, A.G. Characterization of microsatellite markers in peach [*Prunus persica* (L.) Batsch]. *Theor. Appl. Genet.* **2000**, *101*, 421–428. [CrossRef]
36. Cipriani, G.; Lot, G.; Huang, W.G.; Marrazzo, M.T.; Peterlunger, E.; Testolin, R. AC/GT and AG/CT microsatellite repeats in peach [*Prunus persica* (L) Batsch]: Isolation, characterisation and cross-species amplification in *Prunus*. *Theor. Appl. Genet.* **1999**, *99*, 65–72. [CrossRef]
37. The International Peach Genome Initiative; Verde, I.; Abbott, A.G.; Scalabrin, S.; Jung, S.; Shu, S.; Marroni, F.; Zhebentyayeva, T.; Dettori, M.T.; Grimwood, J.; et al. The high-quality draft of peach (*Prunus persica*) identifies unique patterns of genetic diversity, domestication and genome evolution. *Nat. Genet.* **2013**, *45*, 487–494.

38. Micheletti, D.; Dettori, M.T.; Micali, S.; Aramini, V.; Pacheco, I.; Da Silva Linge, C.; Foschi, S.; Banchi, E.; Barreneche, T.; Quilot-Turion, B.; et al. Whole-Genome Analysis of Diversity and SNP-Major Gene Association in Peach Germplasm. *PLoS ONE* **2015**, *10*, e0136803. [CrossRef] [PubMed]
39. Gasic, K.; Linge, C.D.S.; Bianco, L.; Troggio, M.; Rossini, L.; Bassi, D.; Aranzana, M.J.; Arus, P.; Verde, I.; Peace, C.; et al. Development and Evaluation of a 9K SNP Addition to the Peach Ipsc 9K SNP Array V1. *HortScience* **2019**, *54*, S188.
40. Bielenberg, D.G.; Rauh, B.; Fan, S.; Gasic, K.; Abbott, A.G.; Reighard, G.L.; Okie, W.R.; Wells, C.E. Genotyping by Sequencing for SNP-Based Linkage Map Construction and QTL Analysis of Chilling Requirement and Bloom Date in Peach [*Prunus Persica* (L.) Batsch]. *PLoS ONE* **2015**, *10*, e0139406. [CrossRef] [PubMed]
41. Forcada, C.; Guajardo, V.; Chin-Wo, S.R.; Moreno, M.A. Association mapping analysis for fruit quality traits in *Prunus persica* using SNP markers. *Front. Plant Sci.* **2019**, *9*, 2005. [CrossRef]
42. Mas-Gómez, J.; Cantín, C.M.; Moreno, M.; Prudencio, Á.S.; Gómez-Abajo, M.; Bianco, L.; Troggio, M.; Martínez-Gómez, P.; Rubio, M.; Martínez-García, P.J. Exploring genome-wide diversity in the national peach (*Prunus persica*) germplasm collection at CITA (Zaragoza, Spain). *Agronomy* **2021**, *11*, 481. [CrossRef]
43. Cirilli, M.; Baccichet, I.; Chiozzotto, R.; Silvestri, C.; Rossini, L.; Bassi, D. Genetic and phenotypic analyses reveal major quantitativeloci associated to fruit size and shape traits in a non-flat peach collection (*P. persica* L. Batsch). *Hortic. Res.* **2021**, *8*, 232. [CrossRef]
44. Li, W.; Li, Y.; Wang, X.; Zhao, G.; Zhu, G.; Cao, K.; Zang, W.; Wu, J.; Ma, K.; Chen, C.; et al. Genomic analysis provides insights into the westward expansion of domesticated peaches in China. *Hortic. Plant J.* **2024**, *10*, 367–375. [CrossRef]
45. Li, X.; Wang, J.; Su, M.; Zhou, J.; Zhang, M.; Du, J.; Zhou, H.; Gan, K.; Jin, J.; Zhang, X. Single Nucleotide Polymorphism Detection for Peach Gummosis Disease Resistance by Genome-Wide Association Study. *Front. Plant Sci.* **2022**, *12*, 763618. [CrossRef]
46. Cao, K.; Zhou, Z.; Wang, Q.; Guo, J.; Zhao, P.; Zhu, G.; Fang, W.; Chen, C.; Wang, X.; Wang, X.; et al. Genome-wide association study of 12 agronomic traits in peach. *Nat. Commun.* **2016**, *7*, 13246. [CrossRef]
47. Tan, Q.P.; Li, S.; Zhang, Y.Z.; Chen, M.; Wen, B.B.; Jiang, S.; Chen, X.D.; Fu, X.L.; Li, D.M.; Wu, H.Y.; et al. Chromosome-level genome assemblies of five *Prunus* species and genome-wide association studies for key agronomic traits in peach. *Hortic. Res.* **2021**, *8*, 213. [CrossRef] [PubMed]
48. Thurow, L.B.; Gasic, K.; Bassols Raseira, M.D.C.; Bonow, S.; Marques Castro, C. Genome-wide SNP discovery through genotyping by sequencing, population structure, and linkage disequilibrium in Brazilian peach breeding germplasm. *Tree Genet. Genomes* **2020**, *16*, 10. [CrossRef]
49. Liu, H.; Cao, K.; Zhu, G.; Fang, W.; Chen, C.; Wang, X.; Wang, L. Genome-wide association analysis of red flesh character based on resequencing approach in peach. *J. Am. Soc. Hortic. Sci.* **2019**, *144*, 209–216. [CrossRef]
50. Li, X.; Wang, J.; Su, M.; Zhang, M.; Hu, Y.; Du, J.; Zhou, H.; Yang, X.; Zhang, X.; Jia, H.; et al. Multiple-Statistical Genome-Wide Association Analysis and Genomic Prediction of Fruit Aroma and Agronomic Traits in Peaches. *Hortic. Res.* **2023**, *10*, uhad117. [CrossRef] [PubMed]
51. Liu, J.; Bao, Y.; Zhong, Y.; Wang, Q.; Liu, H. Genome-wide association study and transcriptome of olecranon-type traits in peach (*Prunus persica* L.) germplasm. *BMC Genom.* **2021**, *22*, 702. [CrossRef]
52. Huang, Z.; Shen, F.; Chen, Y.; Cao, K.; Wang, L. Preliminary identification of key genes controlling peach pollen fertility using genome-wide association study. *Plants* **2021**, *10*, 242. [CrossRef]
53. Elsadr, H. A Genome Wide Association Study of Flowering and Fruit Quality Traits in Peach [(*Prunus persica* (L.) Batsch]. Doctoral Dissertation, University of Guelph, Guelph, ON, Canada, 2016.
54. Meng, G.; Zhu, G.; Fang, W.; Chen, C.; Wang, X.; Wang, L.; Cao, K. Identification of loci for single/double flower trait by combining genome-wide association analysis and bulked segregant analysis in peach (*Prunus persica*). *Plant Breed.* **2019**, *138*, 360–367. [CrossRef]
55. Fu, W.; da Silva Linge, C.; Gasic, K. Genome-wide association study of brown rot (*Monilinia* spp.) tolerance in peach. *Front. Plant Sci.* **2021**, *12*, 635914. [CrossRef]
56. Serrie, M.; Segura, V.; Blanc, A.; Brun, L.; Dlalah, N.; Gilles, F.; Heurtevin, L.; Le-Pans, M.; Signoret, V.; Viret, S.; et al. Investigating the genetic architecture of biotic stress response in stone fruit tree orchards under natural infections with a multi-environment GWAS approach. *bioRxiv* **2024**. [CrossRef]
57. Pronozin, A.Y.; Salina, E.A.; Afonnikov, D.A. GBS-DP: A bioinformatics pipeline for processing data coming from genotyping by sequencing. *Vavilov J. Genet. Breed.* **2023**, *27*, 737. [CrossRef]
58. Chen, S.; Zhou, Y.; Chen, Y.; Gu, J. fastp: An ultra-fast all-in-one FASTQ preprocessor. *Bioinformatics* **2018**, *34*, i884–i890. [CrossRef]
59. Ewels, P.; Magnusson, M.; Lundin, S.; Käller, M. MultiQC: Summarize analysis results for multiple tools and samples in a single report. *Bioinformatics* **2016**, *32*, 3047–3048. [CrossRef] [PubMed]

60. Howe, K.L.; Contreras-Moreira, B.; De Silva, N.; Maslen, G.; Akanni, W.; Allen, J.; Alvarez-Jarreta, J.; Barba, M.; Bolser, D.M.; Cambell, L.; et al. Ensembl Genomes 2020—Enabling non-vertebrate genomic research. *Nucleic Acids Res.* **2020**, *48*, D689–D695. [CrossRef] [PubMed]
61. Li, H. Aligning sequence reads, clone sequences and assembly contigs with BWA-MEM. *arXiv* **2013**, arXiv:1303.3997.
62. Li, H.; Handsaker, B.; Wysoker, A.; Fennell, T.; Ruan, J.; Homer, N.; Marth, G.; Abecasis, G.; Durbin, R. The Sequence alignment/map (SAM) format and SAMtools 1000 Genome Project Data Processing Subgroup. *Bioinformatics* **2009**, *25*, 2078–2079. [CrossRef]
63. Danecek, P.; Bonfeld, J.K.; Liddle, J.; Marshall, J.; Ohan, V.; Pollard, M.O.; Whitwham, A.; Keane, T.; McCarthy, S.A.; Davies, R.M.; et al. Twelve years of SAMtools and BCFtools. *Gigascience* **2021**, *10*, giab008. [CrossRef] [PubMed]
64. Zheng, X.; Gogarten, S.M.; Lawrence, M.; Stilp, A.; Conomos, M.P.; Weir, B.S.; Laurie, C.; Levine, D. SeqArray-a Storage-Efficient High-Performance Data Format for WGS Variant Calls. *Bioinformatics* **2017**, *33*, 2251–2257. [CrossRef]
65. Zhang, C.; Dong, S.S.; Xu, J.Y.; He, W.M.; Yang, T.L. PopLDdecay: A fast and effective tool for linkage disequilibrium decay analysis based on variant call format files. *Bioinformatics* **2019**, *35*, 1786–1788. [CrossRef]
66. Nguyen, L.-T.; Schmidt, H.A.; Von Haeseler, A.; Minh, B.Q. IQ-TREE: A fast and effective stochastic algorithm for estimating maximum-likelihood phylogenies. *Mol. Biol. Evol.* **2015**, *32*, 268–274. [CrossRef]
67. Letunic, I.; Bork, P. Interactive tree of life v2: Online annotation and display of phylogenetic trees made easy. *Nucleic Acids Res.* **2011**, *39*, W475–W478. [CrossRef]
68. Alexander, D.H.; Novembre, J.; Lange, K. Fast model-based estimation of ancestry in unrelated individuals. *Genome Res.* **2009**, *19*, 1655–1664. [CrossRef]
69. Frichot, E.; Francois, O. LEA: An R Package for Landscape and Ecological Association Studies. *Methods Ecol. Evol.* **2015**, *6*, 925–929. [CrossRef]
70. Purcell, S.; Neale, B.; Todd-Brown, K.; Thomas, L.; Ferreira, M.A.R.; Bender, D.; Maller, J.; Sklar, P.; Bakker, P.I.W.; Daly, M.J.; et al. PLINK: A tool set for whole-genome association and population-based linkage analyses. *Am. J. Hum. Genet.* **2007**, *81*, 559–575. [CrossRef]
71. Dodds, P.N.; Rathjen, J.P. Plant immunity: Towards an integrated view of plant pathogen interactions. *Nat. Rev. Genet.* **2010**, *11*, 539–548. [CrossRef] [PubMed]
72. Emanuelli, F.; Lorenzi, S.; Grzeskowiak, L.; Catalano, V.; Stefanini, M.; Troggio, M.; Myles, S.; Martinez-Zapater, J.M.; Zyprian, E.; Moreira, F.M.; et al. Genetic diversity and population structure assessed by SSR and SNP markers in a large germplasm collection of grape. *BMC Plant Biol.* **2013**, *13*, 39. [CrossRef] [PubMed]
73. Li, X.W.; Meng, X.Q.; Jia, H.J.; Yu, M.L.; Ma, R.J.; Wang, L.R.; Cao, K.; Shen, Z.-J.; Niu, L.; Tian, J.-B.; et al. Peach genetic resources: Diversity, population structure and linkage disequilibrium. *BMC Genet.* **2013**, *14*, 84. [CrossRef]
74. Font i Forcada, C.; Oraguzie, N.; Igartua, E.; Moreno, M.A.; Gogorcena, Y. Population structure and marker–trait associations for pomological traits in peach and nectarine cultivars. *Tree Genet. Genomes* **2013**, *9*, 331–349. [CrossRef]
75. Guajardo, V.; Solís, S.; Almada, R.; Saski, C.; Gasic, K.; Moreno, M.Á. Genome-wide SNP identification in *Prunus* rootstocks germplasm collections using Genotyping-by-Sequencing: Phylogenetic analysis, distribution of SNPs and prediction of their effect on gene function. *Sci. Rep.* **2020**, *10*, 1467. [CrossRef]
76. Kim, Y.S.; Kim, S.H.; Sung, S.Y.; Kim, D.S.; Kim, J.B.; Jo, Y.D.; Kang, S.Y. Genetic relationships among diverse spray- and standard-type Chrysanthemum varieties and their derived radio-mutants determined using AFLPs. *Hortic. Environ. Biotechnol.* **2015**, *56*, 498–505. [CrossRef]
77. Singh, N.; Choudhury, D.R.; Singh, A.K.; Kumar, S.; Srinivasan, K.; Tyagi, R.K.; Singh, N.K.; Singh, R. Comparison of SSR and SNP Markers in Estimation of Genetic Diversity and Population Structure of Indian Rice Varieties. *PLoS ONE* **2013**, *8*, e84136. [CrossRef]

Disclaimer/Publisher's Note: The statements, opinions and data contained in all publications are solely those of the individual author(s) and contributor(s) and not of MDPI and/or the editor(s). MDPI and/or the editor(s) disclaim responsibility for any injury to people or property resulting from any ideas, methods, instructions or products referred to in the content.

MDPI AG
Grosspeteranlage 5
4052 Basel
Switzerland
Tel.: +41 61 683 77 34

Horticulturae Editorial Office
E-mail: horticulturae@mdpi.com
www.mdpi.com/journal/horticulturae

Disclaimer/Publisher's Note: The title and front matter of this reprint are at the discretion of the Guest Editors. The publisher is not responsible for their content or any associated concerns. The statements, opinions and data contained in all individual articles are solely those of the individual Editors and contributors and not of MDPI. MDPI disclaims responsibility for any injury to people or property resulting from any ideas, methods, instructions or products referred to in the content.

www.ingramcontent.com/pod-product-compliance
Lightning Source LLC
Chambersburg PA
CBHW062043220926
51850CB00010B/2645